Quantum Physics

Second Edition

Quantum Physics

Second Edition

Stephen Gasiorowicz

University of Minnesota

John Wiley & Sons, Inc.

New York   Chichester   Brisbane   Toronto   Singapore

To Hildy

ACQUISITIONS EDITOR   Cliff Mills
MARKETING MANAGER   Catherine Faduska
SENIOR PRODUCTION EDITOR   Suzanne Magida
COVER DESIGNER   Harry Nolan
MANUFACTURING COORDINATOR   Dorothy Sinclair
ILLUSTRATION COORDINATOR   Jaime Perea

This book was set in 10/12 Palatino by University Graphics and printed and
bound by Hamilton Printing. The cover was printed by New England Book Components, Inc.

Library of Congress Cataloging in Publication Data
Gasiorowicz, Stephen.
    Quantum physics / Stephen Gasiorowicz.—2nd ed.
       p.   cm.
    Includes bibliographical references and index.
    ISBN 0-471-85737-8 (cloth : alk. paper)
    1. Quantum theory.   I. Title.
  QC174.12.G37   1995
530.1'2—dc20   95-20414
                                                    CIP

Printed in the United States of America

10 9

PREFACE TO THE FIRST EDITION

This book is intended to serve as an introduction to quantum physics. In writing it, I have kept several guidelines in mind.

1. First, it is helpful for the development of intuition in any new field of study to start with a base of detailed knowledge about simple systems. I have therefore worked out a number of problems in great detail, so that the insight thus obtained can be used for more complex systems.

2. Every aspect of quantum mechanics has been helpful in understanding some physical phenomenon. I have laid great stress on applications at every stage of the development of the subject. Although no area of quantum physics is totally developed, my intention is to bridge the gap between a modern physics course and the more formal development of quantum mechanics. Thus, many applications are discussed, and I have stressed order-of-magnitude estimates and the importance of numbers.

3. In keeping with the level of the book, the mathematical structure has been kept as simple as possible. New concepts, such as operators, and new mathematical tools necessarily make their appearance. I have dealt with the former more by analogy than by precise definition, and I have minimized the use of new tools insofar as possible.

In approaching quantum theory, I chose to start with wave mechanics and the Schrödinger equation. Although the state-vector approach gets at the essential structure of quantum mechanics more rapidly, experience has shown that the use of more familiar tools, such as differential equations, makes the theory more accessible and the correspondence with classical physics more transparent.

The book probably contains a little more material than can comfortably be covered in one year. The basic material can be covered in one academic quarter. It consists of Chapters 1 to 6, 8, and 9, in which the motivation for a quantum theory, the Schrödinger equation, and the general framework of wave mechanics are covered. A number of simple problems are solved in Chapter 5, and their relevance to physical phenomena is discussed. The generalization to many particles and to three dimensions is developed. The second-quarter materials deals directly with atomic physics problems and uses somewhat more sophisticated tools. Here we discuss operator methods (Chapter 7), angular momentum (Chapter 10), the hydrogen atom (Chapter 12), operators, matrices, and spin (Chapter 14), the addition to angular momenta (Chapter 15), time-independent perturbation theory (Chapter 16), and the real hydrogen atom (Chapter 17). This material prepares the student to cope with a large variety of problems that are discussed during the third and last quarter. These problems include the interaction of charged particles with a magnetic field (Chapter 13), the helium atom (Chapter 18), problems in the radiation of atoms and related topics (Chapters 22 and 23), collision theory (Chapter 24), and the absorption of radiation in matter (Chapter 25). This material is supplemented by a more qualitative discussion of the structure of atoms and molecules (Chapters 19 to 21). The last chapter on elementary particles and their symmetries serves the dual purpose of describing some of the recent advances on that frontier of physics and showing how the basic ideas of quantum theory have found applicability in the domain of very short distances.

Several topics arise naturally as digressions in the development of the subject matter. Instead of lengthening some long chapters, I have placed this material in a separate "Special Topics" section. There, relativistic kinematics, the equivalence principle, the WKB approximation, a detailed treatment of lifetimes, line widths, and scattering resonances, and the Yukawa theory of nuclear forces are discussed. For the same reason, a brief introduction to the Fourier integral, the Dirac delta function, and some formal materials dealing with operators have been placed in mathematical appendices at the end of the book.

I am indebted to my colleagues at the University of Minnesota, especially Benjamin Bayman and Donald Geffen, for many discussions on the subject of quantum mechanics. I am grateful to Eugen Merzbacher, who read the manuscript and made many helpful suggestions for improvements. I also thank my students in the introductory quantum mechanics course that I taught for several years. Their evident interest in the subject led me to the writing of the supplemental notes that later became this book.

Stephen Gasiorowicz

PREFACE TO THE SECOND EDITION

The first edition of *Quantum Physics* was published more than 20 years ago. The guidelines that I set out in the preface to that edition as well as its general approach are ones to which I still subscribe. This edition does not fundamentally differ from the first but improves on it in some important ways.

PEDAGOGY

To help students, I have filled in more steps and have added more discussion of the physics behind the formal results of calculations. I have also added more section headings and subheadings, which allows for a more orderly use by the instructor. The number of applications has increased, in keeping with my aim to present a book on quantum physics.

APPLICATIONS

I have made substantive changes in the large number of applications contained in the text. I have expanded the discussion of the degenerate fermion gas to include a simplified calculation of the radius of a neutron star. In the discussion of the interaction of the electron with a magnetic field, I address Landau levels and briefly address the Integer Quantum Hall Effect.

Changes that affect the later half of this edition include shortening the discussion of molecular structure and omitting the chapter on elementary particle physics. Instead, I have devoted more attention to radiation theory. This includes a discussion of the Einstein A and B coefficients, lasers, the cooling of atoms, the properties of two-level atoms in strong electric fields, and the fascinating subject of the observtion of quantum jumps. In the special topics section, I have replaced a discussion of the Einstein equivalence principle, which had no real place in the text, with a discussion of the density operator. Even though this is not expounded on within the text, its inclusion allows the students to expand their knowledge should they so desire.

This edition includes many more problems than in the first edition and I have created a full and detailed Solutions Manual, the preparation of which, somewhat to my amazement, turned out to be equivalent to writing a second, smaller text.

ACKNOWLEDGMENTS

Over the years I have communicated with users of the first edition. Misprints as well as insoluble problems were pointed out to me. Because these errors were communicated close to 20 years ago (most of the errors were corrected in the second printing), I do not have a full record of the people who wrote to me. I would like to thank everyone who corresponded, including Professors J. S. Tenn and Tom Devlin. I have had good advice from Professors Lowell Brown, Richard Robinett, Ian Gatland, Yuan Ha, and Robert Lourie, and I would like to especially thank my friend and colleague, Earl Peterson, for many useful suggestions.

I have enjoyed my collaboration with the staff at Wiley. My thanks go to Acquisitions Editor, Cliff Mills, and Senior Production Editor, Suzanne Magida, for making this revision a manageable task. I would also like to acknowledge Harry Nolan for designing the cover and Jaime Perea for coordinating the illustration program.

Stephen Gasiorowicz

CONTENTS

THE LIMITS OF CLASSICAL PHYSICS

The end of the nineteenth century and the beginning of the twentieth witnessed a crisis in physics. A series of experimental results required concepts totally incompatible with classical physics. The development of these concepts, in a fascinating interplay of radical conjectures and brilliant experiments, led finally to the *quantum theory*.[1] Our objective in this chapter is to describe the background of this crisis and, armed with hindsight, to expose the new concepts in a manner that, while not historically correct, will make the transition to quantum theory less mysterious for the reader. The new concepts, *the particle properties of radiation, the wave properties of matter*, and *the quantization of physical quantities* will emerge in the phenomena discussed below.

BLACKBODY RADIATION

Classical Blackbody Radiation

When a body is heated, it is seen to radiate. In equilibrium the light emitted ranges over the whole spectrum of frequencies ν, with a spectral distribution that depends both on the frequency or, equivalently, on the wavelength of the light λ, and on the temperature. One may define a quantity $E(\lambda, T)$, the emissive power, as the

---

[1]An interesting account of the development of quantum theory may be found in M. Jammer, *The Conceptual Development of Quantum Mechanics,* Second Edition, American Institute of Physics, New York (1983). See also A. Pais *Subtle is the Lord . . .* Oxford University Press, NY (1982), Ch. 19.

energy emitted at wavelength λ per unit area, per unit time. Theoretical research in the field of thermal radiation began in 1859 with the work of Gustav Kirchhoff, who showed that for a given λ, the ratio of the emissive power E to the absorptivity A, defined as the fraction of incident radiation of wavelength λ that is absorbed by the body, is the same for all bodies. Kirchhoff considered two emitting and absorbing parallel plates and showed from the equilibrium condition that the energy emitted was equal to the energy absorbed (for each λ), that the ratios E/A must be the same for the two plates. Soon thereafter, he observed that for a *black body*, defined as a surface that totally absorbs all radiation that falls on it, so that $A = 1$, the function $E(\lambda, T)$ is a universal function.

To study this function, it is necessary to obtain the best possible source of blackbody radiation. A practical solution to this problem is to consider the radiation emerging from a small hole in an enclosure heated to a temperature T. Given the imperfections in the surface of the inside of the cavity, it is clear that any radiation falling on the hole will have no chance of emerging again. Thus the surface presented by the hole is very nearly "totally absorbing," and consequently the radiation coming from it is indeed "blackbody radiation." Provided the hole is small enough, this radiation will be the same as that which falls on the walls of the cavity. It is therefore necessary to understand the distribution of radiation inside a cavity whose walls are at a temperature T in Kelvin (K).

Kirchhoff showed that the second law of thermodynamics requires that the radiation in the cavity be isotropic, that is, that the flux be independent of direction; that it be homogeneous, that is, the same at all points; and that it be the same in all cavities at the same temperature—all of this for each wavelength.[2] The emissive power may, by simple geometric arguments, be shown to be connected with the energy density $u(\lambda, T)$ inside the cavity (see problem 1). The relation is

$$u(\lambda, T) = \frac{4E(\lambda, T)}{c} \tag{1-1}$$

The energy density is the quantity of theoretical interest, and further understanding of it came in 1894 from the work of Wilhelm Wien, who, again using very general arguments,[3] showed that the energy density had to be of the form

$$u(\lambda, T) = \lambda^{-5} f(\lambda T) \tag{1-2}$$

with f still an unknown function of a single variable. If, as is convenient, one deals instead with the energy density as a function of frequency, $u(\nu, T)$, then it follows from the fact that, with $\lambda = c/\nu$,

$$u(\nu, T) = u(\lambda, T) \left| \frac{d\lambda}{d\nu} \right|$$

$$= \frac{c}{\nu^2} u(\lambda, T) \tag{1-3}$$

---

[2]These matters are discussed in many textbooks on modern physics and statistical physics. References can be found at the end of this chapter and at the end of the book.

[3]Wien considered a perfectly reflecting spherical cavity contracting adiabatically. The redistribution of the energy as a function of λ has to be caused by the Doppler shift on reflection. See Chapter V in F. K. Richtmyer, E. H. Kennard, and J. N. Cooper *Introduction to Modern Physics*, McGraw-Hill, New York, 1969.

that the Wien law reads

$$u(\nu, T) = \nu^3 g\left(\frac{\nu}{T}\right) \qquad (1\text{-}4)$$

The implications of this law, which was confirmed experimentally (Fig. 1-1), are twofold:

1. Given the spectral distribution of blackbody radiation at one temperature, the distribution at any other temperature can be found with the help of the expressions just given.

2. If the function $f(x)$—or, equivalently, the function $g(x)$—has a maximum for some value of $x > 0$, then the wavelength λ_{max} at which the energy density, and hence the emissive power, has its maximum value, has the form

$$\lambda_{max} = \frac{b}{T} \qquad (1\text{-}5)$$

where b is a universal constant. Its value, as determined from the experiments of O. Lummer and E. Pringsheim (1897) is $b = 0.2898$ cm-K.

Wien used a model (of no interest, except to the historian) to predict a form for $g(\nu/T)$. The form was

$$g\left(\frac{\nu}{T}\right) = C e^{-\beta \nu / T} \qquad (1\text{-}6)$$

and, remarkably enough, this form, containing two adjustable parameters, fit the high frequency (low wavelength) data very well. The formula is not, however, in accord with some very general notions of classical physics. J. W. S. Rayleigh, in 1900, derived the result

$$u(\nu, T) = \frac{8\pi \nu^2}{c^3} kT \qquad (1\text{-}7)$$

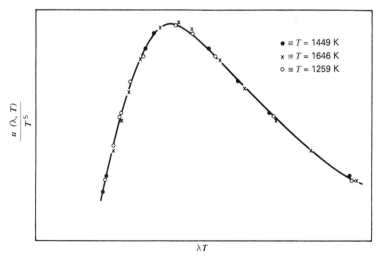

Figure 1-1. Experimental verification of (1.2) in the form of $u(\lambda, T)/T^5$ = a universal function of λT.

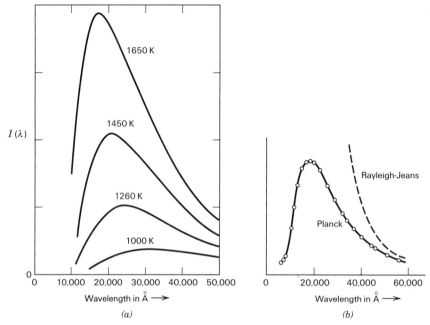

Figure 1-2. (*a*) Distribution of power radiated by a black body at various temperatures. (*b*) Comparison of data at 1600 K with Planck formula and Rayleigh-Jeans formula.

where k is Boltzmann's constant, $k = 1.38 \times 10^{-16}$ erg/K and c is the velocity of light, $c = 3.00 \times 10^{10}$ cm/sec. The ingredients that went into the derivation were (1) the classical law of equipartition of energy, according to which the average energy per degree of freedom for a dynamical system in equilibrium is, in this context,[4] kT, and (2) the calculation of the number of modes (i.e., degrees of freedom) for electromagnetic radiation with frequency in the interval $(\nu, \nu + d\nu)$, confined in a cavity.[5]

The Rayleigh-Jeans law (1-7) (Jeans corrected a minor error in Rayleigh's derivation) does not agree with experiment at high frequencies, where the Wien formula works, though it does fit the experimental curve at low frequencies (Fig. 1-2). The Rayleigh-Jeans law cannot, on general grounds, be correct, since the total energy density (integrated over all frequencies) is predicted to be infinite!

The Planck Distribution and the Quantum of Energy

In 1900, Max Planck found a formula by an ingenious interpolation between the high-frequency Wien formula and the low-frequency Rayleigh-Jeans law. The formula is

$$u(\nu, T) = \frac{8\pi h}{c^3} \frac{\nu^3}{e^{h\nu/kT} - 1} \tag{1-8}$$

---

[4]The equipartition law predicts that the energy per degree of freedom is $kT/2$. For an oscillator—and the modes of the electromagnetic field are simple harmonic oscillators—a contribution of $kT/2$ from the kinetic energy is matched by a like contribution from the potential energy, giving kT.

[5]We will need this result again, and derive it in Chapter 21. The number of modes is $4\pi\nu^2/c^3$, further multiplied by a factor of 2 because transverse electromagnetic waves correspond to two-dimensional harmonic oscillators.

where h, *Planck's constant*, is an adjustable parameter whose numerical value was found to be $h = 6.63 \times 10^{-27}$ erg sec. This law approaches the Rayleigh-Jeans form when $h\nu/kT \ll 1$, and reduces to

$$u(\nu, T) = \frac{8\pi h}{c^3} \nu^3 e^{-h\nu/kT} (1 - e^{-h\nu/kT})^{-1}$$
$$\cong \frac{8\pi h}{c^3} \nu^3 e^{-h\nu/kT}$$

(1-9)

when $h\nu \gg kT$. If we rewrite the formula as a product of the number of modes [we obtain this from (1-7) by dividing the energy density by kT] and another factor that can be interpreted as the average energy per degree of freedom

$$u(\nu, T) = \frac{8\pi\nu^2}{c^3} \frac{h\nu}{e^{h\nu/kT} - 1}$$
$$= \frac{8\pi\nu^2}{c^3} kT \frac{h\nu/kT}{e^{h\nu/kT} - 1}$$

(1-10)

we see that the classical equipartition law is altered whenever the frequencies are not small compared with kT/h. This alteration in the equipartition law shows that the modes have an average energy that depends on their frequency, and that the high frequency modes have a very small average energy. This effective cutoff removes the difficulty of the Rayleigh-Jeans density formula: the total energy in a cavity of unit volume is no longer infinite. We have

$$U(T) = \frac{8\pi h}{c^3} \int_0^\infty d\nu \frac{\nu^3}{e^{h\nu/kT} - 1}$$
$$= \frac{8\pi h}{c^3} \left(\frac{kT}{h}\right)^4 \int_0^\infty \frac{(h\nu/kT)^3 \, d(h\nu/kT)}{e^{h\nu/kT} - 1}$$
$$= \frac{8\pi k^4}{h^3 c^3} T^4 \int_0^\infty dx \frac{x^3}{e^x - 1}$$

(1-11)

The integral can be evaluated,[6] and the result is the Stefan-Boltzmann expression for the total radiation energy per unit volume

$$U(T) = aT^4$$

(1-12a)

with $a = 7.56 \times 10^{-15}$ erg/cm³ K⁴. The general form shown in (1-12a) was derived much earlier using thermodynamic reasoning. The result can also be written in the form of the total emissive power of a black body

$$E(T) = \sigma T^4$$

(1-12b)

with $\sigma = 5.42 \times 10^{-5}$ erg/cm² sec K⁴.

A departure from the pure equipartition law was not entirely unexpected: one consequence of it was the Dulong-Petit law of specific heats, according to which

---

[6] $\int_0^\infty dx \, x^3 (e^x - 1)^{-1} = \int_0^\infty dx \, x^3 e^{-x} \sum_{n=0}^\infty e^{-nx} = \sum_{n=0}^\infty \frac{1}{(n+1)^4} \int_0^\infty dy \, y^3 e^{-y} = 6 \sum_{n=1}^\infty \frac{1}{n^4} = \frac{\pi^4}{15}$

the product of the atomic (or molecular) weight and the specific heat is a constant for all solids; yet departures from the Dulong-Petit predictions were observed as early as 1872.[7] These departures indicated that the specific heat decreased at lower temperatures.[8]

The unqualified success of his formula drove Planck to search for its origin, and within two months he found that he could derive it by assuming that the energy associated with each mode of the electromagnetic field did not vary continuously (with average value kT) but was an integral multiple of some minimum quantum of energy ϵ. Under these circumstances a calculation of the average energy associated with each mode, using the Boltzmann probability distribution in a system of equilibrium at temperature T,

$$P(E) = \frac{e^{-E/kT}}{\sum_E e^{-E/kT}} \tag{1-13}$$

led to

$$\begin{aligned}
\overline{E} &= \sum_E EP(E) \\[1em]
&= \frac{\sum_{n=0}^{\infty} n\epsilon\, e^{-n\epsilon/kT}}{\sum_{n=0}^{\infty} e^{-n\epsilon/kT}} \\[1em]
&= \left. \frac{-\epsilon\, \dfrac{d}{dx} \sum_{n=0}^{\infty} e^{-nx}}{\sum_{n=0}^{\infty} e^{-nx}} \right|_{x=\epsilon/kT} \\[1em]
&= \left. \epsilon\, \frac{e^{-x}}{1 - e^{-x}} \right|_{x=\epsilon/kT} \\[1em]
&= \frac{\epsilon}{e^{\epsilon/kT} - 1}
\end{aligned} \tag{1-14}$$

This agrees with (1-10) provided we make the identification

$$\epsilon = h\nu \tag{1-15}$$

and do not change the number of modes.

Planck argued that for some unknown reason the atoms in the walls of the cavity emitted radiation in "quanta" with energy $nh\nu$ ($n = 1, 2, 3, \ldots$), but consistency demanded, as established by Einstein a few years later, that *electromagnetic radiation behaved as if it consisted of a collection of energy quanta with energy $h\nu$.*[9]

---

[7]According to the equipartition law an assembly of N oscillators (and a lattice of atoms with elastic forces between them may be so viewed) will have energy $3NkT$, the factor 3 coming from the fact that the oscillators in a solid are three-dimensional, rather than two-dimensional as for the radiation field in an enclosure. The specific heat for a mole is obtained by differentiating with respect to T and setting $N = N_0$, Avogadro's number, so that $C_v = 3N_0k = 3R$ where $R = 8.13 \times 10^7$ erg/K.

[8]Specific heats will be discussed very briefly in Chapter 20.

[9]For a given frequency ν there may be any integral number of quanta present, and hence the energy can take on the values, $nh\nu$, with $n = 0, 1, 2, 3, \ldots$.

The energy carried per quantum is extremely small. For light in the optical range, with, say, $\lambda = 6000$ Å,

$$h\nu = h\frac{c}{\lambda} = \frac{6.63 \times 10^{-27} \times 3.00 \times 10^{10}}{6 \times 10^{-5}} \simeq 3.3 \times 10^{-12} \text{ erg}$$

so that the number of light quanta of this wavelength, emitted by a 100-watt source, say, is

$$N = \frac{100 \times 10^7}{3.3 \times 10^{-12}} \cong 3 \times 10^{20} \text{ quanta/sec}$$

With so many quanta present, it is perhaps not surprising that we do not experience the particle nature of light directly; we shall see that on a macroscopic scale no deviations from classical optics are expected. Nevertheless, Planck's interpretation of his formula radically changes our picture of radiation.

The Cosmic Microwave Radiation Background

Blackbody radiation has reappeared on the frontier of fundamental science because of the developments accompanying and following the discovery by A. A. Penzias and R. W. Wilson in 1964 of a cosmic background of radiation in the microwave region. In the late 1940s George Gamow, Ralph Alpher, and Robert Herman studied some of the consequences of the *Big Bang* model of the creation of the universe. Their work, and subsequent calculations by P. J. E. Peebles showed that the present abundance of hydrogen in the universe can be understood only if there was a great deal of radiation present in the very early stages of the universe. The expansion of the universe led to a cooling of the matter and radiation in it, and when a temperature of about 3000 K was reached, the radiation ceased to interact significantly with the matter in the universe because the free electrons were able to combine with the nucleons to form atoms. From that time the universe became transparent to radiation, and the temperature of the radiation fell linearly with the scale of the "box" that contains the radiation, that is, with the size of the universe. The present residue of radiation has been studied in the last few years by NASA's COBE (Cosmic Background Explorer) satellite. As Fig. 1-3 shows, the spectrum is fitted to high accuracy by the blackbody distribution, corresponding to a present temperature of 2.735 K. The cosmic blackbody radiation background confirms the Big Bang scenario, and it yields information on the expansion of the universe, as well as conditions that were obtained at the decoupling time.

Tiny variations in the temperature as a function of direction are consistent with the motion of the solar system relative to the center of the galaxy, combined with the motion of our galaxy in the direction of the Virgo cluster of galaxies, a big clump of mass about 50 million light years away. The speed of this motion is of the order of 370 km/sec and the inhomogeneity may be associated with a Doppler shift associated with the motion. Once this effect was subtracted out, the temperature appeared to be uniform to an accuracy of better than one part in 10^5. This homogeneity actually presented cosmologists with a problem. The blackbody radiation detected from a certain direction in the sky is the radiation from that part of the sky at the decoupling time (of course red-shifted because of the expansion

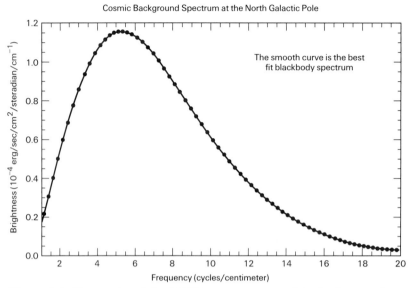

Figure 1-3. The results of the COBE measurements of the background radiation. (Courtesy COBE group and Professor G. Smoot.)

of the universe from that time). The equality of the radiation spectra in very different parts of the sky indicates equality of the temperatures in these parts of the sky at the decoupling time, but such regions are outside each other's horizon of influence (see Fig. 1-4). In 1981 Alan Guth proposed that the very earliest phases of the Big Bang include a period of really rapid exponential growth, which allows

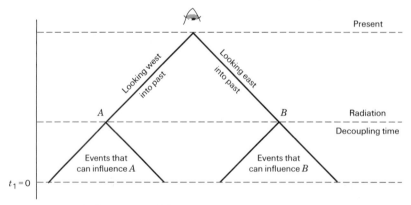

Figure 1-4. The horizon problem: An observer measuring blackbody radiation background by looking east and west sees the effect of conditions at A and B at the decoupling time. In the conventional Big Bang model the equality of temperatures at A and B cannot be understood, since at the time of the Big Bang ($t_1 = 0$) the past lightcones of A and B do not overlap. The inflation scenario assumes that in the very earliest period after the Big Bang the universe underwent an explosive exponential expansion, so that both regions in the past of A and B arise from an earlier, very much smaller region, in which none of the regions were outside of each others' region of influence. The time scale is very much distorted in that the interval between the Big Bang and now is of the order 10^5 larger than the time between the Big Bang and the decoupling time.

one to put the different parts of the sky at the decoupling time into a common origin. The difficulty of the extraordinary homogeneity was somewhat reduced, but it was still difficult to imagine that there would not be some traces of the inhomogeneities that had to be present to seed the coalescence of matter into the earliest galaxies. It was therefore with great relief that cosmologists greeted the announcement from the COBE group that inhomogeneities at the level of 5×10^{-6} in the temperature have been found. It is to be expected that more accurate measurements will continue to contribute to the detailed understanding of the early universe.

THE PHOTOELECTRIC EFFECT

As successful as the Planck formula was, the conclusion from it of the quantum nature of radiation is hardly compelling. An important contribution to its acceptance came from the work of Albert Einstein, who in 1905 used the concept of the quantum nature of light to explain some peculiar properties of metals, when these are irradiated with visible and ultraviolet light.

In 1887, the photoelectric effect was discovered by Heinrich Hertz, who, while engaged in his famous experiments on electromagnetic waves, found that the length of the spark induced in the secondary circuit was reduced when the terminals of the spark gap were shielded from the ultraviolet light coming from the spark in the primary circuit. His observations attracted much interest and the following facts were established by further experiments:

1. When polished metal plates are irradiated, they may emit electrons;[10] they do not emit positive ions.
2. Whether the plates emit electrons depends on the wavelength of the light. In general there will be a threshold that varies from metal to metal: only light with a frequency greater than a given threshold frequency will produce a photoelectric current.
3. The magnitude of the current, when it exists, is proportional to the intensity of the light source.
4. The energy of the photoelectrons is independent of the intensity of the light source but varies linearly with the frequency of the incident light.

Although the existence of the photoelectric effect can be understood within the framework of classical electromagnetic theory, since it was known that there were electrons in metals, and one could imagine them to be accelerated by absorption of radiation, the frequency-dependence of the effect is not comprehensible within that framework. The energy carried by an electromagnetic wave is proportional to the intensity of the source, and frequency has nothing to do with it. Furthermore, a classical explanation of the effect, which would have to involve the concentration of the energy deposited on single photoelectrons, would carry with it an implied time delay between the arrival of the radiation and the departure of the electron, the delay being longer when the intensity is decreased. In fact, no such time delays were ever observed, at least none longer than 10^{-9} sec, even with incident radiation of very low intensity.

---

[10]This was established by an e/m measurement.

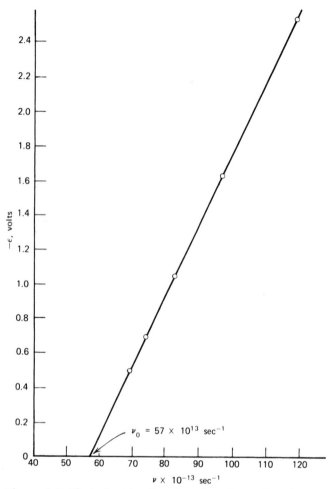

Figure 1-5. Photoelectric effect data showing a plot of retarding potential necessary to stop electron flow from a metal (lithium), or equivalently, electron kinetic energy, as a function of frequency of the incident light. The slope of the line is h/e.

Einstein considered the radiation to consist of a collection of quanta of energy $h\nu$, where ν is the frequency of the light. The absorption of a single quantum by an electron—a process that may take less time than the upper limit previously quoted—increases the electron energy by an amount $h\nu$. Some of the energy must be expended to separate the electron from the metal. This amount, W (called the *work function*), might be expected to vary from metal to metal, but should not depend on the electron energy. The rest is available for the electron kinetic energy, so that on the basis of this picture one expects that the following relation between electron velocity v and light frequency ν

$$\tfrac{1}{2}mv^2 = h\nu - W \tag{1-16}$$

should hold. The threshold effect and the linear relation between electron kinetic energy and the frequency are contained in this formula. The proportionality of the current and the source intensity can also be understood in terms of these light quanta, or *photons*, as they came to be called: a more intense light source emits more photons, and these in turn can liberate more electrons.

Robert A. Millikan carried out extensive experiments and established the correctness of the Einstein formula (Fig. 1-5). What Millikan's and the earlier experiments proved was that sometimes light behaves like a collection of particles, and that these "particles" can act individually, so that it is possible to contemplate the existence of a single photon and ask what its properties are. A by-product of these experiments was information about metals. It was found that W was of the order of several electron volts (1 eV = 1.6×10^{-12} erg), and this could be correlated with other properties of the metals.

THE COMPTON EFFECT

The experiment that provides the most direct evidence for the particle nature of radiation is the so-called Compton effect. Arthur H. Compton discovered that radiation of a given wavelength (in the X-ray region) sent through a metallic foil was scattered in a manner not consistent with classical radiation theory. According to classical theory, the mechanism for the effect is the re-radiation of light by electrons set into forced oscillations by the incident radiation, and this leads to the prediction of intensity observed at an angle θ that varies as $(1 + \cos^2 \theta)$, and does not depend on the wavelength of the incident radiation. Compton found that the radiation scattered through a given angle actually consists of two components: one whose wavelength is the same as that of the incident radiation, the other of wavelength shifted relative to the incident wavelength by an amount that depends on the angle (Fig. 1-6). Compton was able to explain the "modified" component by treating the incoming radiation as a beam of photons of energy $h\nu$, with individual photons scattering elastically off individual electrons. In an elastic collision,

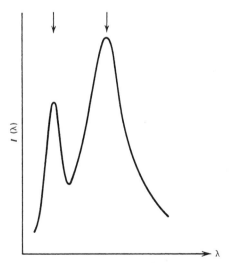

Figure 1-6. The spectrum of radiation scattered by carbon, showing the unmodified line at 0.7078 Å on the left and the shifted line at 0.7314 Å on the right. The former is the wavelength of the primary radiation.

momentum as well as energy must be conserved, and we must first assign a momentum to the photon. By analogy with relativistic particle kinematics we argue that

$$p = \frac{h\nu}{c} \tag{1-17}$$

The argument is that it follows from the relativistic relation between energy and momentum

$$E = [(m_0c^2)^2 + (pc)^2]^{1/2} \tag{1-18}$$

where m_0 is the rest mass of the particle, that the velocity at this momentum is

$$v = \frac{dE}{dp} = \frac{pc^2}{E} = \frac{pc^2}{(m_0^2c^4 + p^2c^2)^{1/2}} \tag{1-19}$$

For a photon this is always c, and hence the *photon rest mass must be zero*. Thus the relation (1-18) becomes

$$E = pc \tag{1-20}$$

which yields (1-17) when we substitute $E = h\nu$. One may also derive (1-20) from consideration of the energy and momentum of an electromagnetic wave, but the analogy argument is simpler.

   Consider, now, a photon with initial momentum **p**, incident upon an electron at rest. After the collision, the photon momentum is **p**', and the electron recoils with momentum **P**. Conservation of momentum yields (Fig. 1-7)

$$\mathbf{p} = \mathbf{p}' + \mathbf{P} \tag{1-21}$$

from which it follows that

$$\mathbf{P}^2 = (\mathbf{p} - \mathbf{p}')^2 = \mathbf{p}^2 + \mathbf{p}'^2 - 2\mathbf{p} \cdot \mathbf{p}' \tag{1-22}$$

Energy conservation reads

$$h\nu + mc^2 = h\nu' + (m^2c^4 + P^2c^2)^{1/2} \tag{1-23}$$

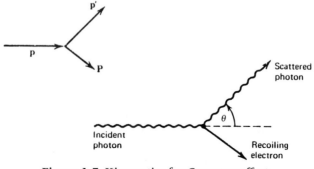

Figure 1-7. Kinematics for Compton effect.

where m is the electron rest mass. Hence

$$m^2c^4 + P^2c^2 = (h\nu - h\nu' + mc^2)^2$$
$$= (h\nu - h\nu')^2 + 2mc^2(h\nu - h\nu') + m^2c^4$$

On the other hand, (1-22) may be rewritten in the form

$$P^2 = \left(\frac{h\nu}{c}\right)^2 + \left(\frac{h\nu'}{c}\right)^2 - 2\frac{h\nu}{c}\cdot\frac{h\nu'}{c}\cos\theta$$

that is,

$$P^2c^2 = (h\nu - h\nu')^2 + 2(h\nu)(h\nu')(1 - \cos\theta) \tag{1-24}$$

where θ is the photon scattering angle. Thus

$$h\nu'(1 - \cos\theta) = mc^2(\nu - \nu')$$

or equivalently

$$\lambda' - \lambda = \frac{h}{mc}(1 - \cos\theta) \tag{1-25}$$

The measurements of the modified component agree very well with this prediction. The unmodified line is due to the scattering by the whole atom; if m is replaced by the mass of the atom, the shift in the wavelength is very small, since an atom is many thousands times more massive than an electron. The quantity h/mc has the dimensions of a length. It is called the Compton wavelength of the electron, and its magnitude is

$$\frac{h}{mc} \cong 2.4 \times 10^{-10} \text{ cm} \tag{1-26}$$

Measurements of the electron recoil were also made, and these are in agreement with the theory. It was furthermore determined by good time resolution coincidence experiments that the outgoing photon and the recoil electron appear simultaneously. There is no question of the correctness of the interpretation of the collision as an ordinary ''billiard ball'' type of collision, that is, of the particlelike behavior of the photon. Since radiation also has wave properties and exhibits interference and diffraction, we might expect some conceptual difficulties. These exist, and we shall discuss them at the end of the chapter.

WAVE PROPERTIES AND ELECTRON DIFFRACTION

In 1923 De Broglie, guided by the analogy of Fermat's principle in optics, and the least-action principle in mechanics, was led to suggest that the dual wave-particle nature of radiation should have its counterpart in a dual particle-wave nature of matter. Thus particles should have wave properties under certain circumstances, and De Broglie suggested an expression for the wavelength associated with the particle.[11] This is given by

$$\lambda = \frac{h}{p} \qquad (1\text{-}27)$$

where h is Planck's constant and p is the momentum of the particle. De Broglie's work attracted much attention, and many people suggested that verification could be obtained by observing electron diffraction.[12] The experimental observation of this effect occurred in experiments of C. J. Davisson and L. H. Germer, who found that in the scattering of electrons by a crystal surface, there was preferential scattering in certain directions.

Figure 1-8 is a simplified picture of what happens. In the scattering of waves by a periodic structure, there will be a phase difference between waves coming from adjacent scattering "planes," whose magnitude is given by $(2\pi/\lambda)\, 2a \sin\, \theta$.

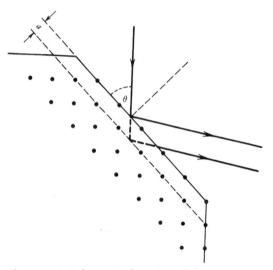

Figure 1-8. Schematic drawing of electron scattering geometry.

There will be constructive interference whenever this phase difference is equal to $2\pi n$, where n is an integer, that is, when

$$\lambda = \frac{2a \sin\, \theta}{n} \qquad (1\text{-}28)$$

The interference pattern observed in electron scattering by Davisson and Germer could be correlated with the preceding formula, provided the association (1-27) was made. This verification constituted a major step in the development of wave mechanics.

The particle diffraction experiments have since been carried out with molecular beams of hydrogen and helium, and with slow neutrons (see Fig. 1-9). Neutron

---

[11]The relation parallels the photon relation $\lambda = c/\nu = hc/h\nu = hc/E = h/p$.

[12]The history of the verification of De Broglie's conjecture can be found in M. Jammer, *loc cit.*

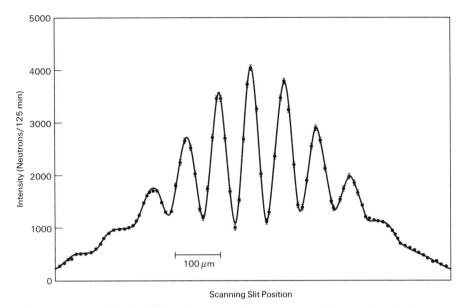

Figure 1-9. Double slit diffraction pattern for neutrons with wavelength $\lambda \approx$ 18.5 Å. (From A. Zeilinger, R. Gahler, C. G. Shull, W. Treimer, and W. Mampe, *Rev. Mod. Phys.* **60:**1067 (1988), by permission.)

diffraction is particularly useful in the study of crystal structure. To get a rough idea of the kind of energies needed for the diffraction experiments, we note that the crystal spacings are of the order of Angstroms. The grating constant in the Davisson-Germer experiment, in which nickel was used, was $a = 2.15$ Å. Hence λ is of the order of 10^{-8} cm, so that $p = h/\lambda \cong 6.6 \times 10^{-19}$ gm cm/sec. Thus for electrons the kinetic energy is $p^2/2m_e = (6.6 \times 10^{-19})^2/(2 \times 0.9 \times 10^{-27}) \cong 2.5 \times 10^{-10}$ ergs, and for neutrons the kinetic energy is $p^2/2m_n = (m_e/m_n) \times$ (electron energy) $\cong (1/1840) \times 2.5 \times 10^{-10}$ ergs $\cong 1.3 \times 10^{-13}$ ergs. In terms of the more convenient electron volt, these energies are approximately 160 eV and 0.08 eV, respectively.

On a macroscopic scale, the wave aspects of particles are beyond our ability to observe them. A droplet 0.1 mm in size, moving at 10 cm/sec will have a *De Broglie wavelength* of $\lambda = 6.6 \times 10^{-27}/4 \times 10^{-5} \cong 1.6 \times 10^{-22}$ cm. Since the "size" of a proton is about 10^{-13} cm, clearly there is no way in which the wave properties of an object of dimensions significantly larger than 10^{-4} cm can be observed. As for the particle properties of radiation, it is the smallness of h that determines the classical properties, in the sense that the dual aspects become apparent only when the product of momentum and dimension is of the order of h. We shall see that the formalism of quantum mechanics describes the situation very well.

THE BOHR ATOM

The Rutherford Planetary Model

The discovery of radioactivity by Henri Becquerel in 1896 provided the tools for an attack on the structure of the atom, which was complementary to the study of the emission of radiation by atoms. Ernest Rutherford was the leading physicist in the study of atomic structure, and he pioneered the use of the particles emitted in radioactive decay as projectiles to study atoms. Experiments carried out in 1908 under Rutherford's guidance by H. W. Geiger and E. Marsden, in which α-particles were aimed at thin foils, showed that a surprisingly large fraction of the α-particles underwent large-angle scattering, a result totally inconsistent with expectations based on the Thomson model of the atom. In the Thomson model, electrons were assumed to be embedded in a distribution of positive charge whose extent determined the atomic radius. Electrons do not deflect α-particles, since they are about 10^4 times less massive. Thus the source of deflection of the α-particles has to be the positive charge, and large-angle deflection implies that the potential at the surface of the charge distribution is large. This in turn implies that the positive charge is limited to a region much smaller than the volume of the atom. Rutherford proposed a new model that accounted for the data. In this model, all the positive charge (and almost all of the mass) is concentrated in a small region in the center of the atom. This positively charged *nucleus* attracts the negatively charged electrons and since the force law has a $1/r^2$ behavior, the *electrons travel in circular or elliptic orbits about the nucleus.*

Although the model gave a good quantitative account of the α-particle scattering data, it faced two insuperable difficulties. Since it implied a periodic motion for the electrons, it could not account for the spectra of radiation from atoms, which did not have the expected harmonic structure (cf. a vibrating string), but instead had the structure

$$\frac{1}{\lambda} = \text{const.} \left(\frac{1}{n_1^2} - \frac{1}{n_2^2} \right) \tag{1-29}$$

where n_1 and n_2 were integers. It also lacked a mechanism for stabilizing atoms: an electron in a circular or elliptic orbit is constantly accelerating, and according to electromagnetic theory, should be radiating. The constant loss of energy would, within a very short time (of the order of 10^{-10} sec) lead to the collapse of the atom, with the electrons plunging into the nucleus.

The Bohr Postulates

Just two years after this model was proposed, Niels Bohr in 1913 advanced a series of postulates, which, while sharply breaking with classical theory, explained the spectral structure and bypassed the stability problem. Bohr proposed that:

1. The electrons move in orbits restricted by the requirement that the angular momentum be an integral multiple of $h/2\pi$, that is, for circular orbits of radius r, the electron velocity v is restricted by

$$mvr = \frac{nh}{2\pi} \tag{1-30}$$

    and furthermore the electrons in these orbits do not radiate in spite of their acceleration. They were said to be in stationary states.

2. Electrons can make discontinuous transitions from one allowed orbit to another, and the change in energy, $E - E'$ will appear as radiation with frequency

$$\nu = \frac{E - E'}{h} \tag{1-31}$$

An atom may absorb radiation by having its electrons make a transition to a higher energy orbit.

The consequences of these postulates are very simply deduced for one-electron atoms such as hydrogen, singly ionized helium, and so on, if we deal with the circular orbits.[13] If the nuclear charge is Ze and that of the electron is $-e$, and if the radius of the orbit is r, then, taking the nuclear mass to be infinite, we balance the Coulomb force against the centrifugal force

$$\frac{Ze^2}{r^2} = \frac{mv^2}{r} \tag{1-32}$$

This, when combined with (1-30) leads to

$$v = \frac{2\pi e^2 Z}{hn} \tag{1-33}$$

and

$$r = \frac{1}{4\pi^2} \frac{n^2 h^2}{Ze^2 m} \tag{1-34}$$

The energy is

$$E = \tfrac{1}{2}mv^2 - \frac{Ze^2}{r} = -\frac{2\pi^2 e^4 Z^2 m}{h^2 n^2} \tag{1-35}$$

which, by postulate (2) immediately leads to the general form (1-29) (Fig. 1-10).

Before evaluating these quantities to obtain an idea of their magnitude, we will introduce some notations that will be very useful. First of all, it is $h/2\pi$ rather than h that appears in most formulas in quantum mechanics. We therefore define

$$\hbar = \frac{h}{2\pi} = 1.0545 \times 10^{-27} \text{ erg sec} \tag{1-36}$$

To keep the expressions for the energy simple, we shall deal with the angular frequency ω, rather than ν, where

$$\omega = 2\pi\nu \tag{1-37}$$

Thus (1-31) reads

$$\omega = \frac{E - E'}{\hbar} \tag{1-38}$$

Similarly, the quantum of radiation carries energy

$$E = \hbar\omega \tag{1-39}$$

---

[13]When elliptical orbits are allowed, a much richer structure emerges. This will be discussed in Chapter 12.

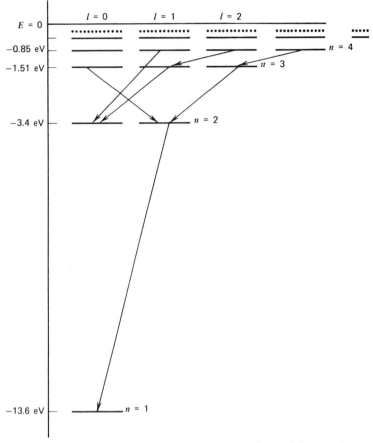

Figure 1-10. Spectrum for hydrogen atom as derived from Bohr atomic model. The existence of the quantum numbers l emerges from a discussion of elliptical orbits. The lines connecting energy levels represent some of the dominant atomic transitions.

It is sometimes convenient to introduce the "reduced wavelength"

$$\lambdabar = \frac{\lambda}{2\pi} = \frac{c}{\omega} \tag{1-40}$$

so that the De Broglie relation reads

$$p = \frac{\hbar}{\lambdabar} \tag{1-41}$$

The Bohr angular momentum quantization condition reads

$$mvr = n\hbar \quad (n = 1, 2, 3, \ldots) \tag{1-42}$$

It is also very convenient to introduce the dimensionless "fine structure constant"

$$\alpha = \frac{e^2}{\hbar c} = \frac{1}{137.0388} \tag{1-43}$$

which we will approximate by 1/137. In terms of these quantities we find the much simpler expressions

$$\frac{v}{c} = \frac{Z\alpha}{n} \qquad r = \frac{n^2}{Z\alpha} \frac{\hbar}{mc} \tag{1-44}$$

and

$$E = -\tfrac{1}{2}mc^2 \frac{(Z\alpha)^2}{n^2} \tag{1-45}$$

Notice that the radius, which has the dimensions of a length, is written in terms of $\hbar/mc$, the reduced Compton wavelength of the electron, and that the energy is written in terms of mc^2. In all atomic calculations we shall express our results in terms of mc^2, $\hbar/mc$, $\hbar/mc^2$, and mc for energy, length, time, and momentum, respectively. Angular momenta will always appear as multiples of $\hbar$.

   Let us now calculate some of the quantities that emerge from the Bohr theory. We calculate

$$mc^2 \cong 0.51 \times 10^6 \text{ eV}$$
$$\cong 0.51 \text{ MeV}$$

$$\frac{\hbar}{mc} \cong 3.9 \times 10^{-11} \text{ cm} \tag{1-46}$$

$$\frac{\hbar}{mc^2} \cong 1.3 \times 10^{-21} \text{ sec}$$

and thus obtain

(a) the radius of the lowest ($n = 1$) Bohr orbit is

$$a_0 = \frac{137}{Z} \frac{\hbar}{mc} = \frac{0.53}{Z} \text{ Å} \tag{1-47}$$

(b) the binding energy of the electron in the lowest Bohr orbit, that is, the energy required to put it in a state with $E = 0$ (corresponding to $n = \infty$) is

$$E = \tfrac{1}{2}mc^2 (Z\alpha)^2 = 13.6Z^2 \text{ eV} \tag{1-48}$$

Thus, for example, a transition from the $n = 1$ state to the $n = 2$ state in hydrogen ($Z = 1$) corresponds to a change in energy of $13.6 (1 - \tfrac{1}{4})$ eV $= 10.2$ eV. The frequency of the emitted radiation can be calculated by converting this into ergs, but it is more convenient to work this out in the form

$$\omega = \frac{mc^2\alpha^2(1 - \tfrac{1}{4})}{2\hbar} = \frac{3\alpha^2}{8} \frac{1}{1.3 \times 10^{-21}} \text{ rad/sec}$$
$$\cong 1.5 \times 10^{16} \text{ rad/sec}$$

Equivalently

$$\lambda = 2\pi \frac{c}{\omega} = \frac{16\pi}{3\alpha^2} \frac{\hbar}{mc}$$

$$\cong 1200 \text{ Å}$$

which lies in the ultraviolet.

The success of the Bohr theory with hydrogenlike atoms gave great impetus to further research on the "Bohr atom." In spite of some extraordinary achievements by Bohr and others, it was clear that the theory was provisional. It said nothing about when the electrons would make their jumps; also the quantization rule was restricted to periodic systems; a more general statement, by Sommerfeld and Wilson,

$$\int_{\substack{\text{closed} \\ \text{path}}} p \, dq = nh \tag{1-49}$$

where p is the momentum associated with the coordinate q, was of no help in treating problems other than those associated with atomic levels of hydrogen.

The quantization of angular momentum held in other situations as well. Its application to elliptic orbits gave a more complete picture of the spectrum of hydrogenlike atoms, and it was directly observed in the experiments of Stern and Gerlach in 1922.

The Correspondence Principle

Niels Bohr made great use of the notion that the quantum theory that he had developed should merge into classical theory in the limit in which classical theory was known to apply. This idea was formulated as the *correspondence principle.* Technically it stated that the classical limit should be reached when the "quantum numbers" are large, for example, for large n in the Bohr atom. Once a consistent theory of quantum phenomena was constructed, it automatically contained classical physics as a limit, but the principle was very helpful in guiding theoretical guesses, and led Heisenberg to the point from which he could make his giant leap to quantum mechanics. To illustrate how the correspondence principle is satisfied by the Bohr atomic model, consider the frequency of the radiation emitted when an electron makes a "jump" from the orbit with quantum number $n + 1$ to the orbit with quantum number n, when n is very large. This is a good domain to ask for the classical limit, since the angular momentum $n\hbar$ is indeed much larger than $\hbar$. Classically an electron moving in a circular orbit with velocity v would be expected to radiate with the frequency of its motion, that is,

$$\nu_{\text{cl}} = \frac{v}{2\pi r} = \frac{Z\alpha c}{n} \frac{Z\alpha mc}{2\pi n^2 \hbar} = \frac{(Z\alpha)^2 mc^2}{2\pi \hbar} \frac{1}{n^3} \tag{1-50}$$

On the other hand, the frequency of the radiation associated with the transition is, according to (1-31),

$$\nu = \frac{\omega}{2\pi} = \frac{1}{2\pi \hbar} \frac{mc^2}{2} (Z\alpha)^2 \left[\frac{1}{n^2} - \frac{1}{(n + 1)^2} \right] \tag{1-51}$$

which approaches ν_{cl} for $n \gg 1$. Note that this is a significant result, since it is only the frequency associated with an $n + 1 \rightarrow n$ transition that corresponds to the fundamental classical frequency. The radiation associated with the jump $n + 2 \rightarrow n$ has no classical counterpart even in the large n limit. We shall see in Chapter 21 that there are no $n + 2 \rightarrow n$ transitions for "circular orbits" in quantum mechanics.

THE WAVE-PARTICLE PROBLEM

The fact that radiation exhibits both wave and particle properties raises a deep conceptual difficulty, as can be seen from the following considerations:

1. Our discussion of the photoelectric effect, in particular the correlation of the number of electrons emitted with the intensity of the radiation, strongly suggests that the intensity of electromagnetic radiation is proportional to the number of photons emitted by the source. Let use now consider a *Gedankenexperiment*[14] in which radiation is diffracted by a two-slit system. Imagine that the intensity of the source is reduced to the point where, on the average, one photon per hour arrives at the screen. Note that we have to deal with entire photons: as the Compton effect as well as the photoelectric effect show, it is not possible to split a photon into parts with frequency ω but energy less than $\hbar\omega$. The decrease of intensity in the incident radiation should not affect the classical diffraction pattern, since, in effect, we are only stretching out the time scale on which the transmission from the source to the photographic plate of a large number of photons takes place. Photons that come to the plate an hour apart clearly cannot be correlated, and we may therefore think about this process one photon at a time. A photon, as a particle, will presumably go through one slit or the other. If we add to our Gedankenexperiment apparatus a small monitor that tells us whether the photon went through slit "1" or slit "2," we can divide the photons into two classes, associated with the two slits. For the first class, we could have closed down slit 2, since the photon did not go through it; for the second class we could have closed down slit 1. We might thus expect that the pattern on the photographic plate should be the same if we repeated the experiment with one slit closed for half the time, and the other slit closed for the other half of the time. This, however, cannot be, since the second experiment does not give an interference pattern. Thus there is an inconsistency that will be traced to the assumption that the presence of the monitor that tells us which slit the photon went through does not affect the experiment. When we discuss the *Heisenberg uncertainty principle,* we shall see that the action of the monitor destroys the interference pattern, so that there is no inconsistency. At this stage it is sufficient to point out that when there is no monitor, each photon acts as a wave, and it does not make sense to ask which slit the photon went through. Presumably, we can still speak of an average intensity of radiation at each slit:

[14]A *Gedankenexperiment* (thought experiment) is one that may be imagined, that is, one that is consistent with the known laws of physics, even though it may not be technically feasible. Thus, measuring the acceleration due to gravity on the surface of the sun is a Gedankenexperiment, whereas measuring the Doppler shift of sunlight as seen from a space ship moving with twice the velocity of light is nonsense.

this must mean that for individual photons we can only speak of a probability of going through one slit or another.

2. The notion of probability must again be invoked in understanding the passage of polarized radiation through an analyzer. As is well known, a beam of radiation of intensity I_0 will be attenuated to $I_0 \cos^2 \alpha$, where α is the angle between the axis of the polarizer and that of the analyzer. In terms of single photons that are *indivisible*, such an attenuation is only explainable if we state that a given photon will either go through or be blocked by the system, with a probability of transmission governed by the construction of the apparatus, that is, by the angle α.

3. In the same way, consider radiation from a distant star. The star is the source of a spherical wave of electromagnetic field excitation, spreading with velocity c. In terms of individual photons it is not sensible to think of the photon as spread thinly over a sphere of radius ct (where t is the time since the photon was emitted), since the collapse of that photon to a single point on a photographic plate, or on the retina of the eye, would violate common sense, if it were "really" happening. We may, however, interpret the spherical distribution as giving us the probability of finding a photon at a given solid angle.

4. Sometimes it is possible to interpret a given experiment both in particle and in wave language, but then the connection between the two complementary descriptions introduces a nonclassical aspect elsewhere. Dicke and Wittke[15] have proposed the following Gedankenexperiment (Fig. 1-11). Consider a cylindrical bird cage with the bars spaced regularly, and spacing

$$a = 2\pi \frac{R}{N}$$

where R is the radius of the cylinder and N is the number of bars. Consider radiation emitted from a source placed on the axis of the cylinder. The bars act as a diffraction grating. If the beam emerges at an angle θ with the original direction, we have maximum intensity if the angle and the wavelength are related by

$$a \sin \theta = n\lambda \qquad (n = 1, 2, 3, \ldots)$$

that is,

$$\lambda = \frac{2\pi R \sin \theta}{Nn} \qquad (1\text{-}52)$$

We could also interpret the intensity peak by assuming that the particles scattered through an angle θ off the bars of the bird cage. The momentum transferred to the cage is $p \sin \theta$ and hence the angular momentum transferred to the cage is

$$L = pR \sin \theta \qquad (1\text{-}53)$$

[15]R. H. Dicke and J. P. Wittke, *Introduction to Quantum Mechanics*, Addison-Wesley, Reading, Mass., 1960.

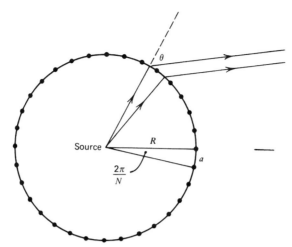

Figure 1-11. End view of Dicke-Wittke "cage" showing equally spaced bars and geometrical quantities connected with it.

If we now make the De Broglie association, $p = 2\pi\hbar/\lambda$ we obtain

$$L = \frac{2\pi\hbar Nn}{2\pi R \sin \theta} \cdot R \sin \theta = Nn\hbar \tag{1-54}$$

that is, angular momentum is quantized! The factor N is associated with the fact that the bird cage looks the same when it is rotated through an angle $2\pi/N$, as will become clear later.

In 1925 the modern theory of quantum mechanics started with the work of Werner Heisenberg, Max Born, Pascual Jordan, Erwin Schrödinger, and Paul Dirac. This theory provides a way of reconciling all of the conflicting concepts at the cost of making us abandon a certain amount of classical thinking. It is one of the joys of being a student of physics to be able to appreciate this beautiful theory and the monumental advances in our understanding the properties of matter that the theory enabled us to make.

Problems

1. Prove the relation (1-1) between the energy density in a cavity and the emissive power. [*Hint:* To do so, look at the figure. The shaded volume element is of magnitude $r^2 \, dr \sin \theta \, d\theta \, d\phi = dV$ where r is the distance to the origin (at the aperture of area dA), θ is the angle with the vertical, and ϕ is the azimuthal angle about the perpendicular axis through the opening. The energy contained in the volume element is dV multiplied by the energy density. The radiation is isotropic, so that what emerges is given by the solid angle $dA \cos \theta / 4\pi r^2$ multiplied by the energy. This is to be integrated over the angles ϕ and θ and, if the flow of radiation in time Δt is wanted, over dr from 0 to $c\Delta t$—the distance from which the radiation will escape in the given time interval.]

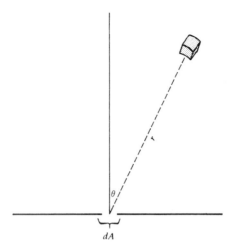

dA

2. Use (1-1) and (1-12) to obtain a formula for the total rate of radiation per unit area of a black body. Assume that the sun radiates as a black body. You are given the radius of the sun $R_\odot = 7 \times 10^{10}$ cm, the average distance of the sun to the earth $d_\odot = 1.5 \times 10^{13}$ cm, and the solar constant, the amount of energy falling on the earth when the sun is overhead 1.4×10^6 ergs/cm^2 sec. Use this information to estimate the surface temperature of the sun.

3. Given (1-8) calculate the energy density in a wavelength interval $\Delta\lambda$. Use your expression to calculate the value of $\lambda = \lambda_{max}$, for which this density is maximal. Show that λ_{max} is of the form b/T, calculate b, and use your estimate of the sun's surface temperature to calculate λ_{max} for solar radiation. [*Hint:* In calculating b you will need the solution x of the equation $(5 - x) = 5e^{-x}$. Solve this graphically or by a successive approximation method, in which you first write $x = 5 - \epsilon$, with $\epsilon \ll 1$.]

4. Ultraviolet light of wavelength 3500 Å falls on a potassium surface. The maximum energy of the photoelectrons is 1.6 eV. What is the work function of potassium?

5. The maximum energy of photoelectrons from aluminum is 2.3 eV for radiation of 2000 Å and 0.90 eV for radiation of 2580 Å. Use these data to calculate Planck's constant and the work function of aluminum.

6. A 100-MeV photon collides with a proton that is at rest. What is the maximum possible energy loss for the photon?

7. A 100-keV photon collides with an electron at rest. It is scattered through 90°. What is its energy after the collision? What is the kinetic energy in eV of the electron after the collision, and what is the direction of its recoil?

8. An electron of energy 100 MeV collides with a photon of wavelength 3×10^7 Å (corresponding to the universal background of blackbody radiation). What is the maximum energy loss suffered by the electron?

9. A beam of X rays is scattered by electrons at rest. What is the energy of the X rays if the wavelength of the X rays scattered at 60° to the beam axis is 0.035 Å?

10. A nitrogen nucleus (mass $\cong 14 \times$ proton mass) emits a photon of energy 6.2 MeV. If the nucleus is initially at rest, what is the recoil energy of the nucleus in eV?

11. Consider a crystal with planar spacing 3.2 Å. What order of magnitude of energies would one need for (a) electrons, (b) helium nuclei (mass $\cong 4 \times$ proton mass) to observe up to three interference maxima?

12. The smallest separation resolvable by a microscope is of the order of magnitude of the wavelength used. What energy electrons would one need in an electron microscope to resolve separations of (a) 150 Å, (b) 5 Å?

13. If one assumes that in a stationary state of the hydrogen atom the electron fits into a circular orbit with an integral number of wavelengths, one can reproduce the results of the Bohr theory. Work this out.

14. The distance between adjacent planes in a crystal are to be measured. If X rays of wavelength 0.5 Å are detected at an angle of 5°, what is the spacing? At what angle will the second maximum occur?

15. Use the Bohr quantization rules to calculate the energy levels for a harmonic oscillator, for which the energy is $p^2/2m + m\omega^2 r^2/2$, that is, the force is $m\omega^2 r$. Restrict yourself to circular orbits. What is the analog of the Rydberg formula? Show that the correspondence principle is satisfied for all values of the quantum number n used in quantizing the angular momentum.

16. Use the Bohr quantization rules to calculate the energy states for a potential given by

$$V(r) = V_0 \left(\frac{r}{a}\right)^k$$

with k very large. Sketch the form of the potential and show that the energy values approach $E_n \simeq Cn^2$.

17. The power, that is, the energy radiated per unit time by an accelerated charge e is classically given by the formula

$$P = \frac{2}{3} \frac{e^2}{c^3} a^2 \text{ erg/sec}$$

where a is the acceleration. In a circular orbit $a = v^2/r$. Calculate the power radiated by an electron in a Bohr orbit characterized by the quantum number n. When n is very large, this should agree with a proper quantum mechanical result according to the correspondence principle.

18. The decay rate for an electron in an orbit may be defined to be the power radiated, P, divided by the energy emitted in the decay. Use the Bohr theory expression for the energy radiated, and the expression for P from Problem 17 to calculate the "correspondence" value of the decay rate when the electron makes a transition from orbit n to orbit $n - 1$. What is the value of this decay rate when $n = 2$? (This will not agree exactly with the true quantum theory result, since the correspondence principle will not hold for such small values of the quantum number.) What is the decay rate when the transition is from an orbit n to an orbit $n - m$? What is the lifetime = (decay rate)$^{-1}$?

19. The classical energy of a plane rotator is given by

$$E = L^2/2I$$

where L is the angular momentum and I is the moment of inertia. Apply the Bohr quantization rules to obtain the energy levels of the rotator. If the Bohr frequency condition is assumed for the radiation in transitions from states labeled by n_1 to states labeled by n_2, show that (a) the correspondence principle holds, and (b) that it implies that only transitions $\Delta n = \pm 1$ should occur.

20. Molecules sometimes behave like rotators. If rotational spectra are characterized by radiation of wavelength of order 10^7 Å, and this is used to estimate interatomic distances in a molecule like H_2, what kind of separations (in Å) are obtained?

References

The topics covered in this chapter are discussed in most modern physics textbooks. For more original discussions see

Eyvind H. Wichmann, *Quantum Physics*, McGraw-Hill, New York, 1969.

Richard P. Feynman, Robert B. Leighton, and Matthew Sands, *The Feynman Lectures on Physics*, Addison-Wesley, Reading, Mass., 1963.

For Bohr's contributions to the development of quantum theory, see Abraham Pais, *Niels Bohr's Times in Physics, Philosophy and Polity*, Oxford University Press, New York, 1991.

WAVE PACKETS AND THE UNCERTAINTY RELATIONS

Quantum mechanics provides us with an understanding of all of the phenomena discussed in Chapter 1. It is indispensable to the understanding of atoms, molecules, atomic nuclei, and aggregates of these. We will approach the study of quantum mechanics through the Schrödinger equation and the appropriate interpretation of its solutions.[1] There is no way of deriving this equation from classical physics, since it lies outside the realm of classical physics. The equation was, in fact, obtained by Erwin Schrödinger by a brilliant guess, based on earlier insights of De Broglie. We will motivate the guess somewhat differently, by seeing how one might try to reconcile the wave and particle properties of electrons. After our presentation of the free particle Schrödinger equation we shall use some general properties of waves to obtain the uncertainty relation between position and momentum, and discuss its implications.

LOCALIZED WAVE PACKETS

It is difficult to think of configurations of particles that somehow simulate wave behavior. This is why the classic diffraction experiments of Fresnel and Young led to the unanimous acceptance of the wave theory of light. On the other hand, it is possible to imagine configurations of waves that are very localized. (A clap of thunder is an example of a superposition of waves leading to an effect localized

---

[1] A different approach can be found in R. P. Feynman, R. B. Leighton, and M. Sands, *The Feynman Lectures on Physics*, Vol. III, Addison-Wesley, Reading, Mass., 1964.

in time at a given location.) Such localized "wave packets" can be achieved by superposing waves with different frequencies in a special way, so that they interfere with each other almost completely outside of a given spatial region. The technical tools for doing this involve Fourier integrals, and Appendix A summarizes them for the reader who is familiar with Fourier series and who does not insist on mathematical rigor.

As an example, consider the function defined by

$$f(x) = \int_{-\infty}^{\infty} dk \, g(k) \, e^{ikx} \tag{2-1}$$

The real part of $f(x)$ is given by $\int_{-\infty}^{\infty} dk \, g(k) \cos kx$, and this is a linear superposition of waves of wavelength $\lambda = 2\pi/k$, since for a given k each wave reproduces itself when x changes to $x + 2\pi/k$. To illustrate such a wave packet, let us choose

$$g(k) = e^{-\alpha(k-k_0)^2} \tag{2-2}$$

The integral can be done: with $k' = k - k_0$ we have

$$f(x) = \int_{-\infty}^{\infty} dk \, g(k) \, e^{i(k-k_0)x} \, e^{ik_0x}$$

$$= e^{ik_0x} \int_{-\infty}^{\infty} dk' \, e^{ik'x} \, e^{-\alpha k'^2}$$

$$= e^{ik_0x} \int_{-\infty}^{\infty} dk' \, e^{-\alpha[k'-(ix/2\alpha)]^2} \, e^{-(x^2/4\alpha)}$$

where in the last step we have completed squares. It is justified to let $k' - (ix/2\alpha) = q$ and still keep the integral along the real axis.[2] Making use of

$$\int_{-\infty}^{\infty} dk \, e^{-\alpha k^2} = \sqrt{\frac{\pi}{\alpha}} \tag{2-3}$$

we obtain

$$f(x) = \sqrt{\frac{\pi}{\alpha}} \, e^{ik_0x} \, e^{-(x^2/4\alpha)} \tag{2-4}$$

The factor e^{ik_0x} is known as a "phase factor," since $|e^{ik_0x}|^2 = 1$. Thus the absolute square of $f(x)$ is

$$|f(x)|^2 = \frac{\pi}{\alpha} e^{-x^2/2\alpha} \tag{2-5}$$

This function peaks at $x = 0$, and depending on the magnitude of α, it represents a broad (α large) or very narrow (α small) wave packet. As it stands, one might be willing to consider $|f(x)|^2$ as a representation of a particle.

---

[2]The reader familiar with the theory of complex variables will have no trouble convincing herself of this.

The width of the packet may be taken to be $2\sqrt{2\alpha}$, since the function falls off to $1/e$ of its peak value. The width of $|f(x)|^2$ and $|g(k)|^2$ are correlated. In our special example, the square of $g(k)$ is a function peaked about k_0 with width $2/\sqrt{2\alpha}$. There is a reciprocity here: a function strongly localized in x is broad in k and vice versa. The product of the two "widths" is

$$\Delta k \, \Delta x \sim \frac{2}{\sqrt{2\alpha}} \cdot 2\sqrt{2\alpha} = 4 \tag{2-6}$$

The exact value of the numerical constant is not important; what matters is that it is independent of α and of order unity. This is a general property of functions that are Fourier transforms of each other (Fig. 2-1). We represent it by the formula

$$\Delta x \, \Delta k \gtrsim O(1) \tag{2-7}$$

where Δx and Δk are the "widths" of the two distributions, and we imply by $O(1)$ that this is a number that may depend on the functions that we are dealing with, but is not significantly different from 1. *It is impossible to make both Δx and Δk small.* This is a general feature of wave packets, but we shall soon see that it has some very deep implications for quantum mechanics.

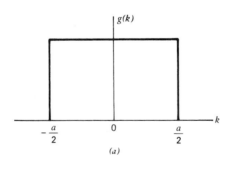

(a)

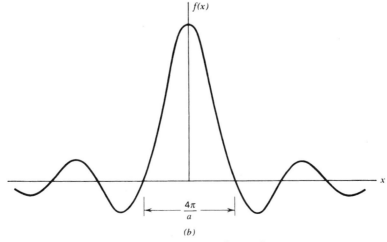

(b)

Figure 2-1. Relation between wave packet and its Fourier transform for a square-shaped wave packet.

THE PROPAGATION OF WAVE PACKETS

In (2-1) we considered a function $f(x)$ that is made up of a continuous superposition of simple waves e^{ikx}. How will such a wave packet propagate in time? The answer to that depends on how the individual waves propagate. In general we shall write for the simple *plane wave* (so called because it only has a spatial variation in x, but not in y or z) the form

$$e^{ikx - i\omega t} \tag{2-8}$$

Here $\omega = 2\pi\nu$ is the angular frequency. The quantity k is related to the wavelength by $k = 2\pi/\lambda$ so that we may write for the simple wave another form

$$e^{2\pi i[(x/\lambda) - \nu t]} \tag{2-9}$$

If we are considering the propagation of a light wave in a vacuum, then there is a simple relation between ν and $1/\lambda$, namely, $\nu = c/\lambda$, so that the simple wave becomes

$$e^{2\pi i(x - ct)/\lambda} = e^{ik(x - ct)}$$

If we now take the superposition, with amplitude $g(k)$ of these simple waves, we get, at time t,

$$f(x, t) = \int_{-\infty}^{\infty} dk \, g(k) \, e^{ik(x - ct)} = f(x - ct) \tag{2-10}$$

This is the same shape that we started with, except that instead of being localized at $x = 0$, it is now localized at $x - ct = 0$. Thus a wave packet of light waves propagates *without distortion* with velocity c, the velocity of light.

We are, however, concerned with waves that are supposed to describe particles, and we may not therefore require that $\omega = kc$. In general ω will be a function of k, so that

$$f(x, t) = \int dk \, g(k) \, e^{ikx - i\omega(k)t} \tag{2-11}$$

For the time being, we do not know what the form of $\omega(k)$ is, but we shall try to determine it from the requirement that $f(x, t)$ resemble a freely moving classical particle.

Let us consider a wave packet that is strongly localized in k-space, about a value k_0. This would correspond to a choice like (2-2) with α large. It is true that this will not represent an $f(x)$ sharply localized in x-space, but our calculation will be easier, and we are, after all, still trying to make intelligent guesses. Since the integral in (2-11) will center around $k = k_0$, we expand $\omega(k)$ about k_0 and assume that $\omega(k)$ is not a very rapidly varying function of k. Thus we write

$$\omega(k) \approx \omega(k_0) + (k - k_0) \left(\frac{d\omega}{dk}\right)_{k_0} + \frac{1}{2}(k - k_0)^2 \left(\frac{d^2\omega}{dk^2}\right)_{k_0} \tag{2-12}$$

The first term is a constant, independent of k. In the second term, the quantity $(d\omega/dk)|_{k_0}$ is the *group velocity*,[3] which describes the propagation of the wave packet. With the notation

$$\left(\frac{d\omega}{dk}\right)_{k_0} = v_g \tag{2-13}$$

and

$$\left(\frac{d^2\omega}{dk^2}\right)_{k_0} = \beta \tag{2-14}$$

and, with $k - k_0 = k'$, the wave packet has the time dependence

$$
\begin{aligned}
f(x, t) &= e^{ik_0 x - i\omega(k_0)t} \int_{-\infty}^{\infty} dk'\, e^{-\alpha k'^2} e^{ik'(x - v_g t)} e^{-ik'^2 \beta t/2} \\
&= e^{ik_0 x - i\omega(k_0)t} \int_{-\infty}^{\infty} dk'\, e^{ik'(x - v_g t)} e^{-(\alpha + i\beta t)k'^2}
\end{aligned}
\tag{2-15}
$$

This is just the integral that led to (2-4), so that replacing x by $x - v_g t$ and α by $\alpha + i\beta t$, we get

$$f(x, t) = e^{i[k_0 x - \omega(k_0)t]} \left(\frac{\pi}{\alpha + i\beta t}\right)^{1/2} e^{-[(x - v_g t)^2/4(\alpha + i\beta t)]} \tag{2-16}$$

and the absolute square of this function is

$$|f(x, t)|^2 = \left(\frac{\pi^2}{\alpha^2 + \beta^2 t^2}\right)^{1/2} e^{-[\alpha(x - v_g t)^2/2(\alpha^2 + \beta^2 t^2)]} \tag{2-17}$$

This represents a wave packet whose peak is traveling with velocity v_g, but it does not have a definite width: the quantity that was α at $t = 0$ now becomes $\alpha + (\beta^2 t^2/\alpha)$, that is, *the packet is spreading*. Since the width is proportional to

$$\left(\alpha + \frac{\beta^2 t^2}{\alpha}\right)^{1/2} = \sqrt{\alpha}\left(1 + \frac{\beta^2 t^2}{\alpha^2}\right)^{1/2}$$

the rate of spreading will be small if α is large, that is, if the packet is spatially large to begin with.

FROM WAVE PACKETS TO THE SCHRÖDINGER EQUATION

The most important result is that if (2-11) is to represent a particle with momentum p and kinetic energy $p^2/2m$, then we must require that

---

[3]The group velocity is a concept discussed in every book that deals with wave propagation. It is the velocity with which signals propagate. See, for example, H. Georgi, *The Physics of Waves*, Prentice-Hall, Englewood Cliffs, N.J., 1993.

$$v_g = \frac{d\omega}{dk} = \frac{p}{m} \tag{2-18}$$

If we further make the association that

$$E = \hbar\omega \tag{2-19}$$

suggested by the quantum relation for radiation, so that

$$\omega = \frac{p^2}{2m\hbar} \tag{2-20}$$

then consistency demands that we make the association

$$k = \frac{2\pi}{\lambda} = \frac{p}{\hbar} \tag{2-21}$$

first derived in a somewhat similar way by De Broglie.

In terms of p, the expression (2-11) can be rewritten in the form[4]

$$\psi(x, t) = \frac{1}{\sqrt{2\pi\hbar}} \int dp \; \phi(p) \; e^{i(px - Et)/\hbar} \tag{2-22}$$

The wave packet $\psi(x, t)$ is a general solution of the partial differential equation

$$i\hbar \frac{\partial \psi(x, t)}{\partial t} = \frac{1}{\sqrt{2\pi\hbar}} \int dp \; \phi(p) \; E \; e^{i(px - Et)/\hbar}$$

$$= \frac{1}{\sqrt{2\pi\hbar}} \int dp \; \phi(p) \; \frac{p^2}{2m} \; e^{i(px - Et)/\hbar} \tag{2-23}$$

$$= -\frac{\hbar^2}{2m} \frac{\partial^2 \psi(x, t)}{\partial x^2}$$

provided, as we have done earlier, we describe the motion of the "particle" in a potential-free region, where $E = p^2/2m$. It is this equation, and its generalization to the case of a particle moving in a potential, that represents the important abstraction from the arguments previously outlined. It should be stressed that the equation represents a guess: there was no justification on the basis of classical physics for the replacement of ω by $E/\hbar$, nor for the replacement of the wave number k by $p/\hbar$.

We must still face the difficulty of the spreading of the wave packets. If we consider a *Gaussian* packet (2-17), we see that no matter how large α is, there will be a time when the spreading will become noticeable. This contradicts experience, which shows very clearly that nuclei, for example, that are very tiny, have not changed during a period of 3×10^9 years (10^{17} sec). We shall see in Chapter 3 that the notions of probability, hinted at in Chapter 1, play a role here, and the spreading really refers to a growing probability that the particle is far from where it was localized at $t = 0$.

---

[4]The numerical factor in front of the integral will be justified later when we assign a physical meaning to $\phi(p)$.

THE UNCERTAINTY RELATIONS

One of the most important qualitative observations that we made in our wave packet discussion is the reciprocity relation between the widths in x- and k-space

$$\Delta k \, \Delta x \gtrsim 1 \qquad (2\text{-}24)$$

If we multiply this by $\hbar$ and use $\hbar k = p$, we obtain the Heisenberg uncertainty relations

$$\Delta p \, \Delta x \gtrsim \hbar \qquad (2\text{-}25)$$

Since the width represents a region in which a particle is likely to be in x-space or in momentum space, (2-25) states that if we try to construct a highly localized wave packet in x-space, then it is impossible to associate a well-defined momentum with it, in contrast with what is taken for granted in classical physics. By the same token, a wave packet characterized by a momentum defined within narrow limits must be spatially very broad. This limitation is one that *quantum mechanics imposes on our use of classical concepts to describe a physical system.* The classical concepts of position and momentum are independent of each other: they refer to different degrees of freedom. In quantum mechanics, as we shall see in more detail in Chapter 6, position and momentum are complementary properties of a system, and the theory does not admit the possibility of an experiment in which both can be established simultaneously. The smallness of $\hbar$ guarantees that only for microscopic systems will the usual notions of classical physics fail. For example, for a dust particle of mass 10^{-4} gm moving with a velocity of 10^{4} cm/sec with an uncertainty in the product of one part in a million implies $\Delta p \sim 10^{-6}$ and thus $\Delta x \sim 10^{-21}$ cm, which is 10^{8} times smaller than the radius of a proton! This is not so for an electron in a Bohr orbit. It we take $\Delta p \sim p \sim mc\alpha$, then $\Delta x \sim \hbar/mc\alpha$, of the order of magnitude of the radii of the orbits.

In what follows we will discuss a number of Gedankenexperiments in which we will show in detail how the wave-particle duality acts to prohibit a violation of the relation (2-25).

(a) *Measurement of position of an electron.* (*The Heisenberg Microscope.*) Consider the experimental setup in Fig. 2-2, whose purpose is to measure the position of an electron. The electrons are in a beam having well-defined momentum p_x and moving in the positive x direction. The microscope (lens + screen) is to be used to see where the electron is located by observing the light that is scattered off the electron. We shine light along the negative x-axis; a particular electron will scatter a particular photon, and the latter recoils through the microscope. The resolution of the microscope, that is, the precision with which the electron can be localized, is known from optics. It is

$$\Delta x \sim \frac{\lambda}{\sin \phi} \qquad (2\text{-}26)$$

where λ is the wavelength of the light. It would appear that by making λ small enough, and by making $\sin \phi$ large, Δx can be made as small as desired. This, we will now show, can only be done at the expense of losing information about the x-component of the electron momentum. Quantum theory tells us that

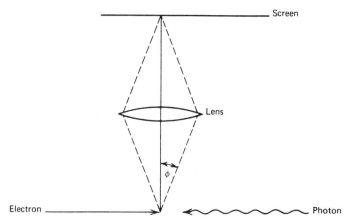

Figure 2-2. Schematic drawing of the Heisenberg microscope for the measurement of electron position.

what registers on the screen behind the lens are really individual photons that got there because they scattered off the electrons. The direction of the photon after scattering is undetermined within the angle subtended by the aperture. Hence the magnitude of the recoil momentum of the electron is uncertain by

$$\Delta p_x \sim 2\,\frac{h\nu}{c}\,\sin\phi \tag{2-27}$$

Hence

$$\Delta p_x\,\Delta x \sim 2\,\frac{h\nu}{c}\,\sin\phi\,\frac{\lambda}{\sin\phi} \sim 4\pi\hbar \tag{2-28}$$

Can we get around this difficulty? After all, the direction of the photon is correlated with its momentum, and if we could somehow measure the recoil of the screen, we could specify the photon (and hence electron) momentum better. True, but once we include the microscope as part of the "observed" system, we must worry about *its* location, since *its* momentum is to be specified. But the microscope, too, must obey the uncertainty relation, and if its momentum is to be specified, its position will be less determined. The final "classical" observation apparatus will always be faced with the indeterminacy.

(b) *The two-slit experiment.* In Chapter 1 we suggested that the interference pattern observed in the passage of an electron[5] through two slits was logically incompatible with our being able to know which slit the electron went through, as such knowledge would imply that the pattern is a superposition of electrons coming from one slit or the other. This, however, cannot give an interference pattern. We may use the uncertainty relation to show that a "monitor" that identifies the slits of passage will destroy the interference pattern. Let the slits

---

[5]We actually discussed photons, but the difficulty is the same for electrons, which are also diffracted.

be separated by a distance a, and let the distance from the slits to the screen be d. The condition for constructive interference is

$$\sin \theta = n \frac{\lambda}{a} \tag{2-29}$$

so that the distance between adjacent maxima on the screen is $d \sin \theta_{n+1} - d \sin \theta_n = d\lambda/a$. Consider a monitor that determines the position of an electron just behind the screen to an accuracy $\Delta y < a/2$, that is, it tells us which slit the electron went through (Fig. 2-3). In doing so, it must impart to the electron a momentum in the y direction (parallel to the screen) whose amount is imprecise, with

$$\Delta p_y > \frac{2h}{a} \tag{2-30}$$

Hence

$$\frac{\Delta p_y}{p} > \frac{2}{a}\frac{h}{p} = \frac{2\lambda}{a} \tag{2-31}$$

Such an uncertainty introduces an indeterminacy in the position of the electron at the screen, whose magnitude is $2\lambda d/a$, at the very least. This, however, is larger than the spacing between maxima, so we conclude that a working monitor will wipe out the interference pattern, and there is no logical contradiction. Conversely, we could, of course, argue that logical consistency demanded that

$$\Delta p_y \, \Delta y > h \tag{2-32}$$

(c) *The "reality" of orbits in the Bohr atom.* As noted in Chapter 1, the Bohr atomic model deals with orbits whose radii are given by $R_n = \hbar n^2/\alpha mc$. Thus an experiment designed to measure the outlines of a given orbit must be such

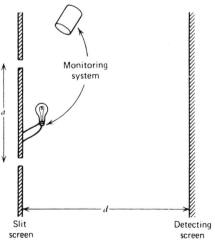

Slit
screen

Detecting
screen

Figure 2-3. The two-slit experiment with monitor.

that a position measurement of the electron in the atom is done with an accuracy

$$\Delta x \ll R_n - R_{n-1} \cong \frac{2\hbar n}{\alpha mc} \tag{2-33}$$

This implies an uncontrollable momentum transfer to the electron that is of magnitude $\Delta p \gg mc\alpha/2n$. This implies an uncertainty in the energy of the electron of magnitude

$$\Delta E \simeq \frac{p\Delta p}{m} \gg \frac{mc\alpha}{n} \cdot \frac{\alpha c}{2n} = \frac{1}{2}\frac{mc^2\alpha^2}{n^2} \tag{2-34}$$

that is, much larger than the binding of the electron in the orbit. Thus such a measurement, as likely as not, will kick the electron out of the orbit, so that no mapping of the orbit is possible.

(d) *The energy–time uncertainty relation.* If we take the relation (2-25) and write it in the form

$$\frac{p\Delta p}{m} \cdot \frac{\Delta xm}{p} \gtrsim \hbar$$

we may interpret the first factor as a measure of the uncertainty in the energy of the system, and the second factor, $\Delta x/v$, as a measure of Δt, an uncertainty in its localizability in time. This suggests the energy–time uncertainty relation

$$\Delta E \, \Delta t \gtrsim \hbar \tag{2-35}$$

Such a relation might also be deduced from the form of the wave packet (2-22) since E and t appear in the same reciprocal relation as p and x, and it is also suggested by the theory of relativity, since space and time, like momentum and energy, are intimately connected with each other.[6] Actually, in nonrelativistic quantum mechanics, space and time play a somewhat different role, and whereas we shall be able to derive (2-25) from the formalism of quantum mechanics, this is not true of (2-35). Nevertheless the energy–time uncertainty relation is as much a part of the qualitative structure of quantum mechanics as (2-25). This can be seen in the special topics Section 4 on "Lifetimes, Line Widths, and Resonance."

Order of Magnitude Estimates

The uncertainty relations may be used to make rough numerical estimates in microscopic physics. Let us illustrate this with several examples, the first of which is the hydrogen atom. If we say that the electron's position inside the atom is unknown, then, if r is its radial coordinate

$$pr \sim \hbar \tag{2-36}$$

---

[6]Both $(ct, \mathbf{r})$ and $E/c, \mathbf{p})$ are *four-vectors* that transform among themselves under Lorentz transformations.

This allows us to express the energy in terms of r:

$$E = \frac{p^2}{2m} - \frac{e^2}{r}$$
$$= \frac{\hbar^2}{2mr^2} - \frac{e^2}{r} \tag{2-37}$$

The minimum value of the energy is obtained from

$$\frac{\partial E}{\partial r} = -\frac{\hbar^2}{mr^3} + \frac{e^2}{r^2} = 0$$

that is,

$$r = \frac{\hbar^2}{me^2} = \frac{\hbar}{mc\alpha} \tag{2-38}$$

and the corresponding value of E is

$$E = -\tfrac{1}{2}mc^2\alpha^2 \tag{2-39}$$

The fact that we obtained the exact value of the energy is, of course, a swindle, since we could equally well have written $pr \sim h$ instead of (2-36), and we would then have obtained a different result. The value of E would, however, have differed from the correct value only by a numerical constant, and the general order of magnitude (that is, $E \sim mc^2/\alpha^2$) would still have been the same. The main point is that in contrast to classical theory, the energy is bounded from below because of the uncertainty principle: an increase in the (negative) potential energy, obtained by decreasing r, that is, localizing the electron closer to the nucleus, carries with it the necessity for increasing the kinetic energy.

As another example, consider the problem of nuclear forces. These have the range of the order of one fermi, that is, 10^{-13} cm. This implies that $p \sim \hbar/r \sim 10^{-14}$ gm cm/sec. The kinetic energy corresponding to this momentum is

$$\frac{p^2}{2M} \sim \frac{10^{-28}}{3.2 \times 10^{-24}} \sim 3 \times 10^{-5} \text{ ergs} \tag{2-40}$$

where M is the nucleon (proton or neutron) mass, which is 1.6×10^{-24} gm. Since the potential that gives rise to the binding must more than compensate for this, we require that

$$|V| \gtrsim 3 \times 10^{-5} \text{ ergs} \gtrsim 20 \text{ MeV} \tag{2-41}$$

Again, this is only a rough order of magnitude, but it does indicate that the potential energy is to be measured in MeV rather than in eV, as in atoms.

Yet another illustration comes from the Yukawa meson theory of nuclear forces. In 1935 Yukawa proposed that the nuclear force arises through the emission of a new quantum, the pi-meson (also called pion), by one of the nucleons, and its absorption by the other. If the mass of the quantum is denoted by μ, then its emission introduces an energy imbalance $\Delta E \sim \mu c^2$, which can only take place for a time $\Delta t \sim \hbar/E \sim \hbar/\mu c^2$. The range corresponding to a particle traveling for this time is of the order of $c\Delta t \sim \hbar/\mu c$. If we take for the range $r_0 = 1.4 \times 10^{-13}$ cm, then we find that

$$\mu c^2 \cong \frac{\hbar c}{r_0} = \frac{10^{-27} \times 3 \times 10^{10}}{1.4 \times 10^{-13}} \text{ ergs}$$

$$\cong 130 \text{ MeV}$$

(2-42)

When the pion was finally discovered, it was found that this estimate was remarkably accurate, since for the pion $\mu c^2 \cong 140$ MeV.

In summary, our tentative attempt to wed wave and particle properties consistent with experiment in the most naive way has led us to an uncertainty in the description of atomic phenomena at the classical level, and this uncertainty is both necessary for a consistent description of (Gedanken) experiments, and in accord with what is observed.

Problems

1. Consider a wave packet defined by (2-1) with $g(k)$ given by

$$g(k) = 0 \qquad k < -K/2$$
$$= N \qquad -K/2 < k < K/2$$
$$= 0 \qquad K/2 < k$$

   (a) Find the form of $f(x)$.
   (b) Find the value of N for which

$$\int_{-\infty}^{\infty} dx |f(x)|^2 = 1$$

   (c) How is this related to the choice of N for which

$$\int_{-\infty}^{\infty} dk |g(k)|^2 = \frac{1}{2\pi}$$

   (d) Show that a reasonable definition of Δx for your answer to (a) yields

$$\Delta k \, \Delta x > 1$$

   independent of the value of K.

2. Given that

$$g(k) = \frac{N}{k^2 + \alpha^2}$$

   calculate the form of $f(x)$. Again, plot the two functions and show that

$$\Delta k \, \Delta x > 1$$

   independent of your choice of α.

3. Consider the problem of the spreading of a Gaussian wave packet for a free particle, where the relation

$$\omega = \frac{\hbar k^2}{2m}$$

holds. Use (2-17) to calculate the fractional change in the size of the wave packet in one second, if

(a) the packet represents an electron, with the wave packet having a size of 10^{-4} cm; 10^{-8} cm.

(b) the packet represents an object of mass 1 gm and has size 1 cm.

It will be convenient to express the width in units of $\hbar/mc$, where m is the mass of the particle represented by the packet.

4. A beam of electrons is to be fired over a distance of 10^4 km. If the size of the initial packet is 1 mm, what will be its size upon arrival, if its kinetic energy is (a) 13.6 eV, (b) 100 MeV?

   [*Caution:* The relation between kinetic energy (K.E.) and momentum is not always K.E. $= p^2/2m!$]

5. The relation between the wavelength and the frequency in a wave guide is given by

$$\lambda = \frac{c}{\sqrt{\nu^2 - \nu_0^2}}$$

What is the group velocity of such waves?

6. For surface tension waves in shallow water, the relation between frequency and wavelength is given by

$$\nu = \left(\frac{2\pi T}{\rho\lambda^3}\right)^{1/2}$$

where T is the surface tension and ρ the density. What is the group velocity of the waves, and its relation to the phase velocity, defined to be $v_p = \lambda\nu$? For gravity waves (deep water), the relation is given by

$$\nu = \left(\frac{g}{2\pi\lambda}\right)^{1/2}$$

What are the group and phase velocities?

7. Use the uncertainty relation to estimate the ground state energy of a harmonic oscillator. The energy is given by

$$E = \frac{p^2}{2m} + \frac{1}{2}m\omega^2 x^2$$

8. Use the uncertainty relation to estimate the ground-state energy of a particle in the potential $V(x) = gx^4$. Check the dimensions of your answer.

9. Nuclei, typically of size 10^{-12} cm, frequently emit electrons, with energies of 1–10 MeV. Use the uncertainty principle to show that electrons of energy 1 MeV could not be contained in the nucleus before the decay.

10. The apparatus sketched below appears to allow a violation of the uncertainty relation. The lateral location can be determined with accuracy $\Delta y \sim a$, and the transverse momentum of the incident beam can be made as small as possible by making L arbitrarily large. Analyze the apparatus in detail, point out the hidden assumptions just made, and show that the uncertainty relation is not violated.

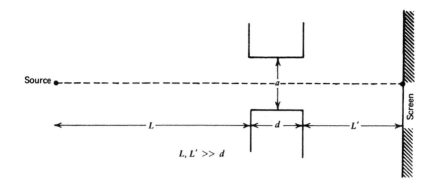

11. What is the energy uncertainty (linewidth, to be expressed in eV) of states with lifetimes of (a) 2.6×10^{-10} sec, (b) 10^{-23} sec, (c) 12 minutes.

12. Monochromatic light with wavelength of 6000 Å passes through a fast shutter that opens for 10^{-9} sec. What will be the spread in wavelengths in the no longer monochromatic light?

References

Wave packets are discussed in a number of textbooks. Most useful at this level are:

S. Borowitz, *Fundamentals of Wave Mechanics*, W. A. Benjamin, New York, 1967.
J. L. Powell and B. Crasemann, *Quantum Mechanics*, Addison-Wesley, Reading, Mass., 1961.
D. Bohm, *Quantum Theory*, Dover Publications Inc., New York (1989).

All textbooks on quantum mechanics necessarily deal with the uncertainty relations. Very thorough discussions can be found in the book by Bohm just cited, and in W. Heisenberg, *The Physical Principles of the Quantum Theory*, Dover Publications, New York, 1930.

Discussions of the uncertainty relations may also be found in any of the more advanced books listed at the end of this book.

THE SCHRÖDINGER WAVE EQUATION AND THE PROBABILITY INTERPRETATION

In this chapter we discuss some properties of the free-particle Schrödinger equation, which we obtained in Chapter 2. We introduce the probability interpretation of the wave function, and are then led to the definition of the momentum in quantum mechanics, and to the Schrödinger equation that describes a particle in a potential $V(x)$. Our starting point is the partial differential equation

$$i\hbar \frac{\partial \psi(x, t)}{\partial t} = - \frac{\hbar^2}{2m} \frac{\partial^2 \psi(x, t)}{\partial x^2} \tag{3-1}$$

which we take to be the correct equation for the description of a free particle.

Inverting the sequence that led to (2-23), we see that the most general solution of this equation is

$$\psi(x, t) = \frac{1}{\sqrt{2\pi\hbar}} \int dp \ \phi(p) \ e^{i[px - (p^2/2m)t]/\hbar} \tag{3-2}$$

[The reason for the normalization factor in front of the integral appears in (3-27).] Before turning to the crucial point of interpreting the meaning of the solution $\psi(x, t)$ of this equation, we draw attention to the fact that the equation is of first order in the time-derivative. This implies that once the initial value of ψ, namely,

$\psi(x, 0)$, is given, its values at all other times can be found. This is evident from the form of the equation as seen by a digital computer[1]

$$\psi(x, t + \Delta t) = \psi(x, t) + \frac{i\hbar}{2m} \frac{\partial^2 \psi(x, t)}{\partial x^2} \Delta t \tag{3-3}$$

or from the form of the most general solution. Given $\psi(x, 0)$, the function $\phi(p)$ may be found from (3-2). The Fourier Integral

$$\psi(x, 0) = \frac{1}{\sqrt{2\pi\hbar}} \int dp \ \phi(p) \ e^{ipx/\hbar} \tag{3-4}$$

may be inverted, and once $\phi(p)$ is known, the solution is known for all values of t. Note that there is no "uncertainty" in the differential equation: once the initial state of the wave packet is specified—and there are, so far, no restrictions on $\psi(x, 0)$—then that wave packet is completely specified at all later times.

THE PROBABILITY INTERPRETATION

In searching for an interpretation for $\psi(x, t)$ we must bear in mind (1) that $\psi(x, t)$ is in general a complex function [e.g., (2-16)], and (2) that the function $|\psi(x, t)|$ is large where the particle is supposed to be, and small elsewhere. It also has associated with it the feature of spreading, discussed in Chapter 2. Almost immediately after the discovery of the Schrödinger equation (which followed the discovery of Quantum Mechanics by Heisenberg in 1925 by only six months), Max Born studied the scattering of a beam of electrons by a target, which suggested to him the proper interpretation of the wave function. He proposed that

$$P(x, t) \ dx = |\psi(x, t)|^2 \ dx \tag{3-5}$$

define the *probability that the particle described by the wave function $\psi(x, t)$ may be found between x and $x + dx$ at time t.* The probability density $P(x, t)$ is real, is large where the particle is supposed to be, and its spreading does not imply that a particular particle is spreading; all it means is that as time goes by, one is less likely to find the particle where one put it at $t = 0$.

For this interpretation to hold, we must require that

$$\int_{-\infty}^{\infty} P(x, t) \ dx = 1 \tag{3-6}$$

since the particle must be somewhere. In a linear equation like (3-1), the solution $\psi(x, t)$ may be multiplied by a constant, and it still remains a solution. Thus (3-6) restricts the solutions $\psi(x, t)$ to a class of functions that are *square integrable*. We shall see below that it is enough to require that

$$\int_{-\infty}^{\infty} dx |\psi(x, 0)|^2 < \infty \tag{3-7}$$

---

[1]For a discrete mesh, $\partial\psi(x, t)/\partial t$ must be replaced by $[\psi(x, t + \Delta t) - \psi(x, t)]/\Delta t$, with Δt small but not vanishing.

that is, the *initial state wave functions must be square integrable*. With an infinite integration interval this means that $\psi(x, 0)$ must go to zero faster than $x^{-1/2}$. We shall also require that the wave functions $\psi(x, t)$ be *continuous* in x.

The Importance of Phases

Since $|\psi(x, t)|^2$ is the physically significant quantity, it would appear that the *phase* of the solution of the equation is somehow unimportant. That is wrong! Since the equation (3-1) is linear, if $\psi_1(x, t)$ and $\psi_2(x, t)$ are solutions, so is

$$\psi(x, t) = \psi_1(x, t) + \psi_2(x, t) \tag{3.8}$$

If $\psi_1(x, t) = R_1 e^{i\theta_1}$ and $\psi_2(x, t) = R_2 e^{i\theta_2}$, with $R_1, R_2, \theta_1, \theta_2$ real, then

$$
\begin{aligned}
|\psi(x, t)|^2 &= |e^{i\theta_1}(R_1 + R_2 e^{i(\theta_2 - \theta_1)})|^2 \\
&= R_1^2 + R_2^2 + 2R_1 R_2 \cos(\theta_1 - \theta_2)
\end{aligned}
\tag{3.9}
$$

which shows that the relative phase *is* important. An overall phase in $\psi(x, t)$ can be ignored; only the relative phase $\theta_1 - \theta_2$ between the two wave functions ψ_1 and ψ_2 appears in $|\psi|^2$.

Notice that the right-hand side of (3-9) is just what one sees in the treatment of superposition of waves. In fact, it is the linearity of the wave functions that leads to the interference pattern that emerges from the cosine in this expression, and when one speaks of "wave behavior" of electrons, or photons, one is primarily speaking of the linearity.

Probability Current

We now show that the condition (3-6), imposed at $t = 0$, holds true for all times. We need (3-1) and its complex conjugate

$$-i\hbar \frac{\partial \psi^*(x, t)}{\partial t} = -\frac{\hbar^2}{2m} \frac{\partial^2 \psi^*(x, t)}{\partial x^2} \tag{3-10}$$

Now

$$
\begin{aligned}
\frac{\partial}{\partial t} P(x, t) &= \frac{\partial \psi^*}{\partial t} \psi + \psi^* \frac{\partial \psi}{\partial t} \\
&= \frac{1}{i\hbar} \left(\frac{\hbar^2}{2m} \frac{\partial^2 \psi^*}{\partial x^2} \psi - \frac{\hbar^2}{2m} \psi^* \frac{\partial^2 \psi}{\partial x^2} \right) \\
&= -\frac{\partial}{\partial x} \left[\frac{\hbar}{2im} \left(\psi^* \frac{\partial \psi}{\partial x} - \frac{\partial \psi^*}{\partial x} \psi \right) \right]
\end{aligned}
$$

If we define the *flux* (or equivalently the *probability current*) by

$$j(x, t) = \frac{\hbar}{2im} \left(\psi^* \frac{\partial \psi}{\partial x} - \frac{\partial \psi^*}{\partial x} \psi \right) \tag{3-11}$$

we see that

$$\frac{\partial}{\partial t} P(x, t) + \frac{\partial}{\partial x} j(x, t) = 0 \tag{3-12}$$

Integrating, we find that

$$\frac{\partial}{\partial t} \int_{-\infty}^{\infty} dx\, P(x, t) = -\int_{-\infty}^{\infty} dx\, \frac{\partial}{\partial x} j(x, t) = 0 \tag{3-13}$$

since for square integrable functions, $j(x, t)$ vanishes at infinity. Incidentally, had we allowed discontinuities in $\psi(x)$, we would have been led to delta-function[2] singularities in the flux, and hence in the probability density, which is unacceptable in a physically observable quantity.

The relation (3-12) is a conservation law. It expresses the fact that a change in the density in a region in x is compensated by a net change in flux into that region

$$\frac{\partial}{\partial t} \int_{a}^{b} dx\, P(x, t) = -\int_{a}^{b} dx\, \frac{\partial}{\partial x} j(x, t)$$
$$= j(a, t) - j(b, t) \tag{3-14}$$

The definition of $P(x, t)$, $j(x, t)$ and the conservation law are maintained if the equation (3-1) is changed to

$$i\hbar \frac{\partial \psi(x, t)}{\partial t} = -\frac{\hbar^2}{2m} \frac{\partial^2 \psi(x, t)}{\partial x^2} + V(x)\, \psi(x, t) \tag{3-15}$$

provided that $V(x)$ is real. This is important, since we will argue later that (3-15) is the Schrödinger equation for a particle in a potential $V(x)$. The generalization to three dimensions is straightforward. Equation 3-15 becomes

$$i\hbar \frac{\partial \psi(x, y, z, t)}{\partial t} = -\frac{\hbar^2}{2m} \left(\frac{\partial^2}{\partial x^2} + \frac{\partial^2}{\partial y^2} + \frac{\partial^2}{\partial z^2} \right) \psi(x, y, z, t)$$
$$+ V(x, y, z)\, \psi(x, y, z, t)$$

that is,

$$i\hbar \frac{\partial \psi(\mathbf{r}, t)}{\partial t} = -\frac{\hbar^2}{2m} \nabla^2 \psi(\mathbf{r}, t) + V(\mathbf{r})\, \psi(\mathbf{r}, t) \tag{3-16}$$

and the generalization of (3-12) reads

$$\frac{\partial}{\partial t} P(\mathbf{r}, t) + \mathbf{\nabla} \cdot \mathbf{j}(\mathbf{r}, t) = 0 \tag{3-17}$$

where

$$P(\mathbf{r}, t) = |\psi(\mathbf{r}, t)|^2 \tag{3-18}$$

---

[2] See Appendix A for a discussion of delta functions.

and

$$\mathbf{j}(\mathbf{r}, t) = \frac{\hbar}{2im} [\psi^*(\mathbf{r}, t)\nabla\psi(\mathbf{r}, t) - \nabla\psi^*(\mathbf{r}, t) \psi(\mathbf{r}, t)] \tag{3-19}$$

Expectation Values and Particle Momentum

Given the probability density $P(x, t)$, expectation values of functions of x may be calculated. In general, we have

$$\langle f(x) \rangle = \int dx \, f(x) \, P(x, t) = \int dx \psi^*(x, t) \, f(x) \, \psi(x, t) \tag{3-20}$$

This integral is only defined if the integral converges. Since there is no particular reason to limit the range of functions $f(x)$ whose expectation value we may wish to calculate, we generalize our earlier statement about the behavior of wave functions at infinity. We shall assume that the wave functions $\psi(x)$ and all of their derivatives vanish sufficiently rapidly at infinity to give us no trouble.

The expression for $\langle f(x) \rangle$ does not help us if we want to calculate the expectation value of the momentum since we do not know how to write momentum in terms of x. We try the following: since classically,

$$p = mv = m \frac{dx}{dt} \tag{3-21}$$

we shall write

$$\langle p \rangle = m \frac{d}{dt} \langle x \rangle = m \frac{d}{dt} \int dx \, \psi^*(x, t) \, x\psi(x, t) \tag{3-22}$$

This yields

$$\langle p \rangle = m \int_{-\infty}^{\infty} dx \left(\frac{\partial \psi^*}{\partial t} x\psi + \psi^* x \frac{\partial \psi}{\partial t} \right)$$

Note that there is no dx/dt under the integral sign. The only quantity that varies with time is $\psi(x, t)$, and it is this variation that gives rise to a change in $\langle x \rangle$ with time. Making use of (3-1) and its complex conjugate, we have

$$\langle p \rangle = \frac{\hbar}{2i} \int_{-\infty}^{\infty} dx \left(\frac{\partial^2 \psi^*}{\partial x^2} x\psi - \psi^* x \frac{\partial^2 \psi}{\partial x^2} \right)$$

Now

$$\frac{\partial^2 \psi^*}{\partial x^2} x\psi = \frac{\partial}{\partial x} \left(\frac{\partial \psi^*}{\partial x} x\psi \right) - \frac{\partial \psi^*}{\partial x} \psi - \frac{\partial \psi^*}{\partial x} x \frac{\partial \psi}{\partial x}$$

$$= \frac{\partial}{\partial x} \left(\frac{\partial \psi^*}{\partial x} x\psi \right) - \frac{\partial}{\partial x} (\psi^*\psi) + \psi^* \frac{\partial \psi}{\partial x}$$

$$- \frac{\partial}{\partial x} \left(\psi^* x \frac{\partial \psi}{\partial x} \right) + \psi^* \frac{\partial \psi}{\partial x} + \psi^* x \frac{\partial^2 \psi}{\partial x^2}$$

Hence the integrand has the form

$$\frac{\partial}{\partial x}\left(\frac{\partial \psi^*}{\partial x}\, x\psi - \psi^* x\, \frac{\partial \psi}{\partial x} - \psi^* \psi\right) + 2\psi^* \frac{\partial \psi}{\partial x}$$

so that

$$\langle p \rangle = \int dx\; \psi^*(x, t)\, \frac{\hbar}{i}\frac{\partial}{\partial x}\, \psi(x, t) \tag{3-23}$$

since the integral of the derivatives vanishes for square integrable functions.

This suggests that the momentum is represented by the *operator*

$$p = \frac{\hbar}{i}\frac{\partial}{\partial x} \tag{3-24}$$

Once we accept this, we get the more general result that

$$\langle f(p) \rangle = \int dx\; \psi^*(x, t) f\left(\frac{\hbar}{i}\frac{\partial}{\partial x}\right)\psi(x, t) \tag{3-25}$$

so that, for example,

$$\langle p^2 \rangle = \int dx\; \psi^*(x, t)\left(-\hbar^2\frac{\partial^2}{\partial x^2}\right)\psi(x, t)$$

The Wave Function in Momentum Space

Armed with this representation we can now discuss the physical significance of $\phi(p)$, which appears in (3-2). First, it is sufficient to consider that equation at $t = 0$, since $\phi(p)$ does not have any time dependence. With

$$\psi(x) = \frac{1}{\sqrt{2\pi\hbar}}\int dp\; \phi(p)\, e^{ipx/\hbar} = \sqrt{\frac{\hbar}{2\pi}}\int dk\; \phi(\hbar k)\, e^{ikx}$$

we find, using the inversion formula for a Fourier integral, that

$$\phi(\hbar k) = \frac{1}{\sqrt{2\pi\hbar}}\int dx\; \psi(x)\, e^{-ikx}$$

that is,

$$\phi(p) = \frac{1}{\sqrt{2\pi\hbar}}\int dx\; \psi(x)\, e^{-ipx/\hbar} \tag{3-26}$$

Now

$$\int dp\ \phi^*(p)\ \phi(p) = \int dp\ \phi^*(p)\ \frac{1}{\sqrt{2\pi\hbar}} \int dx\ \psi(x)\ e^{-ipx/\hbar}$$

$$= \int dx\ \psi(x)\ \frac{1}{\sqrt{2\pi\hbar}} \int dp\ \phi^*(p)\ e^{-ipx/\hbar} \qquad (3\text{-}27)$$

$$= \int dx\ \psi(x)\ \psi^*(x) = 1$$

This result is known as *Parseval's theorem* in the mathematical literature. It states that if a function is normalized to 1, so is its Fourier transform.

Next consider

$$\langle p \rangle = \int dx\ \psi^*(x)\ \frac{\hbar}{i}\ \frac{d\psi(x)}{dx}$$

$$= \int dx\ \psi^*(x)\ \frac{\hbar}{i}\ \frac{d}{dx}\ \frac{1}{\sqrt{2\pi\hbar}} \int dp\ \phi(p)\ e^{ipx/\hbar}$$

$$\qquad\qquad\qquad\qquad\qquad\qquad\qquad (3\text{-}28)$$

$$= \int dp\ \phi(p)\ p\ \frac{1}{\sqrt{2\pi\hbar}} \int dx\ \psi^*(x)\ e^{ipx/\hbar}$$

$$= \int dp\ \phi(p)\ p\phi^*(p)$$

This result, together with (3-27), strongly suggests that $\phi(p)$ should be interpreted as the wave function in momentum space, with $|\phi(p)|^2$ yielding the probability density for finding the particle with momentum p. When $\psi(x, t)$ is a solution of (3-15), we may define $\phi(p, t)$ by

$$\psi(x, t) = \frac{1}{\sqrt{2\pi\hbar}} \int dp\ \phi(p, t)\ e^{ipx/\hbar} \qquad (3\text{-}29)$$

The fact that in general $\phi(p, t)$ has a time dependence does not change (3-27), (3-28), or its interpretation. Lest the reader think that in spite of this symmetry between x- and p-space, $p = (\hbar/i)(\partial/\partial x)$ is an operator, and x is not, we note that x is in fact an operator too. It happens to have a particularly simple form in x-space, but if we want to calculate $\langle f(x) \rangle$ in momentum space, then we can show by methods very similar to the ones used earlier that

$$\langle x \rangle = \int dp\ \phi^*(p, t)\ \left(i\hbar\ \frac{\partial}{\partial p} \right)\ \phi(p, t) \qquad (3\text{-}30)$$

In other words, the operator x has the representation

$$x = i\hbar\ \frac{\partial}{\partial p} \qquad (3\text{-}31)$$

in momentum space.

The following example illustrates some computations for a specific wave function $\psi(x)$.

Example   Consider a particle whose normalized wave function is

$$\psi(x) = 2\alpha\sqrt{\alpha}\, xe^{-\alpha x} \qquad x > 0$$
$$= 0 \qquad\qquad\quad x < 0$$

(a) For what value of x does $P(x) = |\psi(x)|^2$ peak?

(b) Calculate $\langle x \rangle$ and $\langle x^2 \rangle$.

(c) What is the probability that the particle is found between $x = 0$ and $x = 1/\alpha$?

(d) Calculate $\phi(p)$ and use this to calculate $\langle p \rangle$ and $\langle p^2 \rangle$.

(a) The peak in $P(x)$ occurs when $dP(x)/dx = 0$, that is, when

$$\frac{d}{dx}(x^2 e^{-2\alpha x}) = 2x(1 - \alpha x)e^{-2\alpha x} = 0$$

that is, at $x = 1/\alpha$.

(b)

$$\langle x \rangle = \int_0^\infty dx\, x(4\alpha^3 x^2 e^{-2\alpha x}) = \frac{1}{4\alpha}\int_0^\infty dy\, y^3 e^{-y} = \frac{3!}{4\alpha} = \frac{3}{2\alpha}$$

$$\langle x^2 \rangle = \int_0^\infty dx\, x^2\,(4\alpha^3 x^2 e^{-2\alpha x}) = \frac{4!}{8\alpha^2} = \frac{3}{\alpha^2}$$

(c) The desired probability is

$$P = \int_0^{1/\alpha} dx\,(4\alpha^3)x^2\, e^{-2\alpha x} = \frac{1}{2}\int_0^2 dy\, y^2\, e^{-y} = 0.32$$

(d)

$$\phi(p) = \frac{1}{\sqrt{2\pi\hbar}}\int_0^\infty dx\, e^{-ipx/\hbar}\,(2\alpha\sqrt{\alpha})xe^{-\alpha x}$$

$$= \sqrt{\frac{4\alpha^3}{2\pi\hbar}}\frac{d}{d\alpha}\int_0^\infty dx\, e^{-(\alpha + ip/\hbar)x} = -\sqrt{\frac{4\alpha^3}{2\pi\hbar}}\frac{1}{(\alpha + ip/\hbar)^2}$$

From this we can calculate

$$\langle p \rangle = \int_{-\infty}^\infty dp\, p\, |\phi(p)|^2 = \frac{4\alpha^3}{2\pi\hbar}\int_{-\infty}^\infty dp\, \frac{p}{(\alpha^2 + p^2/\hbar^2)^2} = 0$$

$$\langle p^2 \rangle = \frac{4\alpha^3}{2\pi\hbar}\int_{-\infty}^\infty dp\, \frac{p^2}{(\alpha^2 + p^2/\hbar^2)^2} = \frac{8\alpha^3}{2\pi\hbar}\int_0^\infty dp\, \frac{p^2}{(\alpha^2 + p^2/\hbar^2)^2}$$

With the change of variables $p = \hbar\alpha \tan\theta$ we get

$$\langle p^2 \rangle = \frac{4\alpha^2\hbar^2}{\pi}\int_0^{\pi/2} d\theta\, \sin^2\theta = \alpha^2\hbar^2$$

THE SCHRÖDINGER EQUATION FOR A PARTICLE IN A POTENTIAL

We observe that the equation

$$i\hbar \frac{\partial \psi(x, t)}{\partial t} = -\frac{\hbar^2}{2m} \frac{\partial^2 \psi(x, t)}{\partial x^2}$$

may, with the identification $(\hbar/i)(\partial/\partial x) = p_{op}$ be written in the form

$$i\hbar \frac{\partial \psi(x, t)}{\partial t} = \frac{p_{op}^2}{2m} \psi(x, t) \tag{3-32}$$

The operator on the right is just the energy for a free particle. If we generalize this to a particle in a potential, we write

$$i\hbar \frac{\partial \psi(x, t)}{\partial t} = \left[\frac{p_{op}^2}{2m} + V(x) \right] \psi(x, t) \tag{3-33}$$

or, more explicitly

$$i\hbar \frac{\partial \psi(x, t)}{\partial t} = -\frac{\hbar^2}{2m} \frac{\partial^2 \psi(x, t)}{\partial x^2} + V(x) \, \psi(x, t) \tag{3-34}$$

This equation, generalizing (3-1), is the *basic equation of nonrelativistic quantum mechanics,* and it was first proposed by Schrödinger. The Schrödinger equation, just obtained, can also be written in the form

$$i\hbar \frac{\partial \psi(x, t)}{\partial t} = H \, \psi(x, t) \tag{3-35}$$

where H is the *energy operator.* H is commonly called the *Hamiltonian,* because it is an operator version of the classical mechanical Hamiltonian function.

We will find that operators play a central role in quantum mechanics, and in due course we will learn a great deal about their properties. At this point we note a few important ones:

1. In contrast to ordinary numbers, operators do not always commute. If we define

$$[A, B] = AB - BA \tag{3-36}$$

then

$$\begin{aligned} [p, x] \, \psi(x, t) &= \frac{\hbar}{i} \frac{\partial}{\partial x} x\psi(x, t) - x \frac{\hbar}{i} \frac{\partial \psi(x, t)}{\partial x} \\ &= \frac{\hbar}{i} \, \psi(x, t) \end{aligned} \tag{3-37}$$

that is, we have the *commutation relation*

$$[p, x] = \frac{\hbar}{i} \tag{3-38}$$

This leads to an ambiguity in transcribing a classical function $f(x, p)$ into operator form, and we shall adopt the rule that $f(x, p)$ be symmetrized in x and p. Thus

$$xp \rightarrow \tfrac{1}{2}(xp + px)$$
$$x^2p \rightarrow \tfrac{1}{4}(x^2p + 2xpx + px^2) \tag{3-39}$$

and so on. Later we will see that it is the lack of commutativity of x and p that stands behind the uncertainty relations connecting these two variables.

2. The appearance of the operator p, with its i, might lead us to worry about the reality of the expectation value of p. We can, however, check the fact that $\langle p \rangle$ is real. We have

$$
\begin{aligned}
\langle p \rangle - \langle p \rangle^* &= \int dx \ \psi^*(x) \frac{\hbar}{i} \frac{\partial \psi}{\partial x} - \int dx \ \psi(x) \left(-\frac{\hbar}{i} \frac{\partial \psi^*}{\partial x} \right) \\
&= \frac{\hbar}{i} \int dx \left(\psi^* \frac{\partial \psi}{\partial x} + \frac{\partial \psi^*}{\partial x} \psi \right) \\
&= \frac{\hbar}{i} \int dx \frac{\partial}{\partial x} (\psi^* \psi) \\
&= 0
\end{aligned} \tag{3-40}
$$

provided the wave function vanishes at infinity, which it does for a square integrable function. Sometimes one has occasion to use functions that are not square integrable but that have certain periodicity conditions, for example,

$$\psi(x) = \psi(x + L) \tag{3-41}$$

If one restricts oneself to working in the region $0 \leq x \leq L$, then $\hbar/i \ d/dx$ still has a real expectation value, since in (3-40),

$$
\begin{aligned}
\langle p \rangle - \langle p \rangle^* &= \frac{\hbar}{i} \int_0^L dx \frac{\partial}{\partial x} (\psi^*(x) \ \psi(x)) \\
&= \frac{\hbar}{i} |\psi(L)|^2 - \frac{\hbar}{i} |\psi(0)|^2 = 0
\end{aligned} \tag{3-42}
$$

An operator whose expectation value for all admissible wave functions is real is called a *hermitian operator,* and hence p, like x, is a hermitian operator.[3] Thus[4] p^2 is a hermitian operator and, for real $V(x)$, so is

$$H = \frac{p^2}{2m} + V(x) \tag{3-43}$$

---

[3]Some mathematical background on operators is summarized in Appendix B.

[4]From now on we will drop the subscript op on p_{op}. We will use it only when there is danger of confusion with a number described by the letter p.

In summary:

1. The time dependence of wave functions is given by the first-order partial differential equation

$$i\hbar \, \frac{\partial \psi(x, t)}{\partial t} = H \, \psi(x, t) \tag{3-44}$$

where H is the operator $p^2/2m + V(x)$.

2. Wave functions are restricted to square integrable functions.

3. The *probability density* for finding the particle at x is

$$P(x, t) = |\psi(x, t)|^2 \tag{3-45}$$

4. The function $\phi(p, t)$ defined by

$$\psi(x, t) = \frac{1}{\sqrt{2\pi\hbar}} \int dp \, \phi(p, t) \, e^{ipx/\hbar} \tag{3-46}$$

is the wave function in momentum space, and the *probability density* for finding the particle with momentum p is $|\phi(p, t)|^2$.

5. The momentum p and the position x are *operators*, that is, they are quantities that differ from numbers because of their lack of commutativity. In x-space, the momentum operator takes the form

$$p = \frac{\hbar}{i} \, \frac{\partial}{\partial x} \tag{3-47}$$

and in p-space, the x operator takes the form

$$x = i\hbar \, \frac{\partial}{\partial p} \tag{3-48}$$

both consistent with the fundamental commutation relation for x with p,

$$[p, x] = \frac{\hbar}{i} \tag{3-49}$$

We are now ready for a quantitative discussion of quantum mechanics. We have abandoned the notion of a wave packet as representing a particle. This notion was helpful to us in making the Schrödinger equation plausible, but now it is $\psi(x, t)$ and its probabilistic interpretation that tell us where the particle is, without the particle being thought of as "made up out of waves."

Problems

1. Show by explicit calculation that

$$\int_{-\infty}^{\infty} dx \, \psi^*(x) x \psi(x) = \int_{-\infty}^{\infty} dp \, \phi^*(p) \left(i\hbar \, \frac{\partial}{\partial p} \right) \phi(p)$$

[*Hint:* You will need to use what you know about the behavior of $\phi(p)$ at $p = \pm\infty$.]

2. Show that the conservation law (3-11) holds when $\psi(x, t)$ is a solution of the Schrödinger equation with a potential $V(x)$, (3-14), provided that $V(x)$ is real.

3. Suppose that $V(x)$ is complex. Obtain an expression for $\partial P(x, t)/\partial t$ and $d/dt \int dx\, P(x, t)$. For absorption, the last must be negative. What does this tell us about $V(x)$?

4. Given that

$$\psi(x) = \left(\frac{\pi}{\alpha}\right)^{-1/4} e^{-\alpha x^2/2}$$

(a) calculate $\langle x^n \rangle$ for n even. (Why does this vanish for n odd?)
(b) We shall learn later that in quantum mechanics the uncertainty in position may be described by $\Delta x = \sqrt{\langle x^2 \rangle - \langle x \rangle^2}$. Calculate Δx for the $\psi(x)$ given above.

5. (a) Calculate $\phi(p)$ for the system described by the wave function given in Problem 4.
(b) Calculate $\langle p^n \rangle$, and show that this vanishes for n odd.
(c) Assuming that the uncertainty in momentum is given by $\Delta p = \sqrt{\langle p^2 \rangle - \langle p \rangle^2}$, calculate Δp.
(d) Use the result above and the value of Δx calculated in Problem 5 to calculate the value of the product $\Delta p \Delta x$.

6. Calculate $\langle x \rangle$, $\langle x^2 \rangle$, and Δx, as well as $\langle p \rangle$, $\langle p^2 \rangle$, and Δp for the system described by the normalized wave function

$$\psi(x) = \sqrt{\frac{2a^3}{\pi}} \frac{1}{x^2 + a^2}$$

Use these to calculate $\Delta x \Delta p$.
[*Note:* Using contour integrals one can show that

$$\varphi(p) = \sqrt{\frac{a}{\hbar}}\, e^{-a|p|/\hbar}$$

This, or

$$\int_{-\infty}^{\infty} dx \frac{1}{x^2 + a^2} = \pi(a^2)^{-1/2}$$

(and its derivatives w.r.t. a^2) can be used to evaluate the integrals.]

7. Calculate $\langle p \rangle$ and $\langle p^2 \rangle$ for the wave function $\psi(x) = R(x)\, e^{iS(x)/\hbar}$, where $R(x)$ and $S(x)$ are real functions of x.

8. Show that the operator relation

$$e^{ipa/\hbar}\, x e^{-ipa/\hbar} = x + a$$

holds. The operator e^A is defined to be

$$e^A = \sum_{n=0}^{\infty} A^n/n!$$

[*Hint:* Calculate $e^{ipa/\hbar} x e^{-ipa/\hbar} f(p)$ where $f(p)$ is any function of p, and use the representation $x = i\hbar\, d/dp$.]

9. Consider the functions $\psi(\theta)$ of the angular variable θ, restricted to the interval $-\pi \le \theta \le \pi$.

   If the wave functions satisfy the condition $\psi(\pi) = \psi(-\pi)$, show that the operator

$$L = \frac{\hbar}{i}\frac{d}{d\theta}$$

has a real expectation value.

10. Consider $\phi(p)$, the momentum space wave function of a particle. If this function is only defined for positive values of p, what condition must $\phi(p)$ satisfy in order that x have a real expectation value? [Use (3-31).]

Chapter 4

EIGENFUNCTIONS AND EIGENVALUES

Let us consider the time-dependent Schrödinger equation obtained in Chapter 3,

$$i\hbar \frac{\partial \psi(x, t)}{\partial t} = -\frac{\hbar^2}{2m} \frac{\partial^2 \psi(x, t)}{\partial x^2} + V(x)\, \psi(x, t) \tag{4-1}$$

and attempt to solve it by reducing it to a pair of ordinary differential equations in one variable. Write

$$\psi(x, t) = T(t)u(x) \tag{4-2}$$

which implies that

$$i\hbar u(x) \frac{dT(t)}{dt} = \left[-\frac{\hbar^2}{2m} \frac{d^2 u(x)}{dx^2} + V(x)u(x) \right] T(t)$$

Dividing by $u(x)\, T(t)$ we get

$$i\hbar \frac{dT(t)/dt}{T(t)} = \frac{-(\hbar^2/2m)(d^2 u(x)/dx^2) + V(x)\, u(x)}{u(x)} \tag{4-3}$$

This can only be satisfied if both sides are equal to a constant, which we call E. The solution of

$$i\hbar \frac{dT(t)}{dt} = ET(t) \tag{4-4}$$

is

$$T(t) = Ce^{-iEt/\hbar} \tag{4-5}$$

where C is a constant. The other equation is

$$-\frac{\hbar^2}{2m}\frac{d^2u(x)}{dx^2} + V(x)u(x) = Eu(x) \tag{4-6}$$

This equation is frequently called the *time-independent Schrödinger equation*. Its character is really different from that of (4-1). Equation 4-1 describes the time development of $\psi(x, t)$; (4-6) is an *eigenvalue equation*.

EIGENVALUE EQUATIONS

A discussion of eigenvalue equations needs a more careful consideration of operators than was presented in the previous chapter. Most generally, an operator acting on a function maps it into another function. Let us consider some examples

$$\begin{aligned}
Of(x) &= f(x) + x^2 \\
Of(x) &= [f(x)]^2 \\
Of(x) &= f(3x^2 + 1) \\
Of(x) &= [df(x)/dx]^3 \\
Of(x) &= df(x)/dx - 2f(x) \\
Of(x) &= \lambda f(x)
\end{aligned} \tag{4-7}$$

All of these examples share the property that given a function $f(x)$, there is a rule that determines $Of(x)$ for us. There is a special class of operators, called *linear operators* (we denote these operators by L to distinguish them from the general operators O). These have the property that

$$L[f_1(x) + f_2(x)] = Lf_1(x) + Lf_2(x) \tag{4-8}$$

and, with c an arbitrary complex number,

$$Lcf(x) = cLf(x) \tag{4-9}$$

Thus, in our list, the third, fifth, and sixth operators are linear.

A linear operator will map one function into another, as in the example

$$Lf(x) = \frac{df(x)}{dx} - 2f(x)$$

It is instructive to think of the functions as analogous to vectors in a three-dimensional space. Here, the action of an operator is to transform a vector into another vector. In the special case that the vectors are all of unit length, an operator will transform one point on a unit sphere into another. An operator, in this special (but very relevant) example, may be a rotation about an axis (Fig. 4-1). Let the operator be a rotation of, say, 30° about the z-axis. It is easy to visualize what happens to various vectors under this operation. There will be two vectors that have a special property: the unit vectors to the north and south poles will be mapped into themselves under the rotation. This is a special example of an operator equation like (4-6), which may be written as

$$Hu_E(x) = Eu_E(x) \tag{4-10}$$

This equation states that H, the Hamiltonian operator acting on a special class of functions, will give back the function that it is acting on, multiplied by a constant. The constant is called the *eigenvalue*. The solution of the equation depends on E, and we have therefore labeled it with an E. The solution $u_E(x)$ is called the *eigenfunction*, corresponding to the eigenvalue E, of the operator H. We shall see that eigenvalues can form a continuum or be discrete.

Example   Solve the eigenvalue problem $Lf(x) = \lambda f(x)$, where

$$Lf(x) = \frac{\hbar}{i} \frac{df(x)}{dx} - \beta x f(x)$$

in the region $-a \le x \le a$, subject to the boundary condition that $f(a) = f(-a)$.

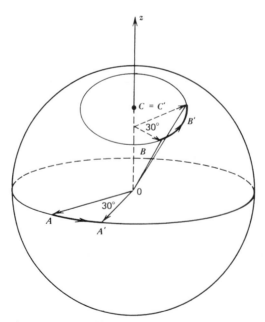

Figure 4-1. An illustration of the operator rotating all vectors by 30° with the vectors lying on the unit sphere: for vectors on the equator $(A \to A')$, at an intermediate latitude $(B \to B')$, and at a pole $(C \to C' = C)$.

Solution

The equation may be rewritten in the form

$$\frac{df(x)}{dx} = \frac{i}{\hbar} (\beta x + \lambda) f(x)$$

that is,

$$\frac{df}{f} = \frac{i}{\hbar} (\beta x + \lambda) \, dx$$

The solution is

$$\ln f(x) = \frac{i}{\hbar} (\beta x^2/2 + \lambda x) + \text{constant}$$

so that

$$f(x) = A \, e^{i\beta x^2/2\hbar + i\lambda x/\hbar}$$

The condition $f(a)$ and $f(-a)$ implies that $e^{i\lambda a} = e^{-i\lambda a}$ or $e^{2i\lambda a} = 1$. Thus the eigenvalues are

$$\lambda = \pm \frac{n\pi}{a} \qquad (n = 0, 1, 2, 3, \ldots)$$

The solution (4-2) is of the form $u_E(x) \, e^{-iEt/\hbar}$. Since (4-1) is a linear equation, a sum of solutions of the preceding form, with permissible values of E, is also a solution. Thus the most general solution of (4-1) is

$$\psi(x, t) = \sum_n C_n u_n(x) e^{-iE_n t/\hbar} + \int dE \, C(E) u_E(x) e^{-iEt/\hbar} \qquad (4\text{-}11)$$

where $u_n(x)$ are the complete set of eigenfunctions corresponding to the discrete eigenvalues E_n, and the $u_E(x)$ are the eigenfunctions corresponding to the continuous eigenvalues E. The C_n are arbitrary constants, and $C(E)$ are arbitrary functions of E. The coefficients are subject to the restriction that $\psi(x, t)$ must be square integrable.

The eigenvalues of the operator H are called the energy eigenvalues, as is suggested by the form of

$$H = \frac{p_{\text{OP}}^2}{2m} + V(x) \qquad (4\text{-}12)$$

Before discussing a very simple but instructive example, we note that the separation of the equation would fail if the potential V depended explicitly on time. We will see later that when this is the case, energy is not a constant of the motion.

THE EIGENVALUE PROBLEM
FOR A PARTICLE IN A BOX

We consider (4-6) with

$$
\begin{aligned}
V(x) &= \infty & x < 0 \\
&= 0 & 0 < x < a \\
&= \infty & a < x
\end{aligned}
\qquad (4\text{-}13)
$$

We see from (4-6) that

$$
\begin{aligned}
u(x) &= 0 & x < 0 \\
&= 0 & a < x
\end{aligned}
\qquad (4\text{-}14)
$$

and inside the box, where $V(x) = 0$, we rewrite (4-6) in the form

$$
\frac{d^2u(x)}{dx^2} + \frac{2mE}{\hbar^2} u(x) = 0
\qquad (4\text{-}15)
$$

First we notice that if $E < 0$, then the equation takes the form

$$
\frac{d^2u(x)}{dx^2} - \kappa^2 u(x) = 0
$$

where $\kappa^2 = 2m|E|/\hbar^2$. The most general solution is some linear combination of $e^{\kappa x}$ and $e^{-\kappa x}$, and although $\sinh \kappa x$ vanishes at $x = 0$, it does not vanish at $x = a$. Thus we cannot have $E < 0$. With positive E, and the notation

$$
k^2 = \frac{2mE}{\hbar^2}
\qquad (4\text{-}16)
$$

(4-15) takes the form

$$
\frac{d^2u(x)}{dx^2} + k^2 u(x) = 0
\qquad (4\text{-}17)
$$

The most general solution is of the form $A \sin kx + B \cos kx$, but the requirement that $u(0) = 0$ limits us to

$$
u(x) = A \sin kx
\qquad (4\text{-}18)
$$

The condition $u(a) = 0$ implies that

$$
ka = n\pi \qquad n = 1, 2, 3, \ldots
\qquad (4\text{-}19)
$$

Thus the energy eigenvalues are

$$
E_n = \frac{\hbar^2 k^2}{2m} = \frac{\hbar^2 \pi^2 n^2}{2ma^2} \qquad n = 1, 2, 3, \ldots
\qquad (4\text{-}20)
$$

It is easy to check that the solutions are normalized if $A = \sqrt{2/a}$, so that

$$u_n(x) = \sqrt{\frac{2}{a}} \sin \frac{n\pi x}{a} \tag{4-21}$$

The solutions have the property that

$$\begin{aligned}
\int_0^a dx \, u_n^*(x) \, u_m(x) &= \int_0^a dx \, \frac{2}{a} \sin \frac{n\pi x}{a} \sin \frac{m\pi x}{a} \\
&= \frac{1}{a} \int_0^a dx \left\{ \cos \frac{(n-m)\pi x}{a} - \cos \frac{(n+m)\pi x}{a} \right\} \\
&= \frac{\sin(n-m)\pi}{(n-m)\pi} - \frac{\sin(n+m)\pi}{(n+m)\pi} \tag{4-22} \\
&= 0 \quad \text{when} \quad n \neq m \\
&= 1 \quad \text{when} \quad n = m
\end{aligned}$$

The conclusion that

$$\int_0^a dx \, u_n^*(x) \, u_m(x) = \delta_{mn} \tag{4-23}$$

implies that eigenfunctions corresponding to different eigenvalues are *orthogonal*. If the eigenfunctions are properly normalized, as they are here, (4-23) is called the *orthonormality condition*. Since the solutions are real, the complex conjugation in that equation is not really necessary, but is inserted for consistency with the more general statement that applies when the eigenfunctions are complex. This will be established in Chapter 6.

We can extract some physical information from the eigensolutions:

1. The state of lowest energy, the *ground state*, is described by $u_1(x)$, and the lowest energy is

$$E_1 = \frac{\pi^2 \hbar^2}{2ma^2} \tag{4-24}$$

Note that classically the lowest energy would be that of a particle at rest in the hole, and for $p = 0$ and $V(x) = 0$, the sum of the kinetic and potential energies would be zero. Here we see the presence of a minimum energy.

2. Since the solutions are real,

$$\langle p \rangle = 0 \tag{4-25}$$

because for any real function $\int dx \, R(x) \, \hbar/i \, (dR(x)/dx)$ is imaginary, which is incompatible with the requirement that $\langle p \rangle = \langle p \rangle^*$, unless $\langle p \rangle = 0$. On the other hand, $\langle p^2 \rangle$ does not vanish. In fact, since inside the box $p^2 = 2mE$, we have

$$\langle p^2 \rangle = 2mE_n = \frac{\hbar^2 \pi^2 n^2}{a^2} \tag{4-26}$$

for the eigenfunction $u_n(x)$.

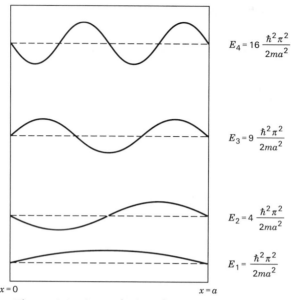

$$E_4 = 16 \frac{\hbar^2 \pi^2}{2ma^2}$$

$$E_3 = 9 \frac{\hbar^2 \pi^2}{2ma^2}$$

$$E_2 = 4 \frac{\hbar^2 \pi^2}{2ma^2}$$

$$E_1 = \frac{\hbar^2 \pi^2}{2ma^2}$$

$x = 0$ $\qquad\qquad$ $x = a$

Figure 4-2. Eigensolutions for particles in a box.

3. The larger the number of nodes in a solution, the higher is its energy, as can be seen in Fig. 4-2. This is understandable, since the kinetic energy grows with the curvature of the solutions. The expectation value of the kinetic energy is

$$\langle K \rangle = -\frac{\hbar^2}{2m} \int dx \; u^*(x) \frac{d^2 u(x)}{dx^2}$$
$$= -\frac{\hbar^2}{2m} \int dx \left[\frac{d}{dx} \left(u^*(x) \frac{du(x)}{dx} \right) - \frac{du^*(x)}{dx} \frac{du(x)}{dx} \right] \tag{4-27}$$

The first term vanishes since $u(x)$ and its derivatives vanish at infinity, so that we are left with

$$\langle K \rangle = \frac{\hbar^2}{2m} \int dx \left| \frac{du(x)}{dx} \right|^2 \tag{4-28}$$

and this is large if $u(x)$ varies a lot.

THE EXPANSION POSTULATE AND ITS PHYSICAL INTERPRETATION

Fourier's theorem states that any function $\psi(x)$ that satisfies the boundary conditions $\psi(0)$ and $\psi(a) = 0$ can be written in the form

$$\psi(x) = \sum C_n \sin \frac{n\pi x}{a} \tag{4-29}$$

Since the eigenfunctions of H for the infinite well are proportional to $\sin n\pi x/a$, we write the preceding expression in terms of the eigenfunctions $u_n(x)$:

$$\psi(x) = \sum A_n u_n(x) \tag{4-30}$$

The orthonormality relation (4-23) can be used to find A_n. We see that

$$\int_0^a dx\, u_m^*(x)\psi(x) = \int_0^a dx\, u_m^*(x) \sum A_n u_n(x)$$

$$= \sum A_n \int_0^a dx\, u_m^*(x)u_n(x) = \sum A_n \delta_{mn}$$

so that

$$A_n = \int_0^a dx\, u_n^*(x)\psi(x) \tag{4-31}$$

As in our discussion of the free wave packet, we can calculate the time development of this arbitrary initial wave function $\psi(x)$. Since each of the eigenfunctions $u_n(x)$ acquires its own time dependence $e^{-iE_n t/\hbar}$, we have quite generally

$$\psi(x,\, t) = \sum A_n u_n(x)e^{-iE_n t/\hbar} \tag{4-32}$$

We may think of the eigenfunctions as the complete collection of orthogonal unit vectors $\mathbf{i}_k$ in some vector space. Any vector $\mathbf{a}$ can be written in the form $\mathbf{a} = a_1\mathbf{i}_1 + a_2\mathbf{i}_2 + \cdots$ and the coefficients a_k can be obtained by using the orthonormality of the unit vectors $\mathbf{i}_m \cdot \mathbf{i}_n = \delta_{mn}$ to get $a_m = \mathbf{a} \cdot \mathbf{i}_m$.

To interpret the coefficients A_n we calculate the expectation value of the energy in an arbitrary state $\psi(x)$. Since inside the box $H = p^2/2m$, and outside the box $\psi(x) = 0$, and since

$$Hu_n(x) = E_n u_n(x)$$

It follows that

$$\langle H \rangle = \int_0^a dx\, \psi^*(x)H\psi(x) = \int_0^a dx\, \psi^*(x)H \sum A_n u_n(x)$$

$$= \sum A_n \int_0^a dx\, \psi^*(x)E_n u_n(x) \tag{4-33}$$

$$= \sum E_n |A_n|^2$$

We also note that

$$\int_0^a dx\, \psi^*(x)\psi(x) = 1 \tag{4-34}$$

implies that

$$1 = \int_0^a dx\, \psi^*(x) \sum A_n u_n(x) = \sum A_n A_n^* = \sum |A_n|^2 \tag{4-35}$$

Equation (4-33) together with the normalization condition (4-35) suggest that $|A_n|^2$, where

$$A_n = \int dx \, u_n^*(x)\psi(x) \tag{4-36}$$

be *interpreted as the probability that a measurement of the energy for the state $\psi(x)$ yields the eigenvalue E_n.* To complete the interpretation of the A_n we must accept the assertion that the eigenvalues E_n are the only possible outcomes of a measurement of the energy, whatever the wavefunction $\psi(x)$ of the system.

If the system is in an eigenstate $u_k(x)$, for example, then the energy measurement will necessarily yield E_k. Since a repeated measurement must yield the same answer (otherwise, how else could we check that the measurement was correct?), we are forced to conclude that once a measurement of the energy is made on a general state $\psi(x)$, and the value E_k results, then this measurement changes the state to $u_k(x)$. Effectively, a measurement in quantum mechanics is a sorting process. When we state that a system is described by a wave function $\psi(x)$, we have in mind a large collection (ensemble) of identical systems, all described by $\psi(x)$. The energy measurements carried out on the members of this collection determine how many of them have energy E_1, how many E_2, and so on. The measurement sorts the systems into sets that, after the measurement, have wave functions $u_1(x)$, $u_2(x)$, and so on.

These statements are not peculiar to the problem of a particle in a box. They hold for more general systems in which there is a potential energy $V(x)$, and also for observables other than the energy, such as the momentum, angular momentum, and so on.

Example   Consider a particle in a box. Its wave function is given by

$$\psi(x) = A(x/a) \qquad\qquad 0 < x < a/2$$
$$\qquad = A(1 - x/a) \qquad a/2 < x < a$$

where $A = \sqrt{12/a}$ so as to satisfy $\int_0^a dx \, |\psi(x)|^2 = 1$. Calculate the probability that a measurement of the energy yields the eigenvalue E_n.

We want to calculate A_n in the expansion

$$A_n = \int_0^a dx \, \psi(x) \sqrt{\frac{2}{a}} \sin\frac{n\pi x}{a}$$

$$= \frac{\sqrt{24}}{a} \left[\int_0^{a/2} dx \left(\frac{x}{a}\right) \sin\frac{n\pi x}{a} + \int_{a/2}^a dx \left(1 - \frac{x}{a}\right) \sin\frac{n\pi x}{a} \right]$$

With the change of variables $\pi x/a = u$ in the first integral and $\pi x/a = \pi - u$ in the second integral, we get

$$A_n = \frac{\sqrt{24}}{\pi} \int_0^{\pi/2} du \, \frac{u}{\pi} \sin nu (1 - (-1)^n)$$

The A_n for n even vanish because of the last factor. The integral is easily calculated, and we get, *for n odd only*

$$A_n = \frac{\sqrt{24}}{\pi} 2 \frac{1}{\pi n^2} (-1)^{n+1}$$

so that

$$|A_n|^2 = \frac{96}{\pi^4 n^4} \qquad \text{for } n \text{ odd}$$

$$= 0 \qquad \text{for } n \text{ even}$$

One can easily check, using the fact that $\Sigma_{\text{all}} \, n^{-4} = \pi^4/90$ and

$$\sum_{\text{all}} n^{-4} = \sum_{\text{even}} n^{-4} + \sum_{\text{odd}} n^{-4} = \sum_{\text{odd}} n^{-4} + (1/16) \sum_{\text{all}} n^{-4}$$

that the sum of all the probabilities is 1:

$$\frac{96}{\pi^4} \sum_{\text{odd}} n^{-4} = \frac{96}{\pi^4} \left(1 - \frac{1}{16}\right) \sum_{\text{all}} n^{-4} = \frac{96}{\pi^4} \cdot \frac{15}{16} \cdot \frac{\pi^4}{90} = 1$$

MOMENTUM EIGENFUNCTION AND THE FREE PARTICLE

The energy operator H is not the only one that has eigenfunctions and eigenvalues. Let us solve the eigenvalue equation for the *momentum operator*

$$p_{\text{op}} u_p(x) = p u_p(x) \tag{4-37}$$

Since $p_{\text{op}} = (\hbar/i)(d/dx)$, this reads

$$\frac{du_p(x)}{dx} = \frac{ip}{\hbar} u_p(x) \tag{4-38}$$

The solution to this equation is

$$u_p(x) = C \, e^{ipx/\hbar} \tag{4-39}$$

with C a constant to be determined by normalization, and the eigenvalue p real, so that the eigenfunction does not blow up at either $+\infty$ or $-\infty$. This is the only constraint on p: we say that p_{op} has a *continuous spectrum*. We might, by analogy with (4-23), expect that the eigenfunctions obey orthonormality conditions. We see that

$$\int_{-\infty}^{\infty} dx \, u_p^*(x) \, u_p(x) = |C|^2 \int_{-\infty}^{\infty} dx \, e^{i(p-p')x/\hbar}$$

$$= 2\pi |C|^2 \hbar \delta(p - p') \tag{4-40}$$

With the choice

$$u_p(x) = \frac{1}{\sqrt{2\pi\hbar}} e^{ipx/\hbar} \tag{4-41}$$

(4-40) reads

$$\int_{-\infty}^{\infty} dx \, u_{p'}^*(x) \, u_p(x) = \delta(p - p') \tag{4-42}$$

This differs from (4-23) only in that the Kroenecker δ_{mn}, appropriate for discrete indices is replaced by a Dirac delta function $\delta(p - p')$ for the continuous indices.

The statement that any wave packet $\psi(x)$ may be expanded in terms of a complete set of eigenfunctions can also be established here. The analog of (4-30) must take into account that we are summing over a continuous index p, so that we write

$$\psi(x) = \int_{-\infty}^{\infty} dp \; \phi(p) \, \frac{e^{ipx/\hbar}}{\sqrt{2\pi\hbar}} \tag{4-43}$$

According to the interpretation implicit in (4-36), $|\phi(p)|^2$, where

$$\phi(p) = \int dx \left(\frac{e^{ipx/\hbar}}{\sqrt{2\pi\hbar}} \right)^{*} \psi(x) \tag{4-44}$$

gives the probability that a measurement of the momentum for an arbitrary packet $\psi(x)$ yields the eigenvalue p. In this way we justify the conjecture made about $\phi(p)$ in Chapter 3.

Example   A particle is in the ground state of a box with sides at $x = 0$ and $x = a$. Suddenly the walls of the box are moved to $\pm\infty$, so that the particle is free. What is the probability that the particle has momentum in the range $(p, p + dp)$? Since a free particle of momentum p has energy $p^2/2m$, which need not equal the ground-state energy, energy is not conserved. How is this possible?

Our initial wave function has the form

$$\psi(x) = \sqrt{\frac{2}{a}} \sin \frac{\pi x}{a} \qquad 0 \leq x \leq a$$

As we saw in (4-44) the probability amplitude that the particle in this state has momentum p is given by

$$\phi(p) = \frac{1}{\sqrt{2\pi\hbar}} \int_{0}^{a} dx \; e^{-ipx/\hbar} \sqrt{\frac{2}{a}} \sin \frac{\pi x}{a}$$

The limits on the integrals come from the condition that $\psi(x)$ vanishes to the left of $x = 0$ and to the right of $x = a$, since that is where the walls initially stood. The integral is easy to evaluate. We get

$$\phi(p) = -e^{-ipa/2\hbar} \frac{2\pi/a}{\sqrt{\pi\hbar a}} \frac{\cos pa/2\hbar}{(\pi/a)^2 - (p/\hbar)^2}$$

so that

$$|\phi(p)|^2 dp = \frac{4\pi}{\hbar a^3} \frac{\cos^2(pa/2\hbar)}{((\pi/a)^2 - (p/\hbar)^2)^2} \, dp$$

Energy is not conserved because there is actually a time dependence in the potential energy. In fact, $V(x)$ changes very rapidly from the walls at $x = 0$ and $x = a$, to walls at $\pm\infty$.

Note that (1) the probability drops very rapidly once p gets to be much larger than $\hbar\pi/a$, and (2) the probability density does not become infinite at $p = \hbar\pi/a$ because the numerator vanishes at that point.

Let us now turn to the free-particle Hamiltonian. When $V(x)$ is zero everywhere, the energy eigenvalue equation reads

$$\frac{d^2u(x)}{dx^2} + k^2u(x) = 0 \tag{4-45}$$

where $k^2 = 2mE/\hbar^2$. The solutions are e^{ikx} and e^{-ikx}, or linear combinations of these, for example, $\cos kx$ and $\sin kx$. The trouble with all of them is that they are not square integrable, since $\int_{-\infty}^{\infty} dx |Ae^{ikx} + Be^{-ika}|^2$ diverges for all values of A and B. There are three ways of getting around this difficulty.

(a) We can consider the problem defined by (4-45) as the limiting case of a particle in a box, with the walls at infinity. To carry this out, we must modify the work in the section on the eigenvalue problem for a particle in a box somewhat. If we were to take the walls of the box at $-a/2$ and $+a/2$, we could enlarge the box by letting $a \to \infty$. This is not such a complicated procedure. All we have to do is shift the origin to the right by $a/2$. Thus if we replace x by $x - a/2$ in our eigenfunctions, we will have accomplished the shift of the walls. The eigenfunctions now become

$$\begin{aligned} u_n(x) &= \sqrt{\frac{2}{a}} \sin \frac{n\pi(x - a/2)}{a} \\ &= \sqrt{\frac{2}{a}} \left\{ \sin \frac{n\pi x}{a} \cos \frac{n\pi}{2} - \cos \frac{n\pi x}{a} \sin \frac{n\pi}{2} \right\} \end{aligned} \tag{4-46}$$

For n even we have $\cos(n\pi/2) = (-1)^{n/2}$ and $\sin(n\pi/2) = 0$. Thus only the first term appears, and we can drop the phase factor $(-1)^{n/2}$. Thus the eigenfunctions for n even are

$$u_n(x) = \sqrt{\frac{2}{a}} \sin \frac{n\pi x}{a} \qquad (n = 2, 4, 6, \ldots) \tag{4-47}$$

and the eigenfunctions for n odd are

$$u_n(x) = \sqrt{\frac{2}{a}} \cos \frac{n\pi x}{a} \qquad (n = 1, 3, 5, \ldots) \tag{4-48}$$

In the limit $a \to \infty$ these eigenfunctions appear not to make any sense. In fact for finite n, they all correspond to energies

$$E_n = \frac{\hbar^2\pi^2 n^2}{2ma^2} \to 0$$

We only get nontrivial solutions when $n \to \infty$ in a way such that

$$k = \frac{n\pi}{a} \tag{4-49}$$

is finite. In that case, we have solutions

$$\sqrt{\frac{2}{a}} \sin kx; \quad \sqrt{\frac{2}{a}} \cos kx \tag{4-50}$$

where the $1/\sqrt{a}$ factors may be kept. They will drop out of the answer to any question we may ask about the system.

(b) We may work with wave packets. A solution of the form

$$\psi(x) = e^{ikx} \tag{4-51}$$

is a special case of (4-43) with

$$\phi(p) = \sqrt{2\pi\hbar} \; \delta(p - \hbar k) \tag{4-52}$$

that is, an infinitely peaked momentum-space distribution. Suppose we replace this limiting $\phi(p)$ by a very sharply peaked function $\sqrt{2\pi\hbar} \; g(p - \hbar k)$. Then e^{ikx} will be replaced by

$$\begin{aligned}
\psi(x) &= \int dp \; e^{ipx/\hbar} \; g(p - \hbar k) \\
&= e^{ikx} \int dq \; e^{iqx/\hbar} \; g(q)
\end{aligned} \tag{4-53}$$

which is a plane wave, e^{ikx}, multiplied by a very broad function of x. We may make this function so broad that it is essentially constant over the region of physical interest. The uncertainty in the momentum will now be of the order of magnitude $\hbar/(\text{size of } x\text{-packet})$, and if the denominator is of macroscopic size, this uncertainty is negligible. We thus satisfy the mathematical requirements without changing any of the physics. The wave-packet description is actually the one that is closest to what really happens physically, since any way of preparing the initial state, for example, firing an electron gun, can never, in practice, create an exact momentum eigenstate.

(c) We can approach the problem by recognizing that the difficulty in normalization stems from the fact that for a wave function like e^{ikx}, the particle is not confined to any region of space, so that the probability of finding it anywhere is zero. If we do not ask questions that involve the probability of finding the particle in any finite region, no problems arise. One way of avoiding the normalization difficulty is to deal with the probability current, or *flux*

$$j(x) = \frac{\hbar}{2im} \left[\psi^*(x) \frac{d\psi(x)}{dx} - \frac{d\psi^*(x)}{dx} \psi(x) \right] \tag{4-54}$$

discussed at the beginning of Chapter 3. For a wave function $Ce^{ipx/\hbar}$, the flux is $|C|^2 \, p/m$; for the wave function $Ce^{-ipx/\hbar}$, the flux is $-|C|^2 \, p/m$. If we note that for a one-dimensional problem, the flux of particles with a density of 1 particle/cm, moving with velocity $v = p/m$ is just v—that is, the number crossing a point $x = x_0$ per second—we see that $|C|^2$ represents the density of particles per cm. Thus (4-41) represents particles with a density $1/2\pi\hbar$ per cm.

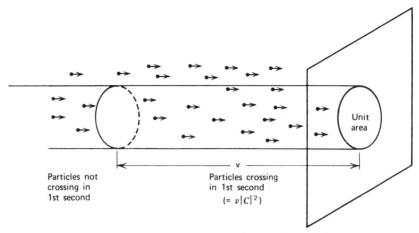

Figure 4-3. The relation between velocity of particles and flux, that is, number of particles crossing a unit area perpendicular to velocity, per unit time.

In three dimensions, with

$$u_\mathbf{p}(\mathbf{r}) = C\, e^{i\mathbf{p}\cdot\mathbf{r}/\hbar} \tag{4-55}$$

the flux will be $|C|^2\, \mathbf{p}/m$, and this corresponds to a flow of particles, with density $|C|^2$ per cm^3 crossing a unit area perpendicular to $\mathbf{p}$, when the particles are moving with velocity $\mathbf{v} = \mathbf{p}/m$ (Fig. 4-3).

Degeneracy

The energy eigenvalue equation (4-45) has two independent solutions, e^{ikx} and e^{-ikx}; equivalently, the pair of real solutions $\cos kx$ and $\sin kx$ is also independent. Whichever pair we choose, we notice that in contrast to the problem of a particle in a box, there are *two* solutions that have the same energy associated with them. This is an example of something that happens quite frequently: *there may be more than one independent eigenfunction that corresponds to the same eigenvalue of a hermitian operator.* When this occurs, we have a *degeneracy*.

In the two cases that we have earlier, the two solutions are orthogonal:

$$\int_{-\infty}^{\infty} dx (e^{-ikx})^* \, e^{ikx} = \int_{-\infty}^{\infty} dx\, e^{2ikx} = 0$$

$$\int_{-\infty}^{\infty} dx\, \sin kx\, \cos kx = 0 \tag{4-56}$$

for $k \neq 0$. It is always possible to make linear combinations such that this is true. Independent of any degeneracy, eigenfunctions corresponding to a given energy $E = \hbar^2 k^2/2m$ are orthogonal to eigenfunctions corresponding to a *different energy*.

What distinguishes the two degenerate eigenfunctions? For the set (e^{-ikx}, e^{ikx}), the difference is that they are eigenfunctions of the momentum operator

$$p_{\text{op}}\, e^{\pm ikx} = \frac{\hbar}{i}\frac{d}{dx}\, e^{\pm ikx} = \pm\hbar k\, e^{\pm ikx} \tag{4-57}$$

corresponding to *different* eigenvalues of the momentum. To answer the same question about the $\sin kx$ and $\cos kx$ solutions, we introduce the notion of *parity*.

PARITY

The eigenfunctions for the free particle ($\sin kx$, $\cos kx$) as well as the eigenfunctions for a particle in a box extending from $-a/2$ to $+a/2$, as given in (4-47) and (4-48) are either even or odd under the interchange $x \to -x$. Suppose we consider the particle in a box that is symmetric about the $x = 0$ point, and suppose that the initial state $\psi(x)$ is even in x. This means that $\psi(x)$ must be of the form

$$\psi(x) = \sum A_n \sqrt{\frac{2}{a}} \cos \frac{n \pi x}{a} \tag{4-58}$$

where the sum extends over $n = 1, 3, 5, \ldots$. At a later time this wave function becomes

$$\psi(x, t) = \sum A_n \sqrt{\frac{2}{a}} \cos \frac{n \pi x}{a} e^{-iE_n t/\hbar} \tag{4-59}$$

Thus the wave function is still even in x at a later time. The same holds for a wave function that is initially odd. Thus for our box, symmetrically centered about $x = 0$, *evenness* and *oddness* are properties that are time independent. We may say that evenness and oddness are *constants of the motion*. Since any constant of the motion is of interest to us, we formalize the discussion somewhat.

We do this by introducing the *parity operator P*, whose rule of operation is to reflect $x \to -x$. Thus for any wave function $\psi(x)$, we have

$$P\psi(x) = \psi(-x) \tag{4-60}$$

For an even wave function we have

$$P\psi^{(+)}(x) = \psi^{(+)}(x) \tag{4-61}$$

and for an odd wave function

$$P\psi^{(-)}(x) = -\psi^{(-)}(x) \tag{4-62}$$

These two equations are eigenvalue equations, and what we have shown is that even functions are eigenfunctions of P with eigenvalue $+1$, while odd functions are eigenfunctions of P with eigenvalue -1. In the problem of the particle in a box, the functions $\cos (n\pi x/a)$ and $\sin (n\pi x/a)$ are not only eigenfunctions of H; *they are simultaneously eigenfunctions of P*.

The eigenvalues ± 1 are the only possible ones. Suppose we have

$$Pu(x) = \lambda u(x) \tag{4-63}$$

Applying P again, we would get

$$P^2 u(x) = P\lambda u(x) = \lambda^2 u(x) \tag{4-64}$$

However, $P^2u(x) = u(x)$, since two reflections should not change anything. Hence $\lambda^2 = 1$, that is, $\lambda = \pm 1$. An arbitrary function $\psi(x)$ can always be written as a sum of an even and an odd function

$$\psi(x) = \tfrac{1}{2}[\psi(x) + \psi(-x)] + \tfrac{1}{2}[\psi(x) - \psi(-x)] \tag{4-65}$$

that is, just as with the eigenfunctions of H discussed in our example, any function can be expanded in terms of the eigenfunctions of this new operator.

The explicit appearance of evenness and oddness came about because we centered the box at $x = 0$. Had we taken it to lie between 0 and a, nothing would have changed, and there would still be symmetry under reflections about $x = a/2$. Such symmetry would, however, be much less apparent. The lesson to be learned here is that in setting up a quantum mechanical problem one should always pay attention to the symmetries in the Hamiltonian, and choose the coordinates in a way that exhibits the symmetries most explicitly. If the box were uneven (Fig. 4-4), no amount of changing coordinates would bring about a symmetry. The important fact is that *the symmetry be in the Hamiltonian*.[1] This may be seen more clearly by asking under what circumstances an even function will remain even for all time.

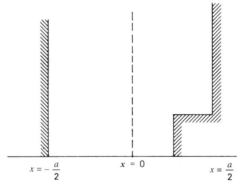

$$x = -\frac{a}{2} \qquad x = 0 \qquad x = \frac{a}{2}$$

Figure 4-4. Box for which there is no symmetry under reflections.

Let

$$\psi(x, 0) = \psi(-x, 0) \equiv \psi^{(+)}(x) \tag{4-66}$$

The time development is given by

$$i\hbar \frac{\partial \psi(x, t)}{\partial t} = H\psi(x, t) \tag{4-67}$$

If we operate with P on this equation, we get

$$i\hbar \frac{\partial}{\partial t} P\psi(x, t) = PH\psi(x, t) \tag{4-68}$$

Under the special circumstances that

$$PH\psi(x, t) = HP\psi(x, t) \tag{4-69}$$

[1]When dealing with the box, we consider the walls as part of the potential, that is, the Hamiltonian. That is why we do not speak of boundary conditions instead of the Hamiltonian.

which holds when H is even under $x \to -x$, that is, when $V(x)$ is an even function (since d^2/dx^2 is even), we have

$$i\hbar \frac{\partial}{\partial t} [P\psi(x, t)] = H[P\psi(x, t)] \tag{4-70}$$

Hence

$$\psi^{(+)}(x, t) = \tfrac{1}{2}(1 + P) \, \psi(x, t) \tag{4-71}$$

and

$$\psi^{(-)}(x, t) = \tfrac{1}{2}(1 - P) \, \psi(x, t) \tag{4-72}$$

separately obey the Schrödinger equation, and do not mix, if the initial state is even (or odd). The condition for the time independence of parity only holds if

$$(PH - HP) \, \psi(x, t) = 0 \tag{4-73}$$

for all possible states, that is, if the operators P and H commute

$$[P, H] = 0 \tag{4-74}$$

This important condition will be seen to be quite general: *any operator that does not have an explicit time dependence and that commutes with the Hamiltonian H is a constant of the motion.* In particular, if the potential changes with time, that is, we have $V(x, t)$, then the energy itself is not a constant of the motion, just as in classical mechanics. Note that when V depends on t, the separation of the equation into an equation for the time dependence and an energy eigenvalue equation is not possible.

    In summary, what differentiates the degenerate eigenfunctions is that they are simultaneous eigenfunctions of another hermitian operator. Both the operators p_{op} and P have the property that they commute with the Hamiltonian $p_{\text{op}}^2/2m$ in this problem. We shall show later that this is a necessary condition for the existence of simultaneous eigenfunctions. For example, p_{op} and P do not commute [since $(\hbar/i)(d/dx)$ changes sign under $x \to -x$], and therefore the eigenfunctions of one of the operators cannot all be simultaneous eigenfunctions of the other.

    We have learned an enormous amount about quantum mechanics from the two simple problems that we have considered. We shall return to these matters in later chapters and generalize them. In Chapter 5 we will again consider some very simple problems, but this time we will concentrate not on the mathematical features, but rather on the physical systems that they are simple models of.

Problems

1. You are given the following operators
   (a) $O_1\psi(x) = x^3\psi(x)$; (b) $O_2\psi(x) = x(d/dx)\psi(x)$; (c) $O_3\psi(x) = \lambda\psi^*(x)$;

   (d) $O_4\psi(x) = e^{\psi(x)}$; (e) $O_5\psi(x) = [d\psi(x)/dx] + a$; (f) $O_6\psi(x) = \displaystyle\int_{-\infty}^{x} dx' (\psi(x')x')$

   Which of these are linear operators?

2. Solve the eigenvalue problem

$$O_6\psi(x) = \lambda\psi(x)$$

What values of the eigenvalue λ lead to square integrable eigenfunctions?

(*Hint:* Differentiate both sides of the equation with respect to x.)

3. Calculate the following commutators: (a) $[O_2, O_6]$; (b) $[O_1, O_2]$. The procedure is to calculate $[A, B]$ by expressing $A(B\psi) - B(A\psi)$ in the form $C\psi$.

4. Calculate

$$\Delta x = \sqrt{\langle x^2 \rangle}$$

for the $u_n(x)$ given in (4-21). Using $\langle p^2 \rangle$ given by (4-28) calculate

$$\Delta p \, \Delta x$$

It is characteristic that for the higher states the uncertainty increases with n.

5. A particle is in the ground state of a box with sides at $x = \pm a$. Very suddenly the sides of the box are moved to $x = \pm b (b > a)$. What is the probability that the particle will be found in the ground state for the new potential? What is the probability that it will be found in the first excited state? In the latter case, the simple answer has a simple explanation. What is it?

6. A particle is known to be localized in the left half of a box with sides at $x = \pm a/2$, with wave function

$$\psi(x) = \sqrt{\frac{2}{a}} \qquad \frac{-a}{2} < x < 0$$

$$= 0 \qquad 0 < x < \frac{a}{2}$$

(a) Will the particle remain localized at later times?
(b) Calculate the probability that an energy measurement yields the ground state energy; the energy of the first excited state.

7. The eigenfunctions for a potential of the form

$$V(x) = \infty \qquad x < 0; \quad x > a$$
$$= 0 \qquad 0 < x < a$$

are of the form

$$u_n(x) = \sqrt{\frac{2}{a}} \sin \frac{n\pi x}{a}$$

Suppose a particle in the preceding potential has an initial normalized wave function of the form

$$\psi(x, 0) = A \left(\frac{\sin \pi x}{a} \right)^5$$

(a) What is the form of $\psi(x, t)$?
(b) Calculate A without doing the integral $\int d\theta \sin^{10}\theta$.

(c) What is the probability that an energy measurement yields E_3, where $E_n = n^2\pi^2\hbar^2/2ma^2$?

{*Hint:* Expand $[e^{i\theta} - e^{-i\theta}]/2i)^5$ in a power series $e^{5i\theta} + \cdots - e^{-5i\theta}$ and recombine into a series of terms involving $\sin 5\theta$ and so on.}

8. Repeat the calculation given in the example on p. 64 for a particle initially in the nth eigenstate. Show that the corresponding probability is given by

$$\frac{2n^2\pi}{\hbar a^3} \frac{1 - (-1)^n \cos pa/\hbar}{[(p/\hbar)^2 - (n\pi/a)^2]^2}$$

Sketch the distribution. Show that it conforms with the uncertainty relation, and that the result is in agreement with the correspondence principle when n is large.

9. A particle in free space is initially in a wave packet described by

$$\psi(x) = \left(\frac{\alpha}{\pi}\right)^{1/4} e^{-\alpha x^2/2}$$

(a) What is the probability that its momentum is in the range $(p, p + dp)$?
(b) What is the expectation value of the energy? Can you give a rough argument, based on the "size" of the wave function and the uncertainty principle, for why the answer should be roughly what it is?

10. The wave function for a particle is given by

$$\psi(x) = Ae^{ikx} + Be^{-ikx}$$

What flux does this represent?

11. What is the flux associated with a particle described by the wave function

$$\psi(x) = u(x) e^{ikx}$$

where $u(x)$ is a real function?

12. Consider the eigenfunctions for a box with sides at $x = \pm a$. Without working out the integral, prove that the expectation value of the quantity

$$x^2p^3 + 3xp^3x + p^3x^2$$

vanishes for all the eigenfunctions.

13. Prove that the parity operator, defined by

$$P\psi(x) = \psi(-x)$$

is a hermitian operator. Also prove that the eigenfunctions of P, corresponding to the eigenvalues $+1$ and -1, are orthogonal.

14. Use parity arguments to show that in the example on p. 62 the A_n for n even must all vanish.

15. Consider the eigenstates of a particle in an infinite box given by (4-21). Calculate $\langle x^2 \rangle$ and $\langle x \rangle$. Use this to calculate the quantity $\Delta x \equiv \sqrt{\langle x^2 \rangle - \langle x \rangle^2}$ for $u_n(x)$. Δx represents the *uncertainty* in the operator x, and it should be close to the naive estimate $\Delta x \approx a$.

References

A detailed discussion of the properties of second-order differential equations as related to quantum mechanics may be found in J. L. Powell and B. Crasemann, *Quantum Mechanics,* Addison-Wesley, Reading, Mass., 1961, and D. S. Saxon, *Elementary Quantum Mechanics,* Holden-Day, San Francisco, 1968.

See also any of the more advanced textbooks listed at the end of the book.

Chapter 5

ONE-DIMENSIONAL POTENTIALS

In this chapter we solve some simple problems of one-dimensional motion. They are of interest because they illustrate some nonclassical effects, and because many physical situations are effectively one-dimensional even though we live in a three-dimensional world.

THE POTENTIAL STEP

For this problem we take (Fig. 5-1) the form of $V(x)$ to be

$$V(x) = 0 \qquad x < 0$$
$$= V_0 \qquad x > 0 \tag{5-1}$$

The Schrödinger equation

$$-\frac{\hbar^2}{2m} \frac{d^2 u(x)}{dx^2} + V(x) \, u(x) = Eu(x) \tag{5-2}$$

takes the form

$$\frac{d^2 u(x)}{dx^2} + \frac{2m}{\hbar^2} [E - V(x)] \, u(x) = 0 \tag{5-3}$$

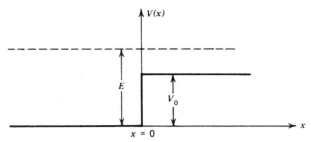

Figure 5-1. The potential step.

We write, as usual

$$\frac{2mE}{\hbar^2} = k^2 \tag{5-4}$$

and we also introduce

$$\frac{2m(E - V_0)}{\hbar^2} = q^2 \tag{5-5}$$

The most general solution of (5-3) for $x < 0$, where $V(x) = 0$ is

$$u(x) = e^{ikx} + R\, e^{-ikx} \tag{5-6}$$

This corresponds to a net flux moving in the positive x direction, of magnitude

$$j = \frac{\hbar}{2im} [(e^{-ikx} + R^* e^{ikx})(ik\, e^{ikx} - ikR\, e^{-ikx}) - \text{complex conjugate}]$$

$$= \frac{\hbar k}{m}(1 - |R|^2) \tag{5-7}$$

We may view e^{ikx} with flux $\hbar k/m$ as an *incoming wave*. If there were no potential, we could choose e^{ikx} as the solution for all x, so that we attribute R to the presence of the potential. This potential gives rise to a reflected wave, $R\, e^{-ikx}$, with a reflected flux $\hbar k |R|^2/m$.

For $x > 0$, we write the solution

$$u(x) = T\, e^{iqx} \tag{5-8}$$

The most general solution for $x > 0$ is a linear combination of e^{iqx} and e^{-iqx}, but a term involving the latter would describe a wave coming from $+\infty$ in the negative direction, and with the "experiment" in which only a wave from the left is sent in, the only wave on the right can be a transmitted wave. The flux corresponding to (5-8) is

$$j = \frac{\hbar q}{m} |T|^2 \tag{5-9}$$

Since there is no time dependence in the problem, the conservation law (3-12) implies that $j(x)$ is independent of x. Hence the flux on the left must be equal to the flux on the right, that is, we expect that

$$\frac{\hbar k}{m}(1 - |R|^2) = \frac{\hbar q}{m}|T|^2 \tag{5-10}$$

The continuity of the wave function implies that

$$1 + R = T \tag{5-11}$$

obtained by matching the two solutions at $x = 0$. In spite of the fact that the potential is discontinuous, the slope of the wave function is also continuous, as can be seen by integrating (5-3) from $-\epsilon$ to $+\epsilon$ (with ϵ arbitrarily small and positive) and using the continuity of the wave function:

$$\left(\frac{du}{dx}\right)_\epsilon - \left(\frac{du}{dx}\right)_{-\epsilon} = \int_{-\epsilon}^{\epsilon} dx \, \frac{d}{dx}\frac{du}{dx}$$
$$= \int_{-\epsilon}^{\epsilon} dx \, \frac{2m}{\hbar^2}[V(x) - E]\,u(x) = 0 \tag{5-12}$$

We note, for future reference, that if the potential contains a term like $V_0\delta(x - a)$, then integration of the equation from $a - \epsilon$ to $a + \epsilon$ gives

$$\left(\frac{du}{dx}\right)_{a+\epsilon} - \left(\frac{du}{dx}\right)_{a-\epsilon} = \frac{2m}{\hbar^2}\int_{a-\epsilon}^{a+\epsilon} dx \, V_0\delta(x - a)\,u(x)$$
$$= \frac{2m}{\hbar^2} V_0\,u(a) \tag{5-13}$$

The continuity of the derivative for our potential implies that

$$ik(1 - R) = iqT \tag{5-14}$$

We can therefore solve (5-11) and (5-14) for R and T to obtain

$$R = \frac{k - q}{k + q}$$
$$T = \frac{2k}{k + q} \tag{5-15}$$

From this we can calculate the reflected and transmitted fluxes:

$$\frac{\hbar k}{m}|R|^2 = \frac{\hbar k}{m}\left(\frac{k - q}{k + q}\right)^2$$
$$\frac{\hbar q}{m}|T|^2 = \frac{\hbar k}{m}\frac{4kq}{(k + q)^2} \tag{5-16}$$

We note the following:

1. In contrast to classical mechanics, according to which a particle going over a potential step would slow down (to conserve energy) but would never be reflected, here we do have a certain fraction of the incident particles reflected. This is, of course, a consequence of the wave properties of the particle; partial reflection of light from an interface between two media is a familiar phenomenon.

2. With the help of (5-16) we can easily check that the conservation law (5-10) is indeed satisfied.

3. For $E \gg V_0$, that is, for $q \to k$ from below, the ratio of the reflected flux to the incident flux, that is, $|R|^2$ approaches zero. This agrees with intuition, which tells us that at very high energies, the presence of the step is but a small perturbation on the propagation of the wave.

4. If the energy E is less than V_0, then q becomes imaginary. If we note that, now the solution for $x > 0$ must be of the form

$$u(x) = T \, e^{-|q|x} \tag{5-17}$$

so as not to blow up at $+\infty$, we see that now

$$|R|^2 = \left(\frac{k - i|q|}{k + i|q|}\right)\left(\frac{k - i|q|}{k + i|q|}\right)^* = 1 \tag{5-18}$$

Thus, as in classical mechanics, there is now total reflection. Note, however, that

$$T = \frac{2k}{k + i|q|} \tag{5-19}$$

does not vanish, and a part of the wave penetrates into the forbidden region. This penetration phenomenon again is characteristic of waves, and we shall see a little later that it permits a "tunneling" through barriers that would totally block particles in a classical description. There is no flux to the right, since $j(x)$ vanishes for a real solution even if the coefficient in front of it is taken to be complex. The phenomenon of total reflection is mathematically identical to what happens to light when it strikes an interface between two media of different refractive indices (going from the larger to the smaller value of n) at an angle larger than the critical angle. The light undergoes total internal reflection, but there is an exponentially decaying electromagnetic field that penetrates into the forbidden region.

5. It is a peculiarity of the idealized sharp potential that the expressions for R and T can be written in terms of E and $(E - V_0)$ only, independent of $\hbar$, which could apparently be set equal to zero. This appears to present us with a paradox, since we associate $\hbar = 0$ with the classical limit. Does this mean that for such a potential one could get partial reflection in the classical limit? There is no paradox, since a condition for the classical limit to hold is that the De Broglie wavelength ($\lambda = 2\pi\hbar/p$) be small compared with the relevant dimensions of the system. In this example, the dimension that is relevant is the size of the

region in which the potential changes from zero to V_0, which is zero in the limiting case. Thus for this idealized potential, there never is a classical region. If we were to round off the potential, then at sufficiently high energies the condition for classical behavior would indeed obtain, but as already noted, at sufficiently high energies there is no reflection.

THE POTENTIAL WELL

We next consider the potential (Fig. 5-2)

$$
\begin{aligned}
V(x) &= 0 & x < -a \\
&= -V_0 & -a < x < a \\
&= 0 & a < x
\end{aligned}
\tag{5-20}
$$

We again write

$$
k^2 = \frac{2mE}{\hbar^2}
\tag{5-21}
$$

and

$$
q^2 = \frac{2m(E + V_0)}{\hbar^2}
\tag{5-22}
$$

We can immediately write down the solutions

$$
\begin{aligned}
u(x) &= e^{ikx} + R\, e^{-ikx} & x < -a \\
u(x) &= A\, e^{iqx} + B\, e^{-iqx} & -a < x < a \\
u(x) &= T\, e^{ikx} & a < x
\end{aligned}
\tag{5-23}
$$

These correspond to an incoming flux $\hbar k/m$ from the left, a reflected flux $\hbar k|R|^2/m$, and a transmitted flux $\hbar k|T|^2/m$ to the right. Inside the well there are waves going in both directions because of the reflections at both discontinuities at $\pm a$. According to flux conservation we should get

$$
\frac{\hbar k}{m} (1 - |R|^2) = \frac{\hbar q}{m} (|A|^2 - |B|^2) = \frac{\hbar k}{m} |T|^2
\tag{5-24}
$$

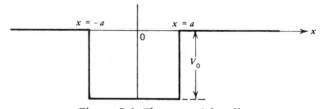

Figure 5-2. The potential well.

Matching wave functions and derivatives gives the four equations

$$e^{-ika} + R e^{ika} = A e^{-iqa} + B e^{iqa}$$

$$ik(e^{-ika} - R e^{ika}) = iq(A e^{-iqa} - B e^{iqa})$$

$$A e^{iqa} + B e^{-iqa} = T e^{ika}$$

$$iq(A e^{iqa} - B e^{-iqa}) = ikT e^{ika}$$

(5-25)

A little algebra yields the results

$$R = i e^{-2ika} \frac{(q^2 - k^2) \sin 2qa}{2kq \cos 2qa - i(q^2 + k^2) \sin 2qa}$$

$$T = e^{-2ika} \frac{2kq}{2kq \cos 2qa - i(q^2 + k^2) \sin 2qa}$$

(5-26)

Again, if $E \gg V_0$, there is practically no reflection, since $q^2 - k^2 \ll 2kq$, and as $E \to 0$, the transmission goes to zero. There is an item of special interest: in the special case that $\sin 2qa = 0$, that is, for the positive energies given by

$$E = -V_0 + \frac{n^2 \pi^2 \hbar^2}{8ma^2} \qquad n = 1, 2, 3, \ldots$$

(5-27)

there is no reflection. This is actually a model of what happens in the scattering of low energy electrons (0.1 eV) by noble gas atoms, for example, neon and argon, in which there is anomalously large transmission. The effect, first observed by Ramsauer and Townsend, is described as a transmission resonance. A more accurate discussion must, of course, involve three-dimensional considerations. In wave language, the effect is due to a destructive interference between the wave reflected at $x = -a$ and the wave reflected once, twice, thrice, . . . , at the edge $x = a$. The resonance condition $2qa = n\pi$, which may be written in the form

$$\lambda = \frac{2\pi}{q} = \frac{4a}{n}$$

(5-28)

is just the one that describes the Fabry-Perot interferometer.

THE POTENTIAL BARRIER

We now consider

$$V(x) = 0 \qquad x < -a$$

$$= V_0 \qquad -a < x < a$$

$$= 0 \qquad a < x$$

(5-29)

We will limit our discussion to energies such that $E < V_0$, that is, energies such that no penetration of the barrier would occur in classical physics (Fig. 5-3). Inside the barrier we have the equation

$$\frac{d^2u(x)}{dx^2} + \frac{2m}{\hbar^2}(E - V_0)\,u(x) = 0$$

that is

$$\frac{d^2u(x)}{dx^2} - \kappa^2 u(x) = 0 \qquad\qquad (5\text{-}30)$$

The general solution

$$u(x) = A\,e^{-\kappa x} + B\,e^{\kappa x} \qquad |x| < a \qquad\qquad (5\text{-}31)$$

is to be matched onto

$$\begin{aligned} u(x) &= e^{ikx} + R\,e^{-ikx} \qquad x < -a \\ &= T\,e^{ikx} \qquad\qquad\quad x > a \end{aligned} \qquad\qquad (5\text{-}32)$$

Actually we need not go through the trouble of solving this since the results can be read off from (5-26) with the substitution

$$q \to i\kappa = i\sqrt{(2m/\hbar^2)(V_0 - E)} \qquad\qquad (5\text{-}33)$$

Thus, for example,

$$T = e^{-2ika}\,\frac{2k\kappa}{2k\kappa \cosh 2\kappa a + i(k^2 - \kappa^2)\sinh 2\kappa a} \qquad\qquad (5\text{-}34)$$

and this implies that

$$|T|^2 = \frac{(2k\kappa)^2}{(k^2 + \kappa^2)^2 \sinh^2 2\kappa a + (2k\kappa)^2} \qquad\qquad (5\text{-}35)$$

There is transmission, even though the energy lies below the top of the barrier. This is a wave phenomenon, and in quantum mechanics it is also one exhibited by particles. This *tunneling* of a particle through a barrier is frequently encountered, and we shall discuss some applications later.

We pause to analyze an apparent difficulty. The wave function does not vanish inside the barrier, and thus there appears to be some probability of finding a particle with negative kinetic energy. How can this make sense? We look to the uncertainty relation to save us from a contradiction with classical physics. An

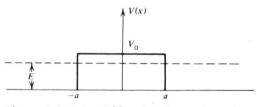

Figure 5-3. Potential barrier. Energy is such that a classical particle would be totally reflected by the barrier.

experiment to study the particle inside the potential barrier must localize it with an accuracy

$$\Delta x \ll 2a \tag{5-36}$$

This measurement will transfer to the particle momentum, with an uncertainty

$$\Delta p \gg \hbar/2a \tag{5-37}$$

This corresponds to a transfer of energy

$$\Delta E \gg \hbar^2/8ma^2 \tag{5-38}$$

In order to observe the negative kinetic energy, this uncertainty must be much less than $|E - V_0|$, so that

$$\frac{\hbar^2 \kappa^2}{2m} \gg E \gg \frac{\hbar^2}{8ma^2} \tag{5-39}$$

This implies that $2\kappa a \gg 1$, and under these circumstances the probability of there being penetration into the barrier, given by

$$|T|^2 \simeq \left(\frac{4k\kappa}{k^2 + \kappa^2} \right)^2 e^{-4\kappa a} \tag{5-40}$$

becomes vanishingly small (for example, for $\kappa a = 10$, $e^{-4\kappa a} = 10^{-18}$).

The approximate expression for the ratio of transmitted flux to incident flux, $|T|^2$, is an extremely sensitive function of the width of the barrier, and of the amount by which the barrier exceeds the incident energy, since

$$\kappa a = \left[\frac{2ma^2}{\hbar^2} (V_0 - E) \right]^{1/2} \tag{5-41}$$

In general, the barriers that occur in physical phenomena are not square, and to discuss some applications, we must first obtain an approximate expression for the transmission coefficient $|T|^2$ through an irregularly shaped barrier. The proper way to do this, given the fact that there is no exact solution available for most potentials, is through the Wentzel–Kramers–Brillouin (WKB) approximation technique.[1] Our discussion will be less mathematical.

We observe that (5-40) consists of a product of two terms, the second of which is by far the more important one. If we write

$$\ln |T|^2 \simeq -2\kappa(2a) + 2 \ln \frac{2(ka)(\kappa a)}{(ka)^2 + (\kappa a)^2}$$

we see that under most circumstances the first term dominates the second for any reasonable size of κa. The procedure we adopt is to treat a smooth, curved barrier as a juxtaposition of square barriers (Fig. 5-4). Since transmission coefficients are

[1]See the WKB approximation in Special Topics section 3.

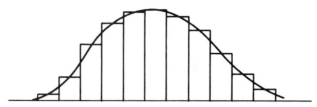

Figure 5-4. Approximation of smooth barrier by a juxta-position of square potential barriers.

multiplicative[2] when they are small (in effect, with most of the flux reflected, the transmission through each slice is an independent, improbable event), we may write, approximately

$$\ln |T|^2 \approx \sum_{\substack{\text{partial} \\ \text{barriers}}} \ln |T_{\substack{\text{partial} \\ \text{barrier}}}|^2$$

$$\approx -2 \sum \Delta x \, \langle \kappa \rangle \tag{5-42}$$

$$\approx -2 \int_{\text{barrier}} dx \, \sqrt{(2m/\hbar^2)[V(x) - E]}$$

In the partial barriers, Δx is the width and $\langle \kappa \rangle$ the average value of κ for that barrier. In the last step a limit of infinitely narrow barriers was taken. It is clear from the expression that the approximation is least accurate near the "turning points" where the energy and potential are nearly equal, since there (5-40) is not a good approximation to (5-35). It is also important that $V(x)$ be a slowly varying function of x, since otherwise the approximation of a curved barrier by a stack of square ones is only possible if the latter are narrow, and there again (5-40) is a poor approximation. A proper treatment, using the WKB approximation includes a discussion of the behavior near the turning points. For most purposes, it is still a fair approximation to write

$$|T|^2 \approx e^{-2\int dx \sqrt{(2m/\hbar^2)[V(x) - E]}} \tag{5-43}$$

with the integration over the region in which the square root is real.

TUNNELING PHENOMENA

The phenomenon of particle tunneling is quite common in atomic and nuclear physics, and we discuss several examples at this point.

Cold Emission

Consider electrons in a metal. As noted in our discussion of the photoelectric effect in Chapter 1, these electrons are held in a metal by a potential, which, to first approximation, may be described by a box of finite depth, as shown in Fig. 5-5a.

---

[2]This statement is only correct for the most important exponential part, as can be seen from the fact that doubling the width will only approximately square the transmission coefficient $|T|^2$.

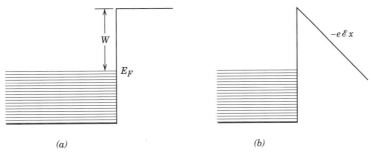

Figure 5-5. (*a*) Electronic energy levels in metal. E_F is the Fermi energy and W is the work function. (*b*) Potential altered by an external electric field.

The electrons are actually stacked up in energy levels that are very dense, since the box is very wide. It is a property of electrons[3] that no more than two of them can occupy any given energy level; thus for the lowest energy state of the metal, all the levels up to a certain energy, called the *Fermi energy* (which depends on the density of free electrons), are filled. When the temperature is above absolute zero, a few electrons are thermally excited to higher levels, but even at room temperature, the number is small. The difference between the Fermi energy and the top of the well is what is required to bring an electron out; it is the *work function* discussed in connection with the photoelectric effect. Electrons can be removed by transferring energy to them, either by photons, or by heating them. They can also be removed by the application of an external electric field $\mathscr{E}$. *Cold emission* occurs because the external field changes the potential seen by an electron from W to $(W - e\mathscr{E}x)$ (Fig. 5-5*b*), if the electron is at the top of the "sea" of levels. The transmission coefficients are

$$|T|^2 = e^{-2\int_0^a dx[2m(W - e\mathscr{E}x)/(\hbar^2)]^{1/2}} \tag{5-44}$$

Since

$$\int dx(A + Bx)^{1/2} = \frac{(A + Bx)^{3/2}}{3B/2}$$

this leads to

$$|T|^2 = e^{-4\sqrt{2/3}\ \sqrt{mW/\hbar^2}(W/e\mathscr{E})} \tag{5-45}$$

The Fowler-Nordheim formula, as (5-45) is called, describes the emission only qualitatively. One effect, which is easily included, is the additional attraction of the electron back to the plate, caused by the image charge. The other effect, much harder to handle, is that there are surface imperfections in the metal surface, which change the electric field locally, and since $\mathscr{E}$ appears in the exponent, this can make a large difference. Incidentally, we see that the exponent may be written in

---

[3]This property of electrons is described by the Pauli exclusion principle, which will be discussed in Chapter 8.

terms of the barrier thickness at the top of the Fermi sea, since that thickness is given by

$$a = \frac{W}{e\mathscr{E}} \tag{5-46}$$

In the last 15 years a version of *cold emission* has found an important application in the *scanning tunneling microscope (STM)*. This tool is based on the extreme sensitivity to the distance in cold emission. The current of electrons induced by an electric potential difference between a metal surface and a very sharp tungsten tip above the surface varies exponentially with the distance to the surface, and this allows for very detailed studies of the topography of the surface. The functioning of the STM depends on some unexpected phenomena and on several technological advances.

First of all, in order to get a 1-Å (or less) resolution, radiation of the order of 10 keV is required by the usual diffraction limit. This does not allow for nondestructive studies of a surface. An electron microscope requires electrons of the order of 150 eV, which is still too large for the scanning of surfaces. As was pointed out by J. A. O'Keefe in 1956, it is possible to bypass the diffraction limit with a new type of microscope. In such a microscope light would shine through a tiny hole onto the object directly in front of the hole, and the transmitted or reflected light would be affected by the object. In such a microscope it is the size of the hole that determines its resolution. The STM operates on this principle. The tungsten tip, to which the potential is applied, acts as the "hole."

A potential difference of 10 V, acting across a spacing of about 5 Å gives rise to very large electric fields, and these fields "pull out" a few atoms from the tip to sharpen it. For example, a tip machined to have a radius of 1000 Å would quite naturally be able to resolve lateral distances to 45 Å because of the sensitivity of the current to the distance from the sampled surface. The tiny atomic protuberances reduce this resolution to 1–2 Å.

The usefulness of the STM depends on the ability of the experimenter to keep the distance from the surface, or alternatively, the current constant. This can now be achieved by means of ceramic (piezoelectric) supports that expand or contract when electric fields are applied to them. Their location can be determined to very high precision by means of interferometers, and thus a constant distance between the tip and the scanned surface can be maintained.

The STM has been used for a study of surfaces of metals and some semiconductors. More recently atoms and molecules have been moved around on surfaces, either by sliding them by means of the tip or by lifting them off the surface by the tip and depositing them at a different location.

Tunneling can also occur if we bring two metal plates close together. Figure 5-6 shows the situation both without a potential difference, and with a potential difference. Without the potential difference, tunneling is not possible because the levels on both sides of the barrier are filled. The effect of even a weak electric field is to change the shape of the barrier a little (Fig. 5-6)—an effect that we can neglect—and to lower the Fermi sea on one side of the barrier. This brings some empty levels in correspondence with the filled ones on the other side of the barrier, and now tunneling can proceed, with transmission coefficient

$$|T|^2 \simeq e^{-2\sqrt{(2mW/\hbar^2)}a} \tag{5-47}$$

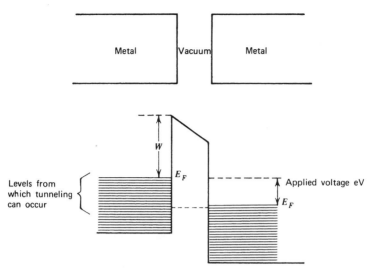

Figure 5-6. Energy diagram for tunneling between two metals separated by vacuum. Tunneling between metals is possible only when there are empty states on the right. Such empty states are created when eV is applied to lower the Fermi level on the right.

Such a factor acts as a resistance. Unfortunately this expression is very sensitive to the gap separation a, and since for a work function of the order of electron volts, the separation has to be of the order of angstroms, it has not proved possible to make metal plates sufficiently flat and parallel. The formula has been applied to the interpretation of currents flowing between two plates with an oxide between them (Ni-NiO-Pb), where the gap can be made as small as 50 Å, and it is in agreement with experiment.

Tunneling in Superconductors

An interesting effect occurs when the metal on the right is in a superconducting state. A characteristic of such a state is that above the Fermi level there is a gap in the level density, that is, there are no allowed states between an energy $E_F - \Delta$ and $E_F + \Delta$ with Δ of the order of 10^{-3} eV compared with the Fermi energy E_F of order 10 eV. These levels do not disappear, but are squeezed up and down, so that the level density just below and just above the gap is very large. If the electric field is small enough, that is, $a\mathscr{E} \leq \Delta/e$, there will be no tunneling, since there is no place for the electrons to go. The qualitative features of the current-voltage relation and the energetics are shown in Fig. 5-7. These features are in good agreement with experiments that were first carried out by I. Giaever in 1960.

Such tunneling experiments may be used to study details of the superconducting state. The gap in the energy level density at the top of the Fermi sea is believed to be due to an attraction between electron pairs, and is the energy required to break such a pair apart. With this picture, one might expect that irradiating a superconductor-oxide-superconductor "sandwich" would break up some pairs, and the single electrons that acquire some kinetic energy from the photons can move into the region of unoccupied levels above the gap, and then tunnel across the gap, increasing the current. This "photon-assisted" tunneling has been ob-

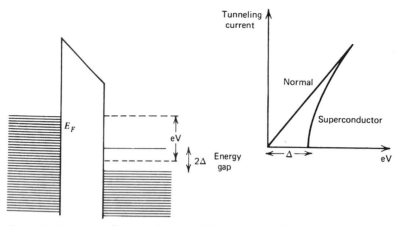

Figure 5-7. Energy diagram for tunneling from metal to superconductor. In contrast to the metal–metal tunneling shown in Figure 5-6, no tunneling is allowed into the energy gap. This affects the current-voltage characteristic as shown.

served. Other examples of tunneling in superconductors involve Josephson junctions.[4]

Tunneling in Nuclear Physics

Tunneling is also important in nuclear physics. Nuclei are very complicated objects, but under certain circumstances it is appropriate to view them as independent particles occupying levels in a potential well. With this picture in mind, the decay of a nucleus into an α-particle (a He nucleus with $Z = 2$) and a daughter nucleus can be described as the tunneling of an α-particle through a barrier caused by the Coulomb potential between the daughter and the α-particle. The α-particle is not viewed as being in a bound state: if it were, the nucleus could not decay. Rather, the α-particle is taken to have positive energy, and its decay is only inhibited by the existence of the barrier.[5]

If we write

$$|T|^2 = e^{-G} \tag{5-48}$$

then

$$G = 2 \left(\frac{2m}{\hbar^2}\right)^{1/2} \int_R^b dr \left(\frac{Z_1 Z_2 e^2}{r} - E\right)^{1/2} \tag{5-49}$$

where R is the nuclear radius[6] and b is the turning point, determined by the vanishing of the integrand (Fig. 5-8); Z_1 is the charge of the daughter nucleus, and

---

[4]A beautiful discussion of superconductivity and of the various Josephson effects can be found in Chapter 21 of volume 3 of R. P. Feynman, R. B. Leighton, and M. Sands, *The Feynman Lectures on Physics,* Addison-Wesley, Reading, Mass., 1965.

[5]If you find it difficult to imagine why a repulsion would keep two objects from separating, think of the inverse process, α capture. It is clear that the barrier will tend to keep the α-particle out.

[6]In fact, early estimations of the nuclear radius came from the study of α-decay. Nowadays one uses the size of the charge distribution as measured by scattering electrons off nuclei to get nuclear radii. It is not clear that the two should be expected to give exactly the same answer.

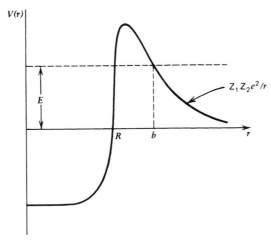

Figure 5-8. Potential barrier for α decay.

Z_2 ($= 2$ here) is the charge of the particle being emitted. The integral can be done exactly

$$\int_R^b dr \left(\frac{1}{r} - \frac{1}{b}\right)^{1/2} = \sqrt{b} \left[\cos^{-1}\left(\frac{R}{b}\right)^{1/2} - \left(\frac{R}{b} - \frac{R^2}{b^2}\right)^{1/2}\right] \tag{5-50}$$

At low energies (relative to the height of the Coulomb barrier at $r = R$), we have $b \gg R$, and then

$$G \simeq 2 \left(\frac{2mZ_1Z_2e^2b}{\hbar^2}\right)^{1/2} \left[\frac{\pi}{2} - \left(\frac{R}{b}\right)^{1/2}\right] \tag{5-51}$$

with $b = Z_1Z_2e^2/E$. If we write for the α-particle energy $E = mv^2/2$, where v is its final velocity, then

$$G \simeq \frac{2\pi Z_1Z_2e^2}{\hbar v} = 2\pi\alpha Z_1Z_2\left(\frac{c}{v}\right) \tag{5-52}$$

The time taken for an α-particle to get out of the nucleus may be estimated as follows: the probability of getting through the barrier on a single encounter is e^{-G}. Thus the number of encounters needed to get through is $n \simeq e^G$. The time between encounters is of the order of $2R/v$, where R is again the nuclear radius, and v is the α velocity inside the nucleus. Thus the lifetime is

$$\tau \simeq \frac{2R}{v} e^G \tag{5-53}$$

The velocity of the α inside the nucleus is a rather fuzzy concept, and the whole picture is very classical, so that the factor in front of the e^G cannot really be predicted without a much more adequate theory. Our considerations do give us an order of magnitude for it. For a 1-MeV α-particle,

$$v = \sqrt{\frac{2E}{m}} = c\sqrt{\frac{2E}{mc^2}} = 3 \times 10^{10-}\sqrt{\frac{2}{4 \times 940}} \simeq 7.0 \times 10^8 \text{ cm/sec}$$

Also, for R we take

$$R \simeq 1.5 \times 10^{-13} \, A^{1/3} \text{ cm} \tag{5-54}$$

and for $A = 216$ we get, for the factor in front, 2.6×10^{-21}. We can also rewrite G in the form

$$G \simeq 4 \frac{Z_1}{\sqrt{E(\text{MeV})}} \tag{5-55}$$

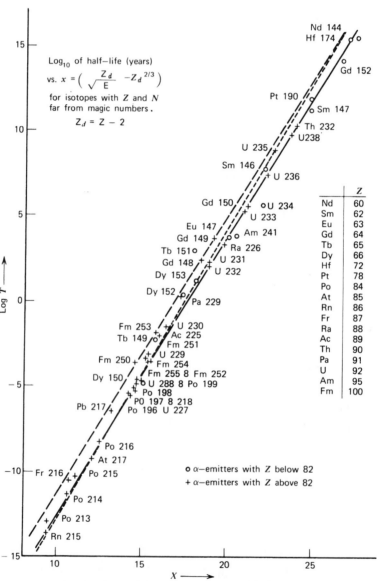

Figure 5-9. Plot of $\log_{10} 1/\tau$ versus $C_2 - C_1 Z_1/\sqrt{E}$ with $C_1 = 1.61$ and a slowly varying $C_2 = 28.9 + 1.6 Z_1^{2/3}$. (From E. K. Hyde, I. Perlman, and G. T. Seaborg, *The Nuclear Properties of the Heavy Elements,* Vol. 1, Prentice-Hall, Englewood Cliffs, N.J. (1964), reprinted by permission.)

so that one predicts, for low energy α's, the straight-line plot

$$\log_{10} \frac{1}{\tau} \simeq \text{const} - 1.73 \frac{Z_1}{\sqrt{E(\text{MeV})}} \qquad (5\text{-}56)$$

with the constant in front of the order of magnitude 27–28 when τ is measured in years instead of seconds, Fig. (5-9) shows that a good fit to the lifetime data of a large number of α emitters is obtained with the formula

$$\log_{10} \frac{1}{\tau} = C_2 - C_1 \frac{Z_1}{\sqrt{E}}$$

where $C_1 = 1.61$ and $C_2 = 28.9 + 1.6 Z_1^{2/3}$. Thus the very simple considerations give a rather remarkable fit to the data.

For larger α energies, the G factor depends on R, and with $R = r_0 A^{1/3}$, one finds that r_0 is a constant, that is, that the notion of a Coulomb barrier taking over the role of the potential beyond the nuclear radius has some validity. Again, simple qualitative considerations explain the data.

The fact that the probability of a reaction (e.g., capture) between nuclei is attenuated by the factor $e^{-2(Z_1 Z_2/\sqrt{E})}$ ($Z_2 = 2$ for α's) implies that at low energies and/or for high Z's, such reactions are rare. That is why all attempts to make thermonuclear reactors concentrate on the burning of hydrogen (actually heavy hydrogen—deuterium).

$$_1H^2 + {}_1H^2 \rightarrow {}_2He^3 + n \qquad (3.27 \text{ MeV})$$

$$_1H^2 + {}_1H^2 \rightarrow {}_1H^3 + p \qquad (4.03 \text{ MeV})$$

$$_1H^2 + {}_1H^3 \rightarrow {}_2He^4 + n \qquad (17.6 \text{ MeV})$$

since reactions involving higher Z elements would require much higher energies, that is, much higher temperatures, with correspondingly greater confinement problems. For the same reason, neutrons are used in nuclear reactors to fission the heavy elements. Protons, at the low energies available, would not be able to get near enough to the nuclei to react with them.

BOUND STATES IN A POTENTIAL WELL

In addition to the solutions for $E > 0$ discussed in the section on the potential well there are, remarkably, also solutions for $E < 0$ provided the potential is negative, that is, $V_0 > 0$ in (5-20). They will turn out to be discrete. Let us write

$$\frac{2mE}{\hbar^2} = -\kappa^2 \qquad (5\text{-}57)$$

The solutions outside the well that are bounded at infinity are

$$u(x) = C_1 e^{\kappa x} \qquad x < -a$$
$$u(x) = C_2 e^{-\kappa x} \qquad a < x \qquad (5\text{-}58)$$

Since we are dealing with real functions, it is more convenient to write the solution inside the well in the form

$$u(x) = A \cos qx + B \sin qx \qquad -a < x < a \qquad (5\text{-}59)$$

Note that

$$q^2 = \frac{2m}{\hbar^2} (V_0 - |E|) > 0 \qquad (5\text{-}60)$$

Matching solutions and derivatives at the edges $x = \pm a$ yields

$$
\begin{aligned}
C_1 e^{-\kappa a} &= A \cos qa - B \sin qa \\
\kappa C_1 e^{-\kappa a} &= q(A \sin qa + B \cos qa) \\
C_2 e^{-\kappa a} &= A \cos qa + B \sin qa \\
-\kappa C_2 e^{-\kappa a} &= -q(A \sin qa - B \cos qa)
\end{aligned}
\qquad (5\text{-}61)
$$

These may be combined to yield

$$
\begin{aligned}
\kappa &= q \, \frac{A \sin qa - B \cos qa}{A \cos qa + B \sin qa} \\
&= q \, \frac{A \sin qa + B \cos qa}{A \cos qa - B \sin qa}
\end{aligned}
\qquad (5\text{-}62)
$$

Together these imply that $AB = 0$, that is, the solutions are either even in x ($B = 0$) or odd in x ($A = 0$). The wave functions are roughly of the shape shown in Fig. 5-10. The ground state, with no nodes, is even. This is a general property of simple systems. The conditions that determine the energy are from (5-62)

$$
\begin{aligned}
\kappa &= q \tan qa \qquad \text{even solutions} \\
\kappa &= -q \cot qa \qquad \text{odd solutions}
\end{aligned}
\qquad (5\text{-}63)
$$

Let us examine these separately.

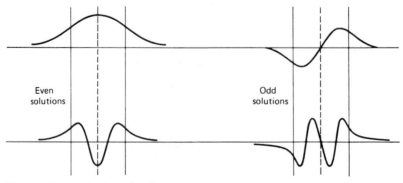

Figure 5-10. Solutions for discrete spectrum in attractive potential well.

(a) The even solutions:

With the notation

$$\lambda = \frac{2mV_0a^2}{\hbar^2}$$

$$y = qa$$ (5-64)

the first of the relations (5-63) reads

$$\frac{\sqrt{\lambda - y^2}}{y} = \tan y$$ (5-65)

If we plot $\tan y$ and $\sqrt{\lambda - y^2}/y$ as functions of y (Fig. 5-11), the points of intersection determine the eigenvalues. These form a discrete set. The larger λ is, the further the curves for $\sqrt{\lambda - y^2}/y$ go, that is, *when the potential is deeper and/or broader, there are more bound states.* The figure also shows that no matter how small λ is, there will always be at least one bound state. This is characteristic of one-dimensional attractive potentials, and is *not* true for three-dimensional potentials, which behave much more like the odd-solution problem that we will discuss later. As λ becomes large, the eigenvalues tend to become equally spaced in y, with the intersection points given approximately by

$$y \simeq (n + \tfrac{1}{2})\pi \qquad n = 0, 1, 2, \ldots$$ (5-66)

This is just the eigenvalue condition for the even solutions of the infinite box centered at the origin (eigenfunctions given in (4-48)). This is as might be expected,

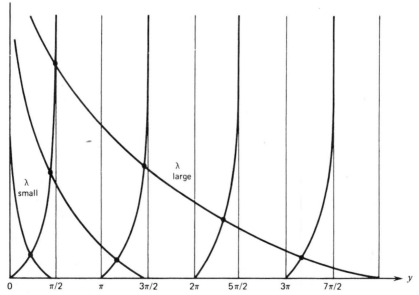

Figure 5-11. Location of discrete eigenvalues for even solutions in square well. The rising curves represent $\tan y$; the falling curves are $\sqrt{\lambda - y^2}/y$ for different values of λ.

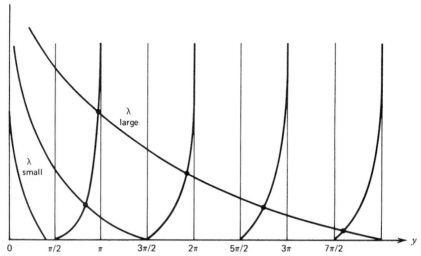

Figure 5-12. Location of discrete eigenvalues for odd solutions in square well. The rising curves represent $-\cot y$; the falling curves are $\sqrt{\lambda - y^2}/y$ for different values of λ. Note that there is no eigenvalue for $\lambda < (\pi/2)^2$.

since for the deep-lying states in the potential, the fact that it is not really infinitely deep does not matter very much.

(b) The odd solutions:

Here the eigenvalue condition reads

$$\frac{\sqrt{\lambda - y^2}}{y} = -\cot y \tag{5-67}$$

Since $-\cot y = \tan (\pi/2 + y)$, the plot in Fig. 5-12 is the same as in Fig. 5-11 with the tangent curves shifted by $\pi/2$. The large λ behavior is more or less the same, with (5-66) replaced by

$$y \simeq n\pi \qquad n = 1, 2, 3, \ldots \tag{5-68}$$

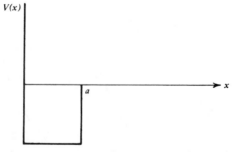

Figure 5-13. Equivalent potential for odd solutions of square well bound state problem.

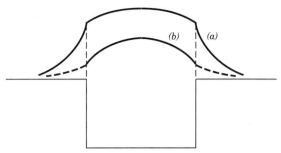

Figure 5-14. (a) Mismatch with E_B too large. (b) Mismatch with E_B too small.

In contrast to the even solutions, there will only be an intersection if $\sqrt{\lambda - \pi^2/4} > 0$, that is, if

$$\frac{2mV_0a^2}{\hbar^2} \geq \frac{\pi^2}{4} \tag{5-69}$$

The odd solutions all vanish at $x = 0$, and hence the bound-state problem for the odd solutions will be the same as for the potential well shown in Fig. 5-13, since in the latter, the condition $u(0) = 0$ would be imposed. We shall see that such conditions are imposed on wave functions in the three-dimensional world.

The detailed calculations that we have carried out support the qualitative understanding of the reason for discrete eigenvalues. These arise because we require the wave functions to vanish at infinity. We see this graphically in Fig. 5-14. The even-parity ground-state wave function inside the potential, of the form $\cos qx$, must tie on continuously to a falling exponential $e^{-\alpha|x|}$ with $\alpha^2 = 2mE_B/\hbar^2$. Large binding means a rapidly falling exponential. Since $q^2 = (2mV_0/\hbar^2 - \alpha^2)$, a large binding energy means that q^2 is small, that is, the wave function is quite flat: thus matching is impossible. As we reduce the trial E_B, the exponential falls less steeply and the wave function inside curves more, so that at some point the matching (continuous slope) becomes possible. If the value of α is lowered beyond this point, the outside curve is too flat to match the more curved inside wave function. For the first excited state, with odd parity, the wave function vanishes at the origin, so that it can only tie onto a falling exponential if it has a chance to turn over inside the potential. The condition that it turns over just enough to tie to a straight line ($\alpha = 0$) is that $\sin qa = 1$, so that $qa = \pi/2$, which corresponds to the condition expressed in (5-69).

DELTA FUNCTION POTENTIALS

We consider a potential $V(x)$ whose spatial behavior is given by $\delta(x)$. Since $\delta(x)$ has the dimensions of a reciprocal length, it is convenient to write for the attractive potential

$$V(x) = -\frac{\hbar^2\lambda}{2ma} \delta(x) \tag{5-70}$$

We have introduced a, an arbitrary quantity with dimensions of a length, so that λ is a dimensionless quantity that we use to characterize the strength of the potential. The equation to be solved is, when $E < 0$,

$$\frac{d^2u(x)}{dx^2} - \kappa^2 u(x) = -\frac{\lambda}{a}\,\delta(x)\,u(x) \tag{5-71}$$

where $\kappa^2 = 2m|E|/\hbar^2$. The solution everywhere, except at $x = 0$, must satisfy the equation $d^2u/dx^2 - \kappa^2 u = 0$, and if it is to vanish at $x \to \pm\infty$, we must have

$$\begin{aligned} u(x) &= e^{-\kappa x} & x > 0 \\ &= e^{\kappa x} & x < 0 \end{aligned} \tag{5-72}$$

The coefficients in front are the same (and here chosen to be unity—we can normalize afterwards) because of the continuity of the wave function. The derivative of the wave function is no longer continuous. As argued before (Eq. 5-13), we have

$$\left(\frac{du}{dx}\right)_{x=0+} - \left(\frac{du}{dx}\right)_{x=0-} = -\frac{\lambda}{a}\,u(0) \tag{5-73}$$

The last relation gives the eigenvalue condition

$$-\kappa - \kappa = -\frac{\lambda}{a}$$

that is

$$\kappa = \frac{\lambda}{2a} \tag{5-74}$$

The double delta function potential is more interesting because it gives a quick way to study the properties of a narrow deep double well, such as shown in Fig. 5-15. Let

$$(2m/\hbar^2)\,V(x) = -\frac{\lambda}{a}\,[\delta(x - a) + \delta(x + a)] \tag{5-75}$$

(Here a in the description of the magnitude of the potential is no longer an arbitrary length, but is connected to the form of the potential.) Because the potential is symmetric under the interchange $x \to -x$, we expect that there will be solutions of definite parity, and we will first consider the even solutions.

1. For the even solution we write

$$\begin{aligned} u(x) &= e^{-\kappa x} & x > a \\ &= A\cosh\kappa x & a > x > -a \\ &= e^{\kappa x} & x < -a \end{aligned} \tag{5-76}$$

and continuity of the wave function gives

$$e^{-\kappa a} = A\cosh\kappa a \tag{5-77}$$

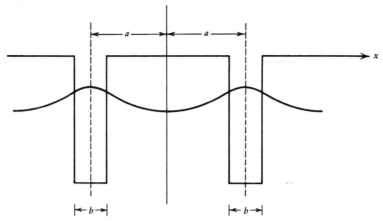

Figure 5-15. Double one-dimensional potential well. The shape of the wave function for a bound state is sketched in. We consider the limiting case in (5-75).

Because of the symmetry, it is sufficient to apply the discontinuity condition for the derivative at $x = a$. Nothing new will come of the application at $x = -a$. We get

$$-\kappa e^{-\kappa a} - \kappa A \sinh \kappa a = -\frac{\lambda}{a} e^{-\kappa a} \tag{5-78}$$

and the eigenvalue condition is

$$\tanh \kappa a = \frac{\lambda}{\kappa a} - 1 \tag{5-79}$$

The preceding may be rewritten in the form

$$\frac{\lambda}{\kappa a} - 1 = \frac{e^{\kappa a} - e^{-\kappa a}}{e^{\kappa a} + e^{-\kappa a}} = \frac{1 - e^{-2\kappa a}}{1 + e^{-2\kappa a}}$$

from which we get

$$e^{-2\kappa a} = \frac{2\kappa a}{\lambda} - 1 \tag{5-80}$$

For λ large we must have $2\kappa a$ slightly larger than λ. If we try $2\kappa a = \lambda + \epsilon$, then, to order ϵ (5-80) reads $e^{-\lambda} \approx \epsilon/\lambda$ so that

$$2\kappa a = \lambda + \lambda e^{-\lambda} \tag{5-81}$$

We can see that for the even solution there is always a single bound state. Figure 5-16 shows a plot of the eigenvalue equation (5-79) with $y = \kappa a$. As the figure shows there is only one intersection point of the curve $\tanh y$ with $(\lambda/y) - 1$. It is obvious that when $y = \lambda$, the right side is zero, whereas $\tanh y > 0$. Thus

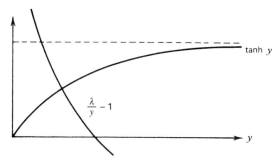

Figure 5-16. Solution of the eigenvalue condition $\tanh y = \lambda/y - 1$.

the intersection point occurs for $y < \lambda$. On the other hand, since $\tanh y < 1$, we must have $(\lambda/y) < 2$ at the intersection point, that is,

$$\kappa > \frac{\lambda}{2a} \tag{5-82}$$

If we compare this with (5-74), we see that the energy for the double well is a *larger negative number*, that is, the energy for the double potential is lower. Note that this is not because somehow the strength of a pair of potentials is larger than that of a single potential, as might be the case if one compared an electron bound to two protons with an electron bound to one proton. The larger binding is there because, as Fig. 5-17 indicates, it is easier to accommodate a sharply dropping exponential to a symmetric function (here $\cosh x$) with a discontinuity

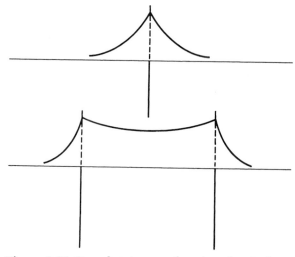

Figure 5-17. Bound state wave functions for single and double delta function attractive potentials. The strength of the potential is measured by the discontinuity in the slopes of the wave function at the potentials. It is the same in all three cases, and shows that for the double potential, a much steeper fall-off to the left and to the right is possible.

in slope as given, than it is to accommodate it to an equally sharply dropping exponential on the other side of the potential. In the real world, a single electron bound to two protons separated by a small distance will have a lower energy than a single proton plus a hydrogen atom far away, even though in the first case there is a more effective repulsion between the protons. Again it is the way in which the wave function can accommodate itself to the geometrical situation that is the dominant effect.

2. The odd solution will have the form

$$
\begin{aligned}
u(x) &= e^{-\kappa x} & x &> a \\
&= A \sinh \kappa x & a &> x > -a \\
&= -e^{\kappa x} & x &< -a
\end{aligned}
\tag{5-83}
$$

Again, because of the antisymmetry, it is sufficient to apply the conditions at $x = a$, say. Continuity of the wave function gives

$$
A \sinh \kappa a = e^{-\kappa a}
\tag{5-84}
$$

and the discontinuity equation reads

$$
-\kappa e^{-\kappa a} - \kappa A \cosh \kappa a = -\frac{\lambda}{a} e^{-\kappa a}
\tag{5-85}
$$

Combining the two yields the eigenvalue condition

$$
\coth \kappa a = \frac{\lambda}{\kappa a} - 1
\tag{5-86}
$$

This equation can be rewritten in the form

$$
\frac{\lambda}{\kappa a} - 1 = \frac{e^{\kappa a} + e^{-\kappa a}}{e^{\kappa a} - e^{-\kappa a}} = \frac{1 + e^{-2\kappa a}}{1 - e^{-2\kappa a}}
$$

Just as for the ground-state eigenvalue, we can find the solution for $\lambda \gg 1$. It is given by changing the sign of $e^{-2\kappa a}$ so that

$$
2\kappa a = \lambda - \lambda e^{-\lambda}
\tag{5-87}
$$

We can show that the odd solution has at most one bound state. Figure 5-18 shows a plot of the reciprocal of the eigenvalue equation, that is, of

$$
\tanh \kappa a = \left(\frac{\lambda}{\kappa a} - 1 \right)^{-1}
$$

with κa denoted by y.

There will only be an intersection if the slope of the former at the origin is larger than that of the second, that is, if

$$
\lambda > 1
\tag{5-88}
$$

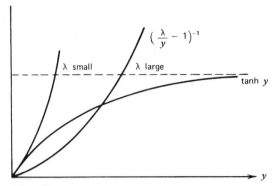

Figure 5-18. Solution of the eigenvalue condition $\tanh y = ((\lambda/y) - 1)^{-1}$.

At $y = \lambda/2$ the term $(\lambda/y - 1)^{-1}$ is already at 1, so that the intersection had to occur for $y < \lambda/2$, that is,

$$\kappa < \frac{\lambda}{2a} \tag{5-89}$$

Thus the odd solution, if there is a bound state, is less strongly bound than the even solution. The wave function, which has to go through zero, is forced to be steep between the wells, and thus can only accommodate to a less rapidly falling exponential. Depending on the size of λ, there may or may not exist an odd parity bound state.

Let us now consider a superposition of the ground state $u_e(x)$, with energy E_e and the excited state $u_o(x)$, with energy E_o (e and o stand for even and odd)

$$\psi(x) = \frac{u_e(x) + \alpha u_o(x)}{\sqrt{1 + \alpha^2}} \tag{5-90}$$

with α chosen so as to make $\int_{-\infty}^{0} dx |\psi(x)|^2$ as small as possible, that is, with the "electron" localized, as far as possible, on the right side. After a time t, the wave function will be

$$\psi(x, t) = (u_e(x) e^{-iE_e t/\hbar} + \alpha u_o(x) e^{-iE_o t/\hbar})/\sqrt{1 + \alpha^2}$$
$$= e^{-iE_e t/\hbar} [u_e(x) + \alpha e^{-i(E_o - E_e)t/\hbar} u_o(x)]/\sqrt{1 + \alpha^2} \tag{5-91}$$

that is, the phase relationship between the two parts will change. In particular, after a time such that

$$e^{-i(E_o - E_e)t/\hbar} = -1 \tag{5-92}$$

the "electron" will be localized on the left side in exactly the same way that it was localized on the right at $t = 0$. Thus there is an oscillatory behavior, which may be described by the electron going back and forth between the two potentials, with frequency

$$\omega = \omega_{oe} = \frac{E_o - E_e}{\hbar} \tag{5-93}$$

For large λ we can use our solutions for the energies to obtain

$$
\begin{aligned}
\omega &= \frac{1}{\hbar} \frac{\hbar^2}{2ma^2} \left(-(\kappa_0 a)^2 - (-(\kappa_e a)^2) \right) \\
&= \frac{\hbar}{8ma^2} \left((\lambda + \lambda e^{-\lambda})^2 - (\lambda - \lambda e^{-\lambda})^2 \right) \\
&= \frac{\hbar \lambda^2}{2ma^2} e^{-\lambda} \\
&= \frac{\hbar}{2ma^2} e^{-\lambda + 2 \ln \lambda}
\end{aligned}
\tag{5-94}
$$

The factor in front has the dimensions of a frequency, and the rest shows the parametric dependence on λ. One can argue that the period of the oscillation just described is approximately equal to the tunneling time across the barrier separating the two wells.[7]

THE KRONIG-PENNEY MODEL

Metals generally have a crystalline structure, that is, the ions are arranged in a way that exhibits a spatial periodicity. This periodicity has an effect on the motion of the free electrons in the metal, and this effect is exhibited in the simple model that we will now discuss.

The periodicity will be built into the potential, for which we require that

$$
V(x + a) = V(x) \tag{5-95}
$$

Since the kinetic energy term $-(\hbar^2/2m)(d^2/dx^2)$ is unaltered by the change $x \to x + a$, the whole *Hamiltonian is invariant under displacements by a*. For the case of zero potential, when the solution corresponding to a given energy $E = \hbar^2 k^2 / 2m$ is

$$
\psi(x) = e^{ikx} \tag{5-96}
$$

the displacement yields

$$
\psi(x + a) = e^{ik(x+a)} = e^{ika} \psi(x) \tag{5-97}
$$

that is, the original solution multiplied by a phase factor, so that

$$
|\psi(x + a)|^2 = |\psi(x)|^2 \tag{5-98}
$$

The observables will therefore be the same at x as at $x + a$, that is, we cannot tell whether we are at x or at $x + a$. In our example we shall also insist that $\psi(x)$ and $\psi(x + a)$ differ only by a phase factor, which need not, however, be of the form e^{ika}.

---

[7]This can be justified for the potentials shown in Fig. 5.15. Our approximate expressions for transmission probabilities are not applicable to delta function potentials.

We digress briefly to discuss this requirement more formally. The invariance of the Hamiltonian under a displacement $x \rightarrow x + a$ can be treated formally as follows. Let D_a be an operator whose rule of operation is that

$$D_a\, f(x) = f(x + a) \tag{5-99}$$

The invariance implies that

$$[H, D_a] = 0 \tag{5-100}$$

We can find the eigenvalues of this operator by noting that

$$D_a\, \psi(x) = \lambda_a \psi(x) \tag{5-101}$$

together with

$$D_{-a}D_a f(x) = D_a D_{-a} f(x) = f(x) \tag{5-102}$$

implies that $\lambda_a \lambda_{-a} = 1$, so that λ_a must be of the form $e^{\sigma a}$. Consider now a simultaneous eigenfunction of H and D_a, $\psi(x)$ and define

$$u(x) = e^{-\sigma x}\, \psi(x) \tag{5-103}$$

Then

$$
\begin{aligned}
D_a u(x) = u(x + a) &= e^{-\sigma(x+a)}D_a\psi(x) \\
&= e^{-\sigma(x+a)}\, e^{\sigma a}\, \psi(x) \\
&= e^{-\sigma x}\, \psi(x) = u(x)
\end{aligned}
\tag{5-104}
$$

Thus $u(x)$ is a periodic function of a obeying

$$u(x + a) = u(x) \tag{5-105}$$

and $\psi(x) = e^{\sigma x}u(x)$. If we now take into account that square integrability requires that the real part of σ must vanish, we get the condition that a simultaneous eigenfunction of H and D_a must be of the form

$$\psi(x) = e^{ix \mathrm{Im}\,\sigma}\, u(x) \tag{5-106}$$

with $u(x) = u(x + a)$. It is convenient to write $\mathrm{Im}\,\sigma = \phi/a$. This statement, known as *Bloch's theorem*, was first used by F. Bloch in the context of quantum mechanics, but it is also known in the mathematical literature as Floquet's theorem.

To simplify the algebra, we will take a series of repulsive delta-function potentials,

$$V(x) = \frac{\hbar^2}{2m} \frac{\lambda}{a} \sum_{n=-\infty}^{\infty} \delta(x - na) \tag{5-107}$$

Away from the points $x = na$, the solution will be that of the free-particle equation, that is, some linear combination of $\sin kx$ and $\cos kx$ (we deal with real functions for simplicity). Let us assume that in the region R_n defined by $(n - 1) a \le x \le na$, we have

$$\psi(x) = A_n \sin k(x - na) + B_n \cos k(x - na) \tag{5-108}$$

and in the region R_{n+1}, defined by $na \le x \le (n + 1) a$ we have

$$\psi(x) = A_{n+1} \sin k[x - (n + 1) a] + B_{n+1} \cos k[x - (n + 1) a] \tag{5-109}$$

Continuity of the wave function implies that $(x = na)$

$$-A_{n+1} \sin ka + B_{n+1} \cos ka = B_n \tag{5-110}$$

and the discontinuity condition (5-13) here reads

$$kA_{n+1} \cos ka + kB_{n+1} \sin ka - kA_n = \frac{\lambda}{a} B_n \tag{5-111}$$

A little manipulation yields

$$\begin{aligned} A_{n+1} &= A_n \cos ka + (g \cos ka - \sin ka) B_n \\ B_{n+1} &= (g \sin ka + \cos ka) B_n + A_n \sin ka \end{aligned} \tag{5-112}$$

where $g = \lambda/ka$.

The requirement from Bloch's theorem that

$$\psi(x + a) = e^{i(x+a)\mathrm{Im}\sigma} u(x + a) = e^{i\phi} e^{ix\mathrm{Im}\sigma} u(x) = e^{i\phi} \psi(x) \tag{5-113}$$

implies that

$$\psi(R_{n+1}) = e^{i\phi} \psi(R_n) \tag{5-114}$$

This is satisfied if

$$\begin{aligned} A_{n+1} &= e^{i\phi} A_n \\ B_{n+1} &= e^{i\phi} B_n \end{aligned} \tag{5-115}$$

When this is inserted into (5-113), we find a consistency condition that reads

$$(e^{i\phi} - \cos ka)(e^{i\phi} - g \sin ka - \cos ka) = \sin ka(g \cos ka - \sin ka)$$

that is,

$$e^{2i\phi} - e^{i\phi}(2 \cos ka + g \sin ka) + 1 = 0$$

Multiplication by $e^{-i\phi}$ gives

$$\cos \phi = \cos ka + \tfrac{1}{2} g \sin ka \tag{5-116}$$

If we take periodic boundary conditions for our "crystal" so that

$$\psi(R_{n+N}) = \psi(R_n) \tag{5-117}$$

then it follows from (5-114) that $e^{iN\phi} = 1$, that is,

$$\phi = \frac{2\pi}{N} m \qquad m = 0, \pm1, \pm2, \ldots \tag{5-118}$$

We used square integrability to show that σ had to be imaginary. If the x-values do not extend to infinity, then we would require that $(e^{\sigma a})^N = 1$, which again would imply that $\sigma a = i\phi$, a pure imaginary number.

We denote ϕ by qa, where q is the wave number of an electron in a box of length Na, with periodic boundary conditions and without any potential, that is, without any ions present. Thus (5-116) should be rewritten in the form

$$\cos qa = \cos ka + \tfrac{1}{2} \lambda \frac{\sin ka}{ka} \tag{5-119}$$

This is a very interesting result, because the left side is always bounded by 1, that is, there are restrictions on the possible ranges of the energy $E = \hbar^2 k^2 / 2m$ that depend on the parameters of our "crystal." Figure 5-19 shows a plot of the function $\cos x + \lambda \sin x / 2x$ as a function of $x = ka$. The horizontal line represents the bounds on $\cos qa$, and the regions of x, for which the curve lies outside the strip, are forbidden regions. Thus there are *allowed energy bands* separated by regions that are forbidden. Note that the onset of a forbidden band corresponds to the condition

$$qa = n\pi \qquad n = \pm1, \pm2, \pm3, \ldots \tag{5-120}$$

This, however, is just the condition for Bragg reflection with normal incidence. The existence of energy gaps can be understood qualitatively. In first approximation the electrons are free, except that there will be Bragg reflection when the waves reflected from successive atoms differ in phase by an integral number of 2π, that is, when (5-120) is satisfied. These reflections give rise to standing waves, with even and odd waves of the form $\cos \pi x / a$ and $\sin \pi x / a$, respectively. The energy levels corresponding to these standing waves are degenerate. Once the attractive interaction between the electrons and the positively charged ions at $x = ma$ (m integer) is taken into account, the even states, peaked at the ion location will move down in energy, and the odd states, peaked in between, will move up in energy. Thus the energy degeneracy is split at $k = n\pi/a$ and this leads to energy gaps, as shown in Fig. 5-19.

The Kronig-Penney model has some relevance to the theory of metals, insulators, and semiconductors if we take into account the fact (to be studied later) that energy levels occupied by electrons cannot accept more electrons. Thus a metal may have an energy band partially filled. If an external field is applied, the electrons are accelerated, and if there are momentum states available to them, the electrons will occupy the momentum states under the influence of the electric field. Insulators have completely filled bands, and an electric field cannot accelerate

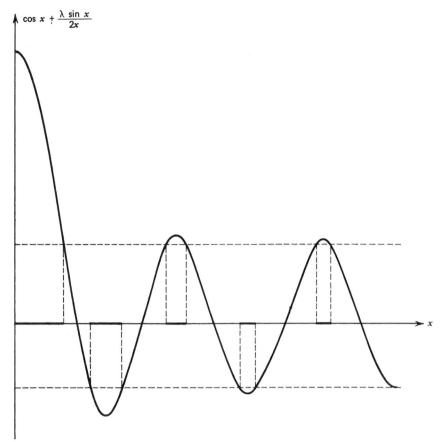

Figure 5-19. Plot of $\cos x + (\lambda/2)(\sin x/x)$ as a function of x. The horizontal lines represent the bounds ± 1. The regions of x for which the curve lines outside the strip are forbidden.

electrons, since there are no neighboring empty states. If the electric field is strong enough, the electrons can "jump" across a forbidden energy gap and go into an empty allowed energy band. This corresponds to the breakdown of an insulator. The semiconductor is an insulator with a very narrow forbidden gap. There, small changes of conditions, such as a rise in temperature, can produce the "jump" and the insulator becomes a conductor.

THE HARMONIC OSCILLATOR

As our last example we consider the harmonic oscillator (Fig. 5-20). In contrast to the examples dealt with until now, the differential equation that needs to be solved is not so trivial, and one reason for discussing this problem is to learn something about the technique for solving such equations.

The classical Hamiltonian is of the form

$$H = \frac{p^2}{2m} + \tfrac{1}{2} kx^2 \tag{5-121}$$

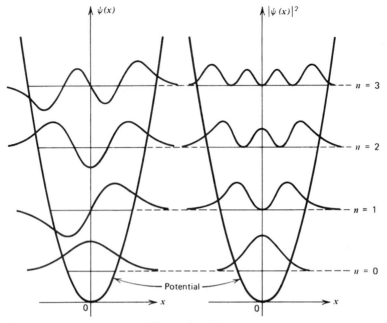

Figure 5-20. Harmonic oscillator eigenfunctions, and probability densities for the lowest four eigenvalues. Note the evenness and oddness properties of the eigenfunctions.

so that the eigenvalue equation is

$$-\frac{\hbar^2}{2m}\frac{d^2u(x)}{dx^2} + \tfrac{1}{2}kx^2u(x) = Eu(x) \qquad (5\text{-}122)$$

We introduce the frequency of the oscillator

$$\omega = \sqrt{k/m} \qquad (5\text{-}123)$$

write

$$\epsilon = \frac{2E}{\hbar\omega} \qquad (5\text{-}124)$$

and change variables to

$$y = \sqrt{\frac{m\omega}{\hbar}}\,x \qquad (5\text{-}125)$$

to finally get the simpler form of the equation

$$\frac{d^2u}{dy^2} + (\epsilon - y^2)\,u = 0 \qquad (5\text{-}126)$$

All quantities that appear are dimensionless.

For any eigenvalue ϵ, as $y^2 \to \infty$, the term involving ϵ is negligible, and we must therefore require that $u(y)$ asymptotically satisfy the equation

$$\frac{d^2 u_0(y)}{dy^2} - y^2 u_0(y) = 0 \tag{5-127}$$

We multiply by $2 du_0/dy$, which allows us to rewrite this in the form

$$\frac{d}{dy}\left(\frac{du_0}{dy}\right)^2 - y^2 \frac{d}{dy}(u_0^2) = 0 \tag{5-128}$$

or, equivalently,

$$\frac{d}{dy}\left[\left(\frac{du_0}{dy}\right)^2 - y^2 u_0^2\right] = -2y u_0^2 \tag{5-129}$$

This simplifies a great deal if we neglect the term on the right side of the equation. We assume that this can be done, and then check that the assumption was correct. If we drop the right side, we find that

$$\frac{du_0}{dy} = (C + y^2 u_0^2)^{1/2}$$

where C is a constant of integration. Since both $u_0(y)$ and du_0/dy must vanish at infinity, we must have $C = 0$. Thus

$$\frac{du_0}{dy} = \pm y u_0 \tag{5-130}$$

whose solution, acceptable at infinity, is

$$u_0(y) = e^{-y^2/2} \tag{5-131}$$

We can now check that $2y u_0^2 = 2y\, e^{-y^2}$ is indeed negligible compared with

$$\frac{d}{dy}(y^2 u_0^2) = \frac{d}{dy}(y^2\, e^{-y^2}) \simeq -4y^3\, e^{-y^2}$$

for large y. If we now introduce a new function $h(y)$, such that

$$u(y) = h(y)\, e^{-y^2/2} \tag{5-132}$$

then the differential equation is easily seen to take the form

$$\frac{d^2 h(y)}{dy^2} - 2y \frac{dh(y)}{dy} + (\epsilon - 1) h(y) = 0 \tag{5-133}$$

This may not seem like much of a simplification, but we have accounted for the behavior at infinity, and we can now look at the behavior near $y = 0$. Let us attempt a power series expansion

$$h(y) = \sum_{m=0}^{\infty} a_m y^m \tag{5-134}$$

When this is inserted into the equation, we find that the coefficients of y^m satisfy the recursion relation

$$(m + 1)(m + 2) a_{m+2} = (2m - \epsilon + 1) a_m \tag{5-135}$$

Thus, given a_0 and a_1, the even and odd series can be generated separately. That they do not mix is a consequence of the invariance of the Hamiltonian under reflections. For arbitrary ϵ, we find that for large m (say $m > N$)

$$a_{m+2} \simeq \frac{2}{m} a_m \tag{5-136}$$

This means that the solution is approximately

$h(y) = $ (a polynomial in y)

$$+ a_N \left[y^N + \frac{2}{N} y^{N+2} + \frac{2^2}{N(N + 2)} y^{N+4} + \frac{2^3}{N(N + 2)(N + 4)} y^{N+6} + \cdots \right]$$

where, for simplicity, we have only taken the even solution. The series may be written in the form

$$a_N y^2 \left(\frac{N}{2} - 1\right)! \left[\frac{(y^2)^{N/2-1}}{(N/2 - 1)!} + \frac{(y^2)^{N/2}}{(N/2)!} + \frac{(y^2)^{N/2+1}}{(N/2 + 1)!} + \cdots \right]$$

If we choose $N = 2k$ for convenience, the series takes the form

$$y^2(k - 1)! \left[\frac{(y^2)^{k-1}}{(k - 1)!} + \frac{(y^2)^k}{k!} + \frac{(y^2)^{k+1}}{(k + 1)!} + \cdots \right]$$

$$= y^2(k - 1)! \left[e^{y^2} - \left\{ 1 + y^2 + \frac{(y^2)^2}{2!} + \cdots + \frac{(y^2)^{k-2}}{(k - 2)!} \right\} \right]$$

which is of the form of a polynomial + a constant $\times y^2 e^{y^2}$. When this is inserted into (5-132), we get a solution that does not vanish at infinity. An acceptable solution can be found if the recursion relation (5-135) terminates, that is, if

$$\epsilon = 2N + 1 \tag{5-137}$$

For that particular value of ϵ the recursion relations yield

$$a_{2k} = (-2)^k \frac{N(N - 2) \cdots (N - 2k + 4)(N - 2k + 2)}{(2k)!} a_0 \tag{5-138}$$

and

$$a_{2k+1} = (-2)^k \frac{(N - 1)(N - 3) \cdots (N - 2k + 3)(N - 2k + 1)}{(2k + 1)!} a_1 \tag{5-139}$$

Thus the results are:

1. There are discrete, equally spaced eigenvalues. Equation (5-137) translates into

$$E = \hbar\omega(n + \tfrac{1}{2}) \tag{5-140}$$

a form that looks familiar, since the relation between energy and frequency is the same as that discovered by Planck for the radiation field modes. This is no accident, since a decomposition of the electromagnetic field into normal modes is essentially a decomposition into harmonic oscillators that are decoupled.

2. The polynomials $h(y)$ are, except for normalization constants, the Hermite polynomials $H_n(y)$, whose properties can be found in any number of textbooks on mathematical physics. We limit ourselves to the following outline of their properties:

$H_n(y)$ satisfy the differential equation

$$\frac{d^2H_n(y)}{dy} - 2y\frac{dH_n(y)}{dy} + 2nH_n(y) = 0 \tag{5-141}$$

They satisfy the following recursion relations

$$H_{n+1} - 2yH_n + 2nH_{n-1} = 0 \tag{5-142}$$

$$H_{n+1} + \frac{dH_n}{dy} - 2yH_n = 0 \tag{5-143}$$

Also

$$\sum_n H_n(y)\frac{z^n}{n!} = e^{2zy-z^2} \tag{5-144}$$

and

$$H_n(y) = (-1)^n e^{y^2} \frac{d^n}{dy^n} e^{-y^2} \tag{5-145}$$

The normalization of the Hermite polynomials is such that

$$\int_{-\infty}^{\infty} dy \, e^{-y^2} H_n(y)^2 = 2^n n! \sqrt{\pi} \tag{5-146}$$

We list a few of the Hermite polynomials below

$$
\begin{aligned}
H_0(y) &= 1 \\
H_1(y) &= 2y \\
H_2(y) &= 4y^2 - 2 \\
H_3(y) &= 8y^2 - 12y \\
H_4(y) &= 16y^4 - 48y^2 + 12 \\
H_5(y) &= 32y^5 - 160y^3 + 129y
\end{aligned}
\tag{5-147}
$$

The orthogonality of eigenfunctions corresponding to different values of n is easily established. The eigenvalue equations

$$\frac{d^2u_n}{dx^2} = \frac{mk}{\hbar^2}x^2u_n - \frac{2mE_n}{\hbar^2}u_n$$

and

$$\frac{d^2u_l^*}{dx^2} = \frac{mk}{\hbar^2}x^2u_l^* - \frac{2mE_l}{\hbar^2}u_l^*$$

when multiplied by u_l^* and u_n, respectively, and the second equation is subtracted from the first one, yields

$$\frac{d}{dx}\left(u_l^*\frac{du_n}{dx} - \frac{du_l^*}{dx}u_n\right) = \frac{2m}{\hbar^2}(E_l - E_n)u_l^*u_n$$

When this equation is integrated over x from $-\infty$ to $+\infty$, the left-hand side vanishes, since the eigenfunctions and their derivatives vanish at $x = \pm\infty$. Thus

$$(E_l - E_n)\int_{-\infty}^{\infty} dx\, u_l^*(x)u_n(x) = 0 \qquad (5\text{-}148)$$

which means that the eigenfunctions for which $E_l \neq E_n$ are orthogonal. The reason for the importance of the harmonic oscillator in quantum mechanics, as in classical mechanics, is that any small perturbation of a system from its equilibrium state will give rise to small oscillations, which are ultimately decomposable into normal modes, that is, independent oscillators.

3. As (5-140) shows, even the lowest state has some energy, the *zero-point energy*. Its presence is a purely quantum mechanical effect, and can be interpreted in terms of the uncertainty principle. It is the zero-point energy that is responsible for the fact that helium does not "freeze" at extremely low temperatures, but remains liquid down to temperatures of the order of 10^{-3} K, at normal pressures. The frequency ω is larger for lighter atoms, which is why the effect is not seen for nitrogen, say. It also depends on detailed features of the interatomic forces, which is why liquid hydrogen does freeze.

Problems

1. Consider an arbitrary potential localized on a finite part of the x-axis. The solutions of the Schrödinger equation to the left and to the right of the potential region are given by

respectively. Show that if we write

$$C = S_{11}A + S_{12}D$$
$$B = S_{21}A + S_{22}D$$

that is, relate the "outgoing" waves to the "ingoing" waves by

$$\begin{pmatrix} C \\ B \end{pmatrix} = \begin{pmatrix} S_{11} & S_{12} \\ S_{21} & S_{22} \end{pmatrix} \begin{pmatrix} A \\ D \end{pmatrix}$$

that the following relations hold

$$|S_{11}|^2 + |S_{21}|^2 = 1$$
$$|S_{12}|^2 + |S_{22}|^2 = 1$$
$$S_{11}S_{12}^* + S_{21}S_{22}^* = 0$$

Use this to show that the *matrix*

$$S = \begin{pmatrix} S_{11} & S_{12} \\ S_{21} & S_{22} \end{pmatrix}$$

and its transpose are unitary.

(*Hint:* Use flux conservation and the possibility that A and D are arbitrary complex numbers.)

2. Calculate the elements of the scattering matrix, S_{11}, S_{12}, S_{21}, and S_{22} for the potential

$$V(x) = 0 \qquad x < -a$$
$$= V_0 \qquad -a < x < a$$
$$= 0 \qquad x < a$$

and show that the general conditions proved in Problem 1 are indeed satisfied.

3. The elements $S_{11} \cdots S_{22}$ are functions of k. Show that

$$S_{11}(-k) = S_{11}^*(k)$$
$$S_{22}(-k) = S_{22}^*(k)$$
$$S_{12}(-k) = S_{21}^*(k)$$

that is, that the matrix has the property

$$S(-k) = S^+(k)$$

4. Consider the odd solution to the potential well [e.g., (5-67)], which can be used as a model for a three-dimensional potential well with zero angular momentum. If the range of the potential is given to be 1.4×10^{-13} cm and the binding energy of a system is -2.2 MeV, and if the mass to be used is 0.8×10^{-24} gm, find the depth of the potential in MeV.

[*Hint:* (1) First, convert distances and masses into units of some mass, so that the range is $d(\hbar/\mu c)$ and the binding energy is of the form $\epsilon(\mu c^2)$. A convenient mass might be the one given. (2) The binding energy is very small, so that it is almost zero. If it were zero, condition (5-69) would yield V_0. Expand about this value.]

5. Without actually solving the Schrödinger equation, set up the solutions so that only the matching of eigenfunctions and their derivatives remain to be done for the following situations:

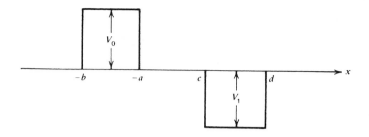

(a) flux $\hbar k/m$ would be incident from the left if the potentials were absent; take $E < V_0$.

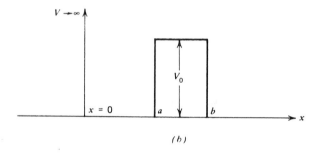

(b)

(b) with flux of magnitude $\hbar k/m$ incident from the right if the potential were absent, $E < V_0$.

6. Show that the conditions for a bound state (5-63) can be obtained by requiring the vanishing of the denominators in (5-26) at $k = i\kappa$. Can you give an argument for why this is not an accident?

7. Consider the scattering matrix for the potential

$$\frac{2m}{\hbar^2} V(x) = \frac{\lambda}{a} \delta(x - b)$$

Show that it has the form

$$\begin{pmatrix} \dfrac{2ika}{2ika - \lambda} & \dfrac{\lambda}{2ika - \lambda} e^{-2ikb} \\[3mm] \dfrac{\lambda}{2ika - \lambda} e^{2ikb} & \dfrac{2ika}{2ika - \lambda} \end{pmatrix}$$

Prove that it is unitary, and that it will yield the condition for bound states when the elements of that matrix become infinite. (This will only occur for $\lambda < 0$.)

8. Calculate α in (5-90), which will localize the particle as far as possible on the right side of the origin.

9. Work out in detail the wave functions for the three lowest eigenfunctions of the harmonic oscillator.

10. Consider the harmonic oscillator potential perturbed by a small cubic term, so that

$$V(x) = \tfrac{1}{2}m\omega^2 \left(x^2 - \frac{1}{a} x^3 \right)$$

If a is large (compared to the characteristic dimension $(\hbar/m\omega)^{1/2}$, estimate how long it takes a particle in the ground state to "leak out" to the region on the far right. Note that with this perturbation alone, there is no lowest energy state, since for large enough x the potential becomes arbitrarily deep.

11. Consider the potential shown below

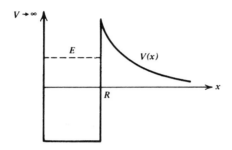

with

$$V(x) = \frac{\hbar^2 l(l + 1)}{2mx^2} \qquad x > R_0$$

Estimate the lifetime of a particle of energy E in this potential. (The outside potential represents a centrifugal barrier in a three-dimensional world.) Express your result in terms of the dimensionless ratio l/kR, where $E = \hbar^2 k^2/2m$, and $l \gg 1$.

12. Consider the Kronig-Penney potential with

$$\lambda = 3\pi$$

(a) Make a detailed plot of

$$\cos x + \frac{\lambda}{2} \frac{\sin x}{x}$$

as a function of $x = ka$.

(b) Show that forbidden energy bands start just above $ka = n\pi$.

(c) Show that the allowed energy bands get narrower as λ increases.

(d) Plot the energy $\hbar^2 k^2/2m$ as a function of q.

13. Consider a potential given by

$$
\begin{aligned}
V(x) &= \infty & x < 0 \\
&= -V_0 & 0 < x < a \\
&= 0 & a < x
\end{aligned}
$$

Assume that a plane wave of momentum $\hbar k$ with flux $\hbar k/m$ is sent in from $+\infty$. (a) Show that the amplitude of the reflected wave can be written in the form $Ce^{i\delta}$. (b) Calculate C and obtain an expression from which you can calculate δ.

14. Consider the following argument: If we have an electron in a potential of width a, then the kinetic energy is, by the uncertainty principle larger than $\hbar^2/2ma^2$. Thus to get a bound state, the potential energy must not only be negative, but it must be larger in magnitude than $\hbar^2/2ma^2$. On the other hand, we have shown, that in one dimension, there is always a bound state, no matter how small V_0 is, provided it is negative. What is wrong with this argument?

15. Consider a potential given by

$$V(x) = \infty \qquad x < 0$$
$$= 0 \qquad x > a$$
$$= a \text{ negative function of } x \text{ in between}$$

Suppose it is known that the interior wave function is such that

$$\frac{1}{u}\frac{du}{dx}\bigg|_{x=a} = f(E)$$

(a) What is the binding energy of a bound state in terms of $f(E_B)$?

(b) Suppose $f(E)$ is a very slowly varying function of E, so that we can take it to be a constant. Calculate the reflected amplitude $R(k)$ in terms of f, if the wave function for $x > a$ has the form $e^{-ikx} + R(k)e^{ikx}$, and check that $|R(k)|^2 = 1$.

16. Consider a particle in the double well shown in the figure. Show that the eigenvalue conditions may be written in the form

$$\tan q(a - b) = \frac{q\alpha(1 + \tanh \alpha b)}{q^2 - \alpha^2 \tanh \alpha b}$$

and

$$\tan q(a - b) = \frac{q\alpha(1 + \coth \alpha b)}{q^2 - \alpha^2 \coth \alpha b}$$

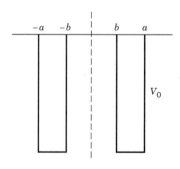

for the even and odd solutions, respectively, where $-E = \hbar^2\alpha^2/2m$ and $E + V_0 = \hbar^2q^2/2m$.

17. Consider the example of Problem 16.
    (a) Show that the eigenvalue conditions approach the single well conditions when $b \to 0$.
    (b) Consider the case where the separation between the well centers becomes large, with the well width fixed. Show that the even and odd eigenvalues approach each other in value. Estimate the energy difference between the lowest lying even and odd eigenvalues.
    (*Hint:* For large z, $\tanh z = 1 - 2e^{-2z}$. Work to lowest order in e^{-2z}.)

18. Prove the *Virial theorem,* which, in one dimension, takes the form

$$\left\langle \frac{p^2}{2m} \right\rangle = \frac{1}{2} \left\langle x \frac{dV}{dx} \right\rangle$$

To do this (a) show that for real wave functions $\psi(x)$,

$$\int_{-\infty}^{\infty} dx \ \psi(x) \ x \ \frac{dV(x)}{dx} \ \psi(x) = -\langle V \rangle + 2 \int_{-\infty}^{\infty} dx \ \frac{d\psi}{dx} \ xV(x)\psi(x)$$

(b) Use the energy eigenvalue equation to prove that

$$2 \int_{-\infty}^{\infty} dx \ \frac{d\psi}{dx} \ xV(x)\psi(x) = E + \frac{\hbar^2}{2m} \int_{-\infty}^{\infty} dx \left(\frac{d\psi}{dx} \right)^2$$

References

The Kronig-Penney model is also discussed in detail in:

E. Merzbacher, *Quantum Mechanics* (2nd edition), John Wiley & Sons, New York, 1970.

For a more detailed discussion of "band theory," see: C. Kittel, *Introduction to Solid State Physics* (6th edition), John Wiley & Sons, New York, 1986, Chapter 9.

For a more complete discussion of barrier penetration, using the WKB approximation, see any of the more advanced textbooks listed at the end of the book.

Chapter 6

THE GENERAL STRUCTURE OF WAVE MECHANICS

In this chapter we pull together and elaborate on the basic principles that were enunciated in connection with some of the specific problems that we solved along the way. Among these are the expansion postulate and its physical meaning, operators, state vectors, degeneracy, and the classical limit. We also introduce the *Dirac notation* in this chapter. This notation will only be used occasionally in the rest of the book, but its compactness merits presentation at this stage of our study of quantum mechanics.

EIGENFUNCTIONS AND EIGENVALUES

The Hamiltonian Operator

The state of a physical system is described by a wave function, which contains all the information about the system. The wave function depends on time, and its temporal development is given by

$$i\hbar \, \frac{\partial \psi(x, t)}{\partial t} = H\psi(x, t) \tag{6-1}$$

The wave function $\psi(x, t)$ is acted on by an *operator* H, the Hamiltonian, which plays a central role in quantum mechanics. The operator H for a simple system of a single particle in a potential V has the form

$$H = \frac{p_{op}^2}{2m} + V(x) \tag{6-2}$$

If $V(x)$ does not have an explicit time dependence, (6-1) can be solved by

$$\psi(x, t) = u_E(x)\, e^{-iEt/\hbar} \tag{6-3}$$

where

$$H u_E(x) = E u_E(x) \tag{6-4}$$

The solutions of this equation $u_E(x)$ are called the *eigenfunctions* of the Hamiltonian, and E are the *eigenvalues*.

In this chapter we concentrate on two important properties of eigenfunctions of H:

1. Eigenfunctions corresponding to different eigenvalues (that is, different values of the constant E) are orthogonal—that is

$$\int dx\; u_E^*(x) u_{E'}(x) = 0 \qquad E \neq E' \tag{6-5}$$

2. The eigenfunctions form a *complete set*—that is—an arbitrary function $\psi(x)$ that is square integrable so that

$$\int dx\; \psi^*(x)\psi(x) < \infty \tag{6-6}$$

may be expanded in terms of eigenfunctions of the Hamiltonian

$$\psi(x) = \sum_E C_E\, u_E(x) \tag{6-7}$$

The spectrum of H may be discrete, as is the case of the infinite hole and for the harmonic oscillator. If the potential $V(x)$ goes to zero as $x \to \infty$, the eigenvalues may be discrete as well as form a continuum. This occurs for an attractive potential, deep enough to form one or more bound states. If this is the case, then it is the total set of continuum and discrete eigenfunctions that are required for the complete set. In that case (6-7) needs to be replaced by

$$\psi(x) = \sum_n C_n u_n(x) + \int dp\; C(p) u_p(x) \tag{6-8}$$

Here the integer n labels the bound states and p labels the continuum states. This labeling is appropriate since for large x the potential energy vanishes and the energy eigenvalue is related to the value of the momentum by $E = p^2/2m$. The eigenfunctions can be multiplied by constants so that they become *normalized*. The orthonormality conditions are

$$\int dx\; u_m^*(x) u_n(x) = \delta_{mn}$$

$$\int dx\; u_q^*(x) u_p(x) = \delta(p - q) \tag{6-9}$$

$$\int dx\; u_n^*(x) u_p(x) = 0$$

The wave function $\psi(x, t)$ can be determined by noting that each eigenfunction has a simple time dependence given by

$$u_E(x, t) = u_E(x) \, e^{-iEt/\hbar} \tag{6-10}$$

It follows from this that

$$\psi(x, t) = \sum_E C_E \, u_E(x) \, e^{-iEt/\hbar} \tag{6-11}$$

which more generally has the form

$$\psi(x, t) = \sum_n C_n u_n(x) e^{-iE_n t/\hbar} + \int dp \, C(p) u_p(x) e^{-ip^2 t/2m\hbar} \tag{6-12}$$

Other Observables

The energy is just one observable attribute of a system. We discussed other observables such as momentum, and will discuss angular momentum in a later chapter. Just as the energy was found to be an eigenvalue of the energy operator H (the Hamiltonian), the momentum was found to be an eigenvalue of the momentum operator p_{op}. The momentum operator eigenvalue equation has the form

$$p_{op} u_p(x) \equiv \frac{\hbar}{i} \frac{d}{dx} u_p(x) = p \, u_p(x) \tag{6-13}$$

The eigenfunctions form a continuum $(-\infty < p < \infty)$, and the eigenfunctions take the form

$$u_p(x) = \frac{1}{\sqrt{2\pi\hbar}} e^{ipx/\hbar} \tag{6-14}$$

The eigenfunctions form an orthonormal set

$$\int_{-\infty}^{\infty} dx \, u_{p_1}^*(x) u_{p_2}(x) = \delta(p_1 - p_2) \tag{6-15}$$

and the expansion theorem is usually written in the form

$$\psi(x) = \int_{-\infty}^{\infty} dp \, \phi(p) u_p(x) \tag{6-16}$$

The momentum operator like the Hamiltonian has *real* eigenvalues.

Operators, all of whose eigenvalues are real, are called hermitian operators. Since all physical observables share that property, they must be described by hermitian operators.

THE EXPANSION POSTULATE
AND THE VECTOR ANALOGY

For an arbitrary observable, which we denote by A, there will be eigenfunctions corresponding to the real eigenvalues a,

$$Au_a(x) = au_a(x) \tag{6-17}$$

The eigenfunctions form an orthogonal set, and they can be normalized so that

$$\int dx \; u_{a_1}^*(x) \; u_{a_2}(x) = \delta_{a_1 a_2} \tag{6-18}$$

Here $\delta_{a_1 a_2}$ is a Kronecker delta for discrete eigenvalues and a Dirac delta function for continuous eigenvalues.

The eigenfunctions $u_a(x)$ also form a complete set, which is equivalent to the statement that an arbitrary (square integrable) function $\psi(x)$ may be expanded in terms of the $u_a(x)$,

$$\psi(x) = \sum_a C_a \, u_a(x) \tag{6-19}$$

It follows from the orthonormality condition that

$$C_a = \int dx \; u_a^*(x)\psi(x) \tag{6-20}$$

The Interpretation of the Expansion Coefficients

The interpretation of the expansion coefficients C_a is the following: If the observable A is measured for a collection of systems each of which is described by the wave function $\psi(x)$, which is normalized to unity, so that

$$\int dx \; \psi^*(x)\psi(x) = 1 \tag{6-21}$$

then

1. The result of any given measurement can only be one of the eigenvalues a.
2. The probability that the eigenvalue a will be found, or, equivalently, the fraction of systems in the collection that will be found to have the eigenvalue a, is $|C_a|^2$.
3. After a measurement on a member of the collection yields a given eigenvalue a_1, for example, then that particular system in the collection must be *projected* by the measurement into the state $u_{a_1}(x)$. It is only in this way that we can be sure that a subsequent measurement of the observable A gives the same result.

A consequence of this interpretation is that the probability that the value of the observable A for a system has any one of the eigenvalues is unity. This means that

$$\sum_a |C_a|^2 = 1 \tag{6-22}$$

This follows from

$$1 = \int dx \; \psi^*(x)\psi(x) = \int dx \left(\sum_a C_a^* u_a^*(x) \right) \psi(x)$$
$$= \sum_a C_a^* C_a$$

This formula implies the following:

$$\sum_a C_a C_a^* = \sum_a \int dx \; u_a^*(x)\psi(x) \int dy \; u_a(y)\psi^*(y)$$

$$= \iint dx dy \psi^*(y)\psi(x) \sum_a u_a(y)u_a^*(x) = 1 \tag{6-23}$$

This, in turn, implies

$$\sum_a u_a(y)u_a^*(x) = \delta(x - y) \tag{6-24}$$

This property of the eigenfunctions is described as the *completeness* relation, and it is equivalent to the statement of the expansion theorem.

The Vector Space Analogy

The expansion theorem may be viewed as a generalization of the expansion of a vector **A** in terms of orthonormal unit vectors in an N-dimensional vector space

$$\mathbf{A} = A_1\mathbf{i}_1 + A_2\mathbf{i}_2 + \cdots + A_N\mathbf{i}_N \tag{6-25}$$

The unit vectors satisfy

$$\mathbf{i}_k \cdot \mathbf{i}_l = \delta_{kl} \tag{6-26}$$

and they are the analogs of the eigenfunctions $u_a(x)$. The coefficients A_n are given by

$$A_n = \mathbf{i}_n \cdot \mathbf{A} \tag{6-27}$$

and these are the analogs of the C_a. We shall often use the language of vector spaces in talking about quantum mechanics. Thus, we shall often refer to the coefficients C_a as the *projections* of $\psi(x)$ along the *basis vectors* $u_a(x)$, and the quantity

$$C_a = \int dx \; u_a^*(x)\psi(x)$$

will often be called the *scalar product* of u_a and ψ.

The analogy between wave functions $\psi(x)$ and N-dimensional vectors is actually quite profound. In both cases we deal with linear spaces: just as a sum of any two vectors in the vector space yields a vector

$$\mathbf{A} + \mathbf{B} = \mathbf{C}$$

so is the sum of two wave functions an acceptable wave function, and in both cases we define a scalar product.

The difference between the vector space of quantum mechanics and a simple N-dimensional vector space is that the vector space in quantum mechanics is continuously infinite-dimensional. Thus the finite, discrete sum in a scalar product $\mathbf{A} \cdot \mathbf{B} = \Sigma_i A_i B_i$ is replaced by an integral in $\int dx\ \phi^*(x)\psi(x)$. This means that in quantum mechanics one has to worry about convergence of integrals, and proving the expansion theorem is very complicated. In mathematical parlance the vector space in quantum mechanics is a *Hilbert Space*. We shall ignore, whenever we can, the mathematical complications, and freely draw on the analogy with an ordinary vector space when explaining quantum mechanics.

Operators and Observables

In vector spaces an operator transforms one vector into another. Linear operators are such that

$$A(\alpha_1\psi_1(x) + \alpha_2\psi_2(x)) = \alpha_1\ A\psi_1(x) + \alpha_2\ A\psi_2(x) \tag{6-28}$$

and if they are to represent observables, they need to be *hermitian*. For hermitian operators

$$\langle A \rangle = \langle A \rangle^*$$

that is, for a state Ψ,

$$\int dx\ \Psi^*(x)A\Psi(x) = \int dx\ (A\Psi(x))^*\Psi(x) \tag{6-29}$$

From this it follows that for hermitian operators

$$\int dx\ \phi^*A\psi = \int dx\ (A\phi)^*\psi \tag{6-30}$$

for any pair of functions ϕ and ψ. To prove this, take

$$\Psi(x) = \phi(x) + \lambda\psi(x) \tag{6-31}$$

where λ is an arbitrary complex constant, then it follows from (6-29) that

$$\int dx\ \Psi^*A\Psi = \int dx\ (\phi^* + \lambda^*\psi^*)A(\phi + \lambda\psi) = \int dx\ \phi^*A\phi \tag{6-32}$$

$$+\ |\lambda|^2 \int dx\ \psi^*A\psi + \lambda \int dx\ \phi^*A\psi + \lambda^* \int dx\ \psi^*A\phi$$

The complex conjugate of this is

$$\int dx\ (A\Psi)^*\Psi = \int dx\ (A\phi)^*\psi + |\lambda|^2 \int dx\ (A\psi)^*\psi$$

$$+\ \lambda^* \int dx\ (A\psi)^*\phi + \lambda \int dx\ (A\phi)^*\psi$$

For a hermitian operator, the left sides of the two equations and the first two terms are equal. Since λ and λ^* are independent parameters (they are complex), we may equate their coefficients separately, and (6-30) follows.

If an operator A is *not* hermitian, we may define the hermitian conjugate operator $A^\dagger$ by

$$\int dx \, (A\phi)^*\psi = \int dx \, \phi^*A^\dagger\psi \tag{6-33}$$

for any pair of wave functions.

DIRAC NOTATION

P.A.M. Dirac introduced a very economical and powerful notation that applies equally well to finite dimensional vector spaces and to Hilbert spaces. We associate with each wave function ψ, a *state vector*, $|\psi\rangle$, called a *ket*. We also associate with each complex conjugate wave function ϕ^* the quantity $\langle\phi|$, called a *bra*. The scalar product of ψ and ϕ^* is associated with the bracket relation (hence the names)

$$\int dx \, \phi^*\psi = \langle\phi|\psi\rangle \tag{6-34}$$

It immediately follows that

$$\langle\phi|\psi\rangle^* = \langle\psi|\phi\rangle \tag{6-35}$$

The expression for the integral involving an operator may be written in two equivalent ways

$$\int dx \, \phi^*A\psi = \langle\phi|A\psi\rangle = \langle\phi|A|\psi\rangle \tag{6-36}$$

The definition of the hermitian conjugate operator reads

$$\langle A\phi|\psi\rangle = \langle\phi|A^\dagger|\psi\rangle \tag{6-37}$$

for any pair of states. If A is a number, rather than an operator, then it can be taken out of the brackets. Thus

$$\langle\phi|a\psi\rangle = a\langle\phi|\psi\rangle \tag{6-38}$$

and

$$\langle a\phi|\psi\rangle = a^*\langle\phi|\psi\rangle \tag{6-39}$$

Eigenvalue equations take the form

$$A|a\rangle = a|a\rangle \tag{6-40}$$

where the state is only labeled by its eigenvalue. How this translates into an x-dependent equation of the type shown in (6-4) will be discussed in Chapter 7. The orthonormality condition in Dirac notation reads

$$\langle a_1 | a_2 \rangle = \delta_{a_1 a_2} \tag{6-41}$$

The expansion theorem reads

$$|\psi\rangle = \sum_a C_a \, |a\rangle \tag{6-42}$$

It follows that on multiplication by a particular eigenstate from the left

$$\langle b|\psi\rangle = \sum_a C_a \, \langle b|a\rangle = \sum_a C_a \, \delta_{ab} = C_b \tag{6-43}$$

With this form we can write

$$\langle\phi|\psi\rangle = \sum_a C_a \, \langle\phi|a\rangle = \sum_a \langle\phi|a\rangle\langle a|\psi\rangle \tag{6-44}$$

Since this is true for all ϕ and ψ, we can "take apart" this relation and write the Dirac-notation analog of (6-24):

$$\sum_a |a\rangle \langle a| = 1 \tag{6-45}$$

We conclude this section by illustrating the use of the Dirac notation to prove the orthogonality of eigenfunctions of hermitian operators that correspond to different eigenvalues. Consider

$$\langle b|A|a\rangle = a \, \langle b|a\rangle \tag{6-46}$$

On the other hand

$$\langle b|A^\dagger|a\rangle = \langle A(b)|a\rangle = b^* \, \langle b|a\rangle \tag{6-47}$$

since $|b\rangle$ is an eigenstate of A with eigenvalue b, and the quantities appear in the *bra*. However, for a hermitian operator we have $A = A^\dagger$. It follows that

$$a\langle b|a\rangle = b^*\langle b|a\rangle \tag{6-48}$$

If we choose $|b\rangle = |a\rangle$, we immediately see that the eigenvalues must be real. Thus $b^* = b$, for example, and (6-48) leads to

$$(a - b)\langle b|a\rangle = 0 \tag{6-49}$$

which is what we set out to prove, since for $a \neq b$, $\langle b|a\rangle$ must vanish.

DEGENERACY AND
SIMULTANEOUS OBSERVABLES

In both the problems discussed in Chapter 4, the particle in the box, and the free particle, we found that the eigenfunctions were simultaneous eigenfunctions of H and another operator, parity in the first case, momentum in the second, and we saw that in both cases the additional operators commuted with H. Let us now examine the general conditions under which this happens.

The eigenfunctions u_a, corresponding to the eigenvalue a of the operator A,

$$Au_a(x) = au_a(x) \tag{6-50}$$

will be simultaneous eigenfunctions of another operator B, when

$$Bu_a(x) = bu_a(x) \tag{6-51}$$

This, however, implies that

$$ABu_a(x) = Abu_a(x) = bAu_a(x) = abu_a(x)$$

and

$$BAu_a(x) = Bau_a(x) = aBu_a(x) = abu_a(x)$$

that is, that

$$(AB - BA)\, u_a(x) = 0 \tag{6-52}$$

If this were to hold for just one u_a, it would not be very interesting, but if it holds for the complete set u_a, then it means that for all square integrable functions $\psi(x) = \Sigma_a\, C_a u_a(x)$,

$$\sum_a C_a(AB - BA)\, u_a(x) = (AB - BA) \sum C_a u_a(x)$$
$$= (AB - BA)\, \psi(x) = 0 \tag{6-53}$$

that is, the operators commute

$$[A, B] = 0 \tag{6-54}$$

Conversely, if we have two hermitian operators A and B that commute, so that (6-54) holds, then

$$ABu_a(x) = BAu_a(x)$$
$$= aBu_a(x) \tag{6-55}$$

that is,

$$A[Bu_a(x)] = a[Bu_a(x)] \tag{6-56}$$

Thus the function $Bu_a(x)$ is also an eigenfunction of A with eigenvalue a. If there is only one eigenfunction of A corresponding to the eigenvalue a, then this implies that $Bu_a(x)$ must be proportional to $u_a(x)$, that is,

$$Bu_a(x) = bu_a(x) \qquad (6\text{-}57)$$

Then $u_a(x)$ is a simultaneous eigenfunction of A and B. This situation, in which the eigenfunctions of A are *not degenerate*, is the one that we saw for the particle in the box. If, on the other hand, there are two eigenfunctions of A corresponding to the eigenvalue a, that is, we have a twofold *degeneracy*

$$\begin{aligned}
Au_a^{(1)}(x) &= au_a^{(1)}(x) \\
Au_a^{(2)}(x) &= au_a^{(2)}(x)
\end{aligned} \qquad (6\text{-}58)$$

a situation illustrated in the free particle example, then we can only assert that $Bu_a^{(1)}(x)$ and $Bu_a^{(2)}(x)$ must be linear combinations of $u_a^{(1)}(x)$ and $u_a^{(2)}(x)$:

$$\begin{aligned}
Bu_a^{(1)}(x) &= b_{11}u_a^{(1)}(x) + b_{12}u_a^{(2)}(x) \\
Bu_a^{(2)}(x) &= b_{21}u_a^{(1)}(x) + b_{22}u_a^{(2)}(x)
\end{aligned} \qquad (6\text{-}59)$$

It is evident, however, that we can take linear combinations of these equations to obtain equations of the type

$$\begin{aligned}
Bv_a^{(1)}(x) &= b_+v_a^{(1)}(x) \\
Bv_a^{(2)}(x) &= b_-v_a^{(2)}(x)
\end{aligned} \qquad (6\text{-}60)$$

For example,

$$\begin{aligned}
B(u_a^{(1)} + \lambda u_a^{(2)}) &= (b_{11} + \lambda b_{21})\,u_a^{(1)} + (b_{12} + \lambda b_{22})\,u_a^{(2)} \\
&= b_\pm(u_a^{(1)} + \lambda u_a^{(2)})
\end{aligned}$$

provided we choose λ such that

$$\frac{b_{12} + \lambda b_{22}}{b_{11} + \lambda b_{21}} = \lambda$$

This is a quadratic equation, and there will be two values of λ, corresponding to the two eigenvalues $b_\pm$. It is more appropriate to denote the simultaneous eigenfunctions of A and B in (6-60) by $u_{ab}^{(1)}(x)$ and $u_{ab}^{(2)}(x)$. Since these correspond to different eigenvalues of the operator B, they will be orthogonal to each other. In practice, for twofold degeneracy, the degenerate eigenfunctions of A, if they are taken to be orthogonal to each other (e.g., e^{ikx} and e^{-ikx} for the free particle case), will automatically be eigenfunctions of B.

Even after finding eigenfunctions of A and then making linear combinations that are eigenfunctions of a commuting operator B, there may still be some degeneracy, that is, there are several eigenfunctions of A and B simultaneously, with the same a and b. This means that there must be a third operator C that commutes with both A and B, and the function can be recombined to be simultaneous eigenfunctions of A, B, and C whose eigenvalues distinguish the degenerate eigenfunctions of A and B. This will go on until there is no more degeneracy. The set of mutually commuting operators A, B, C, ..., M of which our set of functions is a set of common eigenfunctions is called a *complete set of commuting observables*. We have

$$[A, B] = [A, C] = \cdots = [A, M] = 0$$
$$[B, C] = [B, D] = \cdots = [B, M] = 0 \tag{6-61}$$

and so on, with the simultaneous eigenfunctions $u_{ab...m}(x)$:

$$Au_{ab...m}(x) = au_{ab...m}(x)$$
$$Bu_{ab...m}(x) = bu_{ab...m}(x) \tag{6-62}$$
$$Mu_{ab...m}(x) = mu_{ab...m}(x)$$

The state described by $u_{abc...m}(x)$ has definite values of the observables A, B, C, ..., M. This is the largest possible amount of information that we can have about a system all at once. The reason is that if we consider another operator that is not some function of the operators A, B, ..., M (since these commute, such a function is unambiguously defined), then a measurement of it will not give a sharp value for the state $u_{ab...m}(x)$. In general, if two operators do not commute, then a type of uncertainty relation connects the precision with which the two observables can be determined.

UNCERTAINTY RELATIONS

A simple way to define the uncertainty associated with an operator A is to describe it in some way by the fluctuations about the mean value. Let us consider an operator A and some general normalized state described by ψ. Its mean value is just the expectation value $\langle\psi|A|\psi\rangle$, which we abbreviate by the notation $\langle A \rangle$. A measure of the fluctuation is the square of the deviation from the mean. This is how we define the square of the uncertainty $(\Delta A)^2$:

$$(\Delta A)^2 \equiv \langle(A - \langle A\rangle)^2\rangle = \langle A^2 - 2A\langle A\rangle + \langle A\rangle^2\rangle$$
$$= \langle A^2 \rangle - \langle A \rangle^2 \tag{6-63}$$

since the mean value of a number like $\langle A \rangle^2$ is just that number.

We show in Appendix B that for two hermitian operators A and B the uncertainties ΔA and ΔB are correlated if the two operators do not commute. The relation is

$$(\Delta A)^2 \, (\Delta B)^2 \geq \frac{1}{4} \langle i[A,B]\rangle^2 \tag{6-64}$$

where $[A,B] = AB - BA$.

Note that if the state ψ happens to be an eigenstate of one of the operators, A, say, then

$$(\Delta A)^2 = \langle a|A^2|a\rangle - \langle a|A|a\rangle^2$$
$$= a^2\langle a|a\rangle - [a\langle a|a\rangle]^2 = 0 \tag{6-65}$$

as expected, since there is no uncertainty. Thus the left side of (6-64) vanishes. On the right side we also obtain zero since

$$\langle a|[A,B]|a\rangle = \langle a|AB|a_a\rangle - \langle a|BA|a_a\rangle$$
$$= \langle A(a)|B|a\rangle - a\langle a|B|A(a)\rangle$$
$$= a\langle a|B|a\rangle - a\langle a|B|a\rangle = 0$$

For the operators x and p, for which $[x,p] = i\hbar$, it follows that

$$(\Delta x)^2 \, (\Delta p)^2 \geq \frac{\hbar^2}{4} \tag{6-66}$$

Note that in the derivation no use was made of wave properties, x-space or p-space wave functions or of particle-wave duality. Our result depends entirely on the operator properties of the observables A and B.

TIME DEPENDENCE AND THE CLASSICAL LIMIT

Let us now turn to the important question of the classical limit of quantum theory. To do this we must first study the time development of expectation values of operators. In general, the expectation value of an operator changes with time. It may change with time because the operator has an explicit time dependence, for example, the operator $x + pt/m$, and it also changes with time because the expectation value is taken with respect to a wave function that itself changes with time. If we write

$$\langle A \rangle_t = \int \psi^*(x, t) \, A\psi(x, t) \, dx \tag{6-67}$$

then

$$\frac{d}{dt} \langle A \rangle_t = \int \psi^*(x, t) \, \frac{\partial A}{\partial t} \, \psi(x, t) \, dx$$
$$+ \int \frac{\partial \psi^*(x, t)}{\partial t} \, A\psi(x, t) \, dx$$
$$+ \int \psi^*(x, t) \, A \frac{\partial \psi(x, t)}{\partial t} \, dx$$
$$= \left\langle \frac{\partial A}{\partial t} \right\rangle_t + \int \left(\frac{1}{i\hbar} H\psi(x, t) \right)^* A\psi(x, t)$$
$$+ \int \psi^*(x, t) \, A \left(\frac{1}{i\hbar} H\psi(x, t) \right)$$
$$= \left\langle \frac{\partial A}{\partial t} \right\rangle_t + \frac{i}{\hbar} \int \psi^*(x, t) \, HA\psi(x, t) \, dx$$
$$- \frac{i}{\hbar} \int \psi^*(x, t) \, AH\psi(x, t) \, dx$$

that is,

$$\frac{d}{dt} \langle A \rangle_t = \left\langle \frac{\partial A}{\partial t} \right\rangle_t + \frac{i}{\hbar} \langle [H, A] \rangle_t \tag{6-68}$$

In the derivation we made use of the fact that H is a hermitian operator. We observe that if A has no explicit time dependence, then the change of the expectation value *for any state* is

$$\frac{d}{dt} \langle A \rangle_t = \frac{i}{\hbar} \langle [H, A] \rangle_t \tag{6-69}$$

If the operator commutes with H, then its expectation value is always constant, that is, we may say that *the observable is a constant of the motion*. If the Hamiltonian is one of the complete set of commuting observables, then all the others are constants of the motion.

Let us consider successively $A = x$ and $A = p$. We first have

$$\frac{d}{dt} \langle x \rangle = \frac{i}{\hbar} \langle [H, x] \rangle$$

$$= \frac{i}{\hbar} \left\langle \left[\frac{p^2}{2m} + V(x), x \right] \right\rangle$$

Now x commutes with any function of x,

$$[V(x), x] = 0 \tag{6-70}$$

so that we only have to calculate

$$[p^2, x] = p[p, x] + [p, x] p$$
$$= \frac{2\hbar}{i} p \tag{6-71}$$

Thus we obtain

$$\frac{d}{dt} \langle x \rangle = \left\langle \frac{p}{m} \right\rangle \tag{6-72}$$

Next we have

$$\frac{d}{dt} \langle p \rangle = \frac{i}{\hbar} \left\langle \left[\frac{p^2}{2m} + V(x), p \right] \right\rangle$$

$$= -\frac{i}{\hbar} \langle [p, V(x)] \rangle \tag{6-73}$$

since p^2 and p evidently commute. To evaluate the last commutator, we note that

$$pV(x)\,\psi(x) - V(x)\,p\psi(x) = \frac{\hbar}{i} \frac{d}{dx} [V(x)\,\psi(x)] - \frac{\hbar}{i} V(x) \frac{d}{dx} \psi(x)$$

$$= \frac{\hbar}{i} \frac{dV(x)}{dx} \psi(x) \tag{6-74}$$

so that

$$[p, V(x)] = \frac{\hbar}{i} \frac{dV(x)}{dx} \qquad (6\text{-}75)$$

and thus

$$\frac{d}{dt} \langle p \rangle_t = -\left\langle \frac{dV(x)}{dx} \right\rangle_t \qquad (6\text{-}76)$$

We may combine (6-72) and (6-76) to obtain

$$m \frac{d^2}{dt^2} \langle x \rangle_t = -\left\langle \frac{dV(x)}{dx} \right\rangle_t \qquad (6\text{-}77)$$

This looks very much like the equation of motion of a classical point particle in a potential $V(x)$

$$m \frac{d^2 x_{cl}}{dt^2} = -\frac{dV(x_{cl})}{dx_{cl}} \qquad (6\text{-}78)$$

The only thing that keeps us from making the identification

$$x_{cl} = \langle x \rangle \qquad (6\text{-}79)$$

is that

$$\left\langle \frac{dV}{dx} \right\rangle \neq \frac{d}{d\langle x \rangle} V(\langle x \rangle) \qquad (6\text{-}80)$$

Under the circumstances where the preceding inequality becomes an approximate equality, the motion is essentially classical, as was first noted by Ehrenfest. This requires that the potential be a slowly varying function of its argument. If we write

$$F(x) = -\frac{dV(x)}{dx} \qquad (6\text{-}81)$$

then

$$F(x) = F(\langle x \rangle) + (x - \langle x \rangle) F'(\langle x \rangle) + \frac{(x - \langle x \rangle)^2}{2!} F''(\langle x \rangle) + \cdots$$

If the uncertainty $(\Delta x)^2 = \langle (x - \langle x \rangle)^2 \rangle$ is small, and the higher terms in the expansion can be neglected, then we have

$$\begin{aligned} \langle F(x) \rangle &\cong F(\langle x \rangle) + \langle x - \langle x \rangle \rangle F'(\langle x \rangle) \\ &\cong F(\langle x \rangle) \end{aligned} \qquad (6\text{-}82)$$

It is indeed true that even for electrons and other subatomic particles, (6-82) can be valid. For macroscopic fields (6-82) is a good approximation, and this allows us to describe electron or proton orbits in an accelerator by means of classical equations of motion.

Problems

1. If A and B are hermitian operators, prove that (1) the operator of AB is only hermitian if A and B commute, that is, if $AB = BA$, and (2) the operator $(A + B)^n$ is hermitian.

2. Prove that $A + A^\dagger$ and $i(A - A^\dagger)$ are hermitian for any operator, as is $AA^\dagger$.

3. Prove that if H is a hermitian operator, then the hermitian conjugate operator of e^{iH} (defined to be $\sum\limits_{n=0}^{\infty} i^n H^n / n!$) is the operator e^{-iH}.

4. Prove the Schwartz inequality

$$\langle \psi | \psi \rangle \langle \phi | \phi \rangle \geq |\langle \psi | \phi \rangle|^2$$

Note that this is equivalent to $\cos^2 \theta \leq 1$ for three-dimensional vectors.

(*Hint:* Consider $\langle \psi + \lambda\phi | \psi + \lambda\phi \rangle \geq 0$ and calculate the value of λ that minimizes the left-hand side.)

5. The normalized eigenfunctions for

$$
\begin{aligned}
V(x) &= \infty & x < 0 \\
&= 0 & 0 < x < a \\
&= \infty & a < x
\end{aligned}
$$

are $u_n(x) = \sqrt{2/a} \sin (n\pi x/a)$. Show formally that

$$\sum_n \sqrt{\frac{2}{a}} \sin \frac{n\pi x}{a} \sqrt{\frac{2}{a}} \sin \frac{n\pi y}{a} = \delta(x - y) \qquad \text{for} \qquad 0 \leq x, y \leq a$$

6. If A is hermitian, show that $\langle A^2 \rangle \geq 0$.

7. Consider the hermitian operator H that has the property that

$$H^4 = 1$$

What are the eigenvalues of the operator H? What are the eigenvalues if H is not restricted to being Hermitian?

8. An operator is said to be unitary if it has the property that

$$UU^\dagger = U^\dagger U = 1$$

Show that if $\langle \psi | \psi \rangle = 1$ then $\langle U\psi | U\psi \rangle = 1$.

9. Show that if A is hermitian, then e^{iA} is unitary.

10. Show that if the $\{u_a\}$ form an orthonormal complete set, with

$$\langle u_a | u_b \rangle = \delta_{ab}$$

then the set

$$|v_a\rangle = U|u_a\rangle$$

with U unitary is also orthonormal. (The meaning of the preceding is a unitary operator acting on a set of "basis" states yields another set of "basis" states.)

11. Show that

$$\Delta p \, \Delta x \sim \hbar n$$

for a particle in an infinite box in the state characterized by the quantum number n.

12. Use the commutation relations between the momentum p and the position x to obtain the equations describing the time dependence of $\langle x \rangle$ and $\langle p \rangle$ given the Hamiltonians

(a) $H = \dfrac{p^2}{2m} + \tfrac{1}{2}m(\omega_1^2 x^2 + \omega_2 x + \epsilon)$

(b) $H = \dfrac{p^2}{2m} + \tfrac{1}{2}m\omega^2 x^2 - \dfrac{A}{x^2}$

Solve the first set of equations (Hamiltonian (a)).

13. An electron in an oscillating electric field is described by the Hamiltonian operator

$$H = \dfrac{p^2}{2m} - (eE_0 \cos \omega t)\, x$$

Calculate (dx/dt), (dp/dt), and (dH/dt).

References

The general structure of wave mechanics is discussed in all books on quantum mechanics. See, for example, among the more introductory books,

D. Bohm, *Quantum Theory*, Dover Publishers, Inc. New York, 1989.

R. H. Dicke and J. P. Wittke, *Introduction to Quantum Mechanics*, Addison-Wesley, Reading, Mass., 1960.

J. L. Powell and B. Crasemann, *Quantum Mechanics*, Addison-Wesley, Reading, Mass., 1961.

E. Merzbacher, *Quantum Mechanics*, John Wiley & Sons, New York, 1970, is among the more advanced textbooks.

OPERATOR METHODS IN QUANTUM MECHANICS

The discussion of the general structure of wave mechanics placed equal weight on the operators that represent the observables, and on their eigenfunctions. Although the latter were at one point described as analogous to an orthonormal basis of unit vectors in an *N*-dimensional vector space—which would certainly downgrade them in importance—they, rather than the operators, seemed to play the leading role in our discussion of the physical problems in Chapter 5. In this chapter we will show, using a simple example, (a) that one can go very far toward finding the eigenvalue spectrum using the operators alone, and (b) that the description of eigenfunctions as a basis can be made a little more abstract. The latter is important because so far we have only considered functions that depend on *x* or on *p*. We shall see later that there exist observables that cannot be associated with *x*-space in any direct way, and for these a more abstract notion of *eigenstate* must be developed. These remarks will become somewhat clearer in the course of the solution of our example, the *harmonic oscillator problem*.[1]

THE ENERGY SPECTRUM OF THE HARMONIC OSCILLATOR

The Hamiltonian for the harmonic oscillator has the form

$$H = \frac{p^2}{2m} + \tfrac{1}{2}m\omega^2 x^2 \tag{7-1}$$

---

[1]There are few problems that are exactly soluble, whether as differential equations or in operator form. This example is the simplest and thus most suitable for our purposes.

where x and p are operators. We do not insist that p be represented by $(\hbar/i)(d/dx)$. The only vestige of the explicit representation that we obtained in Chapter 3 is the statement of the fundamental commutation relation

$$[p, x] = \frac{\hbar}{i} \tag{7-2}$$

Classically, the Hamiltonian could be factored into

$$H = \omega \left(\sqrt{\frac{m\omega}{2}} \, x - i \frac{p}{\sqrt{2m\omega}} \right) \left(\sqrt{\frac{m\omega}{2}} \, x + i \frac{p}{\sqrt{2m\omega}} \right)$$

but because p and x do not commute, we have

$$\omega \left(\sqrt{\frac{m\omega}{2}} \, x - i \frac{p}{\sqrt{2m\omega}} \right) \left(\sqrt{\frac{m\omega}{2}} \, x + i \frac{p}{\sqrt{2m\omega}} \right)$$

$$= \frac{p^2}{2m} + \frac{m\omega^2}{2} x^2 - \frac{i\omega}{2} (px - xp) \tag{7-3}$$

$$= H - \tfrac{1}{2}\hbar\omega$$

We now introduce a special notation for the operators in terms of which $H - \hbar\omega/2$ is factored. We write

$$A = \sqrt{\frac{m\omega}{2\hbar}} \, x + i \frac{p}{\sqrt{2m\omega\hbar}}$$

$$A^\dagger = \sqrt{\frac{m\omega}{2\hbar}} \, x - i \frac{p}{\sqrt{2m\omega\hbar}} \tag{7-4}$$

with the additional factors of $\sqrt{1/\hbar}$ inserted so as to make the A and $A^\dagger$ dimensionless. Since x and p are hermitian, $A^\dagger$ is indeed the hermitian conjugate of A. We compute

$$[A, A^\dagger] = \left[\sqrt{\frac{m\omega}{2\hbar}} \, x, \, - i \frac{p}{\sqrt{2m\omega\hbar}} \right] + \left[i \frac{p}{\sqrt{2m\omega\hbar}}, \, \sqrt{\frac{m\omega}{2\hbar}} \, x \right] = 1 \tag{7-5}$$

The Hamiltonian takes the form

$$H = \hbar\omega \left(\frac{1}{2} + A^\dagger A \right) \tag{7-6}$$

The simplicity of the Hamiltonian is reflected in the simplicity of the commutation relations of A and $A^\dagger$ with H. We have[2]

$$[H, A] = [\hbar\omega A^\dagger A, A] = \hbar\omega[A^\dagger, A]A = -\hbar\omega A \tag{7-7}$$

[2]We shall make repeated use of the rules for commutators exhibited in Appendix B:

$$[A + B, C] = [A, C] + [B, C] \quad \text{and} \quad [AB, C] = A[B, C] + [A, C] B$$

It is, of course, essential that the order of operators not be disturbed.

Similarly,

$$[H, A^\dagger] = [\hbar\omega A^\dagger A, A^\dagger] = \hbar\omega A^\dagger[A, A^\dagger] = \hbar\omega A^\dagger \qquad (7\text{-}8)$$

Incidentially, it is a useful technical trick in deriving commutation relations involving hermitian adjoint operators to recall that

$$[A, B]^\dagger = (AB - BA)^\dagger = B^\dagger A^\dagger - A^\dagger B^\dagger = [B^\dagger, A^\dagger] \qquad (7\text{-}9)$$

In particular

$$[H, A]^\dagger = [A^\dagger, H] = -[H, A^\dagger]$$
$$= (-\hbar\omega A)^\dagger \qquad (7\text{-}10)$$

from which (7-8) follows.

Let us now write down the eigenvalue equation, which reads

$$Hu_E = Eu_E \qquad (7\text{-}11)$$

In the past, whenever we wrote down such an equation, the implication was that H contained some differential operators like d/dx and that u_E was a function of x. That was appropriate when our operators were specifically tied to the space defined by all square integrable functions of x, but in what we are doing now, we are not being very specific about what our operators operate on. We shall assume that they are defined in some abstract vector space, and relate that abstract vector space to the space of functions of x later. To translate this abstraction into the language that we use to describe the equations, we shall not speak of eigenfunctions but of *eigenstates*, and what we called wave functions or wave packets, we shall now call *state vectors*. Thus the eigenfunction $u_{ab...m}(x)$ of the maximal commuting set of observables can be replaced by the eigenvector or eigenstate of this maximal commuting set, $u_{ab...m}$; the labels $a, b, \ldots, m$ give the values of the eigenvalues of the observables $A, B, \ldots, M$, and this description, without the x, does explicitly show the maximum information content.

Let us now take (7-7) and have it act on u_E:

$$HAu_E - AHu_E = -\hbar\omega Au_E$$

With the help of (7-11) this becomes

$$HAu_E = (E - \hbar\omega)\, Au_E \qquad (7\text{-}12)$$

This equation states that if u_E is an eigenstate of H with eigenvalue E, then Au_E is also an eigenstate of H but with eigenvalue $E - \hbar\omega$, that is, with energy lowered by one unit of

$$\epsilon = \hbar\omega \qquad (7\text{-}13)$$

We may therefore write

$$Au_E = c(E)\, u_{E-\epsilon} \qquad (7\text{-}14)$$

The constant $c(E)$ is necessary, since even if u_E is normalized to 1, Au_E need not be. In our emphasis on separation from x-dependence, the normalization condition that was always written as

$$\int u_E^*(x)\, u_E(x)\, dx = 1$$

is now, using the Dirac notation, written as

$$\langle u_E | u_E \rangle = 1 \tag{7-15}$$

We shall always normalize all eigenstates to 1, unless they belong to the continuum, in which case

$$\langle u_E | u_{E'} \rangle = \begin{cases} \delta(E - E') \\ \text{or } \delta(p - p') \end{cases} \tag{7-16}$$

If we now apply (7-7) to the state $u_{E-\epsilon}$, we find, in exactly the same way, that $Au_{E-\epsilon}$, or, equivalently, $A^2 u_E$ gives a state of energy $E - 2\epsilon$. Thus by repeated application of the operator A to any u_E we can generate states of lower and lower energy. The operator A is appropriately called a *lowering operator*. There is a limit to how many times it can be applied, since it is a consequence of (7-1) that H must always have positive expectation value. For an arbitrary wave function

$$\langle \psi | p^2 | \psi \rangle = \int \psi^*(x)\, p^2 \psi(x)\, dx = \int [p^\dagger \psi(x)]^*\, (p\psi)\, dx$$
$$= \int [p\psi(x)]^* [p\psi(x)]\, dx \tag{7-17}$$
$$= \hbar^2 \int |d\psi(x)/dx|^2\, dx \geq 0$$

which we rewrite in our coordinate-deemphasizing way as

$$\langle \psi | p^2 | \psi \rangle = \langle p^\dagger \psi | p\psi \rangle$$
$$= \langle p\psi | p\psi \rangle \geq 0 \tag{7-18}$$

Similarly, since x is also a hermitian operator

$$\langle \psi | x^2 | \psi \rangle = \langle x^\dagger \psi | x\psi \rangle$$
$$= \langle x\psi | x\psi \rangle \geq 0 \tag{7-19}$$

and all scalar products of vectors with themselves yield the square of their length, that is, a positive number. Thus our lowering procedure must end somewhere, and there is a *ground state*, which we will now denote by u_0, beyond which lowering ends. This must mean that

$$Au_0 = 0 \tag{7-20}$$

The energy of the ground state is

$$Hu_0 = (\hbar\omega A^\dagger A + \tfrac{1}{2}\hbar\omega)\, u_0 = \tfrac{1}{2}\hbar\omega\, u_0 \tag{7-21}$$

Let us apply (7-8) to the ground state:

$$HA^\dagger u_0 - A^\dagger H u_0 = \hbar\omega A^\dagger u_0$$

that is,

$$HA^\dagger u_0 = (\hbar\omega + \tfrac{1}{2}\hbar\omega)\, A^\dagger u_0 \tag{7-22}$$

The energy has been raised by one unit of $\hbar\omega$, and $A^\dagger$ is aptly described as a *raising operator*. We will change our notation a little, namely, label the state by the number of energy units $\epsilon = \hbar\omega$ it has over the ground-state energy $\tfrac{1}{2}\hbar\omega$. Thus we write

$$A^\dagger u_0 = c u_1 \tag{7-23}$$

Note that (7-12) implies that

$$A u_1 = c' u_0 \tag{7-24}$$

so that $A^\dagger$ and A move up and down the same "ladder." All the states may be generated by repeated application of $A^\dagger$ to u_0. One consequence is that the energy spectrum is given by

$$E = (n + \tfrac{1}{2})\hbar\omega \qquad n = 0, 1, 2, \ldots \tag{7-25}$$

We have succeeded in obtaining the energy spectrum without solving any differential equation. We have also a general representation of the eigenvectors

$$u_n = \frac{1}{\sqrt{n!}}\, (A^\dagger)^n\, u_0 \tag{7-26}$$

where we have put in the correct normalization constant. One way of seeing how this constant is obtained is to note that the commutation relation

$$[A, A^\dagger] = 1$$

suggests that we may formally represent A as follows

$$A = \frac{d}{dA^\dagger} \tag{7-27}$$

This allows us to evaluate

$$\langle u_0 | A^m (A^\dagger)^n | u_0 \rangle = \left\langle u_0 \left| \left(\frac{d}{dA^\dagger}\right)^m (A^\dagger)^n \right| u_0 \right\rangle$$

For $m < n$ we get on the right-hand side

$$n(n - 1)(n - 2) \cdots (n - m + 1)\, \langle u_0 | (A^\dagger)^{n-m} | u_0 \rangle$$

and the term in the brackets is easily seen to vanish, since

$$\langle u_0|(A^\dagger)^{n-m}|u_0\rangle = \langle Au_0|(A^\dagger)^{n-m-1}|u_0 \rangle = 0$$

This follows from $Au_0 = 0$. For $m > n$ we write the left-hand side as

$$\langle u_0|A^m(A^\dagger)^n|u_0\rangle = \left\langle u_0\left|A^{m-n}\left(\frac{d}{dA^\dagger}\right)^n(A^\dagger)^n\right|u_0\right\rangle$$

After the differentation we are left with $n!\langle u_0|A^{m-n}|u_0\rangle = 0$. For $m = n$,

$$\langle u_0|A^n(A^\dagger)^n|u_0\rangle = \left\langle u_0\left|\left(\frac{d}{dA^\dagger}\right)^n(A^\dagger)^n\right|u_0\right\rangle = n!$$

Thus the normalization coefficient in (7-26) is the correct one, and we have also shown that

$$\langle u_m|u_n\rangle = 0 \qquad \text{for} \quad m \neq n \tag{7-28}$$

The statement that an arbitrary state vector can be expanded in eigenstates of H now reads in the coordinate independent way

$$\psi = \sum_{n=0}^{\infty} C_n u_n \tag{7-29}$$

and since $\langle u_m|u_n\rangle = \delta_{mn}$, we have

$$C_m = \langle u_m|\psi\rangle \tag{7-30}$$

REPRESENTATIONS OF ABSTRACT STATES: FROM OPERATORS TO THE SCHRÖDINGER EQUATION

We have succeeded in making the point that one can solve for the eigenvalues of the harmonic oscillator using operator methods alone. For this problem all that is needed to specify the eigenstates is the energy, that is, the integer $n = 0, 1, 2, \ldots$ appearing in

$$E = (n + \tfrac{1}{2})\hbar\omega$$

and thus the complete set of commuting observables consists of H alone.[3] Thus the label n on the eigenstate u_n describes its whole content. We would therefore be quite willing to give up the privileged role of the eigenfunction in x-space, $u_n(x)$, except for one point: $u_n(x)$ does provide us with more information in that it gives us the probability density [via $|u_n(x)|^2$] of finding the particle at x. Does this

---

[3]The parity is contained in the label n. States with n even are positive parity states, and those with n odd have negative parity. This follows from the fact that under reflection, A and $A^\dagger$ are odd.

additional content single out the x-space wave function after all? Let us recall the role of the wave function in momentum space $\phi(p)$ that appears in Chapter 3, for example. As the Fourier transform of the x-space function it might have had some claim to a privileged role, but later, in (4-44), for example, we explained that $\phi(p)$ was "merely" an expansion coefficient of an arbitrary $\psi(x)$ in eigenstates of the momentum operator, and that is why its absolute square yielded the probability of finding a momentum p for that state. Similarly, the fact that $|\psi(x)|^2$ yields the probability density of finding x for the position of the system could be interpreted by the statement that $\psi(x)$ is the expansion coefficient of an arbitrary abstract state in eigenstates of the position operator x_{op}. We write the eigenvalue equation abstractly as

$$x_{op}\phi_x = x\phi_x \qquad (7\text{-}31)$$

keeping x as a subscript to stress that it is a label of the eigenstate, just as n is the label for u_n. The spectrum of x_{op}, a hermitian operator, is continuous, so that the expansion theorem, instead of taking a form like (7-29), really reads

$$\psi = \int dx\, C(x)\, \phi_x \qquad (7\text{-}32)$$

Since the eigenstates defined in (7-31) form an orthonormal set,

$$\langle \phi_x | \phi_{x'} \rangle = \delta(x - x') \qquad (7\text{-}33)$$

we can derive

$$C(x) = \langle \phi_x | \psi \rangle \qquad (7\text{-}34)$$

and this quantity is the probability amplitude for finding a particle at x—more specifically, the measurement of the observable x will yield the eigenvalue x with probability $|C(x)|^2$. All we have to do is change the notation, rewriting (7-32) as

$$\psi = \int dx\psi(x)\, \phi_x \qquad (7\text{-}35)$$

to show that the wave function in x-space has no privileged role, and we use it only as a matter of convenience. The basic principles deal with operators and their eigenvectors and eigenvalues in an abstract space, and the rest is a matter of *representation*. The latter is, of course, crucial in obtaining numbers, which is what physics is all about. That is why we will not lay too much stress on the formal structure of the theory, and continue using wave functions. Later we will have to deal with operators that have no classical analog, such as the intrinsic spin of electrons and other particles, and there we will exercise our freedom to use other representations.

It will be convenient for future reference to calculate the amplitude that a position measurement of an eigenstate of momentum u_p yields the value x. This means that we are interested in calculating $\psi(x)$ defined by the equation

$$u_p = \int dx\, \psi(x)\phi_x \qquad (7\text{-}36)$$

By our rules,

$$\psi(x) = \langle \phi_x | u_p \rangle \tag{7-37}$$

We know, however, that the amplitude for finding a particle with momentum p at the point x is given by the free-particle wave function

$$\psi(x) = \frac{e^{ipx/\hbar}}{\sqrt{2\pi\hbar}} \tag{7-38}$$

We thus have the result that

$$\langle x|p \rangle = \frac{e^{ipx/\hbar}}{\sqrt{2\pi\hbar}} \tag{7-39}$$

where we have simplified the notation for $|\phi_x\rangle \to |x\rangle$ and $|u_p\rangle \to |p\rangle$. Note that

$$\int dp \, \langle x|p \rangle \langle p|x' \rangle = \delta(x - x') \tag{7-40}$$

and

$$\int dx \, \langle p|x \rangle \langle x|p' \rangle = \delta(p - p') \tag{7-41}$$

If we generalize the completeness relation (6-45)

$$\sum_a |u_a\rangle\langle u_a| = 1 \tag{7-42}$$

to continuous variables, so that the sum is replaced by an integral over the continuous variable, then

$$\int dp \, |p\rangle\langle p| = 1 \tag{7-43}$$

and

$$\int dx \, |x\rangle\langle x| = 1 \tag{7-44}$$

The equations (7-40) and (7-41) are thus equivalent to

$$\langle x|x' \rangle = \delta(x - x')$$

and $\tag{7-45}$

$$\langle p|p' \rangle = \delta(p - p')$$

respectively. Note that the first of these is just (7-33) in our simplified notation.

How do we get to the Schrödinger equation from our abstract framework? Consider first (7-20), which reads

$$A|u_0\rangle = 0$$

Thus we also have

$$\langle x|A|u_0\rangle = 0$$

that is

$$\left\langle x \left| \sqrt{\frac{m\omega}{2\hbar}}\, x_{\mathrm{op}} + i\, \frac{p_{\mathrm{op}}}{\sqrt{2m\omega\hbar}} \right| u_0 \right\rangle = 0 \tag{7-46}$$

Now the operator x acting on the state $|x\rangle$ gives

$$x_{\mathrm{op}}|x\rangle = x|x\rangle \tag{7-47}$$

so that

$$\langle x|x_{\mathrm{op}}|u_0\rangle = x\langle x|u_0\rangle = xu_0(x) \tag{7-48}$$

$$
\begin{aligned}
\langle x|p_{\mathrm{op}}|u_0\rangle &= \int dp \,\langle x|p_{\mathrm{op}}|p\rangle\langle p|u_0\rangle \\
&= \int dp \, p\langle x|p\rangle\langle p|u_0\rangle \\
&= \int dp \, p\, \frac{e^{ipx/\hbar}}{\sqrt{2\pi\hbar}}\, \langle p|u_0\rangle \\
&= \frac{\hbar}{i}\frac{d}{dx}\int dp\, \frac{e^{ipx/\hbar}}{\sqrt{2\pi\hbar}}\, \langle p|u_0\rangle \\
&= \frac{\hbar}{i}\frac{d}{dx}\int dp\, \langle x|p\rangle\langle p|u_0\rangle \\
&= \frac{\hbar}{i}\frac{d}{dx}\, u_0(x)
\end{aligned}
\tag{7-49}
$$

Thus $A|u_0\rangle = 0$ becomes a differential equation in x-space, which reads

$$\left(m\omega x + \hbar\, \frac{d}{dx} \right) u_0(x) = 0 \tag{7-50}$$

Equation (7-50) is a simple differential equation whose solution is

$$u_0(x) = Ce^{-m\omega x^2/2\hbar} \tag{7-51}$$

The constant C is determined by the requirement that $u_0(x)$ be normalized to unity:

$$1 = C^2 \int_{-\infty}^{\infty} dx\, e^{-m\omega x^2/\hbar} = C^2 \sqrt{\frac{\hbar\pi}{m\omega}}$$

so that

$$C = \left(\frac{m\omega}{\hbar\pi}\right)^{1/4} \tag{7-52}$$

We may also obtain the higher energy states by working out in detail

$$
\begin{aligned}
u_n(x) &= \frac{1}{\sqrt{n!}} (A^\dagger)^n u_0(x) \\
&= \frac{1}{\sqrt{n!}} \left(\frac{m\omega}{\hbar\pi}\right)^{1/4} \left(\sqrt{\frac{m\omega}{2\hbar}}\, x - \sqrt{\frac{\hbar}{2m\omega}}\frac{d}{dx}\right)^n e^{-m\omega x^2/2\hbar}
\end{aligned}
\tag{7-53}
$$

Quite generally

$$H|u_E\rangle = E|u_E\rangle \tag{7-54}$$

written as

$$\langle x|H|u_E\rangle = E\langle x|u_E\rangle \tag{7-55}$$

becomes a differential equation for $\langle x|u_E\rangle = u_E(x)$, of the form

$$Hu_E(x) = Eu_E(x) \tag{7-56}$$

where in H, x_{op} is replaced by x and p_{op} by $\dfrac{\hbar}{i}\dfrac{d}{dx}$.

THE TIME DEPENDENCE OF OPERATORS

We conclude this chapter by discussing the time development of a system in our representation-independent way. The time-dependent Schrödinger equation

$$i\hbar \frac{d\psi(t)}{dt} = H\psi(t) \tag{7-57}$$

is now an operator equation in an abstract space. The $\psi(t)$ is a vector, and it points in a direction that depends on time. The equation can easily be solved. The solution is

$$\psi(t) = e^{-iHt/\hbar}\,\psi(0) \tag{7-58}$$

where $\psi(0)$ is the vector at time $t = 0$ and the operator $e^{-iHt/\hbar}$ is defined by

$$e^{-iHt/\hbar} = \sum_{n=0}^{\infty} \frac{(-iHt/\hbar)^n}{n!} \tag{7-59}$$

The solution (7-58) allows us to describe the change with time of the expectation value of some operator B that does not have any explicit time dependence:

$$\begin{aligned}
\langle B \rangle_t &= \langle \psi(t) | B \psi(t) \rangle \\
&= \langle e^{-iHt/\hbar} \, \psi(0) | B \, e^{-iHt/\hbar} \, \psi(0) \rangle \\
&= \langle \psi(0) | \, e^{iHt/\hbar} \, B \, e^{-iHt/\hbar} \, \psi(0) \rangle \\
&= \langle \psi(0) | B(t) | \psi(0) \rangle \\
&= \langle B(t) \rangle_0
\end{aligned}$$

(7-60)

We used

$$(e^{-iHt/\hbar})^\dagger = e^{iH^\dagger t/\hbar} = e^{iHt/\hbar}$$

(7-61)

along the way, and defined

$$B(t) = e^{iHt/\hbar} \, B \, e^{-iHt/\hbar}$$

(7-62)

What (7-60) says is that the expectation value of a time-independent operator B in a state that varies with time as (7-58) may be written as the expectation value of a time-varying operator $B(t)$ [given by (7-62)] in the time-independent state $\psi(0)$. This is very useful in the formal discussion of quantum mechanics, since it is convenient to set up a basis of orthonormal eigenvectors in the abstract vector space once and for all, and not worry about how the basis vectors change with time. When we do this, we are working in the *Heisenberg picture*, whereas keeping B without time dependence means that we are working in the *Schrödinger picture*. The result is the same, whatever picture we use: this is analogous to the option of describing a rotating body relative to a fixed set of axes, or of describing the body at rest in a rotating coordinate system. The choice is one of convenience. If we do work in the Heisenberg picture, then state vectors are fixed, and we need not refer to them. How an observable varies with time is determined by (7-62), which yields

$$\begin{aligned}
\frac{d}{dt} B(t) &= \frac{i}{\hbar} H e^{iHt/\hbar} \, B \, e^{-iHt/\hbar} - \frac{i}{\hbar} e^{iHt/\hbar} \, BH \, e^{-iHt/\hbar} \\
&= \frac{i}{\hbar} HB(t) - \frac{i}{\hbar} B(t)H \\
&= \frac{i}{\hbar} [H, B(t)]
\end{aligned}$$

(7-63)

a form remarkably like (6-69). That equation was an equation for expectation values, but since its form was independent of the state in which the expectation value was taken, it had to reflect operator properties, and (7-63) shows that explicitly.

For the harmonic oscillator

$$H = \hbar \omega A^\dagger A + \tfrac{1}{2} \hbar \omega$$

and since H is a constant of the motion, we have

$$H = \hbar \omega A^\dagger(t) \, A(t) + \tfrac{1}{2} \hbar \omega$$

(7-64)

We can also show, using (7-62), that

$$[A(t), A^\dagger(t)] = 1 \tag{7-65}$$

Hence (7-7) and (7-8) still have the same form, and we get

$$\frac{d}{dt} A(t) = -i\omega A(t)$$

$$\tag{7-66}$$

$$\frac{d}{dt} A^\dagger(t) = i\omega A^\dagger(t)$$

Thus the time dependence of $A(t)$ and $A^\dagger(t)$ is obtained by solving (7-66), with the result that

$$A(t) = e^{-i\omega t} A(0)$$

$$\tag{7-67}$$

$$A^\dagger(t) = e^{i\omega t} A^\dagger(0)$$

Using the relation (7-4) it is easy to show that

$$p(t) = p(0) \cos \omega t - m\omega x(0) \sin \omega t$$

$$\tag{7-68}$$

$$x(t) = x(0) \cos \omega t + \frac{p(0)}{m\omega} \sin \omega t$$

expressing the operators $x(t)$ and $p(t)$ in terms of the operators $x(0)$ and $p(0)$.

Problems

1. Use the commutation relation (7-5) and the definition of the state u_n given in (7-26) to prove that

$$A u_n = \sqrt{n}\, u_{n-1}$$

(*Hint:* Use induction, that is, shown that if this relation is true for n, it is true for $n + 1$, and establish it directly for $n = 1$.

2. Use the preceding relation to show that if $f(A^\dagger)$ is any polynomial in $A^\dagger$, then

$$A f(A^\dagger)\, u_0 = \frac{df(A^\dagger)}{dA^\dagger}\, u_0$$

Note that representing A in the form

$$A = \frac{d}{dA^\dagger}$$

is consistent with the commutation relation (7-5) and is quite analogous to the representation

$$p = \frac{\hbar}{i} \frac{d}{dx}$$

3. Calculate the form of $\langle u_n|x|u_m\rangle$, and show that it vanishes unless $n = m \pm 1$.

(*Hint:* It is sufficient to calculate $\langle u_n|A|u_m\rangle$ since $\langle u_n|A^\dagger|u_m\rangle = \langle Au_n|u_m\rangle = \langle u_m|A|u_n\rangle^*$. Use the results of Problem 1.)

4. Calculate the form of $\langle u_n|p|u_m\rangle$.

5. Use the results of Problems 3 and 4 to calculate $\langle u_m|px|u_n\rangle$ by calculating $\Sigma_k\langle u_m|p|u_k\rangle\langle u_k|x|u_n\rangle$.

6. Do the same calculation as in Problem 5 to obtain $\langle u_m|xp|u_n\rangle$.

7. Use the results of Problems 5 and 6 to show that

$$\langle u_m|[p, x]|u_n\rangle = \frac{\hbar}{i}\,\delta_{mn}$$

8. Calculate $\langle u_n|x|u_n\rangle$ and $\langle u_n|x^2|u_n\rangle$.

9. Calculate $\langle u_n|p|u_n\rangle$ and $\langle u_n|p^2|u_n\rangle$.

10. Calculate

$$(\Delta x)^2\,(\Delta p)^2$$

where the usual definition for a given state is

$$(\Delta x)^2 = \langle u_n|x^2|u_n\rangle - (\langle u_n|x|u_n\rangle)^2$$

11. Consider a pair of operators A and $A^\dagger$ that satisfy commutation relations

$$[A, A^\dagger] = 1$$

(a) Show that the operator $N = A^\dagger A$ has eigenvalues $n = 0, 1, 2, \ldots$.
(b) Show that the harmonic oscillator Hamiltonian may be written in the form

$$H = \hbar\omega(N + \tfrac{1}{2})$$

12. A state $|\alpha\rangle$ that obeys the equation

$$A|\alpha\rangle = \alpha|\alpha\rangle$$

is called a *coherent state*.

(a) Show that the state $|\alpha\rangle$ may be written in the form

$$|\alpha\rangle = C\,e^{\alpha A^\dagger}|0\rangle$$

(b) Use the result of Problem 2 to obtain C.
(c) Expand the state $|\alpha\rangle$ in a series of eigenstates of the number operator N, $|n\rangle$, and use this to find the probability that the coherent state contains n quanta. The distribution is called a *Poisson distribution*.
(d) Calculate $\langle \alpha|N|\alpha\rangle$, the average number of quanta in the coherent state.

13. Use the general operator equation of motion (7-63) to solve for the time dependence of the operator $x(t)$ given that

$$H = \frac{p^2(t)}{2m} + mgx(t)$$

14. Consider the Hamiltonian describing a one-dimensional oscillator in an external electric field.

$$H = \frac{p^2(t)}{2m} + \tfrac{1}{2}m\omega^2x^2(t) - e\mathscr{E}x(t)$$

Calculate the equation of motion for the operators $p(t)$ and $x(t)$ using (7-63) and the commutation relation

$$[p(t), x(t)] = \frac{\hbar}{i}$$

Show that the equation of motion is just the classical equation of motion. Solve for $p(t)$ and $x(t)$ in terms of $p(0)$ and $x(0)$. Show that

$$[x(t_1), x(t_2)] \neq 0 \qquad \text{for} \quad t_1 \neq t_2$$

This shows that operators that commute at the same time need not commute at different times.

15. Use (7-53) to calculate the eigenfunctions for $n = 1, 2, 3$. (*Note:* Be sure to keep track of the ordering of x and d/dx in the expansion of the binomial series.)

16. Use the results of Problem 2 to show that

$$e^{\lambda A} f(A^\dagger) \, u_0 = f(A^\dagger + \lambda) \, u_0$$

(*Hint:* Expand the exponential in a series, and use the fact that

$$f(x + a) = \sum \frac{a^n}{n!} f^{(n)}(x)$$

$$f^{(n)}(x) = \frac{d^n}{dx^n} f(x)$$

to work out this problem.)

17. Use the results of Problem 16 to establish the operator relation

$$e^{\lambda A} f(A^\dagger) \, e^{-\lambda A} = f(A^\dagger + \lambda)$$

Note that an operator relation must hold when it acts on an arbitrary state. Let an arbitrary state be of the form $g(A^\dagger) \, u_0$. Thus what must be proved is that

$$e^{\lambda A} \, f(A^\dagger) \, e^{-\lambda A} \, g(A^\dagger) \, u_0 = f(A^\dagger + \lambda) \, g(A^\dagger) \, u_0$$

This can also be proved from the general relation

$$e^{\lambda A} \, A^\dagger \, e^{-\lambda A} = A^\dagger + \lambda[A, A^\dagger] + \frac{\lambda^2}{2!} [A, [A, A^\dagger]] + \cdots$$

18. Use the preceding relation to prove that

$$e^{aA + bA^\dagger} = e^{aA} \, e^{bA^\dagger} \, e^{-(1/2)ab}$$

The procedure is the following. Let

$$e^{\lambda(aA + bA^\dagger)} \equiv e^{\lambda a A} \, F(\lambda)$$

Differentiation with respect to λ yields

$$(aA + bA^\dagger) \, e^{\lambda(aA + bA^\dagger)} = aA \, e^{\lambda a A} \, F(\lambda) + e^{\lambda a A} \, \frac{dF}{d\lambda}$$

that is,

$$(aA + bA^\dagger) \, e^{\lambda a A} \, F(\lambda) = aA \, e^{\lambda a A} \, (F\lambda) + e^{\lambda a A} \, \frac{dF}{d\lambda}$$

Use Problem 17 to show that

$$\frac{dF}{d\lambda} = (bA^\dagger - \lambda ab)\, F(\lambda)$$

so that

$$F(\lambda) = e^{\lambda b A^\dagger}\, e^{-(1/2)\lambda^2 ab}$$

19. Use the procedure of Problem 17 to show that

$$e^{\lambda A^\dagger} f(A)\, e^{-\lambda A^\dagger} = f(A - \lambda)$$

Show from this that,

$$e^{aA + bA^\dagger} = e^{bA^\dagger}\, e^{aA}\, e^{(1/2)ab}$$

using the method outlined in Problem 18.

20. Use the preceding result to show that

$$e^{ikx} = e^{ik\sqrt{\hbar/2m\omega}\, A^\dagger}\, e^{ik\sqrt{\hbar/2m\omega}\, A}\, e^{-(\hbar k^2/4m\omega)}$$

Note that

$$x = \sqrt{\frac{\hbar}{2m\omega}}\, (A + A^\dagger)$$

Use this expression to calculate

$$\langle u_0 | e^{ikx} | u_0 \rangle$$

21. Show that the result just obtained is the same as the one obtained from

$$\int_{-\infty}^{\infty} dx\, u_0^*(x)\, e^{ikx}\, u_0(x)$$

References

The material discussed in this chapter is also treated in almost all of the books in the reference list at the end of the book. The student is encouraged to look up some of them, since it is always useful to see the same basic material presented from different points of view.

Chapter 8

N-PARTICLE SYSTEMS

Our discussion of a single particle is easily generalized to an N-particle system. The N particles are described by a wave function $\psi(x_1, x_2, \ldots, x_N)$ that is normalized such that

$$\int \cdots \int dx_1 dx_2 \cdots dx_N |\psi(x_1, x_2, \ldots, x_N)|^2 = 1 \tag{8-1}$$

The interpretation of $|\psi(x_1, x_2, \ldots, x_N)|^2$ is a generalization of the interpretation of $|\psi(x)|^2$, that is, it yields the probability density for finding particle 1 at x_1, particle 2 at $x_2, \ldots$, particle N at x_N. The time development of such a wave function is given by the solution of the differential equation

$$i\hbar \frac{\partial}{\partial t} \psi(x_1, \ldots, x_N; t) = H\psi(x_1, \ldots, x_N; t) \tag{8-2}$$

where the Hamiltonian is again constructed in correspondence with the classical form

$$H = \sum_{i=1}^{N} \frac{p_i^2}{2m_i} + V(x_1, x_2, \ldots, x_N) \tag{8-3}$$

as

$$H = -\hbar^2 \left(\frac{1}{2m_1} \frac{\partial^2}{\partial x_1^2} + \cdots + \frac{1}{2m_N} \frac{\partial^2}{\partial x_N^2} \right) + V(x_1, \ldots, x_N) \tag{8-4}$$

145

The whole formalism of quantum mechanics developed before is easily generalized, with the proviso that operators describing single-particle observables commute when they refer to different particles, for example,

$$[p_i, x_j] = \frac{\hbar}{i} \delta_{ij} \tag{8-5}$$

If there are no external fields, such as the common gravitational field of the
earth, or externally imposed electric or magnetic fields, then the potential energy
can only depend on the relative separation of the particles, that is,

$$V = V(x_1 - x_2, x_1 - x_3, \ldots, x_{N-1} - x_N) \tag{8-6}$$

This must be the case, because in the absence of any external agency that somehow
determines an "origin," the displacement of the whole system should not change
any physical properties of the system. In other words, the form of the potential
(8-6) is a consequence of the invariance of all physically significant quantities
under the transformation

$$x_i \rightarrow x_i + a \tag{8-7}$$

A very important special case of (8-6) is the case of two-body forces, in which case

$$V = \sum_{i>j} V(x_i - x_j) \tag{8-8}$$

The summation is over all indices i and j, subject to the condition $i > j$ to avoid
double counting, and the counting of $i = j$. Actually, in the description of electrons
in an atom, we will be dealing with the common Coulomb potential, as well as
the electron–electron repulsion, and there the nucleus provides an origin. The
potential in that case is a three-dimensional generalization of

$$\sum_{i=1}^{N} W(x_i) + \sum_{i>j} V(x_i - x_j) \tag{8-9}$$

CONSERVATION OF TOTAL MOMENTUM

When there are no external forces, then in classical mechanics the total momentum
is conserved. This follows from the equations of motion

$$m_i \frac{d^2 x_i}{dt^2} = -\frac{\partial}{\partial x_i} V(x_1 - x_2, x_1 - x_3, \ldots, x_{N-1} - x_N) \tag{8-10}$$

a consequence of which is that

$$\frac{d}{dt} \sum_i m_i \frac{dx_i}{dt} = -\sum_i \frac{\partial}{\partial x_i} V(x_1 - x_2, \ldots, x_{N-1} - x_N)$$

$$= 0 \tag{8-11}$$

The reason for the vanishing of the right side of the preceding equation is that for every argument in V, there are equal and opposite contributions that come from $\sum_i \partial/\partial x_i$ acting on it. For example, with $u = x_1 - x_2$, $v = x_1 - x_3$, $w = x_2 - x_3$

$$\left(\frac{\partial}{\partial x_1} + \frac{\partial}{\partial x_2} + \frac{\partial}{\partial x_3}\right) V(u, v, w) = \frac{\partial V}{\partial u} + \frac{\partial V}{\partial v} - \frac{\partial V}{\partial u} + \frac{\partial V}{\partial w} - \frac{\partial V}{\partial v} - \frac{\partial V}{\partial w} = 0$$

Hence

$$P = \sum_i m_i \frac{dx_i}{dt} \tag{8-12}$$

is a constant of the motion.

In quantum mechanics the same conclusion holds. We shall demonstrate it by using the invariance of the Hamiltonian under the transformation (8-7). The invariance implies that both

$$Hu_E(x_1, x_2, \ldots, x_N) = Eu_E(x_1, x_2, \ldots, x_N) \tag{8-13}$$

and

$$Hu_E(x_1 + a, x_2 + a, \ldots, x_N + a) = Eu_E(x_1 + a, x_2 + a, \ldots, x_N + a) \tag{8-14}$$

hold. Let us take a infinitesimal, so that terms of $0(a^2)$ can be neglected. Then

$$u(x_1 + a, \ldots, x_N + a) \simeq u(x_1, \ldots, x_N) + a \frac{\partial}{\partial x_1} u(x_1, \ldots, x_N)$$

$$+ a \frac{\partial}{\partial x_2} u(x_1, \ldots, x_N) + \cdots$$

$$\simeq u(x_1, \ldots, x_N) + a \sum_i \frac{\partial}{\partial x_i} u(x_1, \ldots, x_N)$$

and hence, subtracting (8-13) from 8-14)

$$aH \left(\sum_{i=1}^{N} \frac{\partial}{\partial x_i}\right) u_E(x_1, \ldots, x_N) = aE \left(\sum_{i=1}^{N} \frac{\partial}{\partial x_i}\right) u_E(x_1, \ldots, x_N)$$

$$= a \left(\sum_{i=1}^{N} \frac{\partial}{\partial x_i}\right) Eu_E(x_1, \ldots, x_N) \tag{8-15}$$

$$= a \left(\sum_{i=1}^{N} \frac{\partial}{\partial x_i}\right) Hu_E(x_1, \ldots, x_N)$$

If we now define

$$P = \frac{\hbar}{i} \sum_{i=1}^{N} \frac{\partial}{\partial x_i} \equiv \sum_{i=1}^{N} p_i \tag{8-16}$$

so that P is the total momentum operator, we see that we have demonstrated that

$$(HP - PH)\, u_E(x_1, \ldots, x_N) = 0 \qquad (8\text{-}17)$$

Since the energy eigenstates for N-particles form a complete set of states, in the sense that any function of $x_1, x_2, \ldots, x_N$ can be expanded in terms of all the $u_E(x_1, \ldots, x_N)$ the preceding equation can be translated into

$$[H, P]\, \psi(x_1, \ldots, x_N) = 0 \qquad (8\text{-}18)$$

for all $\psi(x_1, \ldots, x_N)$, that is, into the operator relation

$$[H, P] = 0 \qquad (8\text{-}19)$$

This, however, implies that P, the total momentum of the system, is a *constant of the motion*. This is a very deep consequence of what is really a statement about the nature of space. The statement that there is no origin, that is, that the laws of physics are invariant under displacement by a fixed distance, leads to a conservation law. In relativistic quantum mechanics there are no potentials of the form that we consider here; nevertheless the invariance principle, as stated earlier, still leads to a conserved total momentum.

THE TWO-PARTICLE SYSTEM

Our main interest will be in the *two-particle system*, which we discuss next. For two noninteracting particles we have the simple Hamiltonian

$$H = \frac{p_1^2}{2m_1} + \frac{p_2^2}{2m_2} \qquad (8\text{-}20)$$

We might expect that since the two particles are totally uncorrelated, the probability of finding one at x_1 and the other at x_2 is the product of two independent probabilities

$$P(x_1, x_2) = P(x_1)\, P(x_2) \qquad (8\text{-}21)$$

Thus we expect that the solution of

$$\left(-\frac{\hbar^2}{2m_1} \frac{\partial^2}{\partial x_1^2} - \frac{\hbar^2}{2m_2} \frac{\partial^2}{\partial x_2^2} \right) u(x_1, x_2) = Eu(x_1, x_2) \qquad (8\text{-}22)$$

should be separable into

$$u(x_1, x_2) = \phi_1(x_1)\, \phi_2(x_2) \qquad (8\text{-}23)$$

Substituting this into (8-22) and dividing by $u(x_1, x_2)$ we get

$$\frac{-(\hbar^2/2m_1)(d^2\phi_1(x_1)/dx_1^2)}{\phi_1(x_1)} + \frac{-(\hbar^2/2m_2)(d^2\phi_2(x_2)/dx_2^2)}{\phi_2(x_2)} = E \qquad (8\text{-}24)$$

The two terms in the equation depend on different variables, and that is why we set both of them equal to the constants E_1 and E_2, respectively:

$$E = E_1 + E_2$$

$$-\frac{\hbar^2}{2m_1}\frac{d^2\phi_1(x_1)}{dx_1^2} = E_1\phi_1(x_1) \tag{8-25}$$

$$-\frac{\hbar^2}{2m_1}\frac{d^2\phi_2(x_2)}{dx_2^2} = E_2\phi_2(x_2)$$

The two equations are easily solved, and we get

$$u(x_1, x_2) = C\, e^{ik_1x_1 + ik_2x_2} \tag{8-26}$$

with

$$k_1^2 = \frac{2m_1E_1}{\hbar^2} \qquad k_2^2 = \frac{2m_2E_2}{\hbar^2} \tag{8-27}$$

Let us now rewrite the solution using the coordinates

$$x = x_1 - x_2$$
$$X = \frac{m_1x_1 + m_2x_2}{m_1 + m_2} \tag{8-28}$$

that is, the separation between the particles, and the center-of-mass coordinate. If we write

$$k_1x_1 + k_2x_2 = \alpha(x_1 - x_2) + \beta\frac{m_1x_1 + m_2x_2}{m_1 + m_2}$$

we find that

$$\beta = k_1 + k_2 \equiv K$$
$$\alpha = \frac{m_2k_1 - m_1k_2}{m_1 + m_2} \equiv k$$

so that the solution has the form

$$u(x_1, x_2) = C\, e^{iKX}\, e^{ikx} \tag{8-29}$$

where $K = k_1 + k_2$ is the wave number corresponding to the total momentum, and k is the wave number corresponding to the relative momentum. The first factor represents the motion of the center of mass, and the second factor is the "internal" wave function. The energy may be written as

$$E = \frac{\hbar^2K^2}{2(m_1 + m_2)} + \frac{\hbar^2k^2}{2}\left(\frac{1}{m_1} + \frac{1}{m_2}\right) \tag{8-30}$$

The first factor is the energy of the two-particle system, with mass $m_1 + m_2$ moving freely with the total momentum; the second term is the internal energy. If we introduce the *reduced mass* μ, defined by

$$\frac{1}{\mu} = \frac{1}{m_1} + \frac{1}{m_2} \tag{8-31}$$

then the term is $\hbar^2 k^2/2\mu$, which is effectively a one-particle energy, namely, that of a free particle with mass μ and momentum $\hbar k$.

When the Hamiltonian in (8-20) is altered by the addition of a potential that depends on $x_1 - x_2$ only, then we have

$$\left(-\frac{\hbar^2}{2m_1} \frac{\partial^2}{\partial x_1^2} - \frac{\hbar^2}{2m_2} \frac{\partial^2}{\partial x_2^2} \right) u(x_1, x_2) + V(x_1 - x_2)\, u(x_1, x_2) = Eu(x_1, x_2) \tag{8-32}$$

Using the coordinates defined in (8-28), written in the form

$$x_1 = X + \frac{\mu}{m_1} x$$

$$x_2 = X - \frac{\mu}{m_2} x \tag{8-33}$$

a little algebra shows that the equation takes the form

$$\left(-\frac{\hbar^2}{2(m_1 + m_2)} \frac{\partial^2}{\partial X^2} - \frac{\hbar^2}{2\mu} \frac{\partial^2}{\partial x^2} + V(x) \right) u(x, X) = Eu(x, X) \tag{8-34}$$

If we write

$$u(x, X) = e^{iKX}\, \phi(x) \tag{8-35}$$

we find that the equation for $\phi(x)$ is

$$-\frac{\hbar^2}{2\mu} \frac{d^2\phi(x)}{dx^2} + V(x)\, \phi(x) = \epsilon\phi(x) \tag{8-36}$$

that is, a one-particle Schrödinger equation with reduced mass, and energy

$$\epsilon = E - \frac{\hbar^2 K^2}{2(m_1 + m_2)} \tag{8-37}$$

In Chapter 9 we will obtain the separation in a somewhat more sophisticated way. We now turn to the problem of *identical particles*.

IDENTICAL PARTICLES

There is compelling evidence that electrons are indistinguishable. If this were not so, then the spectrum of an atom, say, helium, would vary from experiment to experiment, depending on "what kind" of electrons were contained in it. No such variation has ever been observed. Similarly, nuclear spectra are always the same, indicating that protons are indistinguishable, as are neutrons. Similar evidence from high energy physics experiments indicates very strongly that other particles, for example, pi-mesons, are also indistinguishable. This is a purely quantum-mechanical property: in classical mechanics it is possible to follow the orbits of all particles (in principle) so that they are never really indistinguishable.

We shall learn that electrons are characterized by an internal quantum number called the *spin*. Thus a complete set of quantum numbers for the description of an electron must include the additional *spin label*, which we shall denote by σ. We shall learn later (starting in Chapter 14) that the spin label σ is *double-valued*, that is, two electrons identical in every respect (except for the spin) can still be distinguished by their σ value. A third electron, with the same quantum numbers as the other two, must have a spin label that is equal to that of at least one of the other electrons, since σ can only have two values, which we shall generally label ($\pm$). The existence of the spin label has a further effect on the consequences of indistinguishability, which we discuss next.

A Hamiltonian for indistinguishable particles must be completely symmetric in the coordinates of the particles. For a two-particle system, if the potential does not depend on the spin labels, the Hamiltonian is

$$H = \frac{p_1^2}{2m} + \frac{p_2^2}{2m} + V(x_1, x_2) \tag{8-38}$$

$$V(x_1, x_2) = V(x_2, x_1) \tag{8-39}$$

We write this symmetry symbolically as

$$H(1, 2) = H(2, 1) \tag{8-40}$$

and it is understood that if the Hamiltonian contains operators that refer to the spins of the two particles, then these too must be included in the labeling "1," "2."

A wave function for an N-particle system, labeled by the total energy, for example, now needs to be labeled by the spin labels $\sigma_1, \sigma_2, \ldots, \sigma_N$. For a two-particle system, the energy eigenvalue equation now reads

$$H(1, 2)u_{E\sigma_1\sigma_2}(1, 2) = Eu_{E\sigma_1\sigma_2}(1, 2) \tag{8-41}$$

Since the labeling does not matter, we may, by interchanging the "1" and "2" labels, write this as

$$H(2, 1)u_{E\sigma_2\sigma_1}(2, 1) = Eu_{E\sigma_2\sigma_1}(2, 1) \tag{8-42}$$

On the other hand, using (8-40) we also have

$$H(1, 2)u_{E\sigma_2\sigma_1}(2, 1) = Eu_{E\sigma_2\sigma_1}(2, 1) \tag{8-43}$$

Let us now use the formal approach that we used in our discussion of parity. We introduce an *exchange operator* P_{12}, which, acting on a state, interchanges the two identical particles that we labeled by 1 and 2. Thus

$$P_{12}u_{E\sigma_1\sigma_2}(1, 2) = u_{E\sigma_2\sigma_1}(2, 1) \tag{8-44}$$

Equation 8-43 may be written as follows

$$
\begin{aligned}
H(1, 2)P_{12}u_{E\sigma_1\sigma_2}(1, 2) &= Eu_{E\sigma_2\sigma_1}(2, 1) \\
&= EP_{12}u_{E\sigma_1\sigma_2}(1, 2) \\
&= P_{12} Eu_{E\sigma_1\sigma_2}(1, 2) \\
&= P_{12} H(1, 2)u_{E\sigma_1\sigma_2}(1, 2)
\end{aligned}
\tag{8-45}
$$

Since this applies for the complete set of $u_{E\sigma_1\sigma_2}(1, 2)$ of simultaneous eigenfunctions of H and the spin operators, we deduce the *operator relation*

$$
[H, P_{12}] = 0
\tag{8-46}
$$

Thus P_{12}, like parity, is a constant of the motion. Further, since two exchanges $1 \rightarrow 2, 2 \rightarrow 1$ bring back the original state, we have

$$
(P_{12})^2 = 1
\tag{8-47}
$$

so that the eigenvalues of P_{12} are ± 1. Just as even and odd functions are eigenfunctions of the parity operator, here the eigenstates are the symmetric and antisymmetric combinations

$$
\psi^{(S)}(1, 2) = \frac{1}{N_{2S}}[\psi(1, 2) + \psi(2, 1)]
$$

$$
\tag{8-48}
$$

$$
\psi^{(A)}(1, 2) = \frac{1}{N_{2a}}[\psi(1, 2) - \psi(2, 1)]
$$

where the N_2 are normalization constants. The fact that P_{12} is a constant of the motion implies that a state that is symmetric at an initial time will always be symmetric, and an antisymmetric state will always be antisymmetric.

THE PAULI PRINCIPLE

It is an important *law of nature* that the symmetry or antisymmetry under the interchange of two particles is a characteristic of the particles, and not something that can be arranged in the preparation of the initial state. The law, which was discovered by Pauli, states that

1. Systems consisting of identical particles of half-odd-integral spin (i.e., spin 1/2, 3/2, . . .) are described by antisymmetric wave functions. Such particles are called *fermions,* and are said to obey Fermi-Dirac statistics.
2. Systems consisting of identical particles of integral spin (spin 0, 1, 2, . . .) are described by symmetric wave functions. Such particles are called *bosons,* and are said to obey Bose-Einstein statistics. We shall primarily be concerned with electrons, protons, and neutrons, which have spin $\frac{1}{2}$, and with spin 0 bosons, which carry no spin labeling.

The law extends to N-particle states. For a system of N identical fermions, the wave function is antisymmetric under the interchange of any pair of particles. For example, a three-particle wave function, properly antisymmetrized, has the form

$$\psi^{(A)}(1, 2, 3) = \frac{1}{N_{3a}} [\psi(1, 2, 3) - \psi(2, 1, 3) + \psi(2, 3, 1)$$
$$- \psi(3, 2, 1) + \psi(3, 1, 2) - \psi(1, 3, 2)] \tag{8-49}$$

whereas the three identical boson wave function has the form

$$\psi^{(S)}(1, 2, 3) = \frac{1}{N_{3S}} [\psi(1, 2, 3) + \psi(2, 1, 3) + \psi(2, 3, 1)$$
$$+ \psi(3, 2, 1) + \psi(3, 1, 2) + \psi(1, 3, 2)] \tag{8-50}$$

It should be stressed that for more than two identical particles, one could in principle have mixed symmetries: for example, the wave function is antisymmetric under the exchange of (1, 2) and (1, 3), but symmetric under the exchange (2, 3). The Pauli principle excludes such mixed symmetry states.

N Fermions in a Potential Well

Let us now consider a very interesting special case, in which N fermions do not interact with each other, but do interact with a common potential. In that case

$$H = \sum_{i=1}^{N} H_i \tag{8-51}$$

where

$$H_i = \frac{p_i^2}{2m} + V(x_i) \tag{8-52}$$

The eigenstates of the one-particle Hamiltonian are denoted by $u_{E\sigma_k}(x_k)$, where

$$H_k\, u_{E\sigma_k}(x_k) = E_k u_{E\sigma_k}(x_k) \tag{8-53}$$

It is understood that for each value of E_k there are two possible values of the spin σ_k label.

A solution of

$$Hu_E(1, 2, \ldots, N) = Eu_E(1, 2, \ldots, N) \tag{8-54}$$

is

$$u_E(1, 2, 3, \ldots, N) = u_{E_1\sigma_1}(x_1)u_{E_2\sigma_2}(x_2), \ldots, u_{E_N\sigma_N}(x_N) \tag{8-55}$$

We shall write this as

$$u_E(1, 2, 3, \ldots, N) = u_{E_1}(x_1)\, u_{E_2}(x_2), \ldots, u_{E_N}(x_N) \tag{8-56}$$

where we have suppressed the σ_i labels that go with the E_i. Also

$$E = E_1 + E_2 + \cdots + E_N \tag{8-57}$$

Our task now is to antisymmetrize (8-56). If there are only two particles, we evidently have

$$u^{(A)}(1, 2) = \frac{1}{\sqrt{2}} [u_{E_1}(x_1) u_{E_2}(x_2) - u_{E_1}(x_2) u_{E_2}(x_1)] \qquad (8\text{-}58)$$

With three particles, the form is

$$\begin{aligned} u^{(A)}(1, 2, 3) = \frac{1}{\sqrt{6}} [&u_{E_1}(x_1) u_{E_2}(x_2) u_{E_3}(x_3) - u_{E_1}(x_2) u_{E_2}(x_1) u_{E_3}(x_3) \\ + &u_{E_1}(x_2) u_{E_2}(x_3) u_{E_3}(x_1) - u_{E_1}(x_3) u_{E_2}(x_2) u_{E_3}(x_1) \\ + &u_{E_1}(x_3) u_{E_2}(x_1) u_{E_3}(x_2) - u_{E_1}(x_1) u_{E_2}(x_3) u_{E_3}(x_2)] \end{aligned} \qquad (8\text{-}59)$$

For N particles, the answer is a determinant, the so-called *Slater determinant:*[1]

$$u^{(A)}(1, 2, \ldots, N) = \frac{1}{\sqrt{N!}} \begin{vmatrix} u_{E_1}(x_1) & u_{E_1}(x_2) & \cdots & u_{E_1}(x_N) \\ u_{E_2}(x_1) & u_{E_2}(x_2) & \cdots & u_{E_2}(x_N) \\ \vdots & & & \\ u_{E_N}(x_1) & u_{E_N}(x_2) & \cdots & u_{E_N}(x_N) \end{vmatrix} \qquad (8\text{-}60)$$

In the preceding three equations we have supressed the label σ_k that goes with each E_k. Clearly the interchange of two particles involves the interchange of two columns in the determinant, and this changes the sign. If two electrons are in the same energy eigenstate, for example, $E_1 = E_2$, and if they are in the same spin state, that is, the spin labels are the same $\sigma_1 = \sigma_2$, then the determinant vanishes when $x_1 = x_2$, that is, the electrons cannot be at the same place. Thus the requirement of antisymmetry introduces an effective interaction between two fermions: qualitatively we see that two particles in the same state tend to stay away from each other, since the joint wave function vanishes when their separation goes to zero. Thus even noninteracting particles behave as if there were a repulsive interaction between them. We will see that a complete set of commuting observables for electrons includes an additional two-valued observable associated with the spin. Thus a state of given energy, angular momentum, parity, and so on can be occupied by two electrons (of opposite spin variable), but by no more than two electrons. This is a restricted version of the *Pauli exclusion principle*.

When Is Antisymmetrization Necessary?

The statement "no two electrons can be in the same quantum state" implies that a wave function for a system containing two electrons must be antisymmetric in the coordinates of the two electrons. The question arises whether we really have to worry about this when we consider a hydrogen atom on earth and another one on the moon. If they are both in the ground state, do they necessarily have to have opposite spin states? What then happens when we consider a third hydrogen atom in its ground state? Intuition tells us that we do not have to worry, and intuition is correct. We can see this by examining whether there is a difference between using the totally uncorrelated wave function for two electrons

$$\psi_a(x_1) \psi_b(x_2) \qquad (8\text{-}61)$$

---

[1]The wave function for N identical bosons is totally symmetric, and the general form is obtained by expanding the determinant in (8-60) and making all the signs positive.

and the antisymmetrized wave function

$$\frac{1}{N} \left(\psi_a(x_1)\, \psi_b(x_2) - \psi_a(x_2)\, \psi_b(x_1) \right) \tag{8-62}$$

The normalization factor N is given by the requirement that

$$\frac{1}{N^2} \int dx_1 \int dx_2 \, |\psi_a(x_1)\psi_b(x_2) - \psi_a(x_2)\psi_b(x_1)|^2 = 1 \tag{8-63}$$

which with

$$\int dx \, |\psi_a(x)|^2 = \int dx \, |\psi_b(x)|^2 = 1 \tag{8-64}$$

leads to

$$N^2 = 2 \left(1 + \left| \int dx \, \psi_a^*(x)\, \psi_b(x) \right|^2 \right)$$

$$\equiv 2(1 + |S_{ab}|^2) \, . \tag{8-65}$$

Suppose we want to calculate the probability that the electron with the a label is in some spatial region R. For the uncorrelated wave function, which is $\Psi(x, y) = \psi_a(x)\psi_b(y)$, the probability density is given by

$$P(R) = \int_R dx \int dy \, |\psi_a(x)|^2 \, |\psi_b(y)|^2 = \int_R dx \, |\psi_a(x)|^2 \tag{8-66}$$

We integrate over the whole range of the coordinates of the b electron, since we do no care where it is.

For the antisymmetrized wave function we have

$$|\Psi(x, y)|^2 = \frac{1}{N^2} \, [\psi_a^*(x)\psi_b^*(y) - \psi_b^*(x)\psi_a^*(y)] \, [\psi_a(x)\psi_b(y) - \psi_b(x)\psi_a(y)]$$

This is to be integrated over the domain R for those variables associated with the label a and over the whole range of coordinates associated with the b label. We thus get

$$P_a(R) = \frac{1}{N^2} \int_R dx \, |\psi_a(x)|^2 \int_{-\infty}^{\infty} dy \, |\psi_b(y)|^2$$

$$+ \frac{1}{N^2} \int_R dy \, |\psi_a(y)|^2 \int_{-\infty}^{\infty} dx \, |\psi_b(x)|^2$$

$$- \frac{1}{N^2} \int_R dx \int_R dy \, [\psi_a^*(x)\psi_b(x)\psi_b^*(y)\psi_a(y) + \psi_b^*(x)\psi_a(x)\,\psi_a^*(y)\psi_b(y)]$$

$$= \frac{2}{N^2} \int_R dx \, |\psi_a(x)|^2 - \frac{2}{N^2} \int_R dx \int_R dy \, \psi_a^*(x)\psi_b(x)\psi_b^*(y)\psi_a(y) \tag{8-67}$$

The interference term has both integrals over the range R, since a occurs in both a wave function with the label. The difference will only be significant if the *overlap integral* $\int_R dx\ \psi_a^*(x)\psi_b(x)$ is important in the region R for the variable x. Since wave functions fall off exponentially for bound states, it is clear that this can only be important if the atoms are very near to each other.

For example, consider a two-electron system, each in a Gaussian wave packet, one centered about the origin and the other about $x = L$. A calculation of the probability of finding an electron in a region R involves the overlap integral of $Ce^{-\beta x^2}$, and $Ce^{-\beta(x-L)^2}$. The overlap integral is of the form

$$C^2 \int_R dx\ e^{-\beta(x^2+(x-L)^2)}$$

and this is easily seen to be proportional to $e^{-\beta L^2/2}$. Thus for L large, the overlap integral vanishes very rapidly, and our intuition that the wave function of an electron under consideration need not be antisymmetrized with any or all other, distant electrons, proves to be correct.

The Pauli exclusion principle must be taken into account in atoms and in molecules, but not in situations where the atoms are separated by significant distances. Even in crystal lattices, where the spacing between the atoms is several angstroms, the overlap is frequently small, and antisymmetrization is not necessary.

Ground State Energy for Free Particles in a Box

An interesting consequence of the Pauli exclusion principle is that the ground state for N electrons in a potential is very different from the ground state for N bosons or N distinguishable particles. Consider, for example, the infinite potential box,

$$
\begin{aligned}
V(x) &= \infty & & x < 0 \\
&= 0 & & 0 < x < b \\
&= \infty & & b < x
\end{aligned}
\tag{8-68}
$$

The solution of the Schrödinger equation that vanishes at $x = 0$ and $x = b$ is given by

$$u_n(x) = \sqrt{\frac{2}{b}}\ \sin\frac{n\pi x}{b} \tag{8-69}$$

with $n = 1, 2, 3, \ldots$, and the energy eigenvalues are

$$E_n = \frac{\hbar^2\pi^2 n^2}{2mb^2} \tag{8-70}$$

For N noninteracting bosons, the ground state has all the particles in the $n = 1$ state, and thus the energy is given by

$$E = N\frac{\hbar^2\pi^2}{2mb^2} \tag{8-71}$$

so that the energy per particle is

$$\frac{E}{N} = \frac{\hbar^2 \pi^2}{2mb^2} \tag{8-72}$$

For N noninteracting fermions the situation is quite different. Only two electrons can go into each of the states $n = 1, 2, 3, \ldots$, so that $N/2$ states are filled. Thus the total energy is given by

$$E = 2 \sum_{n=1}^{N/2} \frac{\hbar^2 \pi^2 n^2}{2mb^2} = \frac{\hbar^2 \pi^2}{mb^2} \frac{N^3}{24} \tag{8-73}$$

In obtaining the last result we have assumed that N is large, so that it does not matter whether the last level is filled with one or two electrons, and we have used

$$\sum_{n=1}^{N/2} n^2 \approx \int_1^{N/2} n^2 \, dn \approx \frac{1}{3} \left(\frac{N}{2} \right)^3$$

Thus the energy per particle

$$\frac{E}{N} = \frac{\hbar^2 \pi^2}{24mb^2} N^2 \tag{8-74}$$

grows with N^2. Equivalently, for a given energy, the number of bosons filling the well is proportional to E, while the number of fermions filling the well is proportional to $E^{1/3}$. The highest level to be filled in the fermion case is the one for which $n = N/2$, and its energy is

$$E_F = \frac{\hbar^2 \pi^2 N^2}{8mb^2} \tag{8-75}$$

The subscript F has been put in because this energy is called the *Fermi energy*. We may write it in terms of the density of fermions, which in the one-dimensional problem is $N/b \equiv \rho$, as

$$E_F = \frac{\hbar^2 \pi^2}{8m} \rho^2 \tag{8-76}$$

We shall return to the significance of these remarks in Chapter 9.

The exclusion principle plays an extremely important role in the structure of atoms. The enormous richness in the variety of chemical properties of the various elements is directly traceable to the fact that only a limited number of electrons can occupy a given energy eigenstate. This will be discussed in Chapter 19.

Problems

1. What is the reduced mass of an electron–proton system? How does it differ from the reduced mass of an electron–deuteron system? What is the reduced mass of a system of two identical particles?

2. Prove that the exchange operator P_{12} is hermitian.

3. Consider two noninteracting electrons in an infinite potential well. What is the ground state wave function if the two electrons are in the *same* spin state?

4. Consider two electrons in the same spin state, interacting with a potential

$$V(|x_1 - x_2|) = -V_0 \qquad |x_1 - x_2| \leq a$$
$$= 0 \qquad \text{elsewhere}$$

What is the lowest energy of the two-electron state, assuming that the total momentum of the two electrons is zero?

[*Hint:* Separate the equation in a manner leading to (8-37) and then apply the Pauli principle.]

5. Consider two identical particles described by the energy operator

$$H = H(p_1, x_1) + H(p_2, x_2)$$

where

$$H(p, x) = \frac{p^2}{2m} + \tfrac{1}{2}m\omega^2 x^2$$

Separate out the center of mass motion, and obtain the energy spectrum for this system. Show that it agrees with that obtained by solving

$$H\psi(x_1, x_2) = E\psi(x_1, x_2)$$

with

$$\psi(x_1, x_2) = u_1(x_1)\, u_2(x_2)$$

Discuss the degeneracy of the energy spectrum.

6. Consider two electrons described by the Hamiltonian

$$H = \frac{p_1^2}{2m} + \frac{p_2^2}{2m} + V(x_1) + V(x_2)$$

where $V(x) = \infty$ for $x < 0$ and for $x > a$; $V(x) = 0$ for $0 < x < a$. Assume that the electrons are in the same spin state, that is, $\sigma_1 = \sigma_2$.
(a) What is the lowest energy of the two-electron state?
(b) What is the energy eigenfunction for this ground state?
(c) What is the energy and the wave function of the first excited state, still keeping $\sigma_1 = \sigma_2$?

7. Consider a two-electron system, in which $\sigma_1 = \sigma_2$, so that we do not need to consider spin. Suppose the electrons are in Gaussian wave packets about $x = a$ and $x = -a$, so that their wave functions are $\sqrt{\pi/\mu}\, e^{-\mu^2(x-a)^2/2}$ and $\sqrt{\pi/\mu}\, e^{-\mu^2(x+a)^2/2}$, respectively. Construct a properly normalized two-electron wave function. Suppose $1/\mu = 0.5$ Å. Estimate for what values of a can one ignore the effects of the Pauli principle to an accuracy of 1 part in 1000.

8. Use the wave function for the two-electron system in Problem 7 to calculate the probability that the separation between the two electrons is in the range $(x, x + dx)$. Also show that the expectation value of the center of mass of the two-electron system is $\langle (x_1 + x_2)/2 \rangle = 0$.

(*Hint:* Write x_1 and x_2 in terms of the center-of-mass variable $X = (x_1 + x_2)/2$ and the separation $x = x_1 - x_2$, and express the wave function in terms of the new variables.)

9. Plot the probability density obtained in Problem 8 as a function of x, for the two cases (i) $1/a = \mu/2$, (ii) $1/a = 2\mu$. Discuss the physics behind your results.

10. Suppose the electrons were replaced by bosons in Problems 7, 8, 9. List the changes in the formulas, plot the probability density as a function of x for the two separations, and explain the difference between the fermion and boson case for the probability density.

References

See any of the references listed at the end of Chapter 6 and also

D. S. Saxon, *Elementary Quantum Mechanics*, Holden-Day, San Francisco, 1968.
D. Park, *Introduction to the Quantum Theory*, (3rd Ed.) McGraw-Hill, New York, 1992.

Chapter 9

THE SCHRÖDINGER EQUATION IN THREE DIMENSIONS I

The Hamiltonian for a single particle moving in three-dimensional space reads

$$H = \frac{P_x^2 + p_y^2 + p_z^2}{2m} + V(x, y, z) \tag{9-1}$$

In this chapter, we concentrate on the special case in which the potential is of the form

$$V(x, y, z) = V_1(x) + V_2(y) + V_3(z) \tag{9-2}$$

The equation

$$-\frac{\hbar^2}{2m} \left(\frac{\partial^2}{\partial x^2} + \frac{\partial^2}{\partial y^2} + \frac{\partial^2}{\partial z^2} \right) u_E(x, y, z)$$
$$+ [V_1(x) + V_2(y) + V_3(z)] u_E(x, y, z) = E u_E(x, y, z) \tag{9-3}$$

is easily seen to be solved by

$$u_E(x, y, z) = u_{\epsilon_1}(x) \, v_{\epsilon_2}(y) \, w_{\epsilon_3}(z) \tag{9-4}$$

where the functions on the right are solutions of

$$
\left[-\frac{\hbar^2}{2m}\frac{d^2}{dx^2} + V_1(x) \right] u_{\epsilon_1}(x) = \epsilon_1 u_{\epsilon_1}(x)
$$

$$
\left[-\frac{\hbar^2}{2m}\frac{d^2}{dy^2} + V_2(y) \right] v_{\epsilon_2}(y) = \epsilon_2 v_{\epsilon_2}(y) \tag{9-5}
$$

$$
\left[-\frac{\hbar^2}{2m}\frac{d^2}{dz^2} + V_3(z) \right] w_{\epsilon_3}(z) = \epsilon_3 w_{\epsilon_3}(z)
$$

and

$$
E = \epsilon_1 + \epsilon_2 + \epsilon_3
$$

FREE PARTICLES IN A BOX

A particularly interesting example is the three-dimensional generalization of the potential hole with infinite walls. If the three-dimensional box is cubical in shape, with side L, then

$$
\begin{aligned}
V_1(x) &= \infty & x < 0 \\
&= 0 & 0 < x < L \\
&= \infty & L < x
\end{aligned} \tag{9-6}
$$

and so on. Thus, the general solution is

$$
u_E(x, y, z) = \left(\frac{2}{L}\right)^{3/2} \sin\frac{n_1 \pi x}{L} \sin\frac{n_2 \pi y}{L} \sin\frac{n_3 \pi z}{L} \tag{9-7}
$$

and

$$
E = \frac{\hbar^2 \pi^2}{2mL^2} (n_1^2 + n_2^2 + n_3^2) \tag{9-8}
$$

Note that there is quite a lot of degeneracy in the problem: there are as many solutions for a given E as there are sets of integers $\{n_1, n_2, n_3\}$ that satisfy (9-8). The degeneracy is usually associated with the existence of mutually commuting operators, and this example is no exception. Here these operations are H_x, H_y, and H_z, defined by

$$
H_x = \frac{p_x^2}{2m} + V_1(x)
$$

$$
H_y = \frac{p_y^2}{2m} + V_2(y) \tag{9-9}
$$

$$
H_z = \frac{p_z^2}{2m} + V_3(z)
$$

with

$$H_x + H_y + H_z = H \qquad (9\text{-}10)$$

EFFECTS OF THE EXCLUSION PRINCIPLE

The potential just considered is so simple that we can use it to discuss the energies of noninteracting electrons (and other identical fermions) in a three-dimensional box. As a first step, it is interesting to ask for the ground-state energy of N non-interacting identical fermions, for example, electrons, in the box of volume L^3. For each triplet of integers, $(1, 1, 1)$, $(2, 1, 1)$, $(1, 2, 1)$, . . . , two electrons can be accommodated. It is easier to ask the question in a different way: How many triplets of integers $\{n_1, n_2, n_3\}$ are there such that E given by (9-8) is less than the energy E_F? Each triplet forms a lattice point in a three-dimensional space, and if there are very many of them, then it is a very good approximation to say that they must lie inside a sphere of radius R given according to (9-8) by

$$n_1^2 + n_2^2 + n_3^2 = R^2 = \frac{2mE_F}{\hbar^2\pi^2} L^2 \qquad (9\text{-}11)$$

That number is given by the volume of the octant of the sphere for which all the n_i are positive (see Fig. 9-1). Thus the number of lattice points is

$$\frac{1}{8} \cdot \frac{4\pi}{3} R^3 = \frac{1}{8} \frac{4\pi}{3} \left(\frac{2mE_F}{\hbar^2\pi^2} L^2 \right)^{3/2} \qquad (9\text{-}12)$$

and hence the number of electrons with energy less than the energy E_F is twice that, that is,

$$N = \frac{\pi}{3} L^3 \left(\frac{2mE_F}{\hbar^2\pi^2} \right)^{3/2} \qquad (9\text{-}13)$$

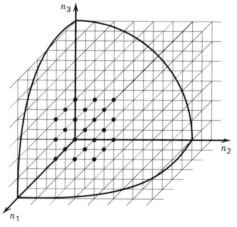

Figure 9-1. Counting of states for independent particle system.

The number of electrons is proportional to the volume of the box L^3, which is to be expected. In terms of the density of electrons,

$$n = \frac{N}{L^3} \tag{9-14}$$

we have

$$E_F = \frac{\hbar^2 \pi^2}{2m} \left(\frac{3n}{\pi}\right)^{2/3} \tag{9-15}$$

The energy E_F is the energy of the most energetic electron in the ground state of a noninteracting electron gas with density n. It is known as the *Fermi energy* and the labeling was chosen accordingly.

To calculate the total energy, the number of lattice points may be written as

$$\frac{1}{8} \int_{|\mathbf{n}| \leq R} d^3\mathbf{n} \tag{9-16}$$

The factor $1/8$ comes from our restriction to positive integers in (9-8); in the integration (9-16) this restriction is removed and must be compensated for by the factor in front. This energy must be doubled, since there are two electrons with the same energy at this lattice point. Hence the total energy is

$$\begin{aligned} E_{tot} &= \frac{\hbar^2 \pi^2}{mL^2} \frac{1}{8} \int \mathbf{n}^2 \, d^3\mathbf{n} \\ &= \frac{\hbar^2 \pi^2}{8mL^2} 4\pi \int_0^R n^4 \, dn \\ &= \frac{\hbar^2 \pi^3}{10mL^2} R^5 \end{aligned} \tag{9-17}$$

Since R is related to the number of electrons by

$$N = 2 \cdot \frac{1}{8} \cdot \frac{4\pi}{3} R^3 \tag{9-18}$$

we finally get

$$E_{tot} = \frac{\hbar^2 \pi^3}{10mL^2} \left(\frac{3N}{\pi}\right)^{5/3} \tag{9-19}$$

If we write this in terms of $n = N/L^3$, we get

$$E_{tot} = \frac{\hbar^2 \pi^3}{10m} \left(\frac{3n}{\pi}\right)^{5/3} L^3 \tag{9-20}$$

The consequences of the Pauli exclusion principle are quite staggering. We will discuss a few of them after the following observations:

(a) The wave number, defined by $E = \hbar^2k^2/2m$ at the top of the "Fermi sea" is given by

$$k_F = (3\pi^2 n)^{1/3} \tag{9-21}$$

Since $k = 2\pi/\lambda$, we get for the De Broglie wavelength

$$\lambda = 2.03n^{-1/3} \tag{9-22}$$

Since $n^{-1/3}$ is approximately the interparticle spacing d, we can state our result in the easily memorized form

$$d = \frac{\lambda_F}{2} \tag{9-23}$$

This is a better than expected mnemonic device. Since the exclusion principle forbids two electrons with identical quantum numbers to be on top of each other, this means that they must be at least a half-wave apart.

(b) If we keep a fixed number of electrons, then (9-19) written in terms of the volume V containing them takes the form

$$E_{\text{tot}} = \frac{\hbar^2\pi^3}{10m}\left(\frac{3N}{\pi}\right)^{5/3} V^{-2/3} \tag{9-24}$$

This result is actually independent of the shape of the volume, if N is very large. We used a cube to do the computations, since that is the simplest way to do the calculation.

DEGENERACY PRESSURE AND ASTROPHYSICAL APPLICATIONS

If the electron gas is compressed, the electrons are pushed closer to each other, and this decreases the De Broglie wavelength, that is, increases the energy. Thus the compression is resisted, and the pressure resisting the compression is called the *degeneracy pressure*. It is given by

$$\begin{aligned} p_{\text{deg}} &= -\frac{\partial E_{\text{tot}}}{\partial V} \\ &= \frac{\hbar^2\pi^3}{15m}\left(\frac{3n}{\pi}\right)^{5/3} \end{aligned} \tag{9-25}$$

The bulk modulus B of a material (the reciprocal of the *compressibility*), is defined by

$$B = -V\frac{\partial p}{\partial V} \tag{9-26}$$

and if we use the degeneracy pressure for p, we find that $B = 5p_{deg}/3$, so that

$$B = \frac{\hbar^2 \pi^3}{9m} \left(\frac{3n}{\pi}\right)^{5/3} \tag{9-27}$$

The use of a degenerate electron gas model for matter gives the correct order of magnitude for the bulk modulus B. For example, for sodium $n = 2.65 \times 10^{22}$ cm^{-3}, and this leads to $B = 9.2 \times 10^{10}$ dynes/cm^2. The experimental value is 6.4×10^{10} dynes/cm^2.

The resistance to compression that originates in the Pauli exclusion principle plays an important role in stellar evolution. Stars "burn" by undergoing a succession of nuclear reactions. Hydrogen converts to helium by the series of reactions

$$^1\text{H} + {}^1\text{H} \rightarrow {}^2\text{H} + e^+ + \nu + 1.44 \text{ MeV}$$

$$^2\text{H} + {}^1\text{H} \rightarrow {}^3\text{He} + \gamma + 5.49 \text{ MeV}$$

$$^3\text{He} + {}^3\text{He} \rightarrow {}^4\text{He} + {}^1\text{H} + {}^1\text{H} + 12.85 \text{ MeV}$$

When all of the hydrogen is converted to helium, the burning stops. Gravitational contraction compresses the helium until it begins to burn in the reaction

$$^4\text{He} + {}^4\text{He} \rightarrow {}^8\text{Be} + \gamma - 95 \text{ keV}$$

$$^8\text{Be} + {}^4\text{He} \rightarrow {}^{12}\text{C} + \gamma + 7.4 \text{ MeV}$$

The variety of nuclear processes gets larger, and the process of synthesis of nuclei is now well understood. At some point, when the star consists primarily of iron, silicon, and neighboring elements, the burning stops. The material then resumes gravitational contraction, and the only barrier to total gravitational collapse is the effect of degeneracy pressure.

The gravitational pressure is easily calculated if we assume that the density of material ρ is independent of radius, and that the shape of the star is spherical. The potential energy of the material in a shell lying between radius r and $r + dr$ is

$$dV_g = -G \frac{(4\pi\rho r^3/3)(4\pi\rho r^2 \, dr)}{r} = -\frac{(4\pi)^2 G\rho^2}{3} r^4 \, dr \tag{9-28}$$

so that the potential energy of the material contained in a sphere of radius R is

$$V_g = -\frac{(4\pi)^2 G\rho^2}{3} \int_0^R r^4 \, dr = -\frac{(4\pi)^2}{15} G\rho^2 R^5 \tag{9-29}$$

We also have the connection between ρ, R, and the stellar mass M. The star consists of N nucleons (in the form of iron, silicon, and so on) each of mass m_n, so that

$$\frac{4\pi}{3} \rho R^3 = M = (Nm_n) \tag{9-30}$$

After a little algebra we get an expression for the gravitational potential energy in terms of the volume of the star V:

$$V_g = -\frac{3}{5}\left(\frac{4\pi}{3}\right)^{1/3} G(Nm_n)^2 V^{-1/3} \tag{9-31}$$

The gravitational pressure is

$$p_g = -\frac{\partial V_g}{\partial V} = -\frac{1}{5}\left(\frac{4\pi}{3}\right)^{1/3} G(Nm_n)^2 V^{-4/3} \tag{9-32}$$

This is opposed by the degeneracy pressure, which, according to (9-25) is

$$p_{\text{deg}} = \frac{\hbar^2\pi^3}{15m_e}\left(\frac{3n}{\pi}\right)^{5/3} = \frac{\hbar^2\pi^3}{15m_e}\left(\frac{3N_e}{\pi}\right)^{5/3} V^{-5/3} \tag{9-33}$$

where N_e is the number of electrons; it is equal to the number of protons in the star. Assuming equal numbers of protons and neutrons, we have $N_e = N/2$.

The two pressures balance, for a given value of N, when

$$\frac{1}{5}\left(\frac{4\pi}{3}\right)^{1/3} G(Nm_n)^2 V^{-4/3} = \frac{\hbar^2\pi^3}{15m_e}\left(\frac{3N_e}{\pi}\right)^{5/3} V^{-5/3}$$

that is, when the radius of the star is R^*,

$$R^* = \left(\frac{3}{4\pi}\right)^{1/3} V^{1/3} = \left(\frac{81\pi^2}{128}\right)^{1/3} \frac{\hbar^2}{Gm_e m_n^2} N^{-1/3} \tag{9-34}$$

For a star of one solar mass,

$$N \approx \frac{2\times 10^{33}\text{ gm}}{1.67\times 10^{-24}\text{ gm}} = 1.2\times 10^{57}$$

and the radius of the degenerate star is $R^* \approx 1.1\times 10^4$ km. The radius of a non-degenerate star, the sun, is $\approx 7\times 10^8$ km!

If the mass is somewhat larger than a solar mass, the average energy of the electrons increases. When the electrons acquire relativistic energies, our expression for the degeneracy pressure changes drastically. In effect, the electron energy is no longer $p^2/2m_e$, but pc. It can be shown (see Problem 1) that in this domain the degeneracy pressure also scales as $V^{-4/3}$ and for a sufficiently large value of N, the gravitational pressure overcomes the degeneracy pressure. As a consequence of this large net pressure, the reaction

$$e^- + p \rightarrow n + \nu$$

takes place. The neutrinos escape, since matter, even degenerate matter, is transparent to them, and we are left with a *neutron star*. The degeneracy pressure of the neutrons, which are also fermions, and thus also obey the exclusion principle, can be calculated in the same way as the electron pressure, except that N_e is replaced by N and m_e by m_n. We now obtain

$$R_n^* = \left(\frac{81\pi^2}{16}\right)^{1/3} \frac{\hbar^2}{Gm_n^3} N^{-1/3} \tag{9-35}$$

For a star with two solar masses, we end up with $R_n^* \approx 10$ km! If the mass (equivalently N) is so large that the neutrons become relativistic, then there is no counterbalance to the huge gravitational pressure, and a *black hole* forms.

Problems

1. Repeat the calculation of the Fermi energy of a gas of fermions by assuming that the fermions are massless, so that the energy-momentum relation is $E = pc$.

2. Calculate the degeneracy of states in a cubic box of volume L^3 as a function of E, that is, calculate the number of states in the interval $(E, E + dE)$, and use this to obtain the density of states of an electron gas, keeping in mind that there are two electrons per energy state.
   (*Hint:* How many $\{n_1, n_2, n_3\}$ are there for which $\Sigma_i n_i^2 = 2mEL^2/\hbar^2\pi^2$?)

3. Calculate the energy spectrum for free electrons in a box of sides a, a, L, with $a \ll L$. Discuss the spacings when $a = 10$ Å and L is 10^{-4} cm.

4. Calculate the energy density of a gas of photons in a cubic box of volume L^3, keeping in mind that there are two photons (two states of polarization) per energy state.

5. Calculate the energy spectrum of a gas of photons in a box of sides a, a, L, and $a \ll L$.

6. Given that the number density of free electrons in copper is 8.5×10^{22} cm^{-3}, calculate (1) the Fermi energy in electron volts; (2) the velocity of an electron moving with kinetic energy equal to the Fermi energy.

7. A nucleus consists of N neutrons and Z protons, with $N + Z = A$. If the radius of the nucleus is given by $R = r_0 A^{1/3}$, with $r_0 = 1.1$ fm (1 fm $= 10^{-13}$ cm), and if the neutron and proton masses are both very nearly 1.6×10^{-24} gm, write expressions for the Fermi energy of the proton "gas" and the neutron "gas," assuming that the protons and the neutrons move freely. What are the Fermi energies if $N = 126$ and $Z = 82$?

8. Consider a neutron gas in its ground state, with mass density ρ varying from 10^{11} to 10^{16} gm cm^{-3}. Calculate the Fermi energy as a function of ρ. Note that at some point the neutron gas becomes relativistic, that is, the relation between energy and momentum is relativistic. In what range of densities should one begin to use the relativistic formula?

Chapter 10

THE SCHRÖDINGER EQUATION IN THREE DIMENSIONS II

THE CENTRAL POTENTIAL

In this chapter we consider the very important case in which the potential $V(x, y, z)$ depends on $r = \sqrt{x^2 + y^2 + z^2}$ only. For a two-particle system, in which the potential depends on the separation between the two particles alone, the Hamiltonian has the form

$$H = \frac{\mathbf{p}_1^2}{2m_1} + \frac{\mathbf{p}_2^2}{2m^2} + V(|\mathbf{r}_1 - \mathbf{r}_2|) \qquad (10\text{-}1)$$

With the usual decomposition into center of mass and relative coordinates

$$\mathbf{R} = \frac{m_1\mathbf{r}_1 + m_2\mathbf{r}_2}{m_1 + m_2}$$

$$\mathbf{r} = \mathbf{r}_1 - \mathbf{r}_2 \qquad (10\text{-}2)$$

the total momentum, and the relative momentum

$$\mathbf{P} = \mathbf{p}_1 + \mathbf{p}_2$$

$$\mathbf{p} = \frac{m_2\mathbf{p}_1 - m_1\mathbf{p}_2}{m_1 + m_2} \qquad (10\text{-}3)$$

the Hamiltonian takes the form

$$H = \frac{\mathbf{P}^2}{2M} + \frac{\mathbf{p}^2}{2\mu} + V(|\mathbf{r}|) \tag{10-4}$$

Here M, the total mass of the system, is given by

$$M = m_1 + m_2 \tag{10-5}$$

and the *reduced mass* μ is given by

$$\mu = \frac{m_1 m_2}{M} \tag{10-6}$$

It is easily checked that

$$[P_i, R_j] = \frac{\hbar}{i} \delta_{ij}$$
$$[p_i, r_j] = \frac{\hbar}{i} \delta_{ij} \tag{10-7}$$

with all other commutators vanishing. Since the potential does not depend on the center-of-mass coordinate $\mathbf{R}$, the total momentum operator $\mathbf{P}$ commutes with H, and we can construct simultaneous eigenfunctions of $\mathbf{P}$ and H. The eigenfunctions of $\mathbf{P}$ are of the form

$$U(\mathbf{P}, \mathbf{R}) = \frac{1}{(2\pi\hbar)^{3/2}} e^{i\mathbf{P}\cdot\mathbf{R}/\hbar} \tag{10-8}$$

An eigenfunction of H will then have the form

$$\psi(\mathbf{R}, \mathbf{r}) = U(\mathbf{P}, \mathbf{R}) \, u_E(\mathbf{r}) \tag{10-9}$$

where

$$\left(\frac{\mathbf{p}^2}{2\mu} + V(r)\right) u_E(\mathbf{r}) = E u_E(\mathbf{r}) \tag{10-10}$$

and $E = E_{\text{tot}} - P^2/2M$, is the internal energy, after the energy of motion of the two-particle system has been subtracted.

We write this equation in the form

$$\left(-\frac{\hbar^2}{2\mu} \nabla^2 + V(r)\right) u_E(\mathbf{r}) = E u_E(\mathbf{r}) \tag{10-11}$$

CONSEQUENCES OF ROTATIONAL INVARIANCE

In this section we show that (10-11) can be separated in such a form that only the radial coordinate r appears in the energy eigenvalue equation. In classical mechanics the separation is based on the use of angular momentum. With

$$\mathbf{L} = \mathbf{r} \times \mathbf{p} \tag{10-12}$$

it follows that

$$\mathbf{L}^2 = (\mathbf{r} \times \mathbf{p}) \cdot (\mathbf{r} \times \mathbf{p}) = r^2 p^2 - (\mathbf{r} \cdot \mathbf{p})^2$$

so that

$$\begin{aligned}
\mathbf{p}^2 &= \frac{1}{r^2} (\mathbf{r} \cdot \mathbf{p})^2 + \frac{1}{r^2} \mathbf{L}^2 \\
&= p_r^2 + \frac{1}{r^2} \mathbf{L}^2
\end{aligned} \tag{10-13}$$

For a central potential (V depends on r alone) the force is radial and exerts no torque on the system. Thus $\mathbf{L}$ is a constant of the motion, and $\mathbf{L}^2$ is just a number. Hence the equation

$$E = \frac{1}{2\mu} \mathbf{p}^2 + V(r)$$

involves radial coordinates only. The same is true in quantum mechanics. In the rest of this section we

1. Identify the angular momentum operators from the requirement that the Hamiltonian is invariant under rotations;
2. Derive the radial equation.

Invariance under Rotations about the z-Axis

Consider the special case of a rotation through an angle θ about the z-axis: with

$$\begin{aligned}
x' &= x \cos \theta - y \sin \theta \\
y' &= x \sin \theta + y \cos \theta
\end{aligned} \tag{10-14}$$

it is easy to see that

$$r' = (x'^2 + y'^2 + z'^2)^{1/2} = (x^2 + y^2 + z^2)^{1/2} = r$$

and

$$\begin{aligned}
\left(\frac{\partial}{\partial x'}\right)^2 + \left(\frac{\partial}{\partial y'}\right)^2 &= \left(\cos \theta \frac{\partial}{\partial x} - \sin \theta \frac{\partial}{\partial y}\right)^2 + \left(\sin \theta \frac{\partial}{\partial x} + \cos \theta \frac{\partial}{\partial y}\right)^2 \\
&= \left(\frac{\partial}{\partial x}\right)^2 + \left(\frac{\partial}{\partial y}\right)^2
\end{aligned}$$

Since the Hamiltonian has an invariance property, we expect a conservation law, as we saw in the case of parity and invariance under displacements. To identify the operators that commute with H, let us consider an infinitesimal rotation about the z-axis. Keeping terms of order θ only so that

$$
\begin{aligned}
x' &= x - \theta y \\
y' &= y + \theta x
\end{aligned}
\tag{10-15}
$$

we require that

$$
Hu_E(x - \theta y, y + \theta x, z) = Eu_E(x - \theta y, y + \theta x, z) \tag{10-16}
$$

If we expand this to first order in θ and subtract from it

$$
Hu_E(x, y, z) = Eu_E(x, y, z) \tag{10-17}
$$

we obtain

$$
H\left(x\frac{\partial}{\partial y} - y\frac{\partial}{\partial x}\right)u_E(x, y, z) = E\left(x\frac{\partial}{\partial y} - y\frac{\partial}{\partial x}\right)u_E(x, y, z) \tag{10-18}
$$

The right side of this equation may be written as

$$
\left(x\frac{\partial}{\partial y} - y\frac{\partial}{\partial x}\right)E\,u_E(x, y, z) = \left(x\frac{\partial}{\partial y} - y\frac{\partial}{\partial x}\right)H\,u_E(x, y, z) \tag{10-19}
$$

If we define

$$
L_z = \frac{\hbar}{i}\left(x\frac{\partial}{\partial y} - y\frac{\partial}{\partial x}\right) = xp_y - yp_x \tag{10-20}
$$

then (10-18) and (10-19) together read

$$
(HL_z - L_zH)u_E(x, y, z) = 0
$$

Since the $u_E(\mathbf{r})$ form a complete set, this implies the operator relation

$$
[H, L_z] = 0 \tag{10-21}
$$

holds. L_z is the z-component of the operator

$$
\mathbf{L} = \mathbf{r} \times \mathbf{p} \tag{10-22}
$$

which is the angular momentum. Had we taken rotations about the x- and y-axes, we would have found, in addition, that

$$
\begin{aligned}
[H, L_x] &= 0 \\
[H, L_y] &= 0
\end{aligned}
\tag{10-23}
$$

Thus the three components of the angular momentum operators commute with the Hamiltonian, that is, the angular momentum is a constant of the motion. This parallels the classical result that central forces imply conservation of the angular momentum.

The Angular Momentum Commutation Relation

We might be tempted to look for simultaneous eigenfunctions of H, L_x, L_y, and L_z, but these do not form a complete set of commuting variables. For example

$$
\begin{aligned}
[L_x, L_y] &= [yp_z - zp_y, zp_x - xp_z] \\
&= [yp_z, zp_x] - [zp_y, zp_x] - [yp_z, xp_z] + [zp_y, xp_z] \\
&= y\,[p_z, z]\,p_x + x[z, p_z]\,p_y \\
&= \frac{\hbar}{i}\,(yp_x - xp_y) \\
&= i\hbar L_z
\end{aligned}
\tag{10-24}
$$

Similarly

$$
[L_y, L_z] = i\hbar L_x \tag{10-25}
$$

$$
[L_z, L_x] = i\hbar L_y \tag{10-26}
$$

The commutation relations (10-24)–(10-26) may be summarized by the formula

$$
\mathbf{L} \times \mathbf{L} = i\hbar \mathbf{L}
$$

A consequence of these commutation relations is that only one component of $\mathbf{L}$ may be chosen with H to form the commuting set of observables. To show this, we suppose we have an eigenfunction of L_x with eigenvalue l_1 that is simultaneously an eigenfunction of L_y with eigenvalue l_2. Then

$$
L_x u = l_1 u
$$

$$
L_y u = l_2 u
$$

This implies that $L_x L_y u = l_1 l_2 u$ and $L_y L_x u = l_2 l_1 u$. Thus (10-24) implies that $L_z u = 0$. However, then

$$
L_y u = \frac{1}{i\hbar}\,(L_z L_x - L_x L_z)u = \frac{1}{i\hbar}\,L_z l_1 u = 0
$$

This implies $l_2 = 0$ and a similar argument shows that $l_1 = 0$. Hence only for $\mathbf{L} = 0$ can there be simultaneous eigenfunctions of all three components of $\mathbf{L}$.

Thus only one component of $\mathbf{L}$ may be chosen with H to form the commuting set of observables. We can do a little better, however, since (10-24)–(10-26) imply that $\mathbf{L}^2$ commutes with all three components of $\mathbf{L}$:

$$
\begin{aligned}
[L_z, \mathbf{L}^2] &= [L_z, L_x^2 + L_y^2 + L_z^2] = [L_z, L_x^2] + [L_z, L_y^2] \\
&= L_x[L_z, L_x] + [L_z, L_x]\,L_x + L_y[L_z, L_y] + [L_z, L_y]\,L_y \\
&= = i\hbar L_x L_y + i\hbar L_y L_x - i\hbar L_y L_x - i\hbar L_x L_y \\
&= 0
\end{aligned}
\tag{10-27}
$$

and so on. We thus choose as our complete set of commuting observables the operators H, L_z (a purely conventional choice), and $\mathbf{L}^2$. We could also have included parity, since the Hamiltonian is manifestly invariant under $x \to -x$, $y \to -y$ and $z \to -z$, but, as we shall see later, specification of $\mathbf{L}^2$ determines the parity.

Separation of Variables in the Schrödinger Equation

In Chapter 11 we will determine the eigenvalues and eigenfunctions of L_z and $\mathbf{L}^2$; here we merely note that their use greatly simplifies the solution of the Schrödinger equation. This follows from a relation derived below:

$$\mathbf{L}^2 = (\mathbf{r} \times \mathbf{p})^2 = [(\mathbf{r} \times \mathbf{p})_x]^2 + [(\mathbf{r} \times \mathbf{p})_y]^2 + [(\mathbf{r} \times \mathbf{p})_z]^2$$

$$= -\hbar^2 \left(y \frac{\partial}{\partial z} - z \frac{\partial}{\partial y} \right)\left(y \frac{\partial}{\partial z} - z \frac{\partial}{\partial y} \right)$$

$$-\hbar^2 \left(z \frac{\partial}{\partial x} - x \frac{\partial}{\partial z} \right)\left(z \frac{\partial}{\partial x} - x \frac{\partial}{\partial z} \right)$$

$$-\hbar^2 \left(x \frac{\partial}{\partial y} - y \frac{\partial}{\partial x} \right)\left(x \frac{\partial}{\partial y} - y \frac{\partial}{\partial x} \right)$$

$$= -\hbar^2 \left[x^2 \left(\frac{\partial^2}{\partial y^2} + \frac{\partial^2}{\partial z^2} \right) + y^2 \left(\frac{\partial^2}{\partial z^2} + \frac{\partial^2}{\partial x^2} \right) \right.$$

$$+ z^2 \left(\frac{\partial^2}{\partial x^2} + \frac{\partial^2}{\partial y^2} \right) - 2xy \frac{\partial^2}{\partial x \partial y} - 2yz \frac{\partial^2}{\partial y \partial z}$$

$$\left. - 2zx \frac{\partial^2}{\partial z \partial x} - 2x \frac{\partial}{\partial x} - 2y \frac{\partial}{\partial y} - 2z \frac{\partial}{\partial z} \right]$$

as a little algebra shows. Similarly

$$(\mathbf{r} \cdot \mathbf{p})^2 = -\hbar^2 \left(x \frac{\partial}{\partial x} + y \frac{\partial}{\partial y} + z \frac{\partial}{\partial z} \right)\left(x \frac{\partial}{\partial x} + y \frac{\partial}{\partial y} + z \frac{\partial}{\partial z} \right)$$

$$= -\hbar^2 \left(x^2 \frac{\partial^2}{\partial x^2} + y^2 \frac{\partial^2}{\partial y^2} + z^2 \frac{\partial^2}{\partial z^2} + 2xy \frac{\partial^2}{\partial x \partial y} + 2yz \frac{\partial^2}{\partial y \partial z} \right. \tag{10-28}$$

$$\left. + 2zx \frac{\partial^2}{\partial z \partial x} + x \frac{\partial}{\partial x} + y \frac{\partial}{\partial y} + z \frac{\partial}{\partial z} \right)$$

The sum of the two yields

$$-\hbar^2 (x^2 + y^2 + z^2)\left(\frac{\partial^2}{\partial x^2} + \frac{\partial^2}{\partial y^2} + \frac{\partial^2}{\partial z^2} \right) + \hbar^2 \left(x \frac{\partial}{\partial x} + y \frac{\partial}{\partial y} + z \frac{\partial}{\partial z} \right) \tag{10-29}$$

We therefore get the identity

$$\mathbf{L}^2 + (\mathbf{r} \cdot \mathbf{p})^2 = r^2 p^2 + i\hbar \mathbf{r} \cdot \mathbf{p} \tag{10-30}$$

Since we are dealing with operators, keeping track of the order of the terms is crucial. It follows from the identity that

$$p^2 = \frac{1}{r^2} \left[\mathbf{L}^2 + (\mathbf{r} \cdot \mathbf{p})^2 - i\hbar \mathbf{r} \cdot \mathbf{p} \right] \tag{10-31}$$

This differs from the classical result (10-13) because **r** and **p** do not commute. With

$$\mathbf{p}^2 = \frac{1}{r^2}\,\mathbf{L}^2 - \hbar^2\frac{1}{r^2}\left(r\frac{\partial}{\partial r}\right)^2 - \hbar^2\frac{1}{r}\frac{\partial}{\partial r} \tag{10-32}$$

the Schrödinger equation takes the form

$$-\frac{\hbar^2}{2\mu}\left[\frac{1}{r^2}\left(r\frac{\partial}{\partial r}\right)\left(r\frac{\partial}{\partial r}\right) + \frac{1}{r}\frac{\partial}{\partial r} - \frac{1}{\hbar^2 r^2}\,\mathbf{L}^2\right]u_E(\mathbf{r}) + V(r)\,u_E(\mathbf{r}) = Eu_E(\mathbf{r}) \tag{10-33}$$

If we work in spherical coordinates (Fig. 10-1), which is the natural thing to do, then the only operator that involves the spherical angles θ and ϕ is $\mathbf{L}^2$. If we therefore pick eigenfunctions of the form

$$u_E(\mathbf{r}) = Y_\lambda(\theta,\,\phi)\,R_{E\lambda}(r) \tag{10-34}$$

where

$$\mathbf{L}^2\,Y_\lambda(\theta,\,\phi) = \lambda\,Y_\lambda(\theta,\,\phi) \tag{10-35}$$

is the eigenvalue equation for $\mathbf{L}^2$, then the equation separates into (10-35) and a purely radial equation in (10-39) below. Our procedure is really no different than

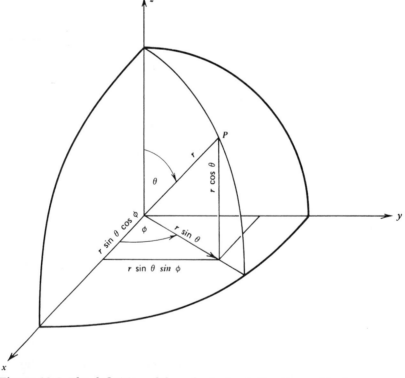

Figure 10-1. The definition of the spherical coordinates used in the text and the relation between the Cartesian coordinates (x, y, z) and the spherical coordinates (r, θ, ϕ).

the conventional separation of variables. It does, however, stress the role of the symmetry in determining the complete commuting set of operators.

We see that λ has the dimensions of $\hbar^2$, and we write it in the form

$$\lambda = l(l + 1)\hbar^2 \tag{10-36}$$

In the next chapter we will prove that $l = 0, 1, 2, 3, \ldots$ and use this fact in the development that follows. We will actually write the angular momentum eigenfunctions as $Y_{lm}(\theta, \phi)$, where the second subscript takes into account that $Y_{lm}(\theta, \phi)$ is simultaneously an eigenfunction of $\mathbf{L}^2$, and L_z, with

$$L_z Y_{lm}(\theta, \phi) = m\hbar Y_{lm}(\theta, \phi) \tag{10-37}$$

We already noted that l is an integer. We will also find that m is an integer with $-l \le m \le l$. Since the Y_{lm} are eigenfunctions of hermitian operators, we expect that the Y_{lm} corresponding to different eigenvalues are orthogonal. We will prove in the next chapter that when properly normalized, an integration over the range of the variables θ, ϕ,

$$\int d\Omega \, Y_{l_1 m_1}(\theta, \phi)^* Y_{l_2 m_2}(\theta, \phi)$$

$$\equiv \int_0^\pi \sin\theta \, d\theta \int_0^{2\pi} d\phi \, Y_{l_1 m_1}(\theta, \phi)^* Y_{l_2 m_2}(\theta, \phi) = \delta_{l_1 l_2} \, \delta_{m_1 m_2} \tag{10-38}$$

THE RADIAL EQUATION

When (10-35) and (10-36) are used in (10-33) and (10-34), we obtain the radial Schrödinger equation

$$-\frac{\hbar^2}{2\mu} \left[\frac{1}{r} \frac{d}{dr} \left(r \frac{d}{dr} \right) + \frac{1}{r} \frac{d}{dr} - \frac{l(l+1)}{r^2} \right] R_{Elm}(r) + V(r) \, R_{Elm}(r) = E R_{Elm}(r) \tag{10-39}$$

We note that there is no dependence on m in the equation. Thus, for a given l, there will always be a $(2l + 1)$-fold degeneracy, since all the possible m-values will have the same energy. We therefore omit the m-label in the radial functions. This equation may be written as

$$\left(\frac{d^2}{dr^2} + \frac{2}{r} \frac{d}{dr} \right) R_{nl}(r) - \frac{2\mu}{\hbar^2} \left[V(r) + \frac{l(l+1)\hbar^2}{2\mu r^2} \right] R_{nl}(r) + \frac{2\mu E}{\hbar^2} R_{nl}(r) = 0 \tag{10-40}$$

where we have replaced the label E by n in the subscripts of the eigenfunction $R_{nl}(r)$. We will examine the solutions to this equation for a variety of potentials restricted by the condition that they go to zero at infinity faster than $1/r$, except for the important special case of the Coulomb potential, which will be discussed in Chapter 12. We will also assume that the potentials are not as singular as $1/r^2$ at the origin, so that

$$\lim_{r \to 0} r^2 V(r) = 0 \tag{10-41}$$

It is sometimes convenient to introduce the function

$$u_{nl}(r) = rR_{nl}(r) \tag{10-42}$$

Since

$$\left(\frac{d^2}{dr^2} + \frac{2}{r}\frac{d}{dr}\right)\frac{u_{nl}(r)}{r} = \frac{1}{r}\frac{d^2}{dr^2}u_{nl}(r) \tag{10-43}$$

it follows that

$$\frac{d^2u_{nl}(r)}{dr^2} + \frac{2\mu}{\hbar^2}\left[E - V(r) - \frac{l(l+1)\hbar^2}{2\mu r^2}\right]u_{nl}(r) = 0 \tag{10-44}$$

This looks very much like a one-dimensional equation, except that

(a) The potential $V(r)$ is altered by the addition of a repulsive centrifugal barrier,

$$V(r) \rightarrow V(r) + \frac{l(l+1)\hbar^2}{2\mu r^2} \tag{10-45}$$

(b) the definition of $u_{nl}(r)$ and the finiteness of the wave function at the origin require that

$$u_{nl}(0) = 0 \tag{10-46}$$

which makes it more like the one-dimensional problem for which $V = +\infty$ in the left-hand region (Fig. 10-2).

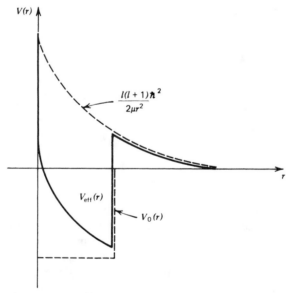

Figure 10-2. Effective potential acting in radial equation for $u = rR(r)$ when the real potential is a square well.

First we consider the radial equation near the origin, dropping all subscripts for convenience. As $r \to 0$, the leading terms in our equation are

$$\frac{d^2u}{dr^2} - \frac{l(l+1)}{r^2} u \simeq 0 \tag{10-47}$$

because the potential does not contribute for small enough r when (10-41) is satisfied. If we make the *Ansatz*

$$u(r) \sim r^s \tag{10-48}$$

we find that the equation will be satisfied, provided that

$$s(s-1) - l(l+1) = 0 \tag{10-49}$$

that is, $s = l + 1$ or $s = -l$. The solution that satisfies the condition $u(0) = 0$, that is, the solution that behaves like r^{l+1}, is called the *regular solution*; the solution that behaves like r^{-1} is the *irregular solution*. In terms of the radial wave function $R(r)$ we have the regular solution behaving like r^l and the ir-regular solution like r^{-l-1}.

For large r we can drop the potential terms, and the equation becomes

$$\frac{d^2u}{dr^2} + \frac{2\mu E}{\hbar^2} u \simeq 0 \tag{10-50}$$

The square integrability condition implies that

$$1 = \int d^3r |\psi(\mathbf{r})|^2 = \int_0^\infty r^2 \, dr \int d\Omega |R_{nl}(r) \, Y_{lm}(\theta, \phi)|^2$$
$$= \int_0^\infty r^2 \, dr |R_{nl}(r)|^2 \tag{10-51}$$

that is,

$$\int_0^\infty dr |u_{nl}(r)|^2 = 1 \tag{10-52}$$

so that the wave function should vanish at infinity. If $E < 0$, so that

$$\frac{2\mu E}{\hbar^2} = -\alpha^2 \tag{10-53}$$

the asymptotic solution is

$$u(r) \sim e^{-\alpha r} \tag{10-54}$$

If $E > 0$, we have solutions that are only normalizable in a box (see discussion in Chapter 4). With

$$\frac{2\mu E}{\hbar^2} = k^2 \tag{10-55}$$

the solution for r large enough so that $V(r)$ is negligible, there will be a linear combination of e^{ikr} and e^{-ikr}, the proper combination being determined by the requirement that the asymptotic solution tie on continuously to the solution that is regular at the origin. We now consider some examples.

THE FREE PARTICLE

In this example $V(r) = 0$, but there is still a centrifugal barrier present. The radial equation (10-40) takes the form

$$\left[\frac{d^2}{dr^2} + \frac{2}{r}\frac{d}{dr} - \frac{l(l + 1)}{r^2} \right] R(r) + k^2 R(r) = 0 \qquad (10\text{-}56)$$

If we introduce the variable $\rho = kr$, we get

$$\frac{d^2 R}{d\rho^2} + \frac{2}{\rho}\frac{dR}{d\rho} - \frac{l(l + 1)}{\rho^2} R + R = 0 \qquad (10\text{-}57)$$

or

$$\frac{d^2 u}{d\rho^2} - \frac{l(l + 1)}{\rho^2} u + u = 0$$

For $l = 0$ the equation for $u(\rho)$ is just $d^2 u/d\rho^2 + u = 0$, so that the solutions are $\sin \rho$ and $\cos \rho$, that is, the regular solution is

$$R_0(\rho) = \frac{\sin \rho}{\rho} \qquad (10\text{-}58)$$

and the irregular solution is

$$R_0(\rho) = \frac{\cos \rho}{\rho} \qquad (10\text{-}59)$$

For general l the solutions can be expressed in terms of simple functions. These are known as *spherical Bessel functions*. The regular solution is $j_l(\rho)$, which may be written in the form

$$j_l(\rho) = (-\rho)^l \left(\frac{1}{\rho}\frac{d}{d\rho} \right)^l \left(\frac{\sin \rho}{\rho} \right) \qquad (10\text{-}60)$$

and the irregular one called the spherical Neumann function $n_l(\rho)$ has the form

$$n_l(\rho) = -(-\rho)^l \left(\frac{1}{\rho}\frac{d}{d\rho} \right)^l \left(\frac{\cos \rho}{\rho} \right) \qquad (10\text{-}61)$$

The first few functions are listed below.

$$j_0(\rho) = \frac{\sin \rho}{\rho} \qquad n_0(\rho) = -\frac{\cos \rho}{\rho}$$

$$j_1(\rho) = \frac{\sin \rho}{\rho^2} - \frac{\cos \rho}{\rho} \qquad n_1(\rho) = -\frac{\cos \rho}{\rho^2} - \frac{\sin \rho}{\rho}$$

$$j_2(\rho) = \left(\frac{3}{\rho^3} - \frac{1}{\rho}\right) \sin \rho - \frac{3}{\rho^2} \cos \rho$$

$$n_2(\rho) = -\left(\frac{3}{\rho^3} - \frac{1}{\rho}\right) \cos \rho - \frac{3}{\rho^2} \sin \rho$$

(10-62)

The combinations that will be of interest for large ρ are the *spherical Hankel functions*

$$h_l^{(1)}(\rho) = j_l(\rho) + i n_l(\rho) \tag{10-63}$$

and

$$h_l^{(2)}(\rho) = [h_l^{(1)}(\rho)]^* \tag{10-64}$$

Again the first few spherical Hankel functions are

$$h_0^{(1)}(\rho) = \frac{e^{i\rho}}{i\rho}$$

$$h_1^{(1)}(\rho) = -\frac{e^{i\rho}}{\rho}\left(1 + \frac{i}{\rho}\right)$$

(10-65)

$$h_2^{(1)}(\rho) = \frac{i\, e^{i\rho}}{\rho}\left(1 + \frac{3i}{\rho} - \frac{3}{\rho^2}\right)$$

Of special interest are

(a) The behavior near the origin: for $\rho \ll l$, it turns out that

$$j_l(\rho) \approx \frac{\rho^l}{1 \cdot 3 \cdot 5 \cdots (2l+1)} \tag{10-66}$$

and

$$n_l(\rho) \simeq -\frac{1 \cdot 3 \cdot 5 \cdots (2l-1)}{\rho^{l+1}} \tag{10-67}$$

For $\rho \gg l$, we have the asymptotic expressions

$$j_l(\rho) \simeq \frac{1}{\rho} \sin\left(\rho - \frac{l\pi}{2}\right) \tag{10-68}$$

and

$$n_l(\rho) \simeq -\frac{1}{\rho} \cos\left(\rho - \frac{l\pi}{2}\right) \tag{10-69}$$

so that

$$h_l^{(1)}(\rho) \simeq -\frac{i}{\rho} e^{i(\rho - l\pi/2)} \tag{10-70}$$

The solution that is regular at the origin is

$$R_l(r) = j_l(kr) \tag{10-71}$$

Its asymptotic form is, using (10-68)

$$R_l(r) \simeq -\frac{1}{2ikr} [e^{-i(kr - l\pi/2)} - e^{i(kr - l\pi/2)}] \tag{10-72}$$

THE INFINITE POTENTIAL WELL

Consider the infinite well in three dimensions. Here

$$\begin{aligned} V(r) &= 0 &&r < a \\ &= \infty &&r > a \end{aligned} \tag{10-73}$$

In this case, writing

$$\frac{2\mu E}{\hbar^2} = k^2 \tag{10-74}$$

the solution that is regular at $r = 0$ is

$$R(r) = Aj_l(kr) \tag{10-75}$$

with the eigenvalues determined by the condition that the solution vanish at $r = a$, that is, by

$$j_l(ka) = 0 \tag{10-76}$$

The roots for a few values of l are:

$l = 0$	1	2	3	4	5
3.14	4.49	5.76	6.99	8.18	9.36
6.28	7.73	9.10	10.42		
9.42					

It follows from (10-68) that for ka large (actually $ka \gg l$) that the roots are given by $ka \approx n\pi + l\pi/2$.

The complete set of simultaneous eigenfunctions of H, $\mathbf{L}^2$, and L_z are

$$u_{nl}(r) = Aj_l(k_{ln}r)Y_{lm}(\theta, \phi) \qquad (10\text{-}77)$$

where the k_{nl} are given by $j_l(k_{nl}a) = 0$, and A is the amplitude required for normalization. Note that

$$\int d^3r \, u_{n'l'}(\mathbf{r})u_{nl}(\mathbf{r}) = 0 \qquad (10\text{-}78)$$

when any of the $n \neq n'$, $l \neq l'$. The Y_{lm} take care of the orthogonality with respect to the quantum numbers l and m. The orthogonality with respect to the label n, which distinguishes different energy eigenvalues for fixed l and m, must surely be there, and in fact, one finds in discussions of Bessel functions that

$$\int_0^1 dt \, j_l(\alpha_m t)j_l(\alpha_n t) = 0 \qquad \text{for} \quad m \neq n \qquad (10\text{-}79)$$

where $j_l(\alpha_n) = 0$. This relation is equivalent to the orthogonality of the radial functions corresponding to different values of the radial quantum number n.

The spectrum of the infinite square well can be described as follows: If the first root for a given l is labeled $n = 1$, the second root $n = 2$, and so on, and if we use the accepted spectroscopic notation for the l values

$$S : l = 0$$
$$P : l = 1$$
$$D : l = 2$$
$$F : l = 3$$
$$G : l = 4$$
$$\cdots$$

then the order in which the levels occur is

$$1S; \; 1P; \; 1D; \; 2S; \; 1F; \; 2P; \; 1G; \; 2D; \; 1H; \; 3S; \; \ldots$$

Suppose we consider a model of the nucleus that consists of protons and neutrons inside such an infinite box. Since neutrons and protons are spin $\frac{1}{2}$ particles, that is, fermions, no more than two neutrons and two protons can occupy a given state. If we concentrate on protons, we observe that in the $1S$ state only *two* protons can appear. In the next level we have $l = 1$, so that there are three states, and hence *six* protons will fill it. For the $1D$ level, with five possible m-values (since $l = 2$), *ten* protons are required to fill this "shell." Thus levels will be filled when the number of protons is 2, 8 ($= 2 + 6$), 18 ($= 2 + 6 + 10$), 20 ($= 18 + 2$), 34 ($= 20 + 14$), 40, 58, 68, 90, 92, 106, . . . , and similarly for the neutrons. A study of real nuclei shows that for the "magic" number of protons and neutrons 2, 8, 20,

28, 50, 82, 126, . . . , these nuclei exhibit special characteristics that can be associated with filled levels, that is, closed shells. The difference between the real "magic" numbers, and those obtained in our primitive model comes about because there is an additional potential that depends on the spin and that shifts the levels about somewhat, thus reordering the numbers. The shell model of the nucleus, when properly constructed, explains many of the properties of nuclei.

It may seem a bit mysterious why treating nuclei as independent particles in a box should provide a good approximation, when we know that the forces between nucleons are very strong. The explanation lies in the exclusion principle. Nuclei in a box interact by colliding with each other. A collision in general leads the nucleon to be scattered into a different quantum state. In the ground state, scattering is inhibited because all the low-lying levels are occupied, and thus cannot serve as available final states.

We next consider another example, analogous to the transmission and reflection by square wells or barriers in the one-dimensional case.

CONTINUUM SOLUTIONS FOR A SQUARE WELL

Consider the potential

$$
\begin{aligned}
V(r) &= -V_0 & r < a \\
&= 0 & r > a
\end{aligned}
\tag{10-80}
$$

Then the radial equation has the form

$$
\frac{d^2R}{dr^2} + \frac{2}{r}\frac{dR}{dr} - \frac{l(l+1)}{r^2} R + \frac{2\mu}{\hbar^2}(V_0 + E) R = 0 \qquad r < a
$$

$$
\frac{d^2R}{dr^2} + \frac{2}{r}\frac{dR}{dr} - \frac{l(l+1)}{r^2} R + \frac{2\mu E}{\hbar^2} R = 0 \qquad r > a
\tag{10-81}
$$

The solution for $r > a$ will be a combination of the regular and irregular solutions of the free field equation

$$
R_l(r) = Bj_l(kr) + Cn_l(kr)
\tag{10-82}
$$

while the solution for $r < a$ must be the regular solution, that is,

$$
R_l(r) = Aj_l(\kappa r)
\tag{10-83}
$$

where

$$
\kappa^2 = \frac{2\mu(E + V_0)}{\hbar^2}
\tag{10-84}
$$

as before.

The matching of $1/R_l\, dR_l/dr$ at $r = a$ gives the ratio C/B from

$$
\kappa \left[\frac{dj_l(\rho)/d\rho}{j_l(\rho)} \right]_{\rho = \kappa a} = k \left[\frac{Bdj_l/d\rho + Cdn_l/d\rho}{Bj_l(\rho) + Cn_l(\rho)} \right]_{\rho = ka}
\tag{10-85}
$$

Asymptotically

$$R_{nl}(r) \approx \frac{B}{2ikr} \left(e^{i(kr - l\pi/2)} - e^{-i(kr - l\pi/2)} \right)$$

$$- \frac{C}{2kr} \left(e^{i(kr - l\pi/2)} + e^{-i(kr - l\pi/2)} \right) \tag{10-86}$$

$$\approx \frac{-C + iB}{2kr} \left[e^{-i(kr - l\pi/2)} + \frac{C + iB}{C - iB} e^{i(kr - l\pi/2)} \right]$$

The factor in front, other than its necessary $1/r$ dependence is irrelevant, since the amplitude is fixed by the normalization condition. The relationship between the two exponentials does have physical significance. First of all, we note that the two terms represent spherical waves, an incoming one and an outgoing one. When there is no potential, $C = 0$, and the factor in front of the outgoing spherical wave is -1. With the potential there, the factor has the property that its absolute square is unity. This is easy to check, since from (10-85) we see that B/C is real. Thus

$$\left| \frac{C + iB}{C - iB} \right|^2 = \frac{1 + iB/C}{1 - iB/C} \times \frac{1 - iB/C}{1 + iB/C} = 1 \tag{10-87}$$

The factor is conventionally written as

$$\frac{C + iB}{C - iB} \equiv -e^{2i\delta_l(k)} \tag{10-88}$$

(equivalently $C/B = -\tan \delta_l(k)$) so that the asymptotic form of the radial solution $R_{nl}(r)$ is

$$R_{nl}(r) \approx \frac{\text{constant}}{r} \sin \left(kr - \frac{l\pi}{2} + \delta_l(k) \right) \tag{10-89}$$

The quantity $\delta_l(k)$ is called a *phase shift* for this reason. Actually, (10-89) is the asymptotic form for *any real potential*. The reason is that the absolute square of the coefficient of the incoming spherical wave must equal the absolute square of

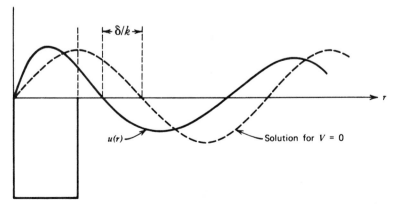

Figure 10-3. Continuum solution $u(r) = rR(r)$ for attractive potential ($l = 0$).

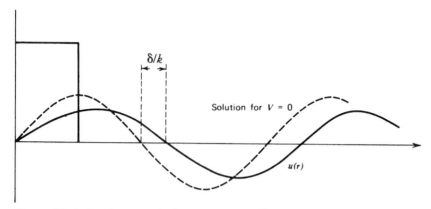

Figure 10-4. Continuum solution $u(r) = rR(r)$ for repulsive potential ($l = 0$).

the coefficient of the outgoing spherical wave. This is just a statement of flux conservation: for a real potential, particles are neither created nor absorbed by the potential.

The actual computation of C/B from (10-85) is tedious, except for $l = 0$. As for the bound state problem, the use $u(r) = rR(r)$ simplifies the calculation greatly. One only needs to match $A \sin \kappa r$ to $B \sin kr + C \cos kr$ at $r = a$ to obtain an expression for $\tan \delta_0$. The results for this case are schematically drawn in Figs. 10-3 and 10-4. They show that an attractive potential tends to "draw in" the wave function, while a repulsive potential tends to push it out. We will return to these matters in Chapter 24, when we discuss collision theory.

THE SQUARE WELL, BOUND STATES

We look for bound state solutions, for which $E < 0$. We write

$$\frac{2\mu}{\hbar^2}(V_0 + E) = \kappa^2$$

$$\frac{2\mu}{\hbar^2}E = -\alpha^2 \tag{10-90}$$

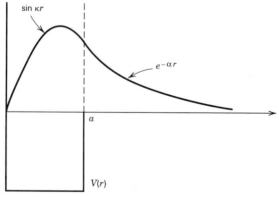

Figure 10-5. The shape of the wave function $u(r) = rR(r)$ for an attractive square well when there exists one bound state ($l = 0$).

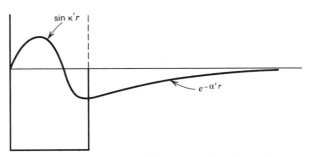

Figure 10-6. The shape of the wave function $u(r) = rR(r)$ for an attractive square well when there exist two bounded states ($l = 0$). Only the wave function for the second bound state is sketched in this figure.

The solution for $r < a$, which must be regular at the origin, is

$$R(r) = Aj_l(\kappa r) \tag{10-91}$$

The solution for $r > a$ must vanish as $r \to \infty$. The second of the equations (10-81) is just the equation for the spherical Bessel function, except that k is replaced by $i\alpha$. The solution that behaves like e^{ikr} now becomes the exponentially falling one, that is, we have

$$R(r) = Bh_l^{(1)}(i\alpha r) \tag{10-92}$$

for $r > a$. The two solutions must match at $r = a$ and so must the derivatives. This leads to the condition

$$\kappa \left[\frac{dj_l(\rho)/d\rho}{j_l(\rho)} \right]_{\rho=\kappa a} = i\alpha \left[\frac{dh_l^{(1)}(\rho)/d\rho}{h_l^{(1)}(\rho)} \right]_{\rho=i\alpha a} \tag{10-93}$$

This is a very complicated transcendental equation involving l, V_0, and E. For $l = 0$ the problem is very simple. In terms of $u(r) = rR(r)$ we again have a situation identical to a one-dimensional potential with $V(x) = \infty$ for $x < 0$. There we know from Chapter 5 [(5-69)] that there will be one or more bound states only if

$$\frac{2mV_0a^2}{\hbar^2} > \frac{\pi^2}{4}$$

Figures 10-5 and 10-6 show the wave functions for the first two bound states for $l = 0$.

Problems

1. Check that P_i, R_i, p_i, r_i obey the commutation relations given in (10-7).

2. Assume that the deuteron (consisting of a neutron and a proton, equal in mass) is a bound state with $l = 0$, and the potential is a square well of range $r_0 = 2.8 \times 10^{-13}$ cm. Given that the binding energy is -2.18 MeV, find the depth of the potential.

[*Hint:* Expand about the case of zero binding energy for which V_0 is given by (5-41)].

3. Calculate the $l = 0$ phase shift for a square well potential. Use the procedure outlined in (10-85) to work out both the attractive and the repulsive potential case. Discuss various limits, such as E large and small, V_0 large and small.

4. Show that for $l = 0$ scattering by a square well of arbitrary range and depth V_0, it is always possible to write the phase shift as an expansion

$$k \cot \delta_0 = -\frac{1}{a} + \frac{r_{\text{eff}} \, k^2}{2} + O(k^4)$$

Obtain an expression for a and r_{eff} in terms of the parameters of the well.

5. Consider a potential of arbitrary shape that vanishes for $r \geq a$. Let the logarithmic derivative of the radial function inside the potential,

$$\left. \frac{1}{R} \frac{dR(r)}{dr} \right|_{r=a} = f_l(E)$$

be a slowly varying function of the energy. Consider $l = 0$.

   (a) If the potential has a bound state with energy, E_B, what is the value of $f_0(E_B)$?
   (b) If $f_0(E)$ is independent of E, what is the phase shift as a function of energy?
   (c) If $f_0(E) = f_0(E_B) + (E - E_B)f_0'$, how does f_0' enter into the phase shift?

It is simpler to work out (b) and (c) in terms of $k \cot \delta_0(k)$, instead of the phase shift, and that is a preferable way to present your results.

6. Consider the potential

$$V(r) = \infty \qquad r < a$$

$$V(r) = 0 \qquad r > a$$

Calculate the $l = 0$ phase shift. What is it for ka very small? What is it for ka very large? Note that this potential is a model for an impenetrable sphere.

7. Consider the eigenvalue condition for a square well potential of range a and depth V_0, for $l = 1$. Obtain an expression from which you could determine the value of V_0 for zero binding energy.

8. Show that for a square well,

$$k^{2l+1} \cot \delta_l(k) \rightarrow \text{constant}$$

as $k \rightarrow 0$.

9. The three-dimensional flux is given by

$$\mathbf{j} = \frac{\hbar}{2i\mu} [\psi^*(\mathbf{r})\nabla\psi(\mathbf{r}) - \nabla\psi^*(\mathbf{r}) \, \psi(\mathbf{r})]$$

Calculate the radial flux integrated over all angles, that is, $\int d\Omega \, \mathbf{i}_r \cdot \mathbf{j}$ for wave functions of the form

$$\psi(\mathbf{r}) = C \frac{e^{\pm ikr}}{r} Y_{lm}(\theta, \phi)$$

with

$$\int d\Omega \, |Y_{lm}(\theta, \phi)|^2 = 1$$

and show that these correspond to outgoing/incoming spherical waves.

10. Show that the flux in the azimuthal direction, $\mathbf{i}_\theta \cdot \mathbf{j}$ is negligible compared to j_r for r very large.

11. Consider the $l = 0$ radial equation for the potential

$$V(r) = V_0[e^{-2(r-r_0)/a} - 2\,e^{-(r-r_0)/a}]$$

(known as the Morse potential). Find the energy eigenvalues by simplifying the differential equation. Do this by defining a new variable $x = Ce^{-r/a}$ with C chosen to simplify the equation as much as possible, and then treating the equation in the manner that the simple harmonic oscillator problem was treated in Chapter 5.

Plot the potential. Show that for a deep, wide potential, the low-lying bound states approximate those of a harmonic oscillator, and explain why this is so.

References

The general properties of second-order differential equations in the context of quantum mechanics are discussed in

J. L. Powell and B. Crasemann, *Quantum Mechanics*, Addison-Wesley, Reading, Mass., 1961.

A comprehensive discussion of such equations may also be found in

P. M. Morse and H. Feshbach, *Methods of Theoretical Physics*, McGraw-Hill, New York, 1953.

ANGULAR MOMENTUM

ANGULAR MOMENTUM OPERATORS IN SPHERICAL VARIABLES

Our task in this chapter is to find the eigenvalues and the eigenfunctions of the operators L_z and $\mathbf{L}^2$. Since the angular momentum has the dimensions of $\hbar$, we may write the eigenvalue equations in the form

$$L_z Y_{lm} = m\hbar Y_{lm}$$
$$\mathbf{L}^2 Y_{lm} = l(l + 1)\, \hbar^2 Y_{lm} \tag{11-1}$$

where m and $l(l + 1)$ are real numbers. The peculiar way of writing the eigenvalue of $\mathbf{L}^2$ will prove its convenience later. There are several ways of proceeding. The conventional way is to write out the operators $\mathbf{L}$ in spherical coordinates. We have

$$x = r \sin \theta \cos \phi$$
$$y = r \sin \theta \sin \phi \tag{11-2}$$
$$z = r \cos \theta$$

so that

$$dx = \sin \theta \cos \phi\, dr + r \cos \theta \cos \phi\, d\theta - r \sin \theta \sin \phi\, d\phi$$
$$dy = \sin \theta \sin \phi\, dr + r \cos \theta \sin \phi\, d\theta + r \sin \theta \cos \phi\, d\phi \tag{11-3}$$
$$dz = \cos \theta\, dr - r \sin \theta\, d\theta$$

These can be solved to give

$$dr = \sin \theta \cos \phi \, dx + \sin \theta \sin \phi \, dy + \cos \theta \, dz$$

$$d\theta = \frac{1}{r} (\cos \theta \cos \phi \, dx + \cos \theta \sin \phi \, dy - \sin \theta \, dz) \qquad (11\text{-}4)$$

$$d\phi = \frac{1}{r \sin \theta} (- \sin \phi \, dx + \cos \phi \, dy)$$

With the help of this equation we can obtain

$$\frac{\partial}{\partial x} = \frac{\partial r}{\partial x} \frac{\partial}{\partial r} + \frac{\partial \theta}{\partial x} \frac{\partial}{\partial \theta} + \frac{\partial \phi}{\partial x} \frac{\partial}{\partial \phi}$$

$$= \sin \theta \cos \phi \frac{\partial}{\partial r} + \frac{1}{r} \cos \theta \cos \phi \frac{\partial}{\partial \theta} - \frac{\sin \phi}{r \sin \theta} \frac{\partial}{\partial \phi}$$

$$\frac{\partial}{\partial y} = \sin \theta \sin \phi \frac{\partial}{\partial r} + \frac{1}{r} \cos \theta \sin \phi \frac{\partial}{\partial \theta} + \frac{\cos \phi}{r \sin \theta} \frac{\partial}{\partial \phi} \qquad (11\text{-}5)$$

$$\frac{\partial}{\partial z} = \cos \theta \frac{\partial}{\partial r} - \frac{\sin \theta}{r} \frac{\partial}{\partial \theta}$$

and thus we finally obtain

$$L_z = \frac{\hbar}{i} \left(x \frac{\partial}{\partial y} - y \frac{\partial}{\partial x} \right) = \frac{\hbar}{i} \frac{\partial}{\partial \phi} \qquad (11\text{-}6)$$

The other two components of the angular momentum are more compactly expressed if we introduce

$$L_\pm = L_x \pm iL_y \qquad (11\text{-}7)$$

$$L_\pm = \frac{\hbar}{i} \left[y \frac{\partial}{\partial z} - z \frac{\partial}{\partial y} \pm i \left(z \frac{\partial}{\partial x} - x \frac{\partial}{\partial z} \right) \right]$$

$$= \frac{\hbar}{i} \left[\pm iz \left(\frac{\partial}{\partial x} \pm i \frac{\partial}{\partial y} \right) \mp i(x \pm iy) \frac{\partial}{\partial z} \right]$$

$$= \pm \hbar r \cos \theta \left(\sin \theta \, e^{\pm i\phi} \frac{\partial}{\partial r} + \frac{1}{r} \cos \theta \, e^{\pm i\phi} \frac{\partial}{\partial \theta} \pm \frac{i \, e^{\pm i\phi}}{r \sin \theta} \frac{\partial}{\partial \phi} \right)$$

$$\mp \hbar r \sin \theta \, e^{\pm i\phi} \left(\cos \theta \frac{\partial}{\partial r} - \frac{\sin \theta}{r} \frac{\partial}{\partial \theta} \right) \qquad (11\text{-}8)$$

Thus

$$L_\pm = \hbar \, e^{\pm i\phi} \left(\pm \frac{\partial}{\partial \theta} + i \cot \theta \frac{\partial}{\partial \phi} \right) \qquad (11\text{-}9)$$

One can then construct the $\mathbf{L}^2$ operator by observing that

$$
\begin{aligned}
L_+L_- &= (L_x + iL_y)(L_x - iL_y) \\
&= L_x^2 + L_y^2 - i[L_x, L_y]
\end{aligned}
\tag{11-10}
$$

so that

$$
\begin{aligned}
\mathbf{L}^2 &= L_z^2 + L_+L_- + i[L_x, L_y] \\
&= L_+L_- + L_z^2 - \hbar L_z
\end{aligned}
\tag{11-11}
$$

In the last line we used (10-24). We thus get a second-order differential operator involving θ and ϕ, and there remains the task of solving the differential equations that (11-1) represents. This is discussed in many textbooks on quantum mechanics or classical electrodynamics. We will proceed algebraically, but digress for a moment to discuss the eigenfunctions of L_z.

THE EIGENFUNCTIONS
AND EIGENVALUES OF L_z

Consider

$$
L_z Y_{lm} = m\hbar Y_{lm}
\tag{11-12}
$$

The equation, using (11-6), reads

$$
\frac{\partial}{\partial \phi} Y_{lm}(\theta, \phi) = im Y_{im}(\theta, \phi)
\tag{11-13}
$$

so that the solution is of the form $Y_{lm}(\theta, \phi) = \Theta_{lm}(\theta)\, \Phi_m(\phi)$, where

$$
\frac{d\Phi_m(\phi)}{d\phi} = im\Phi_m(\phi)
\tag{11-14}
$$

The solution to this, normalized such that

$$
\int_0^{2\pi} d\phi |\Phi_m|^2 = 1
\tag{11-15}
$$

is

$$
\Phi_m(\phi) = \frac{1}{\sqrt{2\pi}}\, e^{im\phi}
\tag{11-16}
$$

It is sometimes argued that since a rotation through 360°, that is, a transformation $\phi \rightarrow \phi + 2\pi$, leaves the system invariant, it is necessary that

$$
e^{2\pi im} = 1
\tag{11-17}
$$

so that m is an integer. This is not quite correct, since the quantities that enter into physical observables are of the type $\int_0^{2\pi} d\phi \, \psi_1^*(\phi) \, A\psi_2(\phi)$, with wave functions $\psi(\phi)$ of the form

$$\psi(\phi) = \sum_{m=-\infty}^{\infty} C_m \frac{e^{im\phi}}{\sqrt{2\pi}} \tag{11-18}$$

If we require that these arbitrary wave functions do not change (except for an overall phase factor) under the transformation $\phi \to \phi + 2\pi$, then we are led to the conclusion that the most general allowed values of m are $m = c + $ integer where c is a constant. It is only if we view the operator L_z as part of the total set (L_x, L_y, L_z) that we can say something about the constant c. We shall argue later that the eigenvalues are distributed symmetrically about zero, so that $c = 0$ or $c = 1/2$, and for the operators considered in this chapter, we shall restrict ourselves to $c = 0$, that is, the condition that m is an integer.

The eigenvalue equation for L_z appears in another context. Consider a classical rotator, rotating in the x-y plane. If the moment of inertia is I, then the energy is

$$E = \frac{L_z^2}{2I} \tag{11-19}$$

and thus the Hamiltonian is

$$H = \frac{L_z^2}{2I} \tag{11-20}$$

The eigenvalues of the Hamiltonian are now immediately seen to be

$$E_m = \frac{\hbar^2 m^2}{2I} \tag{11-21}$$

and the eigenfunctions are $e^{\pm im\phi}$. There is a degeneracy, since H commutes with L_z, and the two eigenfunctions for a given E_m correspond to the two senses of rotation. If we have N particles rigidly fixed on a circle, with equal angles $2\pi/N$ between neighboring particles, and *if the particles are identical*, then the solution of the energy eigenvalue equation

$$H\Phi_E(\phi) = E\Phi_E(\phi) \tag{11-22}$$

will again be $e^{\pm i\lambda\phi}$. The physical system is unaltered under a rotation of $2\pi/N$ radians (or an integral multiple of the angle), and the solutions should reflect this. The same kind of arguments that forced m to be an integer now imply that $\lambda = N \times$ (an integer).[1] The energy is therefore

$$E = \frac{\hbar^2 (Nm)^2}{2I} \tag{11-23}$$

[1]The reader might look back to the Dicke-Wittke Gedanken experiment discussed in Chapter 1.

RAISING AND LOWERING OPERATORS
FOR ANGULAR MOMENTUM

Let us now return to our equations (11-1), and try to obtain the eigenvalues in a manner reminiscent of our treatment of the harmonic oscillator in Chapter 7. The eigenfunctions of the hermitian operators L_z and $\mathbf{L}^2$ will be orthogonal, if the eigenvalues are different, and with proper normalization, we will write

$$\langle Y_{l'm'}|Y_{lm}\rangle = \delta_{ll'}\delta_{mm'} \tag{11-24}$$

Since

$$\langle Y_{lm}|(L_x^2 + L_y^2 + L_z^2)\,Y_{lm}\rangle = \langle L_xY_{lm}|L_xY_{lm}\rangle + \langle L_yY_{lm}|L_yY_{lm}\rangle + m^2\hbar^2 \tag{11-25}$$
$$\geq 0$$

it follows that

$$l(l + 1) \geq 0 \tag{11-26}$$

The operators $L_\pm$ introduced in (11-7) are very useful in what follows, and we shall see that they play the role of raising and lowering operators. First, we already saw that

$$\mathbf{L}^2 = L_+L_- + L_z^2 - \hbar L_z \tag{11-27}$$

In the same way we see that

$$\mathbf{L}^2 = L_-L_+ + L_z^2 + \hbar L_z \tag{11-28}$$

It follows from this, as well as directly from (10-24), that

$$[L_+, L_-] = 2\hbar L_z \tag{11-29}$$

The remaining commutation relations are

$$[L_+, L_z] = [L_x + iL_y, L_z] = -i\hbar L_y - \hbar L_x \tag{11-30}$$
$$= -\hbar L_+$$

and

$$[L_-, L_z] = \hbar L_- \tag{11-31}$$

From the fact that $[\mathbf{L}^2, \mathbf{L}] = 0$, it also follows that

$$[\mathbf{L}^2, L_\pm] = 0$$
$$[\mathbf{L}^2, L_z] = 0 \tag{11-32}$$

This implies that

$$\mathbf{L}^2L_\pm Y_{lm} = L_\pm\mathbf{L}^2Y_{lm} = l(l + 1)\,\hbar^2L_\pm Y_{lm} \tag{11-33}$$

that is, $L_\pm Y_{lm}$ are also eigenfunctions of $\mathbf{L}^2$ with the eigenvalue characterized by l. On the other hand,

$$
\begin{aligned}
L_z L_+ Y_{lm} &= (L_+ L_z + \hbar L_+)\, Y_{lm} \\
&= m\hbar L_+ Y_{lm} + \hbar L_+ Y_{lm} \\
&= \hbar(m + 1)\, L_+ Y_{lm}
\end{aligned}
\tag{11-34}
$$

so that $L_+ Y_{lm}$ is also an eigenfunction of L_z, but with m-value increased by unity. Similarly we can show that

$$
L_z L_- Y_{lm} = \hbar(m - 1)\, L_- Y_{lm}
\tag{11-35}
$$

so that $L_- Y_{lm}$ is an eigenfunction of L_z with m-value lowered by unity. Thus we call $L_\pm$ raising and lowering operators, respectively. We may write

$$
L_\pm Y_{lm} = C_\pm(l, m)\, Y_{l,m\pm 1}
\tag{11-36}
$$

If follows from the hermiticity of L_x and L_y that

$$
L_\pm^\dagger = (L_x \pm iL_y)^\dagger = L_x \mp iL_y = L_\mp
\tag{11-37}
$$

Hence, a consequence of

$$
\langle L_\pm Y_{lm} | L_\pm Y_{lm} \rangle \geq 0
\tag{11-38}
$$

is that

$$
\langle Y_{lm} | L_\mp L_\pm Y_{lm} \rangle \geq 0
\tag{11-39}
$$

and therefore (11-27) and (11-28) imply that

$$
\langle Y_{lm} | (\mathbf{L}^2 - L_z^2 \pm \hbar L_z)\, Y_{lm} \rangle \geq 0
\tag{11-40}
$$

that is,

$$
\begin{aligned}
l(l + 1) &\geq m^2 + m \\
l(l + 1) &\geq m^2 - m
\end{aligned}
\tag{11-41}
$$

Since $l(l + 1) \geq 0$, we can take $l \geq 0$ without loss of generality.[2] Then (11-41) shows that

$$
-l \leq m \leq l
\tag{11-42}
$$

If there is a minimum value of $m(= m_-)$, then for the corresponding eigenstate

$$
L_- Y_{lm-} = 0
\tag{11-43}
$$

---

[2] If we were to find that $l \leq -1$, we would merely define $L = -l - 1$, and replace the old l, with the new, positive L. Nothing would change, since $L(L + 1) = l(l + 1)$.

We may then calculate m_- by using (11-27) and applying it to Y_{lm-}: we get

$$l(l + 1)\, \hbar^2 = m_-^2\hbar^2 - m_-\hbar^2 \tag{11-44}$$

Similarly, if there is a maximum value of $m(=m_+)$, then

$$L_+Y_{lm_+} = 0 \tag{11-45}$$

and an application of (11-28) to the maximum eigenstate gives

$$l(l + 1)\, \hbar^2 = m_+^2\hbar^2 + m_+\hbar^2 \tag{11-46}$$

Hence

$$\begin{aligned} m_- &= -l \\ m_+ &= +l \end{aligned} \tag{11-47}$$

Since the maximum value is to be reached from the minimum value by unit steps (repeated application of L_+), we find (Fig. 11.1): (a) that there are $(2l + 1)$ states, that is, $2l + 1$ is an integer, and (b) that m can take on the values

$$m = -l, -l + 1, -l + 2, \dots, l - 1, l$$

The possibility that l is half-odd integral, that is, $l = 1/2, 3/2, \dots$ will be discussed in Chapter 14 where we discuss *spin*. In this chapter we restrict ourselves to *integral values of l*.

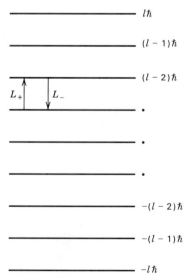

Figure 11-1. Spectrum of the operator L_z for a given value of l.

We may also calculate the coefficients $C_\pm(l, m)$ defined in (11-36). We have

$$
\begin{aligned}
|C_\pm(l, m)|^2 \langle Y_{l,m\pm1}|Y_{l,m\pm1}\rangle &= \langle L_\pm Y_{lm}|L_\pm Y_{lm}\rangle \\
&= \langle Y_{lm}|L_\mp L_\pm Y_{lm}\rangle \\
&= \langle Y_{lm}|(\mathbf{L}^2 - L_z^2 \mp \hbar L_z) Y_{lm}\rangle \\
&= \hbar^2[l(l + 1) - m(m \pm 1)]
\end{aligned}
$$

so that, with a convenient choice of phase, we get

$$
\begin{aligned}
C_+(l, m) &= \hbar \sqrt{l(l + 1) - m(m + 1)} = \hbar \sqrt{(l - m)(l + m + 1)} \\
C_-(l, m) &= \hbar \sqrt{l(l + 1) - m(m - 1)} = \hbar \sqrt{(l + m)(l - m + 1)}
\end{aligned}
\tag{11-48}
$$

THE SPHERICAL HARMONICS

This is as far as operator methods can take us. We shall now use the explicit form of the operators L_z and $L_\pm$ to obtain convenient expressions for the eigenfunctions in terms of the spherical angles θ and ϕ. This development will parallel that of (7-50) to (7-53). We write, as already suggested

$$
Y_{lm}(\theta, \phi) = \Theta_{lm}(\theta) e^{im\phi}
\tag{11-49}
$$

The condition (11-45) reads

$$
\hbar e^{i\phi}\left(\frac{\partial}{\partial\theta} + i \cot\theta \frac{\partial}{\partial\phi}\right) \Theta_{ll}(\theta) e^{il\phi} = \hbar e^{i(l+1)\phi}\left(\frac{\partial}{\partial\theta} - l \cot\theta\right) \Theta_{ll}(\theta) = 0
\tag{11-50}
$$

The solution to this equation is easily found to be

$$
\Theta_{ll}(\theta) = (\sin\theta)^l
\tag{11-51}
$$

The appropriate multiplicative constant will be obtained later from the normalization condition. An arbitrary state is obtained by the lowering procedure

$$
Y_{lm}(\theta, \phi) = C(L_-)^{l-m} (\sin\theta)^l e^{il\phi}
\tag{11-52}
$$

Consider first

$$
\begin{aligned}
L_- Y_{ll}(\theta, \phi) &= \hbar e^{-i\phi}\left(-\frac{\partial}{\partial\theta} + i \cot\theta \frac{\partial}{\partial\phi}\right) (\sin\theta)^l e^{il\phi} \\
&= \hbar e^{i(l-1)\phi}\left(-\frac{\partial}{\partial\theta} - l \cot\theta\right) (\sin\theta)^l
\end{aligned}
$$

Since one can show that for an arbitrary function $f(\theta)$

$$
\left(\frac{d}{d\theta} + l \cot\theta\right) f(\theta) = \frac{1}{(\sin\theta)^l} \frac{d}{d\theta}\left[(\sin\theta)^l f(\theta)\right]
\tag{11-53}
$$

we have obtained

$$Y_{l,l-l} = C' \frac{e^{i(l-1)\phi}}{(\sin\theta)^l} \left(-\frac{d}{d\theta}\right) [(\sin\theta)^l (\sin\theta)^l] \tag{11-54}$$

The next step is the same, except that l is replaced by $l-1$ and the operation in (11-53) acts on the form obtained in (11-54). Thus

$$\begin{aligned} Y_{l,l-2} &= C'' \frac{e^{i(l-2)\phi}}{(\sin\theta)^{l-1}} \left(-\frac{d}{d\theta}\right) \left[(\sin\theta)^{l-1} \frac{1}{(\sin\theta)^l} \left(-\frac{d}{d\theta}\right)(\sin\theta)^{2l}\right] \\ &= C''(-1)^2 \frac{e^{i(l-2)\phi}}{(\sin\theta)^{l-1}} \frac{d}{d\theta}\left[\frac{1}{\sin\theta}\frac{d}{d\theta}(\sin\theta)^{2l}\right] \end{aligned} \tag{11-55}$$

In terms of the variable $u = \cos\theta$, $-1/(\sin\theta)(d/d\theta) = d/du$, and (11-54), (11-55), respectively, read

$$Y_{l,l-l} = C' \frac{e^{i(l-1)\phi}}{(\sin\theta)^{l-1}} \frac{d}{du}[(1-u^2)^l]$$

$$Y_{l,l-2} = C'' \frac{e^{i(l-2)\phi}}{(\sin\theta)^{l-2}} \frac{d^2}{du^2}[(1-u^2)^l] \tag{11-56}$$

The general form is

$$Y_{lm} = C \frac{e^{im\phi}}{(\sin\theta)^m} \left(\frac{d}{du}\right)^{l-m} [(1-u^2)^l] \tag{11-57}$$

The eigenfunctions are to be normalized. Since we are dealing with spherical angles whose range of integration is $0 \le \phi \le 2\pi$, $0 \le \theta \le \pi$ (see Fig. 10.1) and where the integral over the surface of the sphere ($r = $ constant) is

$$\int d\Omega = \int_0^{2\pi} d\phi \int_0^\pi \sin\theta\, d\theta \tag{11-58}$$

we must impose

$$\langle Y_{lm}|Y_{lm}\rangle = 1 = \int_0^{2\pi} d\phi \int_{-1}^1 du |C|^2 \left[\frac{1}{(1-u^2)^{m/2}} \left(\frac{d}{du}\right)^{l-m} (1-u^2)^l\right]^2$$

The integration is tedious. We content ourselves with writing down the appropriately normalized $Y_{lm}(\theta, \phi)$ with the phases that are conventionally used:

$$Y_{lm}(\theta, \phi) = (-1)^m \left[\frac{2l+l}{4\pi}\frac{(l-m)!}{(l+m)!}\right]^{1/2} P_l^m(\cos\theta)\, e^{im\phi} \tag{11-59}$$

for $m \ge 0$, with

$$Y_{l,-m} = (-1)^m Y_{lm}^* \tag{11-60}$$

The associated Legendre polynomials are given by ($m \geq 0$)

$$P_l^m(u) = (-1)^{l+m} \frac{(l+m)!}{(l-m)!} \frac{(1-u^2)^{-m/2}}{2^l l!} \left(\frac{d}{du}\right)^{l-m} (1-u^2)^l \qquad (11\text{-}61)$$

with the value for negative m obtained from

$$P_l^{-m}(u) = (-1)^m \frac{(l-m)!}{(1+m)!} P_l^m(u) \qquad (11\text{-}62)$$

when $m = l$ (or $-l$) we get

$$Y_{ll}(\theta, \phi) = K (\sin \theta)^l e^{il\phi}$$

where K is a constant, so that the probability distribution in polar angle with respect to the z-axis is given by

$$|Y_{ll}|^2 = K^2 (\sin \theta)^{2l} \qquad (11\text{-}63)$$

We observe that for large l this function is strongly confined to the equatorial plane. This is as expected: for $m = l$, L_z has its largest possible value, so that $\mathbf{L}^2 \approx L_z^2$. In the classical limit, in which $l \ggg 1$,

$$\frac{\mathbf{L}^2 - L_z^2}{\mathbf{L}^2} = \frac{1}{l} \to 0 \qquad (11\text{-}64)$$

that is, it is possible to line up the angular momentum in a particular direction (here the z-direction). Such an alignment corresponds to having $\langle L_x^2 \rangle = \langle L_y^2 \rangle = 0$, which is not possible when quantum mechanical effects become important, because of the commutation relations.

We list a few of the eigenfunctions:

$$Y_{0,0} = \frac{1}{\sqrt{4\pi}}$$

$$Y_{1,1} = -\sqrt{\frac{3}{8\pi}} e^{i\phi} \sin \theta$$

$$Y_{1,0} = \sqrt{\frac{3}{4\pi}} \cos \theta$$

$$Y_{2,2} = \sqrt{\frac{15}{32\pi}} e^{2i\phi} \sin^2 \theta \qquad (11\text{-}65)$$

$$Y_{2,1} = -\sqrt{\frac{15}{8\pi}} e^{i\phi} \sin \theta \cos \theta$$

$$Y_{2,0} = \sqrt{\frac{5}{16\pi}} (3 \cos^2 \theta - 1)$$

THE EXPANSION THEOREM

The $Y_{lm}(\theta, \phi)$ form a complete set of orthonormal functions of (θ, ϕ). Thus the usual expansion theorem here implies that any function of θ and ϕ may be expanded in the form

$$f(\theta, \phi) = \sum \sum C_{lm} Y_{lm}(\theta, \phi) \tag{11-66}$$

where

$$C_{lm} = \int d\Omega \; Y^*_{lm}(\theta, \phi) f(\theta, \phi) \tag{11-67}$$

and where the solid angle integration is defined by (11-58). The expansion theorem further states that if $f(\theta, \phi)$ is the angular wave function of some state, normalized such that

$$\int d\Omega \; |f(\theta, \phi)|^2 = 1 \tag{11-68}$$

then $|C_{lm}|^2$ is the probability that simultaneous measurement of $\mathbf{L}^2$ and L_z on the state described by $f(\theta, \phi)$ yields $l(l + 1)\hbar^2$ and $m\hbar$, respectively.

The probability that a measurement of $\mathbf{L}^2$ yields $l(l + 1)\hbar^2$ is

$$P(l) = \sum_{m=-l}^{l} |C_{lm}|^2 \tag{11-69}$$

and the expectation value of L_z is easily seen to be

$$\langle L_z \rangle = \sum_l \sum_{m=-l}^{l} m\hbar |C_{lm}|^2 \tag{11-70}$$

Consider the expansion theorem as it appears in (11-66), written in abstract notation

$$|\psi\rangle = \sum_{l,m} C_{lm}|Y_{lm}\rangle \tag{11-71}$$

With the help of the orthonormality condition

$$\langle Y_{l'm'}|Y_{lm}\rangle = \delta_{ll'}\delta_{mm'} \tag{11-72}$$

we get

$$C_{lm} = \langle Y_{lm}|\psi\rangle \tag{11-73}$$

and when this is inserted into the first equation we get

$$|\psi\rangle = \sum_l \sum_{m=-l}^{l} |Y_{lm}\rangle\langle Y_{lm}|\psi\rangle$$

From this it follows that formally

$$\sum_l \sum_{m=-l}^{l} |Y_{lm}\rangle\langle Y_{lm}| = 1 \qquad (11\text{-}74)$$

where 1 is the unit operator.

We can use the expansion theorem to answer the question often raised by students: What is so special about the z-direction? Cannot we align the angular momentum (as far as possible) with the x-axis? The answer is that indeed this is possible. Such a state, which would be confined near the equatorial plane about the x-axis (in the vicinity of $\phi = \pi/2$) will be a particular linear combination of the $Y_{lm}(\theta, \phi)$. Its physical properties will be exactly the same as the Y_{ll} state.

The Plane Wave in Terms of Spherical Harmonics

The solution of the free particle equation

$$\nabla^2 \psi(\mathbf{r}) + k^2 \psi(\mathbf{r}) = 0$$

can be written in two ways. One is simply the plane wave solution

$$\psi(\mathbf{r}) = e^{i\mathbf{k}\cdot\mathbf{r}} \qquad (11\text{-}75)$$

The other way is to write it as a linear superposition of the partial wave solutions, that is

$$\sum\sum A_{lm} \, j_l(kr) \, Y_{lm}(\theta, \phi) \qquad (11\text{-}76)$$

We may therefore find A_{lm} such that $\psi(\mathbf{r}) = e^{i\mathbf{k}\cdot\mathbf{r}}$ in (11-76). Note that the spherical angles (θ, ϕ) are the coordinates of the vector $\mathbf{r}$ relative to some arbitrarily chosen z-axis. If we define the z-axis by the direction of $\mathbf{k}$ (until now an arbitrary direction), then

$$e^{i\mathbf{k}\cdot\mathbf{r}} = e^{ikr\cos\theta} \qquad (11\text{-}77)$$

Thus the left side of (11-76) has no azimuthal angle, ϕ, dependence, and thus on the right side only terms with $m = 0$ can appear; hence, making use of the fact that

$$Y_{l0}(\theta, \phi) = \left(\frac{2l+1}{4\pi}\right)^{1/2} P_l(\cos\theta) \qquad (11\text{-}78)$$

where the $P_l(\cos\theta)$ are the Legendre polynomials, we get the relation

$$e^{ikr\cos\theta} = \sum_{l=0}^{\infty} \left(\frac{2l+1}{4\pi}\right)^{1/2} A_l j_l(kr) P_l(\cos\theta) \qquad (11\text{-}79)$$

We may use the relation

$$\frac{1}{2} \int_{-1}^{1} d(\cos\theta) \, P_l(\cos\theta) \, P_{l'}(\cos\theta) = \frac{\delta_{ll'}}{2l+1} \qquad (11\text{-}80)$$

which is a direct consequence of the orthonormality relation for the Y_{lm} and (11-78), to obtain

$$A_l j_l(kr) = \frac{1}{2} [4\pi(2l + 1)]^{1/2} \int_{-1}^{1} dz P_l(z) \, e^{ikrz} \tag{11-81}$$

Compare the two sides of the equation as $kr \to 0$. The first term on the left-hand side is

$$A_l \frac{(kr)^l}{1, 3, 5, \ldots, (2l + 1)}$$

and the corresponding power of $(kr)^l$ on the right-hand side has

$$\frac{1}{2} [4\pi(2l + 1)]^{1/2} (ikr)^l \int_{-1}^{1} dz \, P_l(z) z^l / l!$$

The integral can be evaluated by noting that $P_l(z)$ is an lth degree polynomial in z. The coefficient of the leading power, z^l, can be easily obtained from (11-61), as the power of z^l in

$$(-1)^l \frac{1}{2^l l!} \left(\frac{d}{dz}\right)^l (1 - z^2)^l = \frac{2l(2l - 1)(2l - 1) \cdots (l + 1)}{2^l l!} z^l + 0(z^{l-1})$$

We can rewrite this in the form

$$z^l = \frac{2^l l!}{2l(2l - 1)(2l - 2) \cdots (l + 1)} P_l(z) + \text{terms involving } P_{l+1}(z) \text{ and higher.}$$

With the help of (11-80) we finally get

$$A_l \frac{(kr)^l}{1, 3, 5, \ldots, (2l + 1)} =$$
$$\frac{1}{2} [4\pi(2l + 1)]^{1/2} (ikr)^l \frac{1}{l!} \frac{2^l l!}{2l(2l - 1)(2l - 2) \cdots (l + 1)} \frac{2}{2l + 1}$$

What results is the expansion

$$e^{ikr \cos \theta} = \sum_{l=0}^{\infty} (2l + 1) \, i^l j_l(kr) \, P_l(\cos \theta) \tag{11-82}$$

which we will find exceedingly useful in discussions of collision theory.

Problems

1. A molecule consists of two identical atoms, each of which, in its ground state, has spin 0. The molecule has, among its possible excitations, rotational excitations. If only rotations about the z-axis are considered, so that $H = L_z^2/2I$

and the separation between the atoms is considered fixed, what is the rotational spectrum? If the atoms have spin 1/2 and they are both in the same spin state, what is the spectrum?

2. Express the spherical harmonics listed in (11-65) in terms of $x = r \sin\theta \cos\phi$, $y = r \sin\theta$, and $z = r \cos\theta$.

3. Calculate $\langle Y_{lm_1}|L_x|Y_{lm_2}\rangle$ and $\langle Y_{lm_1}|L_y|Y_{lm_2}\rangle$.

4. Calculate $\langle Y_{lm_1}|L_x^2|Y_{lm_2}\rangle$ and $\langle Y_{lm_1}|L_y^2|Y_{lm_2}\rangle$.
   [*Hint:* Use (11-36) and (11-48) to calculate $\langle Y_{lm_1}|L_+^2|Y_{lm_2}\rangle$ and other required quantities.]

5. The Hamiltonian for an axially symmetric rotator is given by

$$H = \frac{L_x^2 + L_y^2}{2I_1} + \frac{L_z^2}{2I_2}$$

What are the eigenvalues of H? Sketch the spectrum, assuming that $I_1 > I_2$.

6. Prove that $\langle L_x^2\rangle = \langle L_y^2\rangle = 0$ is only possible for a state of total angular momentum $l = 0$.
   (*Hint:* Use the completeness relation

$$\sum\sum |Y_{lm}\rangle\langle Y_{lm}| = 1)$$

7. If the quantization axis lies in the x-direction, that is, L_x is the chosen operator, then we may define the point r by angles Θ and Φ, where Θ is the angle that the radius vector $\mathbf{r}$ makes with the x-axis, and Φ is chosen for convenience as the angle that the projection of $\mathbf{r}$ makes onto the (y, z) plane (perpendicular to the x-axis), relative to the y-axis. The spherical harmonics will have the form $Y_{LM}(\Theta, \Phi)$, and these can be expanded in terms of the $Y_{lm}(\theta, \phi)$,

$$Y_{LM}(\Theta, \Phi) = \sum_l \sum_m C_{lm}(L, M)Y_{lm}(\theta, \phi)$$

   (a) Find Θ, Φ in terms of θ, ϕ.
   (b) Consider the wave function for $M = L$, and find out as much as you can about $C_{lm}(L, L)$.

8. Obtain the explicit form of the properly normalized $l = 3$ spherical harmonics, $Y_{33}, Y_{32}, Y_{31}, Y_{30}$.

9. Use the procedure outlined in this chapter to discuss rotations in four dimensions. The generalization of L is now the set of operators that may be written as

$$L_{ij} = -i(x_i\partial_j - x_j\partial_i)$$

where ∂_i stands for $\partial/\partial x_i$, and $(i, j = 1, 2, 3, 4)$. Introduce

$$(J_1, J_2, J_3) = (L_{23}, L_{31}, L_{12})$$

and

$$(K_1, K_2, K_3) = (L_{14}, L_{24}, L_{34})$$

   (a) Find the commutation relations of all six operators among themselves.
   (b) Show that the operators

$$\mathbf{J}^{(+)} = \tfrac{1}{2}(\mathbf{J} + \mathbf{K}); \ \mathbf{J}^{(-)} = \tfrac{1}{2}(\mathbf{J} - \mathbf{K})$$

each obey angular momentum commutation relations and that they commute with each other. Use the final result to determine the maximal set of mutually commuting observables, and thus the quantum numbers that would be used to label an eigenfunction.

10. A particle in a spherically symmetric potential is in a state described by the wave packet

$$\psi(x, y, z) = C(xy + yz + zx) \, e^{-\alpha r^2}$$

What is the probability that a measurement of the square of the angular momentum yields 0? What is the probability that it yields $6\hbar^2$? If the value of l is found to be 2, what are the relative probabilities for $m = 2, 1, 0, -1, -2$?

11. Consider the following model of a perfectly smooth cylinder. It is a ring of equally spaced, identical particles, with mass M/N so that the mass of the ring is M and its moment of inertia is MR^2, with R the radius of the ring. Calculate the possible values of the angular momentum. Calculate the energy eigenvalues. What is the energy difference between the ground state of zero angular momentum, and the first rotational state? Show that this approaches infinity as $N \to \infty$. Contrast this with the comparable energy for a "nicked" cylinder, which lacks the symmetry under the rotation through $2\pi/N$ radians. This example implies that it is impossible to set a perfectly smooth cylinder in rotation, which is consistent with the fact that for a perfectly smooth cylinder such a rotation would be unobservable.

12. Express $\mathbf{L}^2$ in terms of $\partial/\partial\theta$ and $\partial/\partial\phi$. Write down the differential equation obeyed by Θ_{lm} defined in (11-49).

References

This is standard material found in any of the books listed in the bibliography. For a deeper look into the consequences of invariance under rotation, see especially

K. Gottfried, *Quantum Mechanics*, Vol. 1, W. A. Benjamin, New York, 1966.
M. E. Rose, *Elementary Theory of Angular Momentum*, John Wiley & Sons, New York, 1957.

Chapter 12

THE HYDROGEN ATOM

The hydrogen atom is the simplest atom, since it contains only one electron. Thus the Schrödinger equation becomes a one-particle equation after the center-of-mass motion is separated out. We shall deal with hydrogenlike atoms, that is, atoms containing one electron only, but allowing for a nucleus more complicated than a single proton. The potential then is

$$V(r) = -\frac{Ze^2}{r} \tag{12-1}$$

and the radial Schrödinger equation is

$$\left(\frac{d^2}{dr^2} + \frac{2}{r}\frac{d}{dr}\right) R + \frac{2\mu}{\hbar^2}\left[E + \frac{Ze^2}{r} - \frac{l(l+1)\,\hbar^2}{2\mu r^2}\right] R = 0 \tag{12-2}$$

We will concentrate on the bound states, that is, $E < 0$ solutions. It is convenient to make a change of variables,

$$\rho = \left(\frac{8\mu|E|}{\hbar^2}\right)^{1/2} r \tag{12-3}$$

The equation then reads

$$\frac{d^2R}{d\rho^2} + \frac{2}{\rho}\frac{dR}{d\rho} - \frac{l(l+1)}{\rho^2}R + \left(\frac{\lambda}{\rho} - \frac{1}{4}\right) R = 0 \tag{12-4}$$

where we have introduced the dimensionless parameter

$$\lambda = \frac{Ze^2}{\hbar}\left(\frac{\mu}{2|E|}\right)^{1/2} = Z\alpha\left(\frac{\mu c^2}{2|E|}\right)^{1/2} \tag{12-5}$$

The second form makes it easier to compute with it, since $\alpha = 1/137$ and the energy is expressed in units of the rest mass; the first form does, however, make clear that the velocity of light c does not really appear in the equation, that is, that it is strictly a nonrelativistic equation.

THE ENERGY SPECTRUM

We try to solve (12-4) in what is by now a familiar way. First, we extract the large ρ behavior. For large ρ, the only terms that remain in the equation are

$$\frac{d^2R}{d\rho^2} - \frac{1}{4}R \simeq 0 \tag{12-6}$$

and the solution, which behaves properly at infinity, is $R \sim e^{-\rho/2}$. As in our treatment of the harmonic oscillator, we write

$$R(\rho) = e^{-\rho/2}\,G(\rho) \tag{12-7}$$

substitute this into (12-4), and obtain the equation for $G(\rho)$. A little algebra, which we do not reproduce, leads to the equation

$$\frac{d^2G}{d\rho^2} - \left(1 - \frac{2}{\rho}\right)\frac{dG}{d\rho} + \left[\frac{\lambda - 1}{\rho} - \frac{l(l + 1)}{\rho^2}\right]G = 0 \tag{12-8}$$

We now write a power expansion for $G(\rho)$. This takes the form

$$G(\rho) = \rho^l \sum_{n=0}^{\infty} a_n\rho^n \tag{12-9}$$

The fact that $R(\rho)$, and hence $G(\rho)$, behaves like ρ^l at the origin was established in Chapter 10 for all potentials satisfying (10-41). When (12-9) is substituted into the differential equation, we find a relation between various coefficients a_n. The recursion relation is obtained from the differential equation obeyed by

$$H(\rho) = \sum_{n=0}^{\infty} a_n\rho^n \tag{12-10}$$

which is

$$\frac{d^2H}{d\rho^2} + \left(\frac{2l + 2}{\rho} - 1\right)\frac{dH}{d\rho} + \frac{\lambda - 1 - l}{\rho}H = 0 \tag{12-11}$$

as can easily be obtained by substituting $G(\rho) = \rho^l H(\rho)$ into (12-8). We then have

$$\sum_{n=0}^{\infty} \left[n(n-1) a_n \rho^{n-2} + n a_n \rho^{n-1} \left(\frac{2l+2}{\rho} - 1 \right) + (\lambda - 1 - l) a_n \rho^{n-1} \right] = 0$$

(12-12)

that is,

$$\sum_{n=0}^{\infty} \{(n+1)[n a_{n+1} + (2l+2) a_{n+1}] + (\lambda - 1 - l - n) a_n\} \rho^{n-1} = 0$$

Since this must vanish term by term, we get the recursion relation

$$\frac{a_{n+1}}{a_n} = \frac{n+l+1-\lambda}{(n+1)(n+2l+2)}$$

(12-13)

For large n this ratio is

$$\frac{a_{n+1}}{a_n} \simeq \frac{1}{n}$$

(12-14)

and, as for the harmonic oscillator problem, we can show that we do not get a solution $R(\rho)$ that is well behaved at infinity, unless the series in (12-9) terminates. This means that for a given l, for some $n = n_r$ we must have

$$\lambda = n_r + l + 1$$

(12-15)

Let us introduce the *principal quantum number n* defined by

$$n = n_r + l + 1$$

(12-16)

Then, it follows from the fact that $n_r \geq 0$, that

1. $n \geq l + 1$
2. n is an integer
3. the relation

$$\lambda = n$$

implies that

$$E = -\frac{1}{2} \mu c^2 \frac{(Z\alpha)^2}{n^2}$$

(12-17)

a result familiar from the old Bohr model. Notice that it is the *reduced mass* that appears in the expression; this, of course, is not peculiar to the differential equation approach. In the old Bohr theory, too, a proper treatment of the classical orbits, subsequently to be restricted by the quantization of angular momentum condition, would have introduced the reduced mass in the energy formula. The presence of the reduced mass

$$\mu = \frac{mM}{m + M} \tag{12-18}$$

where m is the electron mass, and M the mass of the nucleus, means that the frequencies

$$\omega_{ij} = \frac{E_i - E_j}{\hbar} = \frac{mc^2/2\hbar}{1 + m/M} (Z\alpha)^2 \left(\frac{1}{n_i^2} - \frac{1}{n_j^2} \right) \tag{12-19}$$

differ slightly for different hydrogenlike atoms. In particular, the difference between the hydrogen spectrum and the deuterium spectrum—where M, the nuclear mass, is very close to being twice the proton mass—was responsible for the discovery of deuterium by Urey and collaborators in 1932.

The Degeneracy of the Spectrum

Let us now discuss the degeneracy of the energy spectrum. For $\lambda = 1$, the ground state, we must have $n_r = 0$, $l = 0$. There is a unique ground state. For $\lambda = 2$, we have two possibilities: (i) $n_r = 1$, $l = 0$: here (12-13), written in the form

$$\frac{a_{n+1}}{a_n} = \frac{n - n_r}{(n + 1)(n + 2l + 2)} \tag{12-20}$$

shows that $a_1/a_0 = -1/(1 \times 2)$, so that

$$H(\rho) = a_0 (1 - \rho/2) \tag{12-21}$$

while the angular distribution is spherically symmetric. The other possibility is (ii) $n_r = 0$, $l = 1$. Here the radial wave function is constant, $H(\rho) = a_0$, but the angular part of the wave function involves $Y_{1m}(\theta, \phi)$. The degeneracy is $(2l + 1)$, so that there are three such states. The total degeneracy for $\lambda = n = 2$ is $3 + 1 = 4 = 2^2$.

For $\lambda = 3$, we have the following possibilities: $n_r = 2$, $l = 0$. Here there is one state with $a_2/a_1 = -1/6$, and $a_1/a_0 = -1$, so that the radial function involves

$$H(\rho) = a_0 \left(1 - \rho + \frac{1}{6} \rho^2 \right) \tag{12-22}$$

$n_r = 1$, $l = 1$ has three states, with $H(\rho) = a_0(1 - \rho/4)$, while $n_r = 0$, $l = 2$ has $(2l + 1) = 5$ states with $H(\rho) = a_0$. Thus there are $1 + 3 + 5 = 9 = 3^2$ degenerate states with eigenvalue $\lambda = n = 3$. Quite generally, the degeneracy for $\lambda = n$ is

$$1 + 3 + 5 + \cdots + [2(n - 1) + 1] = n^2 \tag{12-23}$$

A priori we expect a $(2l + 1)$ degeneracy for a radial potential, because the radial Hamiltonian does not depend on L_z, but only on $\mathbf{L}^2$. Here there is an additional degeneracy. This special degeneracy is characteristic of a pure $1/r$ potential. Were the Coulomb potential to be modified by an additional term close in, so that

$$V(r) = -\frac{Ze^2}{r} + \frac{\hbar^2}{2\mu} \frac{g^2}{r^2} \tag{12-24}$$

then the radial equation would be unchanged, except where we had $l(l + 1)/r^2$ we would now have $l^*(l^* + 1)/r^2$, where $l^*(l^* + 1) = l(l + 1) + g^2$, that is $l^* = -1/2 + \sqrt{(l + 1/2)^2 + g^2}$. This would give rise to the energy

$$E = -\frac{1}{2} \mu c^2 \frac{(Z\alpha)^2}{[n_r + 1/2 + \sqrt{(l + 1/2)^2 + g^2}]^2} \tag{12-25}$$

in which, for example $(n_r = 1, l = 2)$ and $(n_r = 2, l = 1)$ no longer are degenerate. The special degeneracy characteristic of the pure $1/r$ potential was formerly called "accidental," since there was no obvious reason for it. This, however, depends on what one means by "obvious." It is already known in classical mechanics that the potential $1/r$ has some special features: the orbits consist of ellipses that maintain their orientation in space, instead of forming precessing orbits (Fig. 12-1). Small modifications of the potential do cause a precession. Such modifications may come from a variety of sources, for example, the perturbations due to other planets, in the Kepler problem. In considering the planetary orbit of Mercury, it was found that after allowance was made for the effects of other planets, a precession of the perihelion in the amount of 42″ per century remained unaccounted for, and this was finally explained by Einstein's general theory of relativity, which predicted just the right amount of $1/r^2$ potential to be added to the Newtonian $1/r$.

In the real hydrogen atom there are small perturbations coming from spin effects and relativistic effects. These will be discussed in Chapter 17. To a very good approximation, however, we have, for a given n, the possible values of $l = 0, 1, 2, \ldots, (n - 1)$, and for each there is the $(2l + 1)$ degeneracy. Thus the total degeneracy is still n^2. Since there are two possible states for the electron because of its spin, the true degeneracy is really $2n^2$. This plays a role in the quantum mechanical explanation of the periodic table.

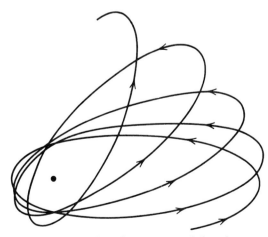

Figure 12-1. Orbits for a potential that does not have the exact $1/r$ form do not close upon themselves and precess as shown here. The orbits remain planar as long as the potential is radial.

THE RADIAL EIGENFUNCTIONS

Let us now return to the differential equation. If we set $\lambda = n$ in the recursion relation (12-13) so that

$$a_{k+1} = \frac{k + l + 1 - n}{(k + 1)(k + 2l + 2)} a_k \tag{12-26}$$

we find that

$$a_{k+1} = (-1)^{k+1} \frac{n - (k + l + 1)}{(k + 1)(k + 2l + 2)} \cdot \frac{n - (k + l)}{k(k + 2l + 1)}$$

$$\cdots \frac{n - (l + 1)}{1 \cdot (2l + 2)} a_0 \tag{12-27}$$

With the help of this we can obtain the power series expansion for $H(\rho)$. Equivalently, we observe that the equation for $H(\rho)$ is that for the *associated Laguerre polynomials*:

$$H(\rho) = L_{n-l-1}^{(2l+1)} (\rho) \tag{12-28}$$

The polynomials are tabulated and their various properties can be found in the mathematical literature.[1]

After conversion back to the radial coordinate r and after normalization, the first few radial functions can be computed. These are listed below. We use

$$a_0 = \frac{\hbar}{\mu c \alpha} \tag{12-29}$$

in the tabulation $R_{nl}(r)$:

$$R_{10}(r) = 2 \left(\frac{Z}{a_0} \right)^{3/2} e^{-Zr/a_0}$$

$$R_{20}(r) = 2 \left(\frac{Z}{2a_0} \right)^{3/2} \left(1 - \frac{Zr}{2a_0} \right) e^{-Zr/2a_0}$$

$$R_{21}(r) = \frac{1}{\sqrt{3}} \left(\frac{Z}{2a_0} \right)^{3/2} \frac{Zr}{a_0} e^{-Zr/2a_0}$$

$$R_{30}(r) = 2 \left(\frac{Z}{3a_0} \right)^{3/2} \left[1 - \frac{2Zr}{3a_0} + \frac{2(Zr)^2}{27a_0^2} \right] e^{-Zr/3a_0} \tag{12-30}$$

$$R_{31}(r) = \frac{4\sqrt{2}}{3} \left(\frac{Z}{3a_0} \right)^{3/2} \frac{Zr}{a_0} \left(1 - \frac{Zr}{6a_0} \right) e^{-Zr/3a_0}$$

$$R_{32}(r) = \frac{2\sqrt{2}}{27\sqrt{5}} \left(\frac{Z}{3a_0} \right)^{3/2} \left(\frac{Zr}{a_0} \right)^2 e^{-Zr/3a_0}$$

[1]An extremely useful book is M. Abramowitz and I. A. Stegun (eds.), *Handbook of Mathematical Functions*, National Bureau of Standards Publication, Washington, D.C., 1964.

The following qualitative features emerge from the sampling of eigensolutions:

(a) The behavior of r^l for small r, which forces the wave function to stay small for a range of radii that increases with l, is a consequence of the centrifugal repulsive barrier that keeps the electrons from coming close to the nucleus.

(b) The relation (12-26) shows that $H(\rho)$ is a polynomial of degree $n_r = n - l - 1$, and thus it has n_r radial nodes (zeros). There will be $n - l$ "bumps" in the probability density distribution

$$P(r) = r^2[R_{nl}]^2 \qquad (12\text{-}31)$$

When, for a given n, l has its largest value $l = n - 1$, then there is only one bump. As (12-30) suggests, and as can be seen from the solution to the differential equation,

$$R_{n,n-1}(r) \propto r^{n-1} e^{-Zr/a_0 n} \qquad (12\text{-}32)$$

Hence $P(r) \propto r^{2n} e^{-2Zr/a_0 n}$ will peak at a value of r determined by

$$\frac{dP(r)}{dr} = \left(2nr^{2n-1} - \frac{2Z}{a_0 n} r^{2n}\right) e^{-2Zr/a_0 n} = 0 \qquad (12\text{-}33)$$

that is, at

$$r = \frac{n^2 a_0}{Z} \qquad (12\text{-}34)$$

which is the Bohr atom value for circular orbits. Smaller values of l give probability distributions with more bumps. One can show that they correspond to elliptical orbits in the large quantum number limit.

(c) Plots of the radial probability density $P(r)$ for finding the electron at a distance r from the origin can be constructed with the help of the wave functions. Figure 12-2 shows the general pattern. We must remember that the wave function also has an angular part, whose absolute square is $P_l^m (\cos \theta)^2$. Plots of the associated Legendre functions $P_l^m (\cos \theta)$ are given in Fig. 12-3. As m increases, the probability density is seen to shift from the z-axis toward the equatorial plane. When $|m| = l$, then $|P_l^l \cos \theta|^2 \propto \sin^{2l} \theta$ as can be read off from (11-63). This function is peaked about $\theta = \pi/2$. As l increases, the width of the peak can be shown to decrease like $l^{-1/2}$, and thus for large quantum numbers we get the classical picture of planar orbits. The finite width of the peak can be understood from the following considerations. When $|m| = l$, we have $L_z^2 = l^2$ and consequently $L_x^2 + L_y^2 = l$. Thus the angular momentum vector can never be perfectly oriented along an axis. Incidentally, the degeneracy in m allows us to orient the "orbit" relative to some other axis, so that there really is no distinguished z-axis. Thus a state that is an eigenstate of L_x with eigenvalue l will be "oriented" in the x-direction. The wave function will now be a linear combination of the $Y_{lm}(\theta, \phi)$, but because of the degeneracy, the energy will be the same as for the z-oriented orbits.

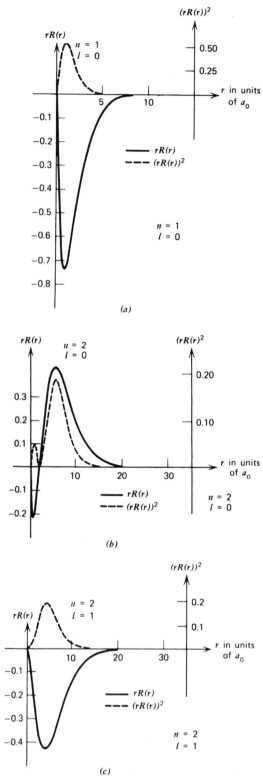

Figure 12-2. The radial wave functions $u(r) = rR(r)$ and the radial probability density function $u^2(r)$ for values of $n = 1, 2, 3$, and the values of l possible. The left abscissa measures $u(r)$ and the right abscissa measures $u^2(r)$. The wave functions are given by solid lines, and the probability distributions by the dashed lines. The ordinate measures r in units of a_0.

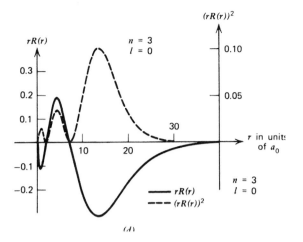

(A)

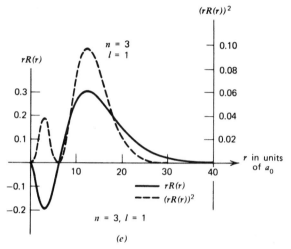

(e)

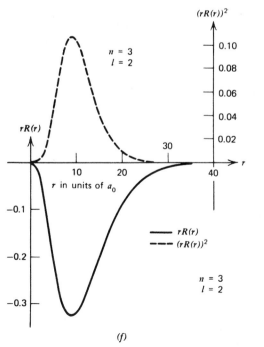

(f)

Figure 12-2. (continued)

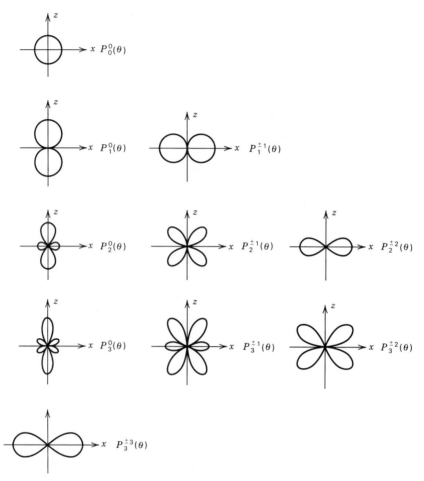

Figure 12-3. Shapes of the associated Legendre polynomials as a function of θ, the angle between the z-axis and the equatorial plane, denoted by the x-axis.

(d) Given the wave functions, we can calculate

$$\langle r^k \rangle = \int_0^\infty dr \; r^{2+k} \, [R_{nl}(r)]^2 \tag{12-35}$$

Some useful expectation values are given below:

$$\langle r \rangle = \frac{a_0}{2Z} \, [3n^2 - l(l + 1)]$$

$$\langle r^2 \rangle = \frac{a_0^2 n^2}{2Z^2} \, [5n^2 + 1 - 3l(l + 1)]$$

$$\left\langle \frac{1}{r} \right\rangle = \frac{Z}{a_0 n^2} \tag{12-36}$$

$$\left\langle \frac{1}{r^2} \right\rangle = \frac{Z^2}{a_0^2 n^3 (l + \frac{1}{2})}$$

$$\left\langle \frac{1}{r^3} \right\rangle = \frac{Z^3}{a_0^3 n^3 l(l + \frac{1}{2})(l + 1)}$$

In electron (or proton) scattering one is also interested in solutions of the Schrödinger equation with a $1/r$ potential for $E > 0$. These involve special functions, the confluent hypogeometric functions. A discussion of these is beyond the scope of this book.

Problems

1. Compare the wavelengths of the $2P \to 1S$ transitions in (1) hydrogen, (2) deuterium (nuclear mass $= 2 \times$ proton mass), (3) positronium (a bound state of an electron and a positron, whose mass is the same as that of an electron).

2. An electron is in the ground state of tritium, for which the nucleus consists of a proton and two neutrons. A nuclear reaction instantaneously changes the nucleus to He^3, that is, two protons and one neutron. Calculate the probability that the electron remains in the ground state of He^3.

3. The relativistic analog of the Schrödinger equation for a spin 0 electron (thus not applicable to the real electron) is the operator version of

$$(E - V)^2 = p^2 c^2 + m^2 c^4$$

that is,

$$\left(\frac{E}{\hbar c} + \frac{Ze^2}{\hbar c} \frac{1}{r} \right)^2 \psi = -\nabla^2 \psi + \left(\frac{mc}{\hbar} \right)^2 \psi$$

(a) Find the radial equation.
(b) Find the eigenvalue spectrum by noting the close relationship of the radial equation obtained in (a) with the radial equation for the hydrogen atom problem.

4. Using the expression for $\langle 1/r \rangle_{n,l}$ calculate the expression for

$$\langle T \rangle_{n,l} = \left\langle \frac{p^2}{2m} \right\rangle_{n,l}$$

for an arbitrary hydrogen atom eigenstate (with Z arbitrary). Show that generally for this potential

$$\langle T \rangle = -\tfrac{1}{2} \langle V \rangle$$

This is a special example of the *Virial theorem*.

5. An electron in the Coulomb field of a proton is in a state described by the wave function

$$\tfrac{1}{6} [4 \psi_{100}(\mathbf{r}) + 3 \psi_{211}(\mathbf{r}) - \psi_{210}(\mathbf{r}) + \sqrt{10}\, \psi_{21-1}(\mathbf{r})]$$

(a) What is the expectation value of the energy?
(b) What is the expectation value of $\mathbf{L}^2$?
(c) What is the expectation value of L_z?

6. An electron in the Coulomb field of a proton is in a state described by the wave function

$$\psi(\mathbf{r}) = \left(\frac{\alpha}{\sqrt{\pi}} \right)^{3/2} e^{-\alpha^2 r^2 / 2}$$

Write out an expression for the probability that it will be found in the ground state of the hydrogen atom?

7. Use the recursion relation to establish (12-32).

8. An electron is in the $n = 2$, $l = 1$, $m = 0$ state of the hydrogen atom. What is its wave function in momentum space?

9. The expectation value of $f(\mathbf{r}, \mathbf{p})$ in any stationary state is a constant. Calculate

$$0 = \frac{d}{dt} \langle \mathbf{r} \cdot \mathbf{p} \rangle = \frac{i}{\hbar} \langle H, \mathbf{r} \cdot \mathbf{p}] \rangle$$

for a Hamiltonian

$$H = \frac{\mathbf{p}^2}{2m} + V(r)$$

and show that

$$\left\langle \frac{\mathbf{p}^2}{m} \right\rangle = \langle \mathbf{r} \cdot \nabla V(r) \rangle$$

Use this to establish the result of Problem 4. Also use this result to calculate $\langle 1/r \rangle$.

10. Use the techniques developed in this chapter to discuss the three-dimensional harmonic oscillator problem, with

$$H = \frac{\mathbf{p}^2}{2m} + \tfrac{1}{2}m\omega^2 r^2$$

Note that the associated Laguerre polynomials also appear in this problem.

11. J. Schwinger points out that the average radial force must vanish for a stationary state. Use this to calculate $\langle n, l|1/r^3|n, l \rangle$.
   Hint: Calculate

$$\left\langle n, l \left| \frac{d}{dr} \left[\frac{\hbar^2 l(l+1)}{2mr^2} - \frac{Ze^2}{r} \right] \right| n, l \right\rangle$$

References

A very thorough discussion of the hydrogenlike atoms is to be found in

E. U. Condon and G. H. Shortley, *The Theory of Atomic Spectra*, Cambridge University Press, Cambridge, England, 1959.

The problem is discussed in every advanced book on quantum mechanics.

INTERACTION OF ELECTRONS WITH ELECTROMAGNETIC FIELD

CLASSICAL THEORY

In Chapter 12 we discussed the interaction of an electron with the static Coulomb field due to a point charge. To generalize this to the interaction with an external magnetic or electric field, we must first review the classical theory. Maxwell's equations in Gaussian units read, in the vacuum,

$$\nabla \cdot \mathbf{B}(\mathbf{r}, t) = 0 \tag{13-1}$$

$$\nabla \times \mathbf{E}(\mathbf{r}, t) + \frac{1}{c} \frac{\partial \mathbf{B}(\mathbf{r}, t)}{\partial t} = 0 \tag{13-2}$$

$$\nabla \cdot \mathbf{E}(\mathbf{r}, t) = 4\pi \rho(\mathbf{r}, t) \tag{13-3}$$

$$\nabla \times \mathbf{B}(\mathbf{r}, t) - \frac{1}{c} \frac{\partial \mathbf{E}(\mathbf{r}, t)}{\partial t} = \frac{4\pi}{c} \mathbf{j}(\mathbf{r}, t) \tag{13-4}$$

where $\rho(\mathbf{r}, t)$ and $\mathbf{j}(\mathbf{r}, t)$ are the charge and current densities that are the sources of the electromagnetic fields $\mathbf{E}(\mathbf{r}, t)$ and $\mathbf{B}(\mathbf{r}, t)$. The conservation of charge equation

$$\frac{\partial \rho(\mathbf{r}, t)}{\partial t} + \nabla \cdot \mathbf{j}(\mathbf{r}, t) = 0 \tag{13-5}$$

is automatically satisfied.

A point electron of mass μ and charge $-e$ obeys the Lorentz force equation

$$\mu \frac{d^2\mathbf{r}}{dt^2} = -e[(\mathbf{E}(\mathbf{r}, t) + \frac{\mathbf{v}}{c} \times \mathbf{B}(\mathbf{r}, t)] \tag{13-6}$$

The transition to quantum mechanics is made by constructing a Hamiltonian for the system. To do this, it is necessary to introduce the *potentials* for the electromagnetic system. The first two Maxwell equations (13-1) and (13-2) are satisfied by defining the vector and scalar potentials $\mathbf{A}(\mathbf{r}, t)$ and $\phi(\mathbf{r}, t)$ such that

$$\mathbf{B}(\mathbf{r}, t) = \mathbf{\nabla} \times \mathbf{A}(\mathbf{r}, t)$$

$$\mathbf{E}(\mathbf{r}, t) = -\frac{1}{c} \frac{\partial \mathbf{A}(\mathbf{r}, t)}{\partial t} - \mathbf{\nabla}\phi(\mathbf{r}, t) \tag{13-7}$$

The equation of motion for the electrons does not involve the potentials $\mathbf{A}$ and ϕ directly. These potentials are not well defined. It is clear that the equation

$$\mathbf{B}(\mathbf{r}, t) = \mathbf{\nabla} \times \mathbf{A}(\mathbf{r}, t)$$

is unchanged if we make the transformation to a new vector potential

$$\mathbf{A}'(\mathbf{r}, t) = \mathbf{A}(\mathbf{r}, t) - \mathbf{\nabla}f(\mathbf{r}, t) \tag{13-8}$$

since $\mathbf{\nabla} \times \mathbf{\nabla}f(\mathbf{r}, t) = 0$. The electric field will be unchanged if, in addition to making the change expressed in (13-8) we change of ϕ to ϕ' given by

$$\phi'(\mathbf{r}, t) = \phi(\mathbf{r}, t) + \frac{1}{c} \frac{\partial f(\mathbf{r}, t)}{\partial t} \tag{13-9}$$

This invariance, known as *invariance under gauge transformations*, allows us the freedom to define the potentials in a variety of ways, to suit our convenience.

The source-dependent pair of equations (13-3) and (13-4) now read

$$-\nabla^2\phi(\mathbf{r}, t) - \frac{1}{c} \frac{\partial}{\partial t} (\mathbf{\nabla} \cdot \mathbf{A}) = 4\pi\rho(\mathbf{r}, t) \tag{13-10}$$

and

$$\mathbf{\nabla} \times (\mathbf{\nabla} \times \mathbf{A}) + \frac{1}{c^2} \frac{\partial^2\mathbf{A}(\mathbf{r}, t)}{\partial t^2} + \frac{1}{c} \frac{\partial}{\partial t} \mathbf{\nabla}\phi = \frac{4\pi}{c} \mathbf{j}(\mathbf{r}, t)$$

which may be rewritten as

$$-\nabla^2\mathbf{A}(\mathbf{r}, t) + \frac{1}{c^2} \frac{\partial^2\mathbf{A}(\mathbf{r}, t)}{\partial t^2} + \mathbf{\nabla} \left(\mathbf{\nabla} \cdot \mathbf{A} + \frac{1}{c} \frac{\partial\phi}{\partial t} \right) = \frac{4\pi}{c} \mathbf{j}(\mathbf{r}, t) \tag{13-11}$$

If the charge distribution is static, that is, $\rho(\mathbf{r})$ is independent of time, it is convenient to choose the gauge such that

$$\mathbf{\nabla} \cdot \mathbf{A}(\mathbf{r}, t) = 0 \tag{13-12}$$

This choice of $f(\mathbf{r}, t)$ is given the name of *Coulomb gauge*. In that case we have

$$-\nabla^2\phi(\mathbf{r}) = 4\pi\rho(\mathbf{r}) \tag{13-13}$$

that is, we have a time-independent scalar potential, and then the equation for $\mathbf{A}(\mathbf{r}, t)$ reads

$$-\nabla^2\mathbf{A}(\mathbf{r}, t) + \frac{1}{c^2}\frac{\partial^2\mathbf{A}(\mathbf{r}, t)}{\partial t^2} = \frac{4\pi}{c}\mathbf{j}(\mathbf{r}, t) \tag{13-14}$$

When the charge distribution is not static, it is more convenient to choose the so-called *Lorentz gauge* for which

$$\nabla \cdot \mathbf{A}(\mathbf{r}, t) + \frac{1}{c}\frac{\partial\phi(\mathbf{r}, t)}{\partial t} = 0 \tag{13-15}$$

This leaves the equation for the vector potential unaltered, but now the scalar potential also obeys a wave equation. A technical point worth noting is that the relation

$$\nabla \times (\nabla \times \mathbf{A}) = -\nabla^2\mathbf{A} + \nabla(\nabla \cdot \mathbf{A})$$

used to obtain (13-11) is only valid in Cartesian coordinates. Thus, $\nabla^2\mathbf{A}(\mathbf{r}, t)$, as it appears, must be calculated in terms of x, y, and z.

The transition to quantum mechanics requires a Hamiltonian formulation of the equation of motion (13-6). In the absence of interaction with the electromagnetic field, it is easily seen that the Hamilton equations

$$\frac{dx_i}{dt} = \frac{\partial H}{\partial p_i}$$
$$\frac{dp_i}{dt} = -\frac{\partial H}{\partial x_i} \tag{13-16}$$

with

$$H = \frac{p^2}{2\mu} + V(r) \tag{13-17}$$

yields

$$\mu\frac{d^2x_i}{dt^2} = -\frac{\partial V}{\partial x_i} = F_i \tag{13-18}$$

The Hamiltonian for the interaction of an electron with an external electromagnetic field, represented by the potentials $(\mathbf{A}(\mathbf{r}, t), \phi(\mathbf{r}, t))$ is given by

$$H = \frac{(\mathbf{p} + (e/c)\mathbf{A}(\mathbf{r}, t))^2}{2\mu} + e\phi(\mathbf{r}, t) \tag{13-19}$$

The Hamilton equations of motion are

$$\frac{dx_i}{dt} = \frac{\partial H}{\partial p_i} = \frac{p_i + (e/c)A_i}{\mu} \qquad (13\text{-}20)$$

and

$$\frac{dp_i}{dt} = -\frac{\partial H}{\partial x_i} = -\frac{e}{\mu c}\left(p_k + \frac{e}{c}A_k\right)\frac{\partial A_k}{\partial x_i} + e\frac{\partial \phi}{\partial x_i} \qquad (13\text{-}21)$$

Thus

$$\mu\frac{d^2 x_i}{dt^2} = \frac{dp_i}{dt} + \frac{e}{c}\left(\frac{\partial A_i}{\partial t} + \frac{\partial A_i}{\partial x_k}\frac{dx_k}{dt}\right)$$

$$= e\frac{\partial \phi}{\partial x_i} + \frac{e}{c}\frac{\partial A_i}{\partial t} - \frac{e}{c}\frac{\partial A_k}{\partial x_i}\frac{dx_k}{dt} + \frac{e}{c}\frac{\partial A_i}{\partial x_k}\frac{dx_k}{dt} \qquad (13\text{-}22)$$

The first two terms are seen to be equal to $-eE_i$, and the second two terms can be checked to equal $-(e/c)(\mathbf{v} \times \mathbf{B})_i$. Thus H given in (13-19) is the correct choice of Hamiltonian.

THE SCHRÖDINGER EQUATION FOR AN ELECTRON IN AN ELECTROMAGNETIC FIELD

The Schrödinger equation takes the form

$$\left[\frac{((\hbar/i)\nabla + (e/c)\mathbf{A}(\mathbf{r}, t))^2}{2\mu} + e\phi(\mathbf{r}, t)\right]\psi(\mathbf{r}, t) = i\hbar\frac{\partial\psi(\mathbf{r}, t)}{\partial t} \qquad (13\text{-}23)$$

where we have replaced the operator $\mathbf{p}$ by $(\hbar/i)\nabla$. Before proceeding with the solution of the energy eigenvalue equation, we need to ask what happens to gauge invariance. If we write the equation in terms of $\mathbf{A}'$ and ϕ', defined in (13-8) and (13-9), the preceding equation takes the form

$$\left[\frac{((\hbar/i)\nabla + (e/c)\mathbf{A}'(\mathbf{r}, t) + (e/c)\nabla f(\mathbf{r}, t))^2}{2\mu} + e\phi'(\mathbf{r}, t) - \frac{e}{c}\frac{\partial f(\mathbf{r}, t)}{\partial t}\right]\psi(\mathbf{r}, t)$$

$$= i\hbar\frac{\partial\psi(\mathbf{r}, t)}{\partial t}$$

which looks like a different equation. It is easy to see that if the transformations (13-8) and (13-9) are accompanied by a phase change in the wave function, $\psi(\mathbf{r}, t) \rightarrow \psi'(\mathbf{r}, t)$, where

$$\psi'(\mathbf{r}, t) = e^{i\Lambda(\mathbf{r}, t)}\psi(\mathbf{r}, t) \qquad (13\text{-}24)$$

then, since

$$\frac{\partial\psi}{\partial t} = \frac{\partial}{\partial t}(e^{-i\Lambda}\psi') = -i\frac{\partial\Lambda}{\partial t}\psi + e^{-i\Lambda}\frac{\partial\psi'}{\partial t}$$

and

$$\frac{\hbar}{i} \nabla \psi = \frac{\hbar}{i} \nabla(e^{-i\Lambda}\psi') = -\hbar\nabla\Lambda\psi - e^{-i\Lambda}\frac{\hbar}{i}\nabla\psi'$$

we get the original equation in terms of $\mathbf{A}'$, ϕ', and ψ' provided

$$\Lambda(\mathbf{r}, t) = \frac{e}{\hbar c} f(\mathbf{r}, t) \tag{13-25}$$

Let us return to the Schrödinger equation. We shall specialize to time-independent fields, so that $\mathbf{A} = \mathbf{A}(\mathbf{r})$, and $\phi = \phi(\mathbf{r})$. In that case we can write

$$\psi(\mathbf{r}, t) = e^{-iEt/\hbar}\psi(\mathbf{r}) \tag{13-26}$$

and

$$\left[\frac{1}{2\mu}\left(\frac{\hbar}{i}\nabla + \frac{e}{c}\mathbf{A}\right) \cdot \left(\frac{\hbar}{i}\nabla + \frac{e}{c}\mathbf{A}\right) + e\phi(\mathbf{r})\right]\psi(\mathbf{r}) = E\psi(\mathbf{r}) \tag{13-27}$$

Equation (13-27) can be written in the form

$$-\frac{\hbar^2}{2\mu}\nabla^2\psi - \frac{ie\hbar}{\mu c}\mathbf{A} \cdot \nabla\psi - \frac{ie\hbar}{2\mu c}(\nabla \cdot \mathbf{A})\psi + \frac{e^2}{2\mu c^2}A^2\psi + e\phi(\mathbf{r})\psi = E\psi \tag{13-28}$$

We now make use of the freedom to choose a gauge function $f(\mathbf{r})$ such that

$$\nabla \cdot \mathbf{A}(\mathbf{r}) = 0 \tag{13-29}$$

to get

$$-\frac{\hbar^2}{2\mu}\nabla^2\psi - \frac{ie\hbar}{\mu c}\mathbf{A} \cdot \nabla\psi + \frac{e^2}{2\mu c^2}A^2\psi + e\phi(\mathbf{r})\psi = E\psi \tag{13-30}$$

THE CONSTANT MAGNETIC FIELD

For a constant uniform magnetic field, $\mathbf{B}$, we may take[1]

$$\mathbf{A} = -\tfrac{1}{2}\mathbf{r} \times \mathbf{B} \tag{13-31}$$

This means that the three components of $\mathbf{A}$ are

$$\mathbf{A} = -\tfrac{1}{2}(yB_z - zB_y, zB_x - xB_z, xB_y - yB_x)$$

and consequently

$$\nabla \times \mathbf{A} = (\tfrac{1}{2}B_x + \tfrac{1}{2}B_x, B_y, B_z)$$
$$= \mathbf{B}$$

---

[1]Note that this choice is not unique, since we may add the gradient of any function to $\mathbf{A}$ without changing $\mathbf{B}$. This choice, however, is very convenient.

Hence the second term in (13-30) becomes

$$\frac{ie\hbar}{2\mu c} \mathbf{r} \times \mathbf{B} \cdot \nabla\psi = -\frac{ie\hbar}{2\mu c} \mathbf{B} \cdot \mathbf{r} \times \nabla\psi$$

$$= \frac{e}{2\mu c} \mathbf{B} \cdot \mathbf{r} \times \frac{\hbar}{i} \nabla\psi = \frac{e}{2\mu c} \mathbf{B} \cdot \mathbf{L}\psi \tag{13-32}$$

and the third term is

$$\frac{e^2}{8\mu c^2} (\mathbf{r} \times \mathbf{B})^2\psi = \frac{e^2}{8\mu c^2} [r^2\mathbf{B}^2 - (\mathbf{r} \cdot \mathbf{B})^2]\psi = \frac{e^2\mathbf{B}^2}{8\mu c^2} (x^2 + y^2)\psi \tag{13-33}$$

if $\mathbf{B}$ is the direction that defines the z-axis. This is of the form of a two-dimensional harmonic oscillator potential.

Let us compare the magnitudes of the two terms. The ratio is estimated with $\langle L_z \rangle$ taken of order $\hbar$ and $\langle x^2 + y^2 \rangle$ of order a_0^2, with a_0 the Bohr radius:

$$\frac{(e^2/8\mu c^2)a_0^2 B^2}{(e/2\mu c)\hbar B} \approx \frac{1}{4} \frac{e^2}{\hbar c} \frac{B}{e/a_0^2} \approx \frac{1}{548} \frac{B}{e/a_0^2}$$

$$\approx \frac{B}{548(4.8 \times 10^{-10})/(0.5 \times 10^{-8})^2} \tag{13-34}$$

$$\approx \frac{B}{9 \times 10^9 \text{ gauss}}$$

Thus in atomic systems, with the kind of fields available in the laboratory, that is, $B \lesssim 10^4$ gauss, the quadratic term is certainly negligible. The term linear in B, compared with the Coulomb potential energy can be estimated in a similar way

$$\frac{(e/2\mu c)\hbar B}{e^2/a_0} \approx \frac{1}{2} \frac{\hbar/\mu c}{e/a_0} B \approx \frac{1}{274} \frac{B}{e/a_0^2} \approx \frac{B}{5 \times 10^9 \text{ gauss}} \tag{13-35}$$

so that the linear term will only slightly perturb the atomic energy levels. The quadratic term can become very important under two conditions: if the magnetic field is very intense; it is believed that fields as large as 10^{12} gauss may exist on the surface of neutron stars, and these would radically alter the structure of atoms.[2] The quadratic term will also be important when we consider the macroscopic motion of an electron in an external field, for example, the motion of an electron in a synchrotron.

THE NORMAL ZEEMAN EFFECT

Let us first consider the linear term alone, and pick the z-direction to coincide with that of $\mathbf{B}$. Then the Hamiltonian with $\mathbf{B} = 0$ is altered by the addition of

$$H_1 = \frac{e}{2\mu c} BL_z \tag{13-36}$$

[2]See R. Cohen, L. Lodenquai, and M. Ruderman, *Phys. Rev. Lett.*, 25, 467 (1970).

If we define the frequency, called the Larmor frequency,

$$\frac{eB}{2\mu c} = \omega_L \tag{13-37}$$

and deal with energy eigenstates that are simultaneously eigenstates of $\mathbf{L}^2$ and L_z, then the extra term (13-36), when acting on an eigenstate, yields a number, namely,

$$H_1 u_{nlm}(\mathbf{r}) = \hbar\omega_L m u_{nlm}(\mathbf{r}) \tag{13-38}$$

where m is the z-component of the angular momentum eigenvalue, with $-l \le m \le l$. Thus the existing energy levels, with their $(2l + 1)$-fold degeneracy are split into $(2l + 1)$ components that are equally spaced, with energies given by

$$E = -\frac{1}{2}\mu c^2 \frac{(Z\alpha)^2}{n^2} + \hbar\omega_L m \tag{13-39}$$

The size of the splitting is

$$\frac{eB\hbar}{2\mu c} = \frac{e\hbar}{2\mu c}\left(\frac{B}{e/a_0^2}\right)\frac{e}{a_0^2}$$

$$= \frac{e^2\hbar}{2\mu c}\left(\frac{\mu c\alpha}{\hbar}\right)^2\left(\frac{B}{e/a_0^2}\right)$$

$$= \left(\tfrac{1}{2}\alpha^2\mu c^2\right)\alpha\,\frac{B}{e/a_0^2}$$

$$= \left(\frac{B}{2.4 \times 10^9}\right) \times 13.6\text{ eV}$$

Since there are selection rules (to be discussed later) according to which only transitions in which the m-value changes by zero or unity are allowed, it turns out that the single line representing a transition with $B = 0$ splits into *three* lines, as can be seen in Fig. 13-1 (p. 222). This effect is the *normal Zeeman effect*. Actually, unless the electron spin state in the atom is one in which the spin is zero, the interaction of the electron spin with the magnetic field changes the pattern predicted earlier. The more common *anomalous Zeeman effect* will be discussed when we have learned about spin.

LARGE MAGNETIC FIELDS AND THE CLASSICAL LIMIT

It is of some interest to discuss the solution of an electron in a constant magnetic field under conditions where the B^2 term is not negligible, and where the Coulomb potential can be neglected. Under those conditions, with $\mathbf{B}$ again chosen to define the z-direction, the Schrödinger equation reads

$$-\frac{\hbar^2}{2\mu}\nabla^2\psi + \frac{eB}{2\mu c}L_z\psi + \frac{e^2B^2}{8\mu c^2}(x^2 + y^2)\psi = E\psi \tag{13-40}$$

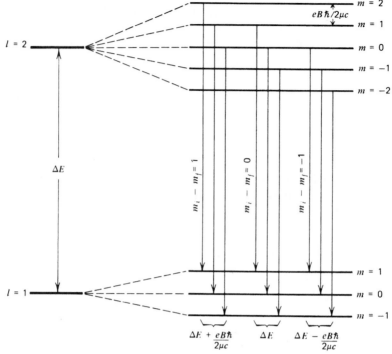

Figure 13-1. Normal Zeeman effect: of the fifteen possible transitions between the $l = 2$ and $l = 1$ states, split by the magnetic field, only nine, corresponding to $\Delta m = m_i - m_f = -1, 0, 1$, occur in the form of three lines.

where we have used (13-30), (13-32), and (13-33). The presence of the "potential" $(x^2 + y^2)$ suggests the use of cylindrical coordinates for the separation of the variables. Writing

$$x = \rho \cos \phi$$
$$y = \rho \sin \phi \tag{13-41}$$

we follow the procedure outlined at the beginning of Chapter 10 to arrive at

$$\frac{\partial}{\partial x} = \cos \phi \, \frac{\partial}{\partial \rho} - \frac{\sin \phi}{\rho} \, \frac{\partial}{\partial \phi}$$
$$\frac{\partial}{\partial y} = \sin \phi \, \frac{\partial}{\partial \rho} + \frac{\cos \phi}{\rho} \, \frac{\partial}{\partial \phi} \tag{13-42}$$

and hence

$$\nabla^2 = \frac{\partial^2}{\partial z^2} + \frac{\partial^2}{\partial \rho^2} + \frac{1}{\rho} \frac{\partial}{\partial \rho} + \frac{1}{\rho^2} \frac{\partial^2}{\partial \phi^2} \tag{13-43}$$

If we now write

$$\psi(\mathbf{r}) = u_m(\rho) \, e^{im\phi} \, e^{ikz} \tag{13-44}$$

we find that the differential equation satisfied by $u_m(\rho)$ is

$$\frac{d^2u}{d\rho^2} + \frac{1}{\rho}\frac{du}{d\rho} - \frac{m^2}{\rho^2}u - \frac{e^2B^2}{4\hbar^2c^2}\rho^2u + \left(\frac{2\mu E}{\hbar^2} - \frac{eB\hbar m}{\hbar^2c} - k^2\right)u = 0 \quad (13\text{-}45)$$

If we introduce the variable

$$x = \sqrt{\frac{eB}{2\hbar c}}\,\rho \quad (13\text{-}46)$$

we can rewrite the equation in the form

$$\frac{d^2u}{dx^2} + \frac{1}{x}\frac{du}{dx} - \frac{m^2}{x^2}u - x^2u + \lambda u = 0 \quad (13\text{-}47)$$

where

$$\lambda = \frac{4\mu c}{eB\hbar}\left(E - \frac{\hbar^2k^2}{2\mu}\right) - 2m \quad (13\text{-}48)$$

It is fairly straightforward to determine that (a) the behavior of $u(x)$ at infinity, determined from

$$\frac{d^2u}{dx^2} - x^2u \approx 0$$

is $u(x) \sim e^{-x^2/2}$, and (b) the behavior of $u(x)$ near $x = 0$, determined from

$$\frac{d^2u}{dx^2} + \frac{1}{x}\frac{du}{dx} - \frac{m^2}{x^2}u \approx 0$$

is $u(x) \sim x^{|m|}$. We thus write

$$u(x) = x^{|m|}\,e^{-x^2/2}\,G(x) \quad (13\text{-}49)$$

and determine the differential equation obeyed by $G(x)$. A little algebra leads to

$$\frac{d^2G}{dx^2} + \left(\frac{2|m| + 1}{x} - 2x\right)\frac{dG}{dx} + (\lambda - 2 - 2|m|)G = 0 \quad (13\text{-}50)$$

This can be brought into the same form as (12-11) if we change variables to

$$y = x^2 \quad (13\text{-}51)$$

The equation then takes the form

$$\frac{d^2G}{dy^2} + \left(\frac{|m| + 1}{y} - 1\right)\frac{dG}{dy} + \frac{\lambda - 2 - 2|m|}{4y}G = 0 \quad (13\text{-}52)$$

We can now proceed as in Chapter 12. Comparison with (12-11) shows that we must have

$$\frac{1}{4} \lambda - \frac{1 + |m|}{2} = n_r \tag{13-53}$$

as an eigenvalue condition, with $n_r = 0, 1, 2, 3, \ldots$. This implies that $E - \hbar^2 k^2 / 2\mu$, the energy with the kinetic energy of the free motion in the z-direction subtracted out, is given by

$$E - \frac{\hbar^2 k^2}{2\mu} = \frac{eB\hbar}{2\mu c} (2n_r + 1 + |m| + m) \tag{13-54}$$

and

$$G(y) = L_{n_r}^{|m|} (y) \tag{13-55}$$

Our discussion of this solution will be confined to the classical limit. To do this, we first review the classical theory. Given the Hamiltonian (13-19), without the scalar potential term, we have

$$\mathbf{v} = \frac{\mathbf{p} + (e/c) \mathbf{A}}{\mu} \tag{13-56}$$

and with $\mathbf{A} = -\frac{1}{2}\mathbf{r} \times B$, we obtain

$$\mu\mathbf{r} \times \mathbf{v} = \mathbf{r} \times \mathbf{p} + \frac{e}{c} \mathbf{r} \times (-\tfrac{1}{2}\mathbf{r} \times \mathbf{B})$$

$$= \mathbf{L} - \frac{e}{2c} [\mathbf{r}(\mathbf{r} \cdot \mathbf{B}) - r^2\mathbf{B}] \tag{13-57}$$

with the help of the identity

$$\mathbf{a} \times (\mathbf{b} \times \mathbf{c}) = \mathbf{b}(\mathbf{a} \cdot \mathbf{c}) - \mathbf{c}(\mathbf{a} \cdot \mathbf{b}) \tag{13-58}$$

We take the z-component of this equation to obtain

$$\mu(\mathbf{r} \times \mathbf{v})_z = L_z + \frac{e}{2c} B (x^2 + y^2)$$

that is,

$$\mu\rho v = L_z + \frac{eB}{2c} \rho^2 \tag{13-59}$$

The expression for the force on the electron

$$\mathbf{F} = -\frac{e}{c} \mathbf{v} \times \mathbf{B} \tag{13-60}$$

yields the relation

$$\frac{\mu v^2}{\rho} = \frac{evB}{c} \tag{13-61}$$

for circular motion. This relation, together with (13-59), after a little algebra, yields,

$$\tfrac{1}{2}\mu v^2 = \frac{eB}{\mu c} L_z \tag{13-62}$$

and

$$\rho = \left[\frac{2c}{eB} L_z\right]^{1/2} \tag{13-63}$$

We now return to the expression for the energy, (13-54). Because of the smallness of $\hbar$, the energy can only be of macroscopic size for reasonable B, if $(2n_r + 1 + |m| + m)$ is very large. We have two cases: (a) If $m < 0$, this implies that n_r is very large. Now n_r determines the degree of the polynomial $L_{n_r}^{|m|}(y)$, that is, the number of the zeros in the function,[3] and if that is very large, the function cannot be large for some small range of y where the classical orbit would be located. (b) If $m > 0$, the coefficient is $(2n_r + 1 + 2m)$, and this can be large, with n_r small, provided that m is large. The energy now is

$$E - \frac{\hbar^2 k^2}{2\mu} \simeq \frac{eB}{\mu c} \hbar m \tag{13-64}$$

in agreement with the classical result. Note that

$$L_z = \hbar m \tag{13-65}$$

is positive, as expected.

We can also show that the radius of the orbit, as determined by the peaking of the radial probability distribution, corresponds to the classical value. Let us take $n_r = 0$. In that case $L_{n_r}^{|m|}(y)$ is just a constant, and the square of the wave function is, according to (13-38),

$$P(x) = x^{2|m|} e^{-x^2} \tag{13-66}$$

This has a maximum where

$$\frac{dP}{dx} = (2|m|x^{2|m|-1} - 2x^{2|m|+1}) e^{-x^2} = 0$$

that is, at

$$x = \sqrt{|m|} \tag{13-67}$$

[3]See the discussion at the beginning of the section on the Degeneracy of the Spectrum in Chapter 12.

which yields

$$\rho = \left(\frac{2c}{eB} \hbar m\right)^{1/2} \tag{13-68}$$

This problem provides a beautiful illustration of the correspondence principle.

LANDAU LEVELS

The choice $\mathbf{A} = (-yB/2, xB/2, 0)$ is not unique. The choice

$$\mathbf{A} = (0, Bx, 0) \tag{13-69}$$

leads to the same magnetic field. It differs by a simple gauge transformation from $\mathbf{A} = -\mathbf{r} \times \mathbf{B}/2$:

$$\left(\frac{-yB}{2}, \frac{xB}{2}, 0\right) = (0, Bx, 0) - \left(\frac{\partial}{\partial x}, \frac{\partial}{\partial y}, \frac{\partial}{\partial z}\right)\left(\frac{yxB}{2}\right) \tag{13-70}$$

With this choice of vector potential, the Hamiltonian operator for an electron in a constant magnetic field takes the form

$$\begin{aligned}
\frac{1}{2\mu}\left(\mathbf{p} - \frac{e}{c}\mathbf{A}\right)^2 &= \frac{1}{2\mu}\left(p_x^2 + \left(p_y + \frac{eB}{c}x\right)^2 + p_z^2\right) \\
&= \frac{1}{2\mu}\left(p_x^2 + p_y^2 + \frac{2eB}{c}xp_y + \left(\frac{eB}{c}\right)^2 x^2 + p_z^2\right)
\end{aligned} \tag{13-71}$$

It is clear that $[H, p_y] = 0$ and $[H, p_z] = 0$, so that we can construct functions that are simultaneous eigenfunctions of p_y, p_z, and H. Let us again choose the state to be an eigenfunction of p_z with eigenvalue zero. If we write the eigenvalue of p_y as $\hbar k$, then the simultaneous eigenfunction takes the form

$$\psi(x, y) = e^{iky}v(x) \tag{13-72}$$

where $v(x)$ is a solution of the equation

$$\frac{1}{2\mu}\left(-\hbar^2\frac{d^2}{dx^2} + \left(\frac{eB}{c}\right)^2\left(x + \frac{\hbar ck}{eB}\right)^2\right)v(x) = Ev(x) \tag{13-73}$$

This is just a harmonic oscillator whose equilibrium point is shifted from $x = 0$ to $-x_0$, where $x_0 = \hbar kc/eB$. We may thus write the solution as

$$\psi(x, y) = e^{ieBx_0y/\hbar c}u(x - x_0) \tag{13-74}$$

where $u(x)$ is the eigensolution of the harmonic oscillator centered at $x = 0$. Comparison with the harmonic oscillator potential $\mu\omega^2 x^2/2$ shows that

$$\omega = \frac{eB}{\mu c} \tag{13-75}$$

and the energy eigenvalues are

$$E_n = \hbar\omega \left(n + \frac{1}{2} \right) \quad n = 0, 1, 2, 3, \ldots \tag{13-76}$$

The energy levels labeled by n are called Landau levels.

If the electron is confined to a strip whose x-dimension is L_1 and y-dimension is L_2, then the boundary condition in the y-direction,

$$\psi(y) = \psi(y + L_2) \tag{13-77}$$

implies that

$$\frac{eBx_0}{\hbar c} L_2 = 2\pi n^* \quad n^* = 0, 1, 2, \ldots \tag{13-78}$$

Since

$$0 < x_0 < L_1 \tag{13-79}$$

we deduce that

$$0 \le n^* \le \frac{eB}{2\pi\hbar c} L_1 L_2 \tag{13-80}$$

We can easily check that $\hbar c/eB$ has the dimensions of an area. (A quick way to see this is that eBv/c, and thus eB has the dimensions of force $= [ML/T^2]$ while $\hbar c$ has the dimensions of $[p][x][L/T] = [ML/T][L][L/T] = [ML^3/T^2]$). We define a *magnetic length* by

$$l_B^2 = \frac{\hbar c}{eB} \tag{13-81}$$

We see that

$$n_{max}^* = \frac{L_1 L_2}{2\pi l_B^2} = (\text{area of sample}/2\pi l_B^2) \tag{13-82}$$

Let us consider what happens when the lowest Landau level ($n = 0$) is filled. A two-dimensional sample of area A can accommodate one electron per energy level (one might wonder whether there are not the usual *two* electrons per energy state, but, as we shall see in the next chapter, the electrons have a magnetic moment associated with the spin, so that the "up" and "down" states of the electron have different energies), so that there is a total of $n_{max}^* = A/2\pi l_B^2$ electrons that fill the lowest Landau level.

INTEGRAL QUANTUM HALL EFFECT

The preceding discussion has relevance to the recently discovered *Integral Quantum Hall Effect* (see Fig. 13-2). We limit ourselves to an idealized description of

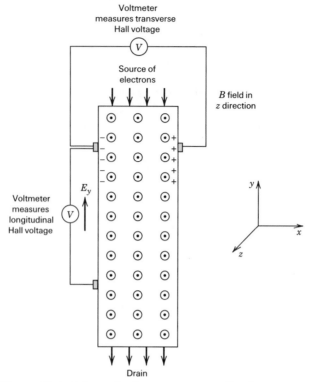

Figure 13-2. Schematics for measuring Hall voltage.

this phenomenon. If an electric field in the y direction is applied to a two-dimensional sample, then electrons will flow in the negative y direction, and the current density is given by

$$j_y = \sigma_0 E_y \tag{13-83}$$

where $\sigma_0 = n_e^2 \tau_0 / m_e^*$. Here n_e is the electron density, m_e^* is the effective electron mass in the material, and τ_0 is a quantity with the dimensions of time, which can be interpreted as the time between collisions of the electron with impurities and other events that cause electrons to lose energy, and not be accelerated indefinitely by the electric field.

If a magnetic field B pointing in the z direction is imposed, then each electron experiences an additional force $\mathbf{F} = -e(\mathbf{v} \times \mathbf{B})/c$. The velocity $\mathbf{v}$ is related to the current density by $\mathbf{j} = -n_e e \mathbf{v}$ so that the the electrons act as if an additional electric field

$$\mathbf{E}' = \frac{(\mathbf{v} \times \mathbf{B})}{c} = \frac{-\mathbf{j} \times \mathbf{B}}{n_e e c}$$

were imposed. Thus in the presence of a magnetic field the current density is given by

$$\mathbf{j} = \sigma_0 \mathbf{E} - \sigma_0 \, \mathbf{j} \times \mathbf{B}/n_e e c \tag{13-84}$$

with **B** pointing in the direction perpendicular to the plane of the sample. The resulting equations

$$j_x = -\frac{\sigma_0 B}{n_e ec} j_y$$

$$j_y = \sigma_0 E_y + \frac{\sigma_0 B}{n_e ec} j_x$$

can be solved in terms of τ_0 to give

$$j_y = \frac{\sigma_0}{1 + (eB\tau_0/m_e^* c)^2} E_y$$

$$j_x = -\frac{n_e ec}{B}\left(1 - \frac{1}{1 + (eB\tau_0/m_e^* c)^2}\right) E_y \tag{13-85}$$

The total number of electrons can be written as fn_{max}^*, which defines f as the ratio of the total number of electrons to the number of Landau states n_{max}^*, so that

$$n_e = \frac{fn_{max}^*}{A} = f\frac{eB}{hc} \tag{13-86}$$

In terms of these, we get

$$\frac{j_y}{E_y} = \sigma_0 \frac{1}{1 + (eB\tau_0/m_e^* c)^2}$$

$$\frac{j_x}{E_y} = -\frac{fe^2}{h} + \frac{n_e ec/B}{1 + (eB\tau_0/m_e^* c)^2} \tag{13-87}$$

The density of electrons and the magnetic field B are under the control of the experimentalist. If n_e is fixed and B is varied, then the ratios in (13-87) can be measured as a function of B. Equivalently, if B is fixed and n_e varied, the ratios can be measured as functions of n_e. It was found by von Klitzing, Dorda, and Pepper in 1980 that the values of B such that $f = 1, 2, 3, \ldots$ lead to (a) vanishing j_y/E_y, and (b) values of $|j_x/E_y| = f(e^2/h)$. A simpleminded explanation of the effect is that when the Landau levels are filled, an electron cannot undergo elastic scattering, since it cannot recoil into another state of the same energy. It cannot be thermally excited into the next Landau level since at low temperatures (≈ 0.1 K) and large magnetic fields ($B \approx 10^5$ gauss) $kT \ll eB\hbar/\mu c$. Thus $\tau_0 \to \infty$, and the observed result follows from (13-87). This is an oversimplified discussion, since it does not take into account a very large number of effects that occur in the nonideal case. For example, some electrons are trapped at imperfections in the crystal lattice, Landau levels are not sharp because of thermal and impurity effects, and electron–electron interactions have been completely neglected. Nevertheless, when all of these complications are taken into account, it is still true that at the critical values of B, j_x/E_y is an integral multiple of e^2/h, to an accuracy of better than one part in 10 million.[4]

---

[4]This precise *quantization* actually follows from gauge invariance, as has been pointed out by R. Laughlin. A presentation of this argument may be found in C. Kittel, *Introduction to Solid State Physics*, Sixth Edition, John Wiley & Sons, Inc. New York (1986).

FLUX QUANTIZATION AND THE
AHARANOV-BOHM EFFECT

Let us return to our discussion following (13-23). In a purely formal way, the equation

$$\frac{1}{2m}\left(\frac{\hbar}{i}\nabla + \frac{e}{c}\mathbf{A}(\mathbf{r})\right)^2 \psi + V(r)\psi(\mathbf{r}) = E\psi(\mathbf{r}) \tag{13-88}$$

can be solved by writing

$$\psi(\mathbf{r}) = e^{-i(e/\hbar c)f(\mathbf{r})}\psi_0(\mathbf{r}) \tag{13-89}$$

where $\psi_0(\mathbf{r})$ is a solution of

$$\frac{1}{2m}\left(\frac{\hbar}{i}\nabla\right)^2 \psi_0 + V(r)\psi_0(\mathbf{r}) = E\psi_0(\mathbf{r}) \tag{13-90}$$

and

$$f(\mathbf{r}) = \int_P^{\mathbf{r}} d\mathbf{r}' \cdot \mathbf{A}(\mathbf{r}') \tag{13-91}$$

is a line integral from a given fixed point P to the point $\mathbf{r}$. The integral only makes sense if $\mathbf{B} = 0$, that is, in a field-free region, since the difference in the integral along two different paths, labeled 1 and 2, is

$$\int_1 d\mathbf{r}' \cdot \mathbf{A}(\mathbf{r}', t) - \int_2 d\mathbf{r}'\mathbf{A}(\mathbf{r}', t) = \oint d\mathbf{r}' \cdot \mathbf{A}(\mathbf{r}', t)$$

$$= \int_S \nabla' \times \mathbf{A}(\mathbf{r}', t) \cdot d\mathbf{S} = \int_S \mathbf{B} \cdot d\mathbf{S} = \Phi \tag{13-92}$$

where we have used Stokes' theorem, and where Φ is the flux of magnetic field through the surface spanned by the two paths (Fig. 13-3). Thus only if $\Phi = 0$ will

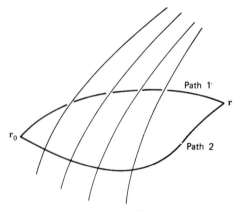

Figure 13-3. The integrals $\int_{r_0}^{r} \mathbf{A}(\mathbf{r}') \cdot d\mathbf{r}'$ along path 1 and path 2 are generally not the same, since the difference is equal to the magnetic flux Φ enclosed by the closed loop.

the phase factor in (13-89) be independent of the choice of path in the line integral. Such an independence is required if we insist that the wave function be single-valued.

If the two paths include flux, then the wave functions of electrons traveling along the two paths will acquire different phases. An interesting consequence is that if an electron moves in a field-free region that is not simply connected, but surrounds a "hole" containing flux Φ, then upon completing a circuit, the electron acquires an additional phase factor $e^{ie\Phi/\hbar c}$. The requirement that the electron wave function be single-valued, so that the phase factor is unity, implies that *the enclosed flux is quantized.*

$$\Phi = \frac{2\pi\hbar c}{e} n \qquad n = 0, \pm 1, \pm 2, \ldots \tag{13-93}$$

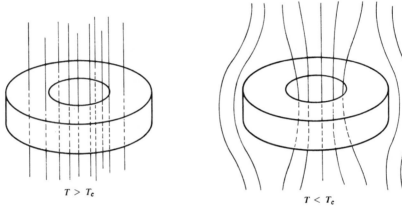

Figure 13-4. A superconductor at temperature $T > T_c$ (the critical temperature) acts like any other metal, and magnetic flux lines can penetrate it. When the temperature is lowered until $T < T_c$, the ring becomes superconducting, and expels magnetic flux lines. Some of these become trapped inside the ring. It is the trapped flux that is found to be quantized.

Such a situation arises in the motion of electrons in a superconducting ring surrounding a region containing flux. The first experiments, done in 1961[5] were based on the following scheme: a ring, made of a superconductor, is placed in an external magnetic field at a temperature above the critical temperature, so that the metal is not superconducting. Since superconductors expel magnetic field lines, except for a thin surface layer, $\mathbf{B} = 0$ inside them. This is the *Meissner effect.*[6] When the ring is cooled below the critical temperature, it becomes superconducting, and magnetic flux is trapped inside the ring (Fig. 13.4). An ingenious measurement of the flux shows that (13-93) holds, with the modification that

$$\Phi = \frac{2\pi\hbar c}{(2e)} n \tag{13-94}$$

[5]B. S. Deaver and W. Fairbank, *Phys. Rev. Lett.,* 7, 43 (1961); R. Döll and M. Nabauer, *ibid.,* 7, 51 (1961).

[6]I strongly recommend Chapter 21 in the *Feynman Lectures on Physics,* Vol. III, for an excellent discussion of these macroscopic manifestations of quantum mechanics.

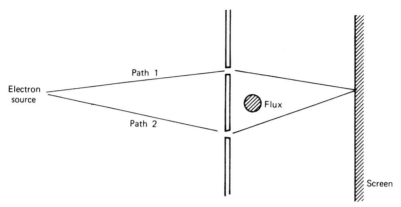

Figure 13-5. Schematic sketch of experiment measuring shift of electron interference pattern by confined magnetic flux.

This is consistent with our present understanding of the phenomenon of superconductivity, according to which, "correlated states" of pairs of electrons (with charge $2e!$) form the fundamental entities that one deals with in the superconductor.

Another manifestation of the dependence of the phase of the wave function on the flux, can, in principle, be seen in an interference experiment (Fig. 13-5) in which a solenoid confining magnetic flux is placed between the slits in a two-slit experiment. The interference pattern at the screen is due to the superposition of two parts of the wave function

$$\psi = \psi_1 + \psi_2 \tag{13-95}$$

where ψ_1 denotes the part of the wave function that describes the electron following path 1, and ψ_2 the part appropriate to path 2. In the presence of the solenoid we have

$$\begin{aligned}\psi &= \psi_1\, e^{ie/\hbar c \int_1 dr\cdot A} + \psi_2\, e^{ie/\hbar c \int_2 dr\cdot A} \\ &= (\psi_1\, e^{ie\Phi/\hbar c} + \psi_2)\, e^{ie/\hbar c \int_2 dr\cdot A}\end{aligned} \tag{13-96}$$

The flux thus causes a *relative change in phase* between ψ_1 and ψ_2, and this will change the interference pattern. This effect, first pointed out by Aharanov and Bohm, has been observed experimentally.

Problems

1. A particle of mass m in a three-dimensional harmonic oscillator of potential energy $m\omega^2 r^2/2$ has a spectrum given by

$$E = \hbar\omega(2n_r + l + 3/2)$$

where n_r is a radial quantum number ($n_r = 0, 1, 2, 3, \ldots$) and l is the orbital angular momentum ($l = 0, 1, 2, 3, \ldots$). Suppose the particle has charge q and the harmonic oscillator is placed in a weak magnetic field **B**. Sketch the spectrum for the three lowest energy states.

2. Consider a *positronium* atom that consists of an electron and a positron (charge $+e$, $m = m_e$) in a hydrogen-like bound state. Write down the Hamiltonian for this system in the presence of a constant external magnetic field, and show that (ignoring the spins of the electron and positron) there is no Zeeman effect.

3. Consider a particle of mass M attached to a rigid massless rod of fixed length R whose other end is fixed at the origin. The rod is free to rotate about its fixed point.

   (a) Give an argument why the Hamiltonian for the system may be written as

$$H = \frac{L^2}{2I} = \frac{(\mathbf{R} \times \mathbf{p})^2}{2I}$$

   with $I = MR^2$.

   (b) If the particle carries charge q, and the rotor is placed in a constant magnetic field $\mathbf{B}$, what is the modified Hamiltonian?

   (c) What is the energy spectrum for small B?

4. Calculate the wavelengths of the three Zeeman lines in the $3D \rightarrow 2P$ transition in hydrogen, when the latter is in a field of 10^4 gauss.

5. Consider the distribution of (13-55)

$$P(x) = x^{2m} e^{-x^2} \qquad (m > 0)$$

   This is known to be peaked at $x = \sqrt{m}$. Show that the width of the peak is approximately $1/2\sqrt{m}$.
   [*Hint:* Let $x = \sqrt{m} \, (1 + \delta)$ and calculate $P(x)/P_{max}$.]

6. Show that for a system described by the Hamiltonian

$$H = \frac{[\mathbf{p} + (e/c) \, \mathbf{A}(\mathbf{r}, \, t)]^2}{2\mu}$$

   the flux $\mathbf{j}$, which satisfies

$$\frac{\partial}{\partial t} \, \psi^*\psi + \nabla \cdot \mathbf{j} = 0$$

   is given by

$$\mathbf{j} = \frac{\hbar}{2i\mu} \left[\psi^*\nabla\psi - \nabla\psi^*\psi + \frac{2ie}{\hbar c} \, \mathbf{A}(\mathbf{r}, \, t)\psi^*\psi \right]$$

7. Consider a charged particle in a magnetic field $\mathbf{B} = (0, 0, B)$ and in a crossed electric field $\mathbf{E} = (E, 0, 0)$. Solve the eigenvalue problem.
   Hint: The proper choice of gauge is important.

8. In this problem we work out an example showing how an enclosed magnetic flux changes the angular momentum of a particle in a region outside the flux tube. Consider a magnetic field confined in a cylindrical region $\rho < a$. Let the flux be Φ. In the region $\rho > a$ there is no magnetic field, and hence the vector potential is of the form

$$\mathbf{A}(\rho,\theta,z) = \nabla\Lambda(\rho,\theta,z)$$

   (a) The choice of gauge $\nabla \cdot \mathbf{A} = 0$ implies that

$$\nabla^2\Lambda = 0$$

Show that a solution of this equation, satisfying (13-92), is

$$\Lambda = \frac{1}{2\pi} \Phi\theta$$

(b) Calculate the angular momentum about the symmetry axis

$$(\mu r \times v)_z = \tilde{L}_z = \left[r \times \left(\frac{\hbar}{i} \nabla + \frac{e}{c} A \right) \right]_z$$

in cylindrical coordinates, and show that for the above Λ it is given by

$$\tilde{L}_z = \frac{\hbar}{i} \frac{\partial}{\partial \theta} + \frac{e}{c} \frac{\Phi}{2\pi}$$

(c) Solve the eigenvalue problem $\tilde{L}_z \psi = \lambda \psi$, and show that single-valuedness of the eigenfunctions leads to flux quantization.

9. Consider an electron confined to a region between two cylinders of radii a and b respectively ($b > a$). (a) Separate the Schrödinger equation in cylindrical coordinates (cf. Eq. 13-43), and show that the equation can be solved in terms of Bessel functions. What are the conditions for the determination of the energy eigenalues? (b) Discuss the degeneracy of the energy eigenfunctions. What is it due to? For Bessel functions, see note below.

Note: The solution of the equation

$$\frac{d^2u}{dz^2} + \frac{1}{z} \frac{du}{dz} + \left(1 - \frac{n^2}{z^2} \right) u = 0$$

with n integral, are known as Bessel functions, for the regular solutions

$$J_n(z) = \left(\frac{z}{2} \right)^n \sum_{l=0}^{\infty} \frac{(iz/2)^{2l}}{l!(n+l)!}$$

and Neumann functions for the irregular solutions

$$N_n(z) = \frac{2}{\pi} J_n(z) \log \frac{\gamma z}{2} - \frac{1}{\pi} \left(\frac{z}{2} \right)^n \sum_{l=0}^{\infty} \frac{(iz/2)^{2l}}{l!(n+l)!} a_{nl}$$

$$- \frac{1}{\pi} \left(\frac{z}{2} \right)^{-n} \sum_{l=0}^{n-1} \frac{(n-l-1)!}{l!} \left(\frac{z}{2} \right)^{2l}$$

$$(\log \gamma = 0.5772\ldots) \qquad a_{nl} = \left(\sum_{m=1}^{l} \frac{1}{m} + \sum_{m=1}^{l+n} \frac{1}{m} \right)$$

They have the asymptotic behavior

$$J_n(z) \sim \left(\frac{2}{\pi z} \right)^{1/2} \cos \left(z - \frac{n\pi}{2} - \frac{\pi}{4} \right) \left[1 + 0 \left(\frac{1}{z^2} \right) \right]$$

$$N_n(z) \sim \left(\frac{2}{\pi z} \right)^{1/2} \sin \left(z - \frac{n\pi}{2} - \frac{\pi}{4} \right) \left[1 + 0 \left(\frac{1}{z^2} \right) \right]$$

A detailed discussion of their properties may be found in any book on the special functions of mathematical physics.

References

The various aspects of electron motion in a magnetic field are very interestingly discussed in

R. P. Feynman, R. B. Leighton, and M. Sands, *The Feynman Lectures on Physics,* Vol. III, Addison-Wesley, Reading, Mass., 1965.

For a detailed discussion of the experiments that unambiguously establish the Aharanov-Bohm effect, and for an excellent discussion of the theory, see M. Peshkin and A. Tonomura, *The Aharanov-Bohm Effect,* Lecture Notes in Physics, Vol. 340, Springer-Verlag, Berlin/New York, 1989.

Chapter 14

OPERATORS, MATRICES, AND SPIN

MATRIX REPRESENTATION OF HARMONIC OSCILLATOR OPERATORS

A proper discussion of atoms is not possible without consideration of the spin of the electron. In spite of the suggestive name, this property of the electron has no classical analog, and, as will soon become evident, it must be treated by somewhat abstract methods. Fortunately we have some preparation for this further departure from a description closely tied to coordinate space, in that we discussed both the harmonic oscillator (Chapter 7) and the angular momentum eigenvalue problem

$$\mathbf{L}^2 Y_{lm} = \hbar^2 l(l + 1)\, Y_{lm}$$
$$L_z Y_{lm} = \hbar m Y_{lm} \tag{14-1}$$

by operator methods. For the harmonic oscillator we found states, defined by

$$u_n = \frac{1}{(n!)^{1/2}} (A^\dagger)^n\, u_0 \tag{14-2}$$

for which

$$H u_n = \hbar\omega(n + \tfrac{1}{2})u_n \tag{14-3}$$

and we could also calculate the action of the raising and lowering operators on u_n,

$$A^\dagger u_n = \sqrt{(n + 1)}\, u_{n+1} \tag{14-4}$$

and

$$A u_n = \sqrt{n}\, u_{n-1} \tag{14-5}$$

We also showed that

$$\langle u_m | u_n \rangle = \delta_{mn} \tag{14-6}$$

a statement that can be made to hold for the eigenstates of any hermitian operator (H here). If we take the scalar product of (14-3) to (14-5) with u_m we find that

$$\langle u_m | H u_n \rangle \equiv \langle u_m | H | u_n \rangle = (n + \tfrac{1}{2})\, \hbar\omega\, \delta_{mn}$$
$$\langle u_m | A^\dagger u_n \rangle \equiv \langle u_m | A^\dagger | u_n \rangle = \sqrt{(n + 1)}\, \delta_{m,n+1} \tag{14-7}$$
$$\langle u_m | A u_n \rangle \equiv \langle u_m | A | u_n \rangle = \sqrt{n}\, \delta_{m,n-1}$$

where we use the more symmetric notation

$$\langle u_i | 0 | u_j \rangle \equiv \langle u_i | 0 u_j \rangle \tag{14-8}$$

These quantities may be arranged in arrays called *matrices*. The conventional notation for a matrix M_{ij} has the first index labeling the row, and the second labeling the column of the array. Thus if we write the scalar product $\langle u_m | H | u_n \rangle$ as H_{mn} we find that

$$H = \hbar\omega \begin{pmatrix} 1/2 & 0 & 0 & 0 & \cdots \\ 0 & 3/2 & 0 & 0 & \cdots \\ 0 & 0 & 5/2 & 0 & \cdots \\ 0 & 0 & 0 & 7/2 & \cdots \\ \vdots & \vdots & \vdots & \vdots & \ddots \end{pmatrix} \tag{14-9}$$

Similarly

$$A^\dagger = \begin{pmatrix} 0 & 0 & 0 & 0 & \cdots \\ \sqrt{1} & 0 & 0 & 0 & \cdots \\ 0 & \sqrt{2} & 0 & 0 & \cdots \\ 0 & 0 & \sqrt{3} & 0 & \cdots \\ \vdots & \vdots & \vdots & \vdots & \end{pmatrix} \tag{14-10}$$

and

$$A = \begin{pmatrix} 0 & \sqrt{1} & 0 & 0 & \cdots \\ 0 & 0 & \sqrt{2} & 0 & \cdots \\ 0 & 0 & 0 & \sqrt{3} & \cdots \\ \vdots & \vdots & \vdots & \vdots & \end{pmatrix} \tag{14-11}$$

We shall call the array $\langle u_m|F|u_n\rangle$, where F is any operator, and the u_i are any complete set, a matrix representation of F in the basis provided by the u_i. This appellation needs some justification. The product of two matrices, for example, satisfies

$$(FG)_{ij} = \sum_n (F)_{in}(G)_{nj} \tag{14-12}$$

and we need to verify this relation for the "matrix representations" of the operators F and G. To do this, let us consider the state Gu_j, and, using completeness, expand it in the form

$$Gu_j = \sum_n C_n u_n \tag{14-13}$$

The coefficients C_n are given by

$$C_n = \langle u_n|G|u_j\rangle \tag{14-14}$$

Hence

$$\langle u_i|FG|u_j\rangle = \langle u_i|F(\sum_n C_n u_n)\rangle$$
$$= \sum_n C_n\langle u_i|F|u_n\rangle \tag{14-15}$$
$$= \sum_n \langle u_i|F|u_n\rangle\langle u_n|G|u_j\rangle$$

which is the same as (14-12), provided we write

$$\langle u_i|F|u_n\rangle = F_{in} \tag{14-16}$$

and so on. We recall that the completeness of the basis vectors $|u_n\rangle$ may be expressed in the form

$$\sum_n |u_n\rangle\langle u_n| = 1 \tag{14-17}$$

The insertion of unity between the two operator F and G in $\langle u_i|FG|u_j\rangle$ in the form of (14-17) immediately yields (14-16).

Further justification for the matrix connection comes from the relation

$$\langle u_m|F|u_n\rangle^* = \langle Fu_n|u_m\rangle = \langle u_n|F^\dagger|u_m\rangle \tag{14-18}$$

which shows that if the operator F is represented by a matrix, then the hermitian conjugate operator $F^\dagger$ will be represented by the hermitian conjugate matrix, since the latter is defined by

$$(F^\dagger)_{nm} = F^*_{mn} \tag{14-19}$$

Note that in our discussion we made no reference to the fact that we started out with eigenstates of the harmonic oscillator Hamiltonian. *The only thing that is special about them is that they diagonalize the matrix representing H.* With another complete set, H would not be diagonal, and reading off its eigenvalues, that is, the matrix elements when it is diagonal, would not be easy.

MATRIX REPRESENTATION OF ANGULAR MOMENTUM OPERATORS

Consider the matrix elements of L_z between different angular momentum states, $\langle l'm'|L_z|lm\rangle$. First of all, we observe that

$$[\mathbf{L}^2, L_z] = 0$$

implies that

$$
\begin{aligned}
0 &= \langle l'm'|[\mathbf{L}^2, L_z]|lm\rangle \\
&= \langle \mathbf{L}^2 l'm'|L_z|lm\rangle - \langle l'm'|L_z|\mathbf{L}^2 lm\rangle \\
&= \hbar^2\{l'(l' + 1) - l(l + 1)\}\langle l'm'|L_z|lm\rangle
\end{aligned}
\tag{14-20}
$$

From this we conclude that if $l' \neq l$, then $\langle l'm'|\, L_z|lm\rangle$ vanishes. Thus L_z, and similarly $L_\pm$, only have matrix elements between states that have the same total angular momentum quantum numbers. If we thus stay with fixed l, that is, with states in which only the m-value is variable, then, with an abbreviated notation, the second of the relations (14-1) reads

$$\langle lm'|L_z|lm\rangle = \hbar m\, \delta_{m'm} \tag{14-21}$$

Furthermore (11-36) with (11-48) implies that

$$\langle lm'|L_\pm|lm\rangle = \hbar[l(l + 1) - m(m \pm 1)]^{1/2}\, \delta_{m',m\pm1} \tag{14-22}$$

This leads to the matrix representations

$$L_z = \hbar \begin{pmatrix} 1 & 0 & 0 \\ 0 & 0 & 0 \\ 0 & 0 & -1 \end{pmatrix} \tag{14-23a}$$

$$L_+ = \hbar \begin{pmatrix} 0 & \sqrt{2} & 0 \\ 0 & 0 & \sqrt{2} \\ 0 & 0 & 0 \end{pmatrix} \tag{14-23b}$$

and

$$L_- = \hbar \begin{pmatrix} 0 & 0 & 0 \\ \sqrt{2} & 0 & 0 \\ 0 & \sqrt{2} & 0 \end{pmatrix} \tag{14-23c}$$

for the $l = 1$ angular momentum operators. The rows and columns are labeled with $m = 1, 0, -1$ in order left to right and top to bottom. It is easy to check that the matrices satisfy the commutation relations. For example,

$$[L_+,L_-] = \hbar \begin{pmatrix} 0 & \sqrt{2} & 0 \\ 0 & 0 & \sqrt{2} \\ 0 & 0 & 0 \end{pmatrix} \begin{pmatrix} 0 & 0 & 0 \\ \sqrt{2} & 0 & 0 \\ 0 & \sqrt{2} & 0 \end{pmatrix} - \hbar^2 \begin{pmatrix} 0 & 0 & 0 \\ \sqrt{2} & 0 & 0 \\ 0 & \sqrt{2} & 0 \end{pmatrix} \begin{pmatrix} 0 & \sqrt{2} & 0 \\ 0 & 0 & \sqrt{2} \\ 0 & 0 & 0 \end{pmatrix}$$

$$= \hbar^2 \begin{pmatrix} 2 & 0 & 0 \\ 0 & 2 & 0 \\ 0 & 0 & 0 \end{pmatrix} - \hbar^2 \begin{pmatrix} 0 & 0 & 0 \\ 0 & 2 & 0 \\ 0 & 0 & 2 \end{pmatrix} = 2\hbar^2 \begin{pmatrix} 1 & 0 & 0 \\ 0 & 0 & 0 \\ 0 & 0 & -1 \end{pmatrix} = 2\hbar L_z \quad (14\text{-}24)$$

General relations between states can also be written in matrix representation. Consider, for example, a relation like

$$\psi = A\phi \tag{14-25}$$

If we take the scalar product of this with any member of a complete set u_i, we have

$$\langle u_i|\psi\rangle = \langle u_i|A\phi\rangle \tag{14-26}$$

Furthermore, the insertion of the unit operator, in the form (14-17) between A and ϕ yields

$$\langle u_i|\psi\rangle = \sum_n \langle u_i|A|u_n\rangle\langle u_n|\phi\rangle \tag{14-27}$$

If we write $\langle u_n|\phi\rangle$ as a column vector α_n

$$\langle u_n|\phi\rangle \rightarrow \begin{pmatrix} \langle u_1|\phi\rangle \\ \langle u_2|\phi\rangle \\ \langle u_3|\phi\rangle \\ \vdots \end{pmatrix} \equiv \begin{pmatrix} \alpha_1 \\ \alpha_2 \\ \alpha_3 \\ \vdots \end{pmatrix} \tag{14-28}$$

and similarly

$$\langle u_n|\psi\rangle \rightarrow \begin{pmatrix} \langle u_1|\psi\rangle \\ \langle u_2|\psi\rangle \\ \langle u_3|\psi\rangle \\ \vdots \end{pmatrix} \equiv \begin{pmatrix} \beta_1 \\ \beta_2 \\ \beta_3 \\ \vdots \end{pmatrix} \tag{14-29}$$

then the matrix representation of (14-25) is

$$\beta_i = \sum_n A_{in}\alpha_n \tag{14-30}$$

Thus matrices represent operators, and column vectors represent states. The scalar product $\langle\phi|u_n\rangle = \langle u_n|\phi\rangle^*$ is written conventionally in the form of a row

$$\langle\phi|u_n\rangle \rightarrow (\alpha_1^*, \alpha_2^*, \alpha_3^*, \ldots) \tag{14-31}$$

so that the scalar product $\langle\phi|\psi\rangle$, for example, can be written as

$$\langle\phi|\psi\rangle = \sum_n \langle\phi|u_n\rangle\langle u_n|\psi\rangle$$

$$= \sum_n \alpha_n^* \beta_n \tag{14-32}$$

An eigenvalue equation is a special case of (14-25). It reads

$$A\phi = a\phi \tag{14-33}$$

and it reads

$$\sum_n A_{in}\alpha_n = a\alpha_i \tag{14-34}$$

in matrix form. This is equivalent to

$$\begin{pmatrix} A_{11}-a & A_{12} & A_{13} & \cdots \\ A_{21} & A_{22}-a & A_{23} & \cdots \\ A_{31} & A_{32} & A_{33}-a & \cdots \\ \vdots & \vdots & \vdots & \end{pmatrix}\begin{pmatrix} \alpha_1 \\ \alpha_2 \\ \alpha_3 \\ \vdots \end{pmatrix} = 0 \tag{14-35}$$

and there will be a nontrivial solution of this equation only if the determinant of the matrix vanishes

$$\det|A_{in} - a\delta_{in}| = 0 \tag{14-36}$$

This is a good way of finding eigenvalues (and eigenvectors) for operators represented by finite matrices, but for infinite matrices this is unfortunately not so simple.

THE SPIN OPERATOR AND ITS MATRIX REPRESENTATION

It is indeed fortunate that there is an alternative to representing operators by functions and differentials, since not all operators can be represented in that way. The simplest example is that corresponding to the angular momentum $l = \frac{1}{2}$. Equations (11-51) and (11-60) tell us that

$$Y_{1/2,\pm 1/2} = C_{\pm}\sqrt{\sin\theta}\, e^{\pm i\phi/2} \tag{14-37}$$

and (11-54) allows us to compute

$$L_- Y_{1/2,1/2} \propto \frac{\cos\theta}{\sqrt{\sin\theta}}\, e^{-i\phi/2} \tag{14-38}$$

This, however, is not proportional to $Y_{1/2,-1/2}$. This indicates problems with extending the established rules to $l = \frac{1}{2}$, and we thus turn to matrix representations.[1] Instead of talking about $l = \frac{1}{2}$, we shall talk about spin, $s = \frac{1}{2}$, reserving the letter

---

[1]It has been pointed out that $Y_{1/2,\pm 1/2}$ gives rise to a probability current from one pole of the sphere ($\theta = 0$) to the other ($\theta = \pi$), with the two poles acting as sources and sinks of probability, respectively.

l for the orbital angular momentum associated with $\mathbf{r} \times \mathbf{p}$. The spin operators are S_x, S_y, and S_z, and they are defined by their commutation relations

$$[S_x, S_y] = i\hbar S_z \tag{14-39}$$

and so on. We wish to represent them by 2×2 matrices. Equation (14-21) yields

$$S_z = \hbar \begin{pmatrix} 1/2 & 0 \\ 0 & -1/2 \end{pmatrix} \tag{14-40}$$

and (14-22) gives

$$S_+ = \hbar \begin{pmatrix} 0 & 1 \\ 0 & 0 \end{pmatrix} \qquad S_- = \hbar \begin{pmatrix} 0 & 0 \\ 1 & 0 \end{pmatrix} \tag{14-41}$$

We may write this representation as

$$\mathbf{S} = \tfrac{1}{2}\hbar\boldsymbol{\sigma} \tag{14-42}$$

where

$$\sigma_x = \begin{pmatrix} 0 & 1 \\ 1 & 0 \end{pmatrix} \qquad \sigma_y = \begin{pmatrix} 0 & -i \\ i & 0 \end{pmatrix} \qquad \sigma_z = \begin{pmatrix} 1 & 0 \\ 0 & -1 \end{pmatrix} \tag{14-43}$$

are the *Pauli matrices*. They satisfy the commutation relations

$$[\sigma_x, \sigma_y] = 2i\sigma_z \tag{14-44}$$

and so on, as they must, to satisfy (14-39), and they also satisfy

$$\sigma_x^2 = \sigma_y^2 = \sigma_z^2 = \begin{pmatrix} 1 & 0 \\ 0 & 1 \end{pmatrix} \equiv 1 \tag{14-45}$$

The Pauli matrices also anticommute:

$$\begin{aligned} \sigma_x\sigma_y &= -\sigma_y\sigma_x \\ \sigma_z\sigma_x &= -\sigma_x\sigma_z \\ \sigma_y\sigma_z &= -\sigma_z\sigma_y \end{aligned} \tag{14-46}$$

These relations are peculiar to the spin $\tfrac{1}{2}$ representations and do not hold for the $l = 1$ matrices, for example.

The eigenstates of S_z will be represented by a two-component column vector, which we call *spinor*. To find these eigenspinors, we solve

$$S_z \begin{pmatrix} u \\ v \end{pmatrix} = \pm\tfrac{1}{2}\hbar \begin{pmatrix} u \\ v \end{pmatrix} \tag{14-47}$$

that is,

$$\begin{pmatrix} 1 & 0 \\ 0 & -1 \end{pmatrix}\begin{pmatrix} u \\ v \end{pmatrix} = \pm\begin{pmatrix} u \\ v \end{pmatrix}$$

or

$$\begin{pmatrix} u \\ -v \end{pmatrix} = \pm \begin{pmatrix} u \\ v \end{pmatrix} \tag{14-48}$$

The plus eigensolution has $v = 0$, and the minus eigensolution has $u = 0$. We thus write

$$\chi_+ = \begin{pmatrix} 1 \\ 0 \end{pmatrix} \qquad \chi_- = \begin{pmatrix} 0 \\ 1 \end{pmatrix} \tag{14-49}$$

for the eigenspinors corresponding to spin up $[S_z = +(1/2)\hbar]$ and spin down $[S_z = -(1/2)\hbar]$, respectively.

An arbitrary spinor can be expanded in this complete set

$$\begin{pmatrix} \alpha_+ \\ \alpha_- \end{pmatrix} = \alpha_+ \begin{pmatrix} 1 \\ 0 \end{pmatrix} + \alpha_- \begin{pmatrix} 0 \\ 1 \end{pmatrix} \tag{14-50}$$

and the expansion postulate yields the interpretation that $|\alpha_+|^2$ and $|\alpha_-|^2$, when properly normalized, so that

$$|\alpha_+|^2 + |\alpha_-|^2 = 1 \tag{14-51}$$

yield the probabilities that a measurement of S_z on the state $\begin{pmatrix} \alpha_+ \\ \alpha_- \end{pmatrix}$ yields $+(1/2)\hbar$ and $-(1/2)\hbar$, respectively.

It is not necessary to keep S_z diagonal. If we look for the eigenstates of the operator $S_x \cos \phi + S_y \sin \phi$, we must solve

$$(S_x \cos \phi + S_y \sin \phi)\begin{pmatrix} u \\ v \end{pmatrix} = \tfrac{1}{2}\hbar\lambda \begin{pmatrix} u \\ v \end{pmatrix} \tag{14-52}$$

that is,

$$\begin{pmatrix} 0 & \cos \phi - i \sin \phi \\ \cos \phi + i \sin \phi & 0 \end{pmatrix}\begin{pmatrix} u \\ v \end{pmatrix} = \lambda \begin{pmatrix} u \\ v \end{pmatrix}$$

This implies that

$$\begin{aligned} v \, e^{-i\phi} &= \lambda u \\ u \, e^{i\phi} &= \lambda v \end{aligned} \tag{14-53}$$

Taking the products of the left- and right-hand sides in the two equations we find that

$$uv(\lambda^2 - 1) = 0 \tag{14-54}$$

Hence

$$\lambda = \pm 1 \tag{14-55}$$

The eigenvector corresponding to $\lambda = 1$ satisfies

$$v = e^{i\phi} u$$

so that the normalized form is

$$\frac{1}{\sqrt{2}} \begin{pmatrix} 1 \\ e^{i\phi} \end{pmatrix}$$

We take advantage of the fact that we can multiply a state vector by an arbitrary phase factor, which we choose to be $e^{-i\phi/2}$. This yields

$$u_+ = \frac{1}{\sqrt{2}} \begin{pmatrix} e^{-i\phi/2} \\ e^{i\phi/2} \end{pmatrix} \tag{14-56}$$

and similarly, the eigenstate corresponding to $\lambda = -1$ can be written in the form

$$u_- = \frac{1}{\sqrt{2}} \begin{pmatrix} e^{-i\phi/2} \\ -e^{i\phi/2} \end{pmatrix} \tag{14-57}$$

which is easily seen to be orthogonal to u_+:

$$u_+^* u_- = \frac{1}{2} (e^{i\phi/2}, e^{-i\phi/2}) \begin{pmatrix} e^{-i\phi/2} \\ -e^{i\phi/2} \end{pmatrix} \tag{14-58}$$
$$= 0$$

respectively. It is interesting to observe that if we change ϕ to $\phi + 2\pi$ *the solutions change sign.* This is characteristic of odd half-integer spin wave functions (fermion states); although this does not violate quantum mechanics, since -1 is just a phase factor, it does mean that no classical macroscopic wave packet can be constructed that has odd half-integral angular momentum.

Given an arbitrary state α, the expectation value of **S** can be calculated. We have

$$\langle \alpha | \mathbf{S} | \alpha \rangle = \sum_i \sum_j \langle \alpha | i \rangle \langle i | \mathbf{S} | j \rangle \langle j | \alpha \rangle$$

or, equivalently,

$$(\alpha_+^*, \alpha_-^*) \, \mathbf{S} \begin{pmatrix} \alpha_+ \\ \alpha_- \end{pmatrix}$$

Thus

$$\langle S_x \rangle = (\alpha_+^*, \alpha_-^*) \tfrac{1}{2}\hbar \begin{pmatrix} 0 & 1 \\ 1 & 0 \end{pmatrix} \begin{pmatrix} \alpha_+ \\ \alpha_- \end{pmatrix}$$

$$= \tfrac{1}{2}\hbar (\alpha_+^*, \alpha_-^*) \begin{pmatrix} \alpha_- \\ \alpha_+ \end{pmatrix} = \tfrac{1}{2}\hbar (\alpha_+^* \alpha_- + \alpha_-^* \alpha_+)$$

$$\langle S_y \rangle = \tfrac{1}{2}\hbar (\alpha_+^*, \alpha_-^*) \begin{pmatrix} 0 & -i \\ i & 0 \end{pmatrix} \begin{pmatrix} \alpha_+ \\ \alpha_- \end{pmatrix} \tag{14-59}$$

$$= \tfrac{1}{2}\hbar (\alpha_+^*, \alpha_-^*) \begin{pmatrix} -i\alpha_- \\ i\alpha_+ \end{pmatrix} = -\frac{i\hbar}{2} (\alpha_+^* \alpha_- - \alpha_-^* \alpha_+)$$

$$\langle S_z \rangle = \tfrac{1}{2}\hbar (\alpha_+^*, \alpha_-^*) \begin{pmatrix} \alpha_+ \\ -\alpha_- \end{pmatrix} = \tfrac{1}{2}\hbar (|\alpha_+|^2 - |\alpha_-|^2)$$

Note that all of these are real, as expected for hermitian operators.

THE INTRINSIC MAGNETIC MOMENT
OF SPIN 1/2 PARTICLES

We shall see later that the spin of an electron appears in the Hamiltonian for the hydrogen atom, for example, coupled to the orbital angular momentum. When an electron is localized at a crystal lattice site, for example, it is often possible to treat the spin as the only degree of freedom that the electron possesses. The electron will have an intrinsic magnetic dipole moment by virtue of its spin, and that magnetic moment[2] is

$$\mathbf{M} = -\frac{eg}{2mc}\,\mathbf{S} \tag{14-60}$$

where g, the gyromagnetic ratio, is very close to 2,

$$g = 2\left(1 + \frac{\alpha}{2\pi} + \cdots\right) = 2.0023192 \tag{14-61}$$

and m is the electron mass and α is the fine structure constant. For such a localized electron, the Hamiltonian in the presence of an external magnetic field B is just the potential energy

$$H = -\mathbf{M} \cdot \mathbf{B} = \frac{eg\hbar}{4mc}\,\boldsymbol{\sigma} \cdot \mathbf{B} \tag{14-62}$$

The Schrödinger equation for the state $\psi(t) = \begin{pmatrix} \alpha_+(t) \\ \alpha_-(t) \end{pmatrix}$ is

$$i\hbar\,\frac{d\psi(t)}{dt} = \frac{eg\hbar}{4mc}\,\boldsymbol{\sigma} \cdot \mathbf{B}\psi(t) \tag{14-63}$$

If $\mathbf{B}$ is taken to define the z-axis, and if we write

$$\psi(t) = \begin{pmatrix} \alpha_+(t) \\ \alpha_-(t) \end{pmatrix} = e^{-i\omega t}\begin{pmatrix} \alpha_+ \\ \alpha_- \end{pmatrix} \tag{14-64}$$

then the equation becomes

$$\hbar\omega\begin{pmatrix} \alpha_+ \\ \alpha_- \end{pmatrix} = \frac{eg\hbar B}{4mc}\begin{pmatrix} 1 & 0 \\ 0 & -1 \end{pmatrix}\begin{pmatrix} \alpha_+ \\ \alpha_- \end{pmatrix} \tag{14-65}$$

The solutions correspond to different frequencies ω. We have, for $\omega = egB/4mc$, $\begin{pmatrix} \alpha_+ \\ \alpha_- \end{pmatrix} = \begin{pmatrix} 1 \\ 0 \end{pmatrix}$, and for $\omega = -(egB/4mc)$, $\begin{pmatrix} \alpha_+ \\ \alpha_- \end{pmatrix} = \begin{pmatrix} 0 \\ 1 \end{pmatrix}$. Thus, if the initial state is

$$\psi(0) = \begin{pmatrix} a \\ b \end{pmatrix} \tag{14-66}$$

---

[2] A "classical" electron moving in a circle with angular momentum $\mathbf{L}$ will form a current loop whose magnetic moment is $\mathbf{M} = -e\mathbf{L}/2mc$. Since the spin is a purely quantum-mechanical variable, one can argue (14-60) only by analogy. For its justification one needs the relativistic Dirac equation from which the value $g = 2$ also emerges. The corrections to $g = 2$ come from quantum electrodynamics. The nonclassical aspects of spin were pointed out by its discoverers, S. Goudsmit and G. Uhlenbeck (1925).

then the state at a later time will be

$$\psi(t) = \begin{pmatrix} a\,e^{-i\omega t} \\ b\,e^{i\omega t} \end{pmatrix} \qquad \omega = \frac{geB}{4mc} \tag{14-67}$$

Suppose that at $t = 0$ the spin is an eigenstate of S_x with eigenvalue $+(1/2)\,\hbar$, that is, it "points in the x direction." This means that

$$\tfrac{1}{2}\hbar \begin{pmatrix} 0 & 1 \\ 1 & 0 \end{pmatrix}\begin{pmatrix} a \\ b \end{pmatrix} = \tfrac{1}{2}\hbar \begin{pmatrix} a \\ b \end{pmatrix}$$

that is, $\begin{pmatrix} a \\ b \end{pmatrix} = \dfrac{1}{\sqrt{2}} \begin{pmatrix} 1 \\ 1 \end{pmatrix}$. Then, at a later time

$$\begin{aligned}
\langle S_x \rangle &= \tfrac{1}{2}\hbar\,\frac{1}{\sqrt{2}}\,(e^{i\omega t},\, e^{-i\omega t})\begin{pmatrix} 0 & 1 \\ 1 & 0 \end{pmatrix}\frac{1}{\sqrt{2}}\begin{pmatrix} e^{-i\omega t} \\ e^{i\omega t} \end{pmatrix} \\[2mm]
&= \frac{\hbar}{4}\,(e^{i\omega t},\, e^{-i\omega t})\begin{pmatrix} e^{i\omega t} \\ e^{-i\omega t} \end{pmatrix} = \frac{\hbar}{2}\,\cos 2\omega t
\end{aligned} \tag{14-68}$$

Similarly

$$\begin{aligned}
\langle S_y \rangle &= \tfrac{1}{2}\hbar\,\frac{1}{\sqrt{2}}\,(e^{i\omega t},\, e^{-i\omega t})\begin{pmatrix} 0 & -i \\ i & 0 \end{pmatrix}\frac{1}{\sqrt{2}}\begin{pmatrix} e^{-i\omega t} \\ e^{i\omega t} \end{pmatrix} \\[2mm]
&= \frac{\hbar}{4}\,(-i\,e^{2i\omega t} + i\,e^{-2i\omega t}) \\[2mm]
&= \frac{\hbar}{2}\,\sin 2\omega t
\end{aligned} \tag{14-69}$$

Thus the spin precesses about the direction of B, with frequency

$$2\omega = \frac{egB}{2mc} \approx \frac{eB}{mc} \equiv \omega_c \tag{14-70}$$

the *cyclotron frequency*. This precession occurs if the spin initially points at an arbitrary angle θ with respect to the z-axis. For a magnetic field of the order of 10^4 gauss (1 T),

$$\omega_c = \frac{(4.8 \times 10^{-10} \text{ e.s.u.})(10^4 \text{ gauss})}{(0.9 \times 10^{-27} \text{ g})(3 \times 10^{10} \text{ cm/sec})} \approx 1.8 \times 10^{11} \text{ rad/sec}$$

which is quite a large frequency.

PARAMAGNETIC RESONANCE

In a solid the gyromagnetic factor g of an electron is affected by the nature of the forces acting in the solid. A knowledge of g provides very useful constraints on what these forces could be, and it is therefore important to be able to measure g. This can be done by the *paramagnetic resonance method*. The principle of the method is the following: We have a magnetic field pointing in the z-direction, and the electron spin precesses about that direction. How fast does it do so? If we could introduce a magnetic field that is perpendicular to the z-axis and rotates with the spin, then the field would "see" an electron spin at rest. The component of the electron spin that points in the x-y plane would preferentially align in a direction opposite to the magnetic field to reach the minimum energy state. For those electrons not already aligned in the minimum energy direction, a transition to the lowest energy will take place, and in the process, energy—in the form of radiation—is given up. This can be detected.

It is not practical to have a magnetic field rotating with a frequency of the order of 10^{11} radians per second. If, however, we have a magnetic field that points in the x-direction, say, and oscillates with a frequency ω, it may be viewed as a superposition of a field rotating in the x-y plane clockwise with frequency ω, and a field rotating counterclockwise with the same frequency, with the phase arranged so that the net effect is in the x-direction. (This is analogous to obtaining a linear polarization out of the sum of two circular polarizations.) Only one of the components will travel in the same direction as the precessing spin. The other component will move in a direction opposite to the spin precession, and its effect on the electron spin averages out to zero.

Consider an electron whose only degrees of freedom are the spin states, under the influence of a large magnetic field B_0 pointing in the z direction, and constant in time, and a small oscillating field $B_1 \cos \omega t$, pointing in the x direction. The Schrödinger equation now reads

$$i\hbar \frac{d}{dt}\begin{pmatrix} a(t) \\ b(t) \end{pmatrix} = \frac{eg\hbar}{4mc}\begin{pmatrix} B_0 & B_1 \cos \omega t \\ B_1 \cos \omega t & -B_0 \end{pmatrix}\begin{pmatrix} a(t) \\ b(t) \end{pmatrix} \tag{14-71}$$

or, with

$$\omega_0 = \frac{egB_0}{4mc} = \tfrac{1}{2}\omega_c \qquad \omega_1 = \frac{egB_1}{4mc} \tag{14-72}$$

$$i\frac{da(t)}{dt} = \omega_0 a(t) + \omega_1 \cos \omega t \, b(t)$$

$$i\frac{db(t)}{dt} = \omega_1 \cos \omega t \, a(t) - \omega_0 b(t) \tag{14-73}$$

Let

$$A(t) = a(t)\, e^{i\omega_0 t}$$
$$B(t) = b(t)\, e^{-i\omega_0 t} \tag{14-74}$$

These satisfy the equations

$$i\frac{dA(t)}{dt} = \omega_1 \cos \omega t \, B(t)\, e^{i\omega_c t}$$

$$\approx \tfrac{1}{2}\omega_1\, e^{i(\omega_c - \omega)t} B(t)$$

$$i\frac{dB(t)}{dt} = \omega_1 \cos \omega t \, A(t)\, e^{-i\omega_c t} \tag{14-75}$$

$$\approx \tfrac{1}{2}\omega_1\, e^{-i(\omega_c - \omega)t}\, A(t)$$

In obtaining these, we made an approximation. We wrote

$$\cos \omega t \, e^{i\omega_c t} = \tfrac{1}{2} \left[e^{i(\omega_c + \omega)t} + e^{i(\omega_c - \omega)t} \right]$$
$$\approx \tfrac{1}{2} e^{i(\omega_c - \omega)t}$$

Since we will be interested in values of $\omega = \omega_c$, and since both are large, the term that has been dropped oscillates very rapidly, and we may expect that its contribution averages to zero. A more detailed treatment supports this observation. We can eliminate $B(t)$:

$$B(t) = \frac{2i}{\omega_1} \frac{dA(t)}{dt} e^{-i(\omega_c - \omega)t} \tag{14-76}$$

and use this to obtain a second-order differential equation for $A(t)$:

$$\frac{d^2 A(t)}{dt^2} - i(\omega_c - \omega) \frac{dA(t)}{dt} + \frac{\omega_1^2}{4} A(t) = 0 \tag{14-77}$$

A trial solution is

$$A(t) = A(0) \, e^{i\lambda t} \tag{14-78}$$

When this is inserted into (14-77), the roots of the equation

$$-\lambda^2 + (\omega_c - \omega) \lambda + \frac{\omega_1^2}{4} = 0$$

that is,

$$\lambda_{\pm} = \frac{\omega_c - \omega \pm \sqrt{(\omega_c - \omega)^2 + \omega_1^2}}{2} \tag{14-79}$$

determine λ.

The most general solution is

$$A(t) = A_+ \, e^{i\lambda_+ t} + A_- \, e^{i\lambda_- t} \tag{14-80}$$

and hence

$$B(t) = -\frac{2}{\omega_1} e^{-i(\omega_c - \omega)t} (\lambda_+ A_+ \, e^{i\lambda_+ t} + \lambda_- A_- \, e^{i\lambda_- t}) \tag{14-81}$$

This finally yields

$$a(t) = e^{-i\omega_c t/2} (A_+ \, e^{i\lambda_+ t} + A_- \, e^{i\lambda_- t})$$

$$b(t) = -\frac{2}{\omega_1} e^{-i(\omega_c/2 - \omega)t} (\lambda_+ A_+ \, e^{i\lambda_+ t} + \lambda_- A_- \, e^{i\lambda_- t}) \tag{14-82}$$

If at $t = 0$, the electron spin points in the positive z direction, then $a(0) = 1$ and $b(0) = 0$, that is,

$$A_+ + A_- = 1$$

$$\lambda_+ A_+ + \lambda_- A_- = 0$$

so that

$$A_+ = \frac{\lambda_-}{\lambda_- - \lambda_+}$$

$$A_- = -\frac{\lambda_+}{\lambda_- - \lambda_+}$$

(14-83)

The probability that at some later time t the spin points in the negative z direction is $|b(t)|^2$:

$$|b(t)|^2 = \frac{4}{\omega_1^2} \left| \frac{\lambda_+ \lambda_-}{\lambda_- - \lambda_+} e^{i\lambda_+ t} - \frac{\lambda_+ \lambda_-}{\lambda_- - \lambda_+} e^{i\lambda_- t} \right|^2$$

$$= \frac{\omega_1^2/4}{(\omega_c - \omega)^2 + \omega_1^2} \left| 1 - e^{-i(\lambda_+ - \lambda_-)t} \right|^2$$

(14-84)

$$= \frac{\omega_1^2}{(\omega_c - \omega)^2 + \omega_1^2} \frac{1 - \cos \sqrt{(\omega_c - \omega)^2 + \omega_1^2}\, t}{2}$$

This quantity is small, since $\omega_1 \ll \omega$, ω_c. When the frequency of the field B_1 is "tuned" to match ω_c, then the probability becomes

$$|b(t)|^2 \rightarrow \frac{1 - \cos \omega_1 t}{2}$$

(14-85)

that is, it approaches unity. Since the energy of the "up" state is different from that of the "down" state, such an energy difference, absorbed from the external field, signals the resonance frequency, so that ω_c, and hence g can be measured with great precision.

Problems

1. If the ground-state vector for the harmonic oscillator is given by

$$u_0 = \begin{pmatrix} 1 \\ 0 \\ 0 \\ \vdots \end{pmatrix}$$

use (14-2) and (14-10) to calculate u_1, u_2, u_3. What is the general pattern? Satisfy yourself that

$$\langle u_m | u_n \rangle = \delta_{mn}$$

2. Given a vector

$$\psi = \frac{1}{\sqrt{6}} \begin{pmatrix} 1 \\ 2 \\ 1 \\ 0 \\ \vdots \end{pmatrix}$$

calculate with the harmonic oscillator operators (14-9), (14-10), (14-11) the quantities

(a) $\langle H \rangle$.

(b) $\langle x^2 \rangle$, $\langle x \rangle$, $\langle p^2 \rangle$, $\langle p \rangle$.

(c) Use this to calculate $\Delta p \, \Delta x$.

[*Note:* The expression for p and x in terms of A and $A^\dagger$ can be obtained from (7-4).]

3. Calculate the top left 4×4 corner of the matrix representation of x^4 for the harmonic oscillator.

4. Use (14-21) and (14-22) to calculate the matrix representation of L_x, L_y, and L_z for angular momentum $3/2$. Check that the commutation relations

$$[L_x, L_y] = i\hbar \, L_z$$

and so on are satisfied.

5. You are given the Hamiltonian

$$H = \frac{1}{2I_1} L_x^2 + \frac{1}{2I_2} L_y^2 + \frac{1}{2I_3} L_z^2$$

Find the eigenvalues of H (a) when the angular momentum of the system is 1; (b) when the angular momentum of the system is 2.

Note: The matrix representations of L_x, L_y, L_z for angular momentum 2 are obtainable from

$$L_z = \hbar \begin{pmatrix} 2 & 0 & 0 & 0 & 0 \\ 0 & 1 & 0 & 0 & 0 \\ 0 & 0 & 0 & 0 & 0 \\ 0 & 0 & 0 & -1 & 0 \\ 0 & 0 & 0 & 0 & -2 \end{pmatrix}$$

$$L_+ = \hbar \begin{pmatrix} 0 & 2 & 0 & 0 & 0 \\ 0 & 0 & \sqrt{6} & 0 & 0 \\ 0 & 0 & 0 & \sqrt{6} & 0 \\ 0 & 0 & 0 & 0 & 2 \\ 0 & 0 & 0 & 0 & 0 \end{pmatrix} \qquad L_- = (L_+)^\dagger$$

6. Calculate the eigenvalues of the matrix

$$H = \begin{pmatrix} 8 & 4 & 6 \\ 4 & 14 & 4 \\ 6 & 4 & 8 \end{pmatrix}$$

What are the eigenvectors?

7. Consider a spin $\frac{1}{2}$ system represented by the normalized state vector

$$\begin{pmatrix} \cos \alpha \\ \sin \alpha e^{i\beta} \end{pmatrix}$$

What is the probability that a measurement of S_y yields $-\hbar/2$?

8. Show that for a state of angular momentum 1, the matrices $\mathbf{L} \cdot \mathbf{n}$, where $\mathbf{n}$ is an arbitrary unit vector, satisfy a polynomial equation of the form

$$\sum \alpha_k (\mathbf{L} \cdot \mathbf{n})^k = 0$$

What is the form of that polynomial? Can you generalize that to arbitrary angular momentum l?

9. For angular momentum 1, we can use the $Y_{lm}(\theta, \phi)$ as eigenstates, and differential operators for the $\mathbf{L}$, as discussed in Chapter 10. Show the correspondence by calculating

$$\int \sin\theta \, d\theta d\phi \, Y^*_{1k}(\theta, \phi) \, L_+ \, Y_{1m}(\theta, \phi)$$

and comparing it with the matrix element $(L_+)_{km}$.

10. Consider an angular momentum 1 system, represented by the state vector

$$u = \frac{1}{\sqrt{26}} \begin{pmatrix} 1 \\ 4 \\ -3 \end{pmatrix}$$

What is the probability that a measurement of L_x yields the value 0?

11. Consider a system of angular momentum 1: What are the eigenfunctions and eigenvalues of the operator $L_x L_y + L_y L_x$?

12. Consider a system of spin 1/2. What are the eigenvalues and eigenvectors of the operator $S_x + S_y$? Suppose a measurement of this operator is made, and the system is found to be in the state corresponding to the larger eigenvalue. What is the probability that a measurement of S_z yields $\hbar/2$?

13. The equation for the rate of change of an operator in the Heisenberg picture is given by (7-62). Consider the operators $S_x(t), \ldots$. What are the equations of motion of these operators, if the Hamiltonian is given by

$$H = \frac{eg}{2mc} \mathbf{S}(t) \cdot \mathbf{B}$$

and the commutation relations are $[S_x(t), S_y(t)] = i\hbar S_z(t)$, and so on. If $\mathbf{B} = (0, 0, B)$, solve for $\mathbf{S}(t)$ in terms of $\mathbf{S}(0)$.

14. A spin 1/2 object is in an eigenstate of S_x with eigenvalue $+\hbar/2$ at time $t = 0$. At that time it is placed in a magnetic field $\mathbf{B} = (0, 0, B)$ in which it is allowed to precess for a time T. At that instant the magnetic field is very rapidly rotated in the y direction, so that its components are $(0, B, 0)$. After another time interval T a measurement of S_x is carried out. What is the probability that the value $\hbar/2$ will be found?

15. Work out the behavior of a spin 1 particle in an external magnetic field. Choose $\mathbf{B} = (0, 0, B)$ and take the initial state to be an eigenstate of

$$\mathbf{S} \cdot \mathbf{n} = S_x \sin \theta \cos \phi + S_y \sin \theta \sin \phi + S_z \cos \theta$$

with eigenvalues $\hbar, 0, -\hbar$ in succession.
[*Hint:* Use the matrix representations given by (14-23a) to (14-23c).]

16. The energy for an electron of mass μ, in a magnetic field $\mathbf{B} = \mathbf{k}B$, with zero momentum in the z direction, is given by

$$E = \frac{eB\hbar}{2\mu c} (2n + 1 + |m| + m) \text{ with } m = 0, \pm 1, \pm 2, \pm 3, \ldots$$

(a) How is the expression for the energy modified when account is taken of the fact that the electron has spin $1/2$?

(b) Suppose B is very large. Sketch the energy spectrum, *including the effect of spin* for the lowest four energy states. For each of the levels that you draw, carefully list the values of the quantum numbers that correspond to that energy, that is, n, m and S_z for the electron.

References

The material on spin is standard, and discussions may be found in all of the books listed at the end of this volume.

Chapter 15

THE ADDITION OF ANGULAR MOMENTA

THE ADDITION OF TWO SPINS

Suppose we have two electrons, whose spins are described by the operators S_1 and S_2. Each of these sets of operators satisfies the standard angular momentum commutation relations

$$[S_{1x}, S_{1y}] = i\hbar S_{1z}$$

and so on,

$$[S_{2x}, S_{2y}] = i\hbar S_{2z} \tag{15-1}$$

and so on, but the two sets of operators commute with each other, since the degrees of freedom associated with different particles are independent, that is,

$$[\mathbf{S}_1, \mathbf{S}_2] = 0 \tag{15-2}$$

Let us now define the total spin $\mathbf{S}$ by

$$\mathbf{S} = \mathbf{S}_1 + \mathbf{S}_2 \tag{15-3}$$

The commutation relations obeyed by the components of $\mathbf{S}$ are

$$\begin{aligned}
[S_x, S_y] &= [S_{1x} + S_{2x}, S_{1y} + S_{2y}] \\
&= [S_{1x}, S_{1y}] + [S_{2x}, S_{2y}] \\
&= i\hbar(S_{1z} + S_{2z}) = i\hbar S_z
\end{aligned} \tag{15-4}$$

and so on. We are therefore justified in calling **S** the total *spin*. We may now determine the eigenvalues and eigenfunctions of $\mathbf{S}^2$ and S_z.

The two-spin system actually has four states. If we denote the spinor of the first electron by $\chi_\pm^{(1)}$, so that

$$S_1^2\chi_\pm^{(1)} = \tfrac{1}{2}(\tfrac{1}{2} + 1)\,\hbar^2\chi_\pm^{(1)}$$
$$S_{1z}\chi_\pm^{(1)} = \pm\tfrac{1}{2}\hbar\chi_\pm^{(1)} \tag{15-5}$$

and similarly for the spinor $\chi_\pm^{(2)}$ of the second electron, then the four states are

$$\chi_+^{(1)}\chi_+^{(2)},\ \chi_+^{(1)}\chi_-^{(2)},\ \chi_-^{(1)}\chi_+^{(2)},\ \chi_-^{(1)}\chi_-^{(2)} \tag{15-6}$$

The eigenvalues of S_z for the four states are

$$S_z\chi_\pm^{(1)}\chi_\pm^{(2)} = (S_{1z} + S_{2z})\chi_\pm^{(1)}\chi_\pm^{(2)}$$
$$= (S_{1z}\chi_\pm^{(1)})\,\chi_\pm^{(2)} + \chi_\pm^{(1)}(S_{2z}\chi_\pm^{(2)})$$

that is,

$$S_z\chi_+^{(1)}\chi_+^{(2)} = \hbar\chi_+^{(1)}\chi_+^{(2)}$$
$$S_z\chi_+^{(1)}\chi_-^{(2)} = S_z\chi_-^{(1)}\chi_+^{(2)} = 0 \tag{15-7}$$
$$S_z\chi_-^{(1)}\chi_-^{(2)} = -\hbar\chi_-^{(1)}\chi_-^{(2)}$$

There are two states with m-value 0. One might expect that one linear combination of them will form an $S = 1$ state, to form a triplet with the $m = 1$ and $m = -1$ states, and the orthogonal combination will form a singlet $S = 0$ state. To check this expectation, let us construct the lowering operator

$$S_- = S_{1-} + S_{2-} \tag{15-8}$$

and apply this to the $m = 1$ state. This should give us the $m = 0$ state that belongs to the $S = 1$ triplet, aside from a coefficient in front. Indeed, using the fact that

$$S_-^{(i)}\chi_+^{(i)} = \hbar\chi_-^{(i)} \tag{15-9}$$

which can be established by noting that

$$\tfrac{1}{2}\hbar\left[\begin{pmatrix} 0 & 1 \\ 1 & 0 \end{pmatrix} - i\begin{pmatrix} 0 & -i \\ i & 0 \end{pmatrix}\right]\begin{pmatrix} 1 \\ 0 \end{pmatrix} = \hbar\begin{pmatrix} 0 \\ 1 \end{pmatrix} \tag{15-10}$$

we get

$$S_-\chi_+^{(1)}\chi_+^{(2)} = (S_{1-}\chi_+^{(1)})\,\chi_+^{(2)} + \chi_+^{(1)}S_{2-}\chi_+^{(2)}$$
$$= \hbar\chi_-^{(1)}\chi_+^{(2)} + \hbar\chi_+^{(1)}\chi_-^{(2)} \tag{15-11}$$
$$= \sqrt{2}\hbar\,\frac{\chi_+^{(1)}\chi_-^{(2)} + \chi_-^{(1)}\chi_+^{(2)}}{\sqrt{2}}$$

The linear combination has been normalized, and the compensating factor in front, $\sqrt{2}\hbar$, agrees with what one would expect from (11-36) and (11-48) with $l = m = 1$. If we now apply S_- to this linear combination, and note that

$$S_-^{(i)}\chi_-^{(i)} = 0 \tag{15-12}$$

we get

$$\begin{aligned} S_- \frac{\chi_+^{(1)}\chi_-^{(2)} + \chi_-^{(1)}\chi_+^{(2)}}{\sqrt{2}} &= \frac{\hbar}{\sqrt{2}} (\chi_-^{(1)}\chi_-^{(2)} + \chi_-^{(1)}\chi_-^{(2)}) \\ &= \sqrt{2}\hbar\chi_-^{(1)}\chi_-^{(2)} \end{aligned} \tag{15-13}$$

as we should, for an angular momentum state $S = 1$. The remaining state, constructed to be orthogonal to (15-11) and properly normalized, is

$$\frac{1}{\sqrt{2}} (\chi_+^{(1)}\chi_-^{(2)} - \chi_-^{(1)}\chi_+^{(2)}) \tag{15-14}$$

and because it has no partners, we conjecture that it is an $S = 0$ state. In order to check this, we compute $\mathbf{S}^2$ for the two states

$$X_\pm = \frac{1}{\sqrt{2}} (\chi_+^{(1)}\chi_-^{(2)} \pm \chi_-^{(1)}\chi_+^{(2)}) \tag{15-15}$$

We have

$$\begin{aligned} \mathbf{S}^2 &= (\mathbf{S}_1 + \mathbf{S}_2)^2 = \mathbf{S}_1^2 + \mathbf{S}_2^2 + 2\mathbf{S}_1 \cdot \mathbf{S}_2 \\ &= \mathbf{S}_1^2 + \mathbf{S}_2^2 + 2S_{1z}S_{2z} + S_{1+}S_{2-} + S_{1-}S_{2+} \end{aligned} \tag{15-16}$$

First of all,

$$\begin{aligned} \mathbf{S}_1^2 X_\pm &= \frac{1}{\sqrt{2}} (\chi_-^{(2)}\mathbf{S}_1^2\chi_+^{(1)} \pm \chi_+^{(2)}\mathbf{S}_1^2\chi_-^{(1)}) \\ &= \tfrac{3}{4}\hbar^2 X_\pm \end{aligned} \tag{15-17}$$

and similarly

$$\mathbf{S}_2^2 X_\pm = \tfrac{3}{4}\hbar^2 X_\pm \tag{15-18}$$

Next, we calculate

$$2S_{1z}S_{2z}X_\pm = 2(\tfrac{1}{2}\hbar)(-\tfrac{1}{2}\hbar) X_\pm = -\tfrac{1}{2}\hbar^2 X_\pm \tag{15-19}$$

Finally

$$\begin{aligned} (S_{1+}S_{2-} + S_{1-}S_{2+}) X_\pm &= \frac{1}{\sqrt{2}} (S_{1+}\chi_+^{(1)}S_{2-}\chi_-^{(2)} + S_{1-}\chi_+^{(1)}S_{2+}\chi_-^{(2)} \\ &\pm S_{1+}\chi_-^{(1)}S_{2-}\chi_+^{(2)} \pm S_{1-}\chi_-^{(1)}S_{2+}\chi_+^{(2)}) \end{aligned}$$

which, with the help of (15-9) and (15-12) yields

$$(S_{1+}S_{2-} + S_{1-}S_{2+}) X_{\pm} = \pm\hbar^2 X_{\pm} \qquad (15\text{-}20)$$

Thus

$$\mathbf{S}^2 X_{\pm} = \hbar^2(\tfrac{3}{4} + \tfrac{3}{4} - \tfrac{1}{2} \pm 1) X_{\pm} = \begin{pmatrix} 2 \\ 0 \end{pmatrix} \hbar^2 X_{\pm}$$

$$= \hbar^2 S(S + 1) X_{\pm} \qquad (15\text{-}21)$$

with $S = 1$ and 0 corresponding to the $\pm$ states.

What we have shown is that the totality of the four states of two spin 1/2 particles may be recombined into a triplet and into a singlet total spin state. It is important to note that the two descriptions are entirely equivalent. In one case we have as our complete set of commuting observables $\mathbf{S}_1^2$, $\mathbf{S}_2^2$, S_{1z}, and S_{2z}. In the other case we have as our complete set of commuting observables $\mathbf{S}^2$, S_z, $\mathbf{S}_1^2$, $\mathbf{S}_2^2$. By the expansion theorem, any function can be expanded in terms of a complete set of eigenstates. *What we have demonstrated here is the expansion of the eigenstates of the second set of observables in terms of the complete set of states of the first set of observables.* This is quite analogous to the expression of the eigenstates of the hydrogen atom in terms of the eigenstates of the momentum operator, in which the coefficients (the analogs of the $1/\sqrt{2}$'s here) are the momentum-space wave functions. It is a simple exercise to invert the process and to find the products of the $\chi^{(1)}\chi^{(2)}$ in terms of triplet and singlet combinations.

In physical problems it frequently happens that to first approximation two sets of completely commuting observables are equally useful in the construction of eigenstates. In next approximation, when additional terms in the Hamiltonian are taken into account, only one of these sets remains useful. A simple example occurs in low energy nuclear physics.

In early studies of the potential $V(r)$ that describes the interaction between neutrons and protons at low energies, it became clear that the strength of the interaction depended on whether the two interacting particles were in a total spin $S = 1$ state or a spin $S = 0$ state. For example, a deuteron had $S = 1$, while the corresponding $S = 0$ state of a neutron and a proton did not bind. This can be described in terms of a spin-dependent potential. Suppose we have

$$V(r) = V_1(r) + \frac{1}{\hbar^2} \mathbf{S}_1 \cdot \mathbf{S}_2 V_2(\mathbf{r}) \qquad (15\text{-}22)$$

we can easily see that S_{1z} and S_{2z} do not commute with the second term, so that the eigenstates of H containing this potential cannot just be simple products of eigenstates of S_{1z} and S_{2z}. If we observe, however, that

$$\mathbf{S}_1 \cdot \mathbf{S}_2 = \tfrac{1}{2}(\mathbf{S}^2 - \mathbf{S}_1^2 - \mathbf{S}_2^2) \qquad (15\text{-}23)$$

so that this term can be replaced by the eigenvalue, when acting on an eigenfunction of $\mathbf{S}^2$, $\mathbf{S}_1^2$, and $\mathbf{S}_2^2$, then

$$V(r) = V_1(r) + \frac{1}{2} V_2(r) \left[S(S + 1) - \frac{3}{2} \right]$$

$$= V_1(r) + \frac{1}{4} \begin{pmatrix} 1 \\ -3 \end{pmatrix} V_2(r) \begin{cases} S = 1 \\ S = 0 \end{cases} \qquad (15\text{-}24)$$

Such a spin-dependent potential is actually observed in the neutron–proton system. The bound state is an $S = 1$ state—this is the deuteron—but there is also an unbound $S = 0$ state. This implies that $V_1 - (3/4)V_2$ is a less attractive potential than $V_1 + (1/4)V_2$. This is only possible if $V_2(r) \neq 0$.

The spin singlet wave function (15-14) implies that, if in a measurement electron (2) is found in an "up" state, then electron (1) *must be* in a spin "down" state. Although the electrons are identical, we may consider a singlet state in which the electrons are moving to the right and to the left with equal and opposite momenta, so that the two-electron system is still at rest in the center of mass. Thus electron (2) could be the one moving to the right, and electron (1) the one moving to the left, and saying that the right one is in an "up" state has a well-defined meaning.

We may ask a more interesting question. Suppose a measurement of S_x is made on electron (2) and it is found that the eigenvalue is $\hbar/2$, that is, electron (2) is in an "up" state along the x-axis. What will a measurement of S_x on electron (1) produce? Since the two electrons are separated by a large distance, one might think that either $+\hbar/2$ of $-\hbar/2$ might be found, perhaps with equal probability, on the grounds that the information about the result of a "projection" of electron (2) into a particular eigenstate of S_x cannot propagate with infinite speed to affect the measurement on electron (1). This is indeed what one would be led to expect if one were to accept certain criteria of what a *complete* physical theory should have, as articulated in a paper by A. Einstein, N. Rosen, and B. Podolsky.[1] On the other hand, quantum mechanics maintains that the two-spin system is described by a single wave function in which the spins are correlated. A measurement of part of the system, the S_x of one electron, combined with the knowledge that the system is in a spin singlet state, is really a measurement of the whole wavefunction. Thus if S_x of electron (2) yields $\hbar/2$, then a measurement of S_x of electron (1) must yield $-\hbar/2$. To see this formally, note that the eigenstates $\chi_\pm$ can be decomposed into eigenstates of S_x, which we denote by $\xi_\pm$. We saw [see (14-56 and (14-57) with $\phi = 0$] that

$$\xi_\pm = \frac{1}{\sqrt{2}}(\chi_+ \pm \chi_-)$$

or equivalently

$$\chi_\pm = \frac{1}{\sqrt{2}}(\xi_+ \pm \xi_-)$$

Let us substitute this into

$$\psi = \frac{1}{\sqrt{2}}(\chi_+^{(1)}\chi_-^{(2)} - \chi_-^{(1)}\chi_+^{(2)})$$

We get

$$\psi = (1/\sqrt{2})^{(3)}[(\xi_+^{(1)} + \xi_-^{(1)})(\xi_+^{(2)} - \xi_-^{(2)}) - (\xi_+^{(1)} - \xi_-^{(1)})(\xi_+^{(2)} + \xi_-^{(2)})]$$
$$= \frac{1}{\sqrt{2}}(\xi_+^{(1)}\xi_-^{(2)} - \xi_-^{(1)}\xi_+^{(2)}) \tag{15-25}$$

---

[1]This is nicely discussed in D. Bohm, *Quantum Theory* and from a more modern point of view, connected with the researches of J. S. Bell, in J. J. Sakurai *Modern Quantum Mechanics*. See also the Afterword in David J. Griffiths, *Introduction to Quantum Mechanics*, Prentice Hall, Englewood Cliffs, N.J. 1995

which clearly indicates that if electron (2) is in the "up" state, then electron (1) must be in the "down" state.

THE ADDITION OF SPIN 1/2 AND ORBITAL ANGULAR MOMENTUM

Much more important for future applications is the combination of a spin with an orbital angular momentum. Since $\mathbf{L}$ depends on spatial coordinates and $\mathbf{S}$ does not, they commute

$$[\mathbf{L}, \mathbf{S}] = 0 \tag{15-26}$$

It is therefore evident that the components of the total angular momentum $\mathbf{J}$, defined by

$$\mathbf{J} = \mathbf{L} + \mathbf{S} \tag{15-27}$$

will satisfy the angular momentum commutation relations.

In asking for linear combinations of the Y_{lm} and the $\chi_{\pm}$ that are eigenstates of

$$J_z = L_z + S_z \tag{15-28}$$

and

$$\begin{aligned}
\mathbf{J}^2 &= \mathbf{L}^2 + \mathbf{S}^2 + 2\mathbf{L} \cdot \mathbf{S} \\
&= \mathbf{L}^2 + \mathbf{S}^2 + 2L_z S_z + L_+ S_- + L_- S_+
\end{aligned} \tag{15-29}$$

we are again looking for the expansion coefficients of one complete set of eigenfunctions in terms of another set of eigenfunctions.

Let us consider the linear combination

$$\psi_{j,m+1/2} = \alpha Y_{lm}\chi_+ + \beta Y_{l,m+1}\chi_- \tag{15-30}$$

It is, by construction, an eigenfunction of J_z with eigenvalue $(m + \frac{1}{2})\hbar$. We now determine α and β such that it is also an eigenfunction of J^2. We shall make use of the fact that

$$\begin{aligned}
L_+ Y_{lm} &= [l(l + 1) - m(m + 1)]^{1/2} \hbar Y_{l,m+1} \\
&= [(l + m + 1)(l - m)]^{1/2} \hbar Y_{l,m+1} \\
L_- Y_{lm} &= [(l - m + 1)(l + m)]^{1/2} \hbar Y_{l,m-1} \\
S_+ \chi_+ &= S_- \chi_- = 0 \qquad S_{\pm}\chi_{\mp} = \hbar\chi_{\pm}
\end{aligned} \tag{15-31}$$

Then

$$\begin{aligned}
J^2\psi_{j,m+1/2} = \alpha\hbar^2 \{ & l(l + 1) \, Y_{lm}\chi_+ + \tfrac{3}{4}Y_{lm}\chi_+ + 2m(\tfrac{1}{2}) \, Y_{lm}\chi_+ \\
& + [(l - m)(l + m + 1)]^{1/2} \, Y_{l,m+1}\chi_-\} + \beta\hbar^2\{l(l + 1) \, Y_{l,m+1}\chi_- \\
& + \tfrac{3}{4}Y_{l,m+1}\chi_- + 2(m + 1)(-\tfrac{1}{2}) \, Y_{l,m+1}\chi_- \\
& + [(l - m)(l + m + 1)]^{1/2} \, Y_{lm}\chi_+\}
\end{aligned} \tag{15-32}$$

This will be of the form

$$\hbar^2 j(j + 1)\, \psi_{j,m+1/2} = \hbar^2 j(j + 1)(\alpha Y_{lm}\chi_+ + \beta Y_{l,m+1}\chi_-) \tag{15-33}$$

provided that

$$\alpha[l(l + 1) + \tfrac{3}{4} + m] + \beta[(l - m)(l + m + 1)]^{1/2} = j(j + 1)\, \alpha$$
$$\beta[l(l + 1) + \tfrac{3}{4} - m - 1] + \alpha[(l - m)(l + m + 1)]^{1/2} = j(j + 1)\, \beta \tag{15-34}$$

This requires that

$$(l - m)(l + m + 1) = [j(j + 1) - l(l + 1) - \tfrac{3}{4} - m]$$
$$\times\, [j(j + 1) - l(l + 1) - \tfrac{3}{4} + m + 1]$$

which evidently has two solutions,

$$j(j + 1) - l(l + 1) - \tfrac{3}{4} = \begin{cases} -l - 1 \\ l \end{cases} \tag{15-35}$$

that is,

$$j = \begin{cases} l - \tfrac{1}{2} \\ l + \tfrac{1}{2} \end{cases} \tag{15-36}$$

For $j = l + 1/2$, we get, after a little algebra

$$\alpha = \sqrt{\frac{l + m + 1}{2l + 1}} \qquad \beta = \sqrt{\frac{l - m}{2l + 1}} \tag{15-37}$$

(Actually we just get the ratio; these are already normalized forms.) Thus

$$\psi_{l+1/2,m+1/2} = \sqrt{\frac{l + m + 1}{2l + 1}}\, Y_{lm}\chi_+ + \sqrt{\frac{l - m}{2l + 1}}\, Y_{l,m+1}\chi_- \tag{15-38}$$

We can guess that the $j = l - 1/2$ solution must have the form

$$\psi_{l-1/2,m+1/2} = \sqrt{\frac{l - m}{2l + 1}}\, Y_{lm}\chi_+ - \sqrt{\frac{l + m + 1}{2l + 1}}\, Y_{l,m+1}\chi_- \tag{15-39}$$

in order to be orthogonal to the $j = l + 1/2$ solution.

GENERAL RULES FOR ADDITION
OF ANGULAR MOMENTA, AND IMPLICATIONS
FOR IDENTICAL PARTICLES

These two examples illustrate the general features that are involved in the addition of angular momenta: If we have the eigenstates $Y_{l_1 m_1}^{(1)}$ of $\mathbf{L}_1^2$ and L_{1z}, and the eigenstates $Y_{l_2 m_2}^{(2)}$ of $\mathbf{L}_2^2$ and L_{2z}, then we can form $(2l_1 + 1)(2l_2 + 1)$ product wave functions

$$Y_{l_1 m_1}^{(1)} Y_{l_2 m_2}^{(2)} \quad \begin{Bmatrix} -l_1 \le m_1 \le l_1 \\ -l_2 \le m_2 \le l_2 \end{Bmatrix} \tag{15-40}$$

These can be classified by the eigenvalue of

$$J_z = L_{1z} + L_{2z} \tag{15-41}$$

which is $m_1 + m_2$, and which ranges from a maximum value of $l_1 + l_2$ down to $-l_1 - l_2$. As in the simple cases discussed earlier, different linear combinations of functions with the same m value will belong to different values of j. In the following table we list the possible combinations for the special example of $l_1 = 4$, $l_2 = 2$. We shall use the simple abbreviation (m_1, m_2) for $Y_{l_1 m_1}^{(1)} Y_{l_2 m_2}^{(2)}$.

m-value	m_1, m_2 combinations	number
6	(4, 2)	1
5	(4, 1) (3, 2)	2
4	(4, 0) (3, 1) (2, 2)	3
3	(4, −1) (3, 0) (2, 1) (1, 2)	4
2	(4, −2) (3, −1) (2, 0) (1, 1) (0, 2)	5
1	(3, −2) (2, −1) (1, 0) (0, 1) (−1, 2)	5
0	(2, −2) (1, −1) (0, 0) (−1, 1) (−2, 2)	5
−1	(1, −2) (0, −1) (−1, 0) (−2, 1) (−3, 2)	5
−2	(0, −2) (−1, −1) (−2, 0) (−3, 1) (−4, 2)	5
−3	(−1, −2) (−2, −1) (−3, 0) (−4, 1)	4
−4	(−2, −2) (−3, −1) (−4, 0)	3
−5	(−3, −2) (−4, −1)	2
−6	(−4, −2)	1

There are a total of 45 combinations, consistent with $(2l_1 + 1)(2l_2 + 1)$.

The highest state has total angular momentum $l_1 + l_2$ as can easily be checked by applying J^2 to $Y_{l_1 l_1}^{(1)} Y_{l_2 l_2}^{(2)}$:

$$\begin{aligned} J^2 Y_{l_1 l_1}^{(1)} Y_{l_2 l_2}^{(2)} &= (\mathbf{L}_1^2 + \mathbf{L}_2^2 + 2L_{1z}L_{2z} + L_{1+}L_{2-} + L_{1-}L_{2+}) \, Y_{l_1 l_1}^{(1)} Y_{l_2 l_2}^{(2)} \\ &= \hbar^2 [l_1(l_1 + 1) + l_2(l_2 + 1) + 2l_1 l_2] \, Y_{l_1 l_1}^{(1)} Y_{l_2 l_2}^{(2)} \\ &= \hbar^2 (l_1 + l_2)(l_1 + l_2 + 1) \, Y_{l_1 l_1}^{(1)} Y_{l_2 l_2}^{(2)} \end{aligned} \tag{15-42}$$

This is $j = 6$ in the example discussed in the table. Successive applications of

$$J_- = L_{1-} + L_{2-} \tag{15-43}$$

will pick out one linear combination from each row in the table. These will form the 13 states that belong to $j = 6$. When this is done, there remains a single state with $m = 5$, two with $m = 4, \ldots$, one with $m = -5$. It is extremely plausible, and can, in fact, be checked, that the $m = 5$ state belongs to $j = 5$. Again successive applications of J_- pick out another linear combination from each row in the table, forming 11 states that belong to $j = 5$. Repetition of this procedure shows that we get, after this, sets that belong to $j = 4, j = 3$, and finally $j = 2$. The multiplicities add up to 45:

$$13 + 11 + 9 + 7 + 5 = 45$$

We shall not work out the details of this decomposition, as it is beyond the scope of this book. We merely state the results.

(a) The products $Y^{(1)}_{l_1 m_1} Y^{(2)}_{l_2 m_2}$ can be decomposed into eigenstates of $\mathbf{J}^2$, with eigenvaues $j(j + 1)\, \hbar^2$, where j can take on the values

$$j = l_1 + l_2, l_1 + l_2 - 1, \ldots, |l_1 - l_2| \tag{15-44}$$

We can verify that the multiplicities check in (15-44): if we sum the number of states, we get $(l_1 \geq l_2)$

$$[2(l_1 + l_2) + 1] + [2(l_1 + l_2 - 1) + 1] + \cdots + [2(l_1 - l_2) + 1]$$

$$= \sum_{n=0}^{2l_2} [2(l_1 - l_2 + n) + 1] \tag{15-45}$$

$$= (2l_2 + 1)(2l_1 + 1)$$

(b) It is possible to generalize (15-38) and (15-39) to give the Clebsch-Gordan series

$$\psi_{jm} = \sum_{m1} C(jm;\, l_1 m_1 l_2 m_2)\ Y^{(1)}_{l_1 m_1} Y^{(2)}_{l_2 m_2} \tag{15-46}$$

The coefficients $C(jm; l_1 m_1 l_2 m_2)$ are called Clebsch-Gordan coefficients, and they have been tabulated for many values of the arguments. We calculated the coefficients for $l_2 = 1/2$, and summarize (15-37) and (15-38) in the table below. Note that $m = m_1 + m_2$, so that the m in (15-37) and (15-38) is really m_1 below.

$$C(jm; l_1 m_1, 1/2, m_2)$$

	$m_2 = 1/2$	$m_2 = -1/2$
$j = l_1 + 1/2$	$\sqrt{\dfrac{l_1 + m + 1/2}{2l_1 + 1}}$	$\sqrt{\dfrac{l_1 - m + 1/2}{2l_1 + 1}}$
$j = l_1 - 1/2$	$-\sqrt{\dfrac{l_1 - m + 1/2}{2l_1 + 1}}$	$\sqrt{\dfrac{l_1 + m + 1/2}{2l_1 + 1}}$

Another useful table is

$$C(jm; l_1 m_1, 1, m_2)$$

	$m_2 = 1$	$m_2 = 0$	$m_2 = -1$
$j = l_1 + 1$	$\sqrt{\dfrac{(l_1 + m)(l_1 + m + 1)}{(2l_1 + 1)(2l_1 + 2)}}$	$\sqrt{\dfrac{(l_1 - m + 1)(l_1 + m + 1)}{(2l_1 + 1)(l_1 + 1)}}$	$\sqrt{\dfrac{(l_1 - m)(l_1 - m + 1)}{(2l_1 + 1)(2l_1 + 2)}}$
$j = l_1$	$-\sqrt{\dfrac{(l + m)(l_1 - m + 1)}{2l_1(l_1 + 1)}}$	$\dfrac{m}{\sqrt{l_1(l_1 + 1)}}$	$\sqrt{\dfrac{(l_1 - m)(l_1 + m + 1)}{2l_1(2l_1 + 1)}}$
$j = l_1 - 1$	$\sqrt{\dfrac{(l_1 - m)(l_1 - m + 1)}{2l_1(2l_1 + 1)}}$	$-\sqrt{\dfrac{(l_1 - m)(l_1 + m)}{l_1(2l_1 + 1)}}$	$\sqrt{\dfrac{(l_1 + m)(l_1 + m + 1)}{2l_1(2l_1 + 1)}}$

A final comment is in order. We noted, when discussing identical particles, that a system of two electrons (or more generally, two fermions) must be in a state that is antisymmetric under the interchange of the two particles. This interchange involves not only the exchange of the spatial coordinates, but also of the spin labels. For a system of two identical spin $1/2$ particles, the $S = 1$ triplet of states

$$\chi_+^{(1)} \chi_+^{(2)}$$
$$\frac{1}{\sqrt{2}} (\chi_+^{(1)} \chi_-^{(2)} + \chi_-^{(1)} \chi_+^{(2)}) \qquad (15\text{-}47)$$
$$\chi_-^{(1)} \chi_-^{(2)}$$

is symmetric under spin label interchange, while the $S = 0$ (singlet)

$$\frac{1}{\sqrt{2}} (\chi_+^{(1)} \chi_-^{(2)} - \chi_-^{(1)} \chi_+^{(2)}) \qquad (15\text{-}48)$$

is antisymmetric. Thus for a triplet state, the spatial wave function must be anti-symmetric, and for a singlet state, it must be symmetric. The spatial wave function of a two-particle system in their center-of-mass system is of the general form

$$u(\mathbf{r}) = R_{nlm}(r) \, Y_{lm}(\theta, \phi) \qquad (15\text{-}49)$$

An interchange of the coordinates of the two particles is equivalent to the change

$$r \rightarrow r$$
$$\theta \rightarrow \pi - \theta \qquad (15\text{-}50)$$
$$\phi \rightarrow \phi + \pi$$

Thus the radial function remains unchanged. However, under this transformation

$$Y_{lm}(\theta \ \phi) \rightarrow Y_{lm}(\pi - \theta, \phi + \pi)$$
$$= (-1)^l \, Y_{lm}(\theta, \phi) \qquad (15\text{-}51)$$

Thus triplet states must have odd orbital angular momentum l, and singlet states must have even orbital angular momentum. We shall see an application of this when we discuss the states of helium.

SOME COMMENTS ON PARITY

Note that the same argument can be applied to checking the properties of the Y_{lm} under inversion. The transformation $x \rightarrow -x$, $y \rightarrow -y$, and $z \rightarrow -z$ is identical to (15-50). Thus we see that a particle that is in an orbital angular momentum state will have its wave function changed by $(-1)^l$. Even l orbital states are also even-parity states, and odd l orbital states are odd-parity states. It should be noted, however, that the particles themselves can have an intrinsic parity. We can define the intrinsic parity of the electron and the proton and neutron to be even. In that case, the parity of a hydrogen $l = 1$ state, for example, is odd, while the parity of the ground state is even.

In relativistic quantum mechanics one can show that the intrinsic parity of an antiparticle of a fermion is opposite to that of the fermion. Thus the e^+ has negative intrinsic parity, and hence the ground state of positronium for which $l = 0$, has *negative* parity.

An interesting application of these remarks occurs in elementary particle physics. One of the first unstable elementary particles to be discovered was the π meson predicted by Yukawa. This particle, which plays an important role in nuclear forces, comes in three charge states π^+, π^0, π^-. It was found to have spin 0, and the question arose whether the wave function of a pion—as this meson came to be called—was even or odd under reflection, assuming that the known particles, the proton and the neutron, had positive intrinsic parity. The following experiment was suggested.

Consider the capture of a π^- by a deuteron. A slow pion in liquid deuterium loses energy by a variety of mechanisms, until it finally ends up in the lowest Bohr orbit about the (pn) nucleus, and is then captured through the action of the nuclear forces. In the nuclear reaction

$$\pi^- + d \rightarrow n + n$$

the angular momentum is 1; the pion has zero spin, the orbital angular momentum is zero in the lowest Bohr state, so that the only contribution is the angular momentum of the deuteron, which is 1. The two neutrons must therefore be in an angular momentum 1 state. If the total spin of the two neutrons is 0, then the orbital angular momentum must be 1. If the total spin of the two-neutron state is 1, then orbital angular momentum 0, 1, and 2 is possible, since adding two angular momenta of one unit each can yield 0, 1, and 2, and adding one unit to two units of angular momentum can yield 3, 2, and 1. However, a singlet state of two identical fermions must have even angular momentum, and is thus excluded. A triplet state must have odd orbital angular momentum, and this is possible if the orbital angular momentum is 1. Such a state, however, has odd parity by (15-51), and hence the pion must have odd parity. In terms of the spectroscopic notation, which we shall use, where a state is labeled according to

$$^{2S+1}L_j \tag{15-52}$$

the two neutron states, from the total class of states $^1S_0$, $^1P_1$, $^1D_2$, $^1F_3$, ..., $^3S_1$, $^3P_2$, $^3P_1$, $^3P_0$, $^3D_3$, $^3D_2$, $^3D_1$, $^3F_4$, $^3F_3$, $^3F_2$, ..., are restricted to $^1S_0$, $^1D_2$, ..., $^3P_{2,1,0}$, $^3F_{4,3,2}$, ... by the Fermi-Dirac statistics argument, and of these there is only one state, the $^3P_1$ state, that has angular momentum 1.

Problems

1. Work out the generalization of (15-38) and (15-39) to the addition of orbital angular momentum L to spin 1.

   (a) Find the eigenstates of $\mathbf{S}^2$ and S_z, where

   $$S_z = \hbar \begin{pmatrix} 1 & 0 & 0 \\ 0 & 0 & 0 \\ 0 & 0 & -1 \end{pmatrix}$$

   (b) If these eigenstates are labeled ξ_{+1}, ξ_0, and ξ_{-1}, find the action of S_+ and S_- on these states.

   (c) Calculate the effect of

   $$\mathbf{J}^2 = \mathbf{L}^2 + \mathbf{S}^2 + 2L_z S_z + L_+ S_- + L_- S_+$$

   on combinations like

   $$\psi_{jm+1} = \alpha Y_{lm}\xi_1 + \beta Y_{l,m+1}\xi_0 + \gamma Y_{l,m+2}\xi_{-1}$$

   (d) Determine the relations between α, β, and γ obtained from

   $$\mathbf{J}^2 \Psi_{j,m} = \hbar^2 j\, (j + 1)\, \Psi_{j,m}$$

2. Find the analog of (15-47) for two spin 1 particles, which can combine to form spin 2, 1, and 0 states. Use the notation $\xi_{+1}^{(i)}$, $\xi_0^{(i)}$, $\xi_{-1}^{(i)}$ for the one-particle spin vectors.

3. A deuteron has spin 1. What are the possible spin and total angular momentum states of two deuterons in an arbitrary angular momentum state L? Do not forget the symmetrization rules.

4. A particle of spin 1 moves in a central potential of the form

   $$V(r) = V_1(r) + \frac{\mathbf{S} \cdot \mathbf{L}}{\hbar^2} V_2(r) + \frac{(\mathbf{S} \cdot \mathbf{L})^2}{\hbar^4} V_3(r)$$

   What are the values of $V(r)$ in the states $J = L + 1$, L, and $L - 1$?

5. Consider the discussion of the determination of the parity of the π^-. Suppose that π^- had spin 1, but was still captured in an $L = 0$ orbital state in the reaction

   $$\pi^- + d \rightarrow 2n$$

   What are the possible two-neutron states? Which states are allowed if the π^- had negative parity?

6. Suppose that π^- has spin 0 and negative parity, but is captured in the reaction

   $$\pi^- + d \rightarrow 2n$$

   from the P orbit. Show that the two neutrons must be in a singlet state.

7. The Hamiltonian of a spin system is given by

   $$H = A + \frac{B\mathbf{S}_1 \cdot \mathbf{S}_2}{\hbar^2} + \frac{C(S_{1z} + S_{2z})}{\hbar}$$

Find the eigenvalues and eigenfunctions of the system of two particles, (a) when both particles have spin $1/2$; (b) when one of the particles has spin $1/2$ and the other has spin 1. Assume in (a) that the two particles are identical.

8. Consider two spin $1/2$ particles, whose spins are described by the Pauli operators $\boldsymbol{\sigma}_1$ and $\boldsymbol{\sigma}_2$. Let $\hat{\mathbf{e}}$ be the unit vector connecting the two particles and define the operator

$$S_{12} = 3(\boldsymbol{\sigma}_1 \cdot \hat{\mathbf{e}})(\boldsymbol{\sigma}_2 \cdot \hat{\mathbf{e}}) - \boldsymbol{\sigma}_1 \cdot \boldsymbol{\sigma}_2$$

Show that if the two particles are in an $S = 0$ state (singlet) then

$$S_{12}X_{\text{singlet}} = 0$$

Show that for a triplet state

$$(S_{12} - 2)(S_{12} + 4)\, X_{\text{triplet}} = 0$$

(*Hint:* Choose $\hat{e}$ along the z-axis.)

9. In a low energy neutron–proton system (which has zero orbital angular momentum) the potential energy is given by

$$V(r) = V_1(r) + V_2(r) \left(3\, \frac{(\boldsymbol{\sigma}_1 \cdot \mathbf{r})(\boldsymbol{\sigma}_2 \cdot \mathbf{r})}{r^2} - \boldsymbol{\sigma}_1 \cdot \boldsymbol{\sigma}_2 \right) + V_3(r)\boldsymbol{\sigma}_1 \cdot \boldsymbol{\sigma}_2$$

where $\mathbf{r}$ is the vector connecting the two particles. Calculate the potential energy for the neutron–proton system

(a) In the spin singlet state.

(b) In the triplet state.

10. Consider two electrons in a spin singlet state.

(a) If a measurement of the spin of one of the electrons shows that it is in a state with $s_z = 1/2$, what is the probability that a measurement of the z-component of the spin of the other electron yields $s_z = 1/2$?

(b) If a measurement of the spin of one of the electrons shows that it is in a state with $s_y = 1/2$, what is the probability that a measurement of the x-component of the spin yields $s_x = -1/2$ for the second electron?

(c) If electron (1) is in a state described by $\cos \alpha_1 \chi_+ + \sin \alpha_1 e^{i\beta_1}\chi_-$ and electron (2) is in a state described by $\cos \alpha_2 \chi_+ + \sin \alpha_2 e^{i\beta_2}\chi_-$, what is the probability that the two-electron state is in a triplet state?

References

The material discussed here is also discussed in one way or another in every textbook on quantum mechanics. Many details can be found in

M. E. Rose, *Elementary Theory of Angular Momentum*, John Wiley & Sons, New York, 1957.

Chapter 16

TIME-INDEPENDENT PERTURBATION THEORY

PERTURBATION THEORY FOR NONDEGENERATE STATES

There are few potentials $V(r)$ for which the Schrödinger equation is exactly solvable, and we have already discussed most of them. We must therefore develop approximation techniques to obtain the eigenvalues and eigenfunctions for potentials that do not lead to exactly soluble equations. In this chapter we discuss perturbation theory. We assume that we have found the eigenvalues and the complete set of normalized eigenfunctions for a Hamiltonian H_0,

$$H_0\phi_n = E_n^0\phi_n \tag{16-1}$$

and we ask for the eigenvalues and eigenfunctions for the Hamiltonian

$$H = H_0 + \lambda H_1 \tag{16-2}$$

that is, for the solutions of

$$(H_0 + \lambda H_1)\,\psi_n = E_n\psi_n \tag{16-3}$$

We will express the desired quantities as power series in λ. The question of convergence of the series will not be discussed. Frequently one can show that the series cannot be convergent, and yet the first few terms, when λ is small, do

266

properly describe the physical system. We will assume that as $\lambda \to 0$, $E_n \to E_n^0$ and $\psi_n \to \phi_n$.

Since the ϕ_i form a complete set, we may expand ψ_n in a series involving all the ϕ_i. We write

$$\psi_n = N(\lambda) \left\{ \phi_n + \sum_{k \neq n} C_{nk}(\lambda) \, \phi_k \right\} \tag{16-4}$$

The factor $N(\lambda)$ is there to allow us to normalize the ψ_n. We have the freedom to choose the phase of ψ_n, and we choose it such that the coefficient of ϕ_n in the expansion is real and positive. Since we require that $\psi_n \to \phi_n$ as $\lambda \to 0$, we have

$$N(0) = 1$$
$$C_{nk}(0) = 0 \tag{16-5}$$

More generally, we have

$$C_{nk}(\lambda) = \lambda C_{nk}^{(1)} + \lambda^2 C_{nk}^{(2)} + \cdots \tag{16-6}$$

and

$$E_n = E_n^0 + \lambda E_n^{(1)} + \lambda^2 E_n^{(2)} + \cdots \tag{16-7}$$

The Schrödinger equation then reads

$$(H_0 + \lambda H_1) \left\{ \phi_n + \sum_{k \neq n} \lambda C_{nk}^{(1)} \phi_k + \sum_{k \neq n} \lambda^2 C_{nk}^{(2)} \phi_k + \cdots \right\}$$
$$= (E_n^0 + \lambda E_n^{(1)} + \lambda^2 E_n^{(2)} + \cdots) \left\{ \phi_n + \sum_{k \neq n} \lambda C_{nk}^{(1)} \phi_k + \sum_{k \neq n} \lambda^2 C_{nk}^{(2)} \phi_k + \cdots \right\} \tag{16-8}$$

Note that the normalization factor $N(\lambda)$ does not appear in this linear equation. Identifying powers of λ yields a series of equations. The first one is

$$H_0 \sum_{k \neq n} C_{nk}^{(1)} \phi_k + H_1 \phi_n = E_n^0 \sum_{k \neq n} C_{nk}^{(1)} \phi_k + E_n^{(1)} \phi_n \tag{16-9}$$

Using $H_0 \phi_k = E_k^0 \phi_k$ we obtain

$$E_n^{(1)} \phi_n = H_1 \phi_n + \sum_{k \neq n} (E_k^0 - E_n^0) \, C_{nk}^{(1)} \phi_k \tag{16-10}$$

If we now take a scalar product with ϕ_n, and use the orthonormality condition

$$\langle \phi_k | \phi_l \rangle = \delta_{kl} \tag{16-11}$$

we obtain

$$\lambda E_n^{(1)} = \langle \phi_n | \lambda H_1 | \phi_n \rangle \tag{16-12}$$

This is a *very important formula*. It states that the first-order energy shift for a given state is just the expectation value of the perturbing potential in that state. If the change in the potential is of a definite sign, then the energy shift will have the same sign. The explicit form of

$$\lambda E_n^{(1)} = \int d^3r \; \phi_n^*(\mathbf{r}) \; \lambda H_1(\mathbf{r}) \; \phi_n(\mathbf{r}) \tag{16-13}$$

shows that for the shift to be significant, both the potential change and the probability density $|\phi_n(\mathbf{r})|^2$ must be large.

If we take the scalar product of (16-10) with ϕ_m, for $m \neq n$, then

$$\langle \phi_m | H_1 | \phi_n \rangle + (E_m^0 - E_n^0) \, C_{nm}^{(1)} = 0$$

that is,

$$\lambda C_{nk}^{(1)} = \frac{\langle \phi_k | \lambda H_1 | \phi_n \rangle}{E_n^0 - E_k^0} \tag{16-14}$$

The numerator is the matrix element of H_1 in the basis of states in which H_0 is diagonal. This formula is used in the next equation, which comes from the identification of terms proportional to λ^2:

$$H_0 \sum_{k \neq n} C_{nk}^{(2)} \phi_k + H_1 \sum_{k \neq n} C_{nk}^{(1)} \phi_k = E_n^0 \sum_{k \neq n} C_{nk}^{(2)} \phi_k + E_n^{(1)} \sum_{k \neq n} C_{nk}^{(1)} \phi_k + E_n^{(2)} \phi_n \tag{16-15}$$

Taking the scalar product with ϕ_n yields

$$E_n^{(2)} = \sum_{k \neq n} \langle \phi_n | H_1 | \phi_k \rangle \, C_{nk}^{(1)} = \sum_{k \neq n} \frac{\langle \phi_n | H_1 | \phi_k \rangle \langle \phi_k | H_1 | \phi_n \rangle}{E_n^0 - E_k^0}$$
$$= \sum_{k \neq n} \frac{|\langle \phi_k | H_1 | \phi_n \rangle|^2}{E_n^0 - E_k^0} \tag{16-16}$$

The last line follows from the hermiticity of H_1:

$$\langle \phi_n | H_1 | \phi_k \rangle = \langle \phi_k | H_1 | \phi_n \rangle^* \tag{16-17}$$

This, too, is a very important formula, especially since the first-order shift frequently vanishes on grounds of symmetry. We may interpret the formula as follows: the second-order energy shift is the sum of terms, whose strength is given by the square of the matrix element connecting the given state ϕ_n to all other states by the perturbing potential, weighted by the reciprocal of the energy difference between the states. We can draw several conclusions from the formula.

(a) If ϕ_n is the *ground state*, that is, the state of lowest energy, then the denominator in the sum is always negative, and hence (16-16) is always negative.

(b) All other things being equal, that is, if all the matrix elements of H_1 are of roughly the same order of magnitude (which is the kind of guess one would make without more specific knowledge), then nearby levels have a bigger effect on the second-order energy shift than distant ones have.

(c) If an important level "k"—important in the sense of lying nearby, or of $\langle \phi_k | H_1 | \phi_n \rangle$ being large—lies above the given level "n," then the second-order shift is downwards; if it lies below, the shift is upward. We speak of this as a tendency of levels to repel each other.

An expression for $C_{nk}^{(2)}$ may be obtained from (16-15) by taking the scalar product with ϕ_m, $m \neq n$, but we shall not require this formula. Also $N(\lambda)$ can be determined from

$$\langle \psi_n | \psi_n \rangle = N^2(\lambda) \left\{ 1 + \lambda^2 \sum_{k \neq n} |C_{nk}^{(1)}|^2 + \cdots \right\}$$
$$= 1$$

(16-18)

It is therefore 1 to first order in λ. Hence, to first order in λ, we may write

$$\psi_n = \phi_n + \sum_{k \neq n} \frac{\langle \phi_k | \lambda H_1 | \phi_n \rangle}{E_n^0 - E_k^0} \phi_k$$

(16-19)

a formula that is sometimes useful.

DEGENERATE PERTURBATION THEORY

The preceding development needs modification when there is degeneracy, since, on the face of it, the denominator involving energy differences could vanish. The difficulty is associated with the fact that, instead of a unique ϕ_n, there is a finite set of ϕ_n^0, all of which have the same energy E_n^0. This set can be made orthonormal with respect to the label "i," because, as we have seen in Chapter 4, this label can be associated with the eigenvalues of some other, simultaneously commuting, hermitian operators. We thus choose the set of $\phi_n^{(i)}$ such that

$$\langle \phi_m^{(j)} | \phi_n^{(i)} \rangle = \delta_{mn} \delta_{ij}$$

(16-20)

The natural way to take the degeneracy into account is to replace (16-4) by an expression that involves linear combinations of the degenerate eigenfunctions of H_0:

$$\psi_n = N(\lambda) \left\{ \sum_i \alpha_i \phi_n^{(i)} + \lambda \sum_{k \neq n} C_{nk}^{(1)} \sum_i \beta_i \phi_k^{(i)} + \cdots \right\}$$

(16-21)

The coefficients α_i, β_i, ... will have to be determined. When the preceding is substituted into the Schrödinger equation (16-3), we get, to first order in λ,

$$H_0 \sum_{k \neq n} C_{nk}^{(1)} \sum_i \beta_i \phi_k^{(i)} + H_1 \sum_i \alpha_i \phi_n^{(i)} = E_n^{(1)} \sum_i \alpha_i \phi_n^{(i)} + E_n^0 \sum_{k \neq n} C_{nk}^{(1)} \sum_i \beta_i \phi_k^{(i)}$$

(16-22)

Taking the scalar product with $\phi_n^{(j)}$ gives the first-order shift equation

$$\sum_i \alpha_i \langle \phi_n^{(j)} | \lambda H_1 | \phi_n^{(i)} \rangle = \lambda E_n^{(1)} \alpha_j$$

(16-23)

This is a finite-dimensional eigenvalue problem. For example, if there is a twofold degeneracy, and if we use the notation

$$\langle \phi_n^{(j)} | H_1 | \phi_n^{(i)} \rangle = h_{ji} \tag{16-24}$$

this equation reads

$$h_{11}\alpha_1 + h_{12}\alpha_2 = E_n^{(1)} \alpha_1$$
$$h_{21}\alpha_1 + h_{22}\alpha_2 = E_n^{(1)} \alpha_2 \tag{16-25}$$

The eigenvalues and the α_i can be determined from this equation, if we add the condition that

$$\sum_i |\alpha_i|^2 = 1 \tag{16-26}$$

We do not bother with the determination of the β_i, since we shall only use degenerate perturbation theory for the first-order energy eigenvalues in our applications. If it so happens that $h_{ij} = 0$ for $i \neq j$, that is, the matrix h_{ij} is diagonal, then the first-order shifts are just the diagonal elements of this matrix. This will happen when the perturbation H_1 commutes with the operator whose eigenvalues the "i" labels represent. For example, in the hydrogen atom, there is a degeneracy associated with the eigenvalues of L_z, that is, all m-values have the same energy. If it happens that

$$[H_1, L_z] = 0 \tag{16-27}$$

and if we choose our $\phi_n^{(i)}$ to be eigenfunctions of L_z, then h_{ij} will be diagonal. To see this, note that with

$$L_z \phi_n^{(i)} = \hbar m^{(i)} \phi_n^{(i)} \tag{16-28}$$

$$\langle \phi_n^{(j)} | [H_1, L_z] | \phi_n^{(i)} \rangle = \langle \phi_n^{(j)} | H_1 L_z - L_z H_1 | \phi_n^{(i)} \rangle \tag{16-29}$$

$$= \hbar (m^{(i)} - m^{(j)}) h_{ji}$$

$$= 0$$

that is, (16-27) implies

$$h_{ji} = 0 \qquad \text{for} \qquad m^{(i)} \neq m^{(j)} \tag{16-30}$$

Some of these features will be illustrated in the example below; others will appear later in our discussion of the real hydrogen atom.

THE STARK EFFECT

To illustrate the application of perturbation theory to a real problem, we will consider the effect of an external electric field on the energy levels of a hydrogen-like atom. This is the *Stark effect*. The unperturbed hamiltonian is

$$H_0 = \frac{\mathbf{p}^2}{2\mu} - \frac{Ze^2}{r} \tag{16-31}$$

whose eigenfunctions we denote by $\phi_{nlm}(\mathbf{r})$. The perturbing potential is

$$\lambda H_1 = e\,\mathscr{E} \cdot \mathbf{r} = e\,\mathscr{E}z \tag{16-32}$$

where $\mathscr{E}$ is the electric field. The quantity $e\mathscr{E}$ will play the role of the parameter λ. The energy shift of the ground state, which is nondegenerate, is given by the expression

$$E_{100}^{(1)} = e\,\mathscr{E}\langle\phi_{100}|z|\phi_{100}\rangle = e\,\mathscr{E}\int d^3r|\phi_{100}(r)|^2\,z \tag{16-33}$$

This integral vanishes, since the square of the wave function is always an even function under parity, and the perturbing potential is an odd function under reflections. Thus for the ground state there is no energy shift that is linear in the electric field $\mathscr{E}$. Classically, a system that has an electric dipole moment, $\mathbf{d}$, will experience an energy shift of magnitude $-\mathbf{d} \cdot \mathscr{E}$. Thus (16-33) shows that in its ground state the atom has no permanent dipole moment. The parity argument can be used whenever the unperturbed Hamiltonian is invariant under reflection, and it can be generalized to the statement that *systems in nondegenerate states cannot have permanent dipole moments.* The statement of nondegeneracy is important: it is only then that the states are also eigenstates of the parity operator, and then $|\phi(r)|^2$ is even, and the expectation value of z vanishes.

Many molecules do have permanent dipole moments, and it is often said that this is because the ground states are degenerate. The expectation value of z in a state like $\alpha\psi_+ + \beta\psi_-$, where the subscripts indicate the parity, certainly does not vanish, and a state like the preceding will be degenerate with its space-inverted state $\alpha\psi_+ - \beta\psi_-$ if the two states $\psi_+\ \psi_-$ have the same energy. This explanation is not quite correct. The reason is that the lowest-lying states are never quite degenerate. Consider, for example, a molecule like ammonia, NH_3. Its structure is tetrahedral, with the three H nuclei forming an equilateral triangle. The N can be at a position (determined by the condition that the energy is minimum) either "above" or "below" the triangle. The even and odd linear combinations of these two states do not have quite the same energy, though the energy difference is very tiny (-10^{-4} eV), because of the large barrier between the "above" and "below" locations. Thus, strictly speaking, the ground state is nondegenerate. However, if $\mathscr{E}d$, where

$$d = e\int \psi_{\text{above}}^*\,z\psi_{\text{above}} = -e\int \psi_{\text{below}}^*\,z\psi_{\text{below}} \tag{16-34}$$

is much larger than the tiny splitting, then the energy shift will be linear in the electric field, and the molecule will behave as if it had an electric dipole moment. This will be seen more explicitly in the discussion leading up to (16-56).

Let us look at the second-order term. It reads

$$E_{100} = e^2\mathscr{E}^2\left\{\sum_{nlm}\frac{|\langle\phi_{nlm}|z|\phi_{100}\rangle|^2}{E_1 - E_n} + \sum_k\frac{|\langle\phi_k|z|\phi_{100}\rangle|^2}{E_1 - \hbar^2k^2/2m}\right\} \tag{16-35}$$

The reason for the second term is that in (16-16) we must sum over a complete set of eigenstates of H_0. For atoms, these consist of the bound states ϕ_{nlm} as well as the continuum states in which the electron has a positive energy. We label the continuum states by $\mathbf{k}$, where k is related to the positive kinetic energy by $E = \hbar^2 k^2 / 2m$. This sum is very difficult to evaluate directly, since it involves an integral over k, which appears in the rather complicated continuum solutions of the Coulomb problem (it can be done by a trick, which we shall not describe). What we can do, however, is to estimate the value of E_{100} by finding an upper bound. Let us rewrite (16-35) in a more symbolic form

$$E_{100} = e^2 \mathscr{E}^2 \sum_E \frac{\langle \phi_{100}|z|\phi_E \rangle \langle \phi_E|z|\phi_{100} \rangle}{E_1 - E} \tag{16-36}$$

where the complete set is now labeled by ϕ_E. Because E_1 is the ground-state energy, and all energies lie above it, we have

$$\frac{1}{E_1 - E} \leq \frac{1}{E_1 - E_2} \tag{16-37}$$

so that

$$E_{100} \leq \frac{e^2 \mathscr{E}^2}{E_1 - E_2} \sum_E \langle \phi_{100}|z|\phi_E \rangle \langle \phi_E|z|\phi_{100} \rangle \tag{16-38}$$

Since for a complete set

$$\sum_E |\phi_E \rangle \langle \phi_E| = 1 \tag{16-39}$$

We get

$$E_{100} < \frac{e^2 \mathscr{E}^2}{E_1 - E_2} \sum_E \langle \phi_{100}|z^2|\phi_{100} \rangle \tag{16-40}$$

This is easy to evaluate. Since the ground-state wave function is spherically symmetric,

$$\langle \phi_{100}|z^2|\phi_{100} \rangle = \langle \phi_{100}|y^2|\phi_{100} \rangle = \langle \phi_{100}|x^2|\phi_{100} \rangle$$
$$= \tfrac{1}{3}\langle \phi_{100}|r^2|\phi_{100} \rangle = a_0^2 \tag{16-41}$$

where the last step follows from (12-36). Hence

$$E_{100} < \frac{8e^2 \mathscr{E}^2 a_0^2}{3mc^2 \alpha^2} = \frac{8}{3}\mathscr{E}^2 a_0^3 \tag{16-42}$$

Note that $\int d^3r\, \mathscr{E}^2$ is an energy, so that dimensional analysis would mandate a form (constant)$\mathscr{E}^2 a_0^3$, since a_0 is the only length in the problem. The exact second-order calcuation yields 2.25 for the coefficient of $\mathscr{E}^2 a_0^3$. For hydrogenlike atoms we must make the replacement $a_0 \rightarrow a_0/Z$.

If we differentiate the energy shift with respect to the electric field, we get an expression for the dipole moment

$$d = -\frac{\partial E_{100}}{\partial \mathscr{E}} = -\frac{9}{2}\mathscr{E}a_0^3 \tag{16-43}$$

The dipole moment is proportional to the electric field $\mathscr{E}$, that is, the dipole moment is induced. The *polarizability*, defined by

$$P = \frac{d}{\mathscr{E}} \tag{16-44}$$

is given by $P = 4.5a_0^3$.

To illustrate degenerate perturbation theory, we calculate the first-order (linear in $\mathscr{E}$) Stark effect for the $n = 2$ states of the hydrogen atom. For the unperturbed system there are really four $n = 2$ states that have the same energy. These are

$$\phi_{200} = (2a_0)^{-3/2}\, 2 \left(1 - \frac{r}{2a_0}\right) e^{-r/2a_0}\, Y_{00}$$

$$\phi_{211} = (2a_0)^{-3/2}\, 3^{-1/2} \left(\frac{r}{a_0}\right) e^{-r/2a_0}\, Y_{11}$$

$$\phi_{210} = (2a_0)^{-3/2}\, 3^{-1/2} \left(\frac{r}{a_0}\right) e^{-r/2a_0}\, Y_{10} \tag{16-45}$$

$$\phi_{2,1,-1} = (2a_0)^{-3/2}\, 3^{-1/2} \left(\frac{r}{a_0}\right) e^{-r/2a_0}\, Y_{1,-1}$$

The $l = 0$ state has even parity, and the $l = 1$ state has odd parity. We want to solve an equation like (16-23) and, on the face of it, four equations are involved. If we note, however, that (1) the perturbing potential (that is, z) commutes with L_z so that it only connects states with the same m-value, and (2) parity forces us to consider only terms in which the perturbing potential must connect $l = 1$ to $l = 0$ terms, that is,

$$\langle \phi_{2,1,\pm1}|z|\phi_{2,1,\pm1}\rangle = 0 \tag{16-46}$$

then the matrix in (16-23) is only a 2×2 matrix. The equation reads

$$e\mathscr{E} \begin{pmatrix} \langle\phi_{200}|z|\phi_{200}\rangle & \langle\phi_{200}|z|\phi_{210}\rangle \\ \langle\phi_{210}|z|\phi_{200}\rangle & \langle\phi_{210}|z|\phi_{210}\rangle \end{pmatrix} \begin{pmatrix} \alpha_1 \\ \alpha_2 \end{pmatrix} = E^{(1)} \begin{pmatrix} \alpha_1 \\ \alpha_2 \end{pmatrix} \tag{16-47}$$

The diagonal elements are zero, because of parity, and the off-diagonal elements are equal, since they are complex conjugates of each other, and each may be chosen to be real. We have

$$\langle\phi_{200}|z|\phi_{210}\rangle = \int_0^\infty r^2\, dr (2a_0)^{-3}\, e^{-r/a_0}\, \frac{2r}{\sqrt{3}a_0}\left(1 - \frac{r}{2a_0}\right) r$$

$$\cdot \int d\Omega\, Y_{00}^*(\sqrt{4\pi/3}\; Y_{10})\, Y_{10} \tag{16-48}$$

$$= -3a_0$$

and hence (16-47) becomes

$$\begin{pmatrix} -E^{(1)} & -3e\mathscr{E}a_0 \\ -3e\mathscr{E}a_0 & -E^{(1)} \end{pmatrix} \begin{pmatrix} \alpha_1 \\ \alpha_2 \end{pmatrix} = 0 \tag{16-49}$$

The eigenvalues of this are

$$E^{(1)} = \pm 3e\mathscr{E}a_0 \qquad\qquad (16\text{-}50)$$

and the corresponding eigenstates, when properly normalized, are

$$\frac{1}{\sqrt{2}} \begin{pmatrix} 1 \\ -1 \end{pmatrix} \text{ and } \frac{1}{\sqrt{2}} \begin{pmatrix} 1 \\ 1 \end{pmatrix}$$

respectively. Thus the linear Stark effect for the $n = 2$ states yields a splitting of degenerate levels as shown in Fig. 16-1.

It is possible to understand the splitting pattern in Fig. 16-1 in a qualitative way. The electron charge distribution for $\phi_{2,1,\pm1}$ is given by $e|\phi_{2,1,\pm1}|^2$ and the angular dependence is given by $\sin^2 \theta$. The charge is therefore distributed into lobes in the x-y plane (see Fig. 12-3), cylindrically symmetric about the z-axis. There is no dipole moment, since the positive charge is at the origin, and thus there is no $\mathbf{d} \cdot \mathbf{E}$ term. On the other hand, the linear combinations $\phi_{200} \pm \phi_{210}$ give rise to charge distributions of the form $A + B \cos^2\theta + C \cos \theta$. The term with the plus sign shifts the electron charge distribution in the positive z direction, so that the dipole moment points in the positive z direction, parallel to the E field. Thus $-\mathbf{d} \cdot \mathbf{E}$ is negative, and the energy is lowered. The term with the minus sign has its energy raised, in qualitative agreement with the result of our calculation.

There are some general comments that can be abstracted from the calculations just concluded.

(a) The states in the presence of the electric field are no longer eigenstates of $\mathbf{L}^2$. In the preceding case, for example, we found that the states that diagonalize the perturbation were equal mixtures of $l = 0$ and $l = 1$, though they are still eigenstates of L_z. The reason is that the perturbation changes the Hamiltonian, so that it no longer commutes with $\mathbf{L}^2$. This can be worked out in detail, but it is really evident that the external field specifies a preferred direction, so that the physical system is no longer invariant under arbitrary rotations. It is still invariant under rotations about the preferred axis, here the z direction, and hence L_z is still a good constant of the motion.

(b) Quite generally, whenever there is a perturbation that does not conserve some quantity (for example, $\mathbf{L}^2$ here), then the states that "diagonalize" the new

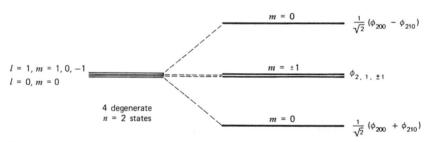

Figure 16-1. Pattern of Stark splitting of hydrogen atom in $n = 2$ state. The fourfold degeneracy is partly lifted by the perturbation. The $m = \pm1$ states remain degenerate and are not shifted in the Stark effect.

Hamiltonian in any approximation, are superpositions of states with different values of the previously conserved quantum numbers, and thus degenerate levels will be split.

(c) We can summarize the procedure in degenerate perturbation theory in matrix language as follows. If H_0 is diagonal, but H_1 is not, then, since H_0 and H_1 do not commute, it is not possible to diagonalize H_1 by itself, without "un-diagonalizing" H_0. One must work with

$$H = H_0 + H_1$$

as a whole. If we work with a subset of degenerate states, all of which are eigenstates of H_0 *with the same eigenvalue,* then, as far as these states are concerned, H_0 is not merely diagonal, but it is proportional to the unit matrix. Since H_1 (and everything else) commutes with the unit matrix, one can diagonalize H_1 by itself, without affecting H_0.

The hydrogenlike atoms considered here were somewhat idealized. As we will see in Chapter 17, there are small relativistic and spin-orbit coupling effects that actually remove some of the degeneracies. Does this mean that we never really need to use degenerate perturbation theory? Actually, even if, say, ϕ_{200} and ϕ_{210} do not have exactly the same energy, it may still be sensible to take some linear combination of them in the perturbation expansion. If we have, for example

$$H_0 \, \phi_{200} = (E_2^0 - \Delta) \, \phi_{200}$$
$$H_0 \, \phi_{210} = (E_2^0 + \Delta) \, \phi_{210} \tag{16-51}$$

with Δ small, then the Schrödinger equation, with the linear combinations, reads

$$(H_0 + \lambda H_1)\left(\alpha_1 \phi_{200} + \alpha_2 \phi_{210} + \lambda \sum_{n \neq 2} C_n \phi_n \right)$$
$$= E \left(\alpha_1 \phi_{200} + \alpha_2 \phi_{210} + \lambda \sum_{n \neq 2} C_n \phi_n \right) \tag{16-52}$$

Taking the scalar product with ϕ_{200} and ϕ_{210}, respectively, leads to the following equation to order λ:

$$\begin{pmatrix} E_2^0 - \Delta - \langle \phi_{200}|\lambda H_1|\phi_{200}\rangle & \langle \phi_{200}|\lambda H_1|\phi_{210}\rangle \\ \langle \phi_{210}|\lambda H_1|\phi_{200}\rangle & E_2^0 + \Delta - \langle \phi_{210}|\lambda H_1|\phi_{210}\rangle \end{pmatrix} \begin{pmatrix} \alpha_1 \\ \alpha_2 \end{pmatrix} = E \begin{pmatrix} \alpha_1 \\ \alpha_2 \end{pmatrix} \tag{16-53}$$

If we write

$$\langle \phi_{200}|\lambda H_1|\phi_{210}\rangle = \langle \phi_{210}|\lambda H_1|\phi_{200}\rangle = a\lambda \tag{16-54}$$

we must find the eigenvalues of the matrix

$$\begin{pmatrix} E_2^0 - \Delta & \lambda a \\ \lambda a & E_2^0 + \Delta \end{pmatrix} \tag{16-55}$$

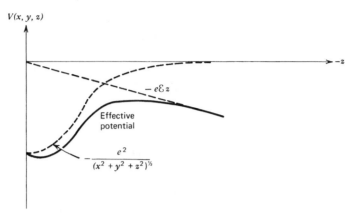

Figure 16-2. Schematic picture of potential energy as a function of z with x and y held fixed. The dotted line represents the Coulomb potential, the dashed line the potential energy due to the external field, and the solid line the total potential.

and these are

$$E = E_2^0 \pm \sqrt{a^2\lambda^2 + \Delta^2} \qquad (16\text{-}56)$$

(In the preceding we have used $\langle\phi_{200}|H_1|\phi_{200}\rangle = \langle\phi_{210}|H_1|\phi_{210}\rangle = 0$.) We see that when $\Delta \gg a\lambda$, we get a "quadratic" effect only. This corresponds to no degeneracy. When $\Delta \ll a\lambda$ we get the result of the form (16-50). In the intermediate region, the preceding, more careful treatment is necessary. Furthermore, when the new linear combinations are used, then in second-order perturbation theory there no longer appear very tiny energy differences in the denominators. We do not discuss this in detail, but this is not difficult to establish.

As a final comment we point out two apparently contradictory facts. (1) The predictions of perturbation theory concerning the Stark effect are borne out very well by experiment, and (2) the perturbation series evidently diverges, since the perturbing potential $e\mathcal{E}z$ grows without bound as z becomes very large, no matter how small $e\mathcal{E}$ is. The question arises whether one has any right to believe in the accuracy of the first few terms of a mathematically divergent series, since it is well known that a mathematically divergent series can be rearranged to give entirely different expansions. The answer lies in the physics and not in the mathematics of the problem. The reason for the divergence can be seen in Fig. 16-2, which gives a rough picture of the total potential for x, y fixed. It appears that there is a barrier created for the bound electron. This barrier is ultimately penetrable, even though for small $e\mathcal{E}$ it is very broad. What the mathematical divergence of the series is responding to is the possibility that the electron in the ground state, for example, has a finite (although very, very small) probability of being sufficiently far away from the nucleus, where the external electric field is stronger than the Coulomb field, and the electron is carried away by the electric field. Thus the new "shifted" energy levels of the hydrogen atom are no longer stationary states, but rather metastable states. If the field is weak, however, they may be stable on a time scale of the age of the universe,[1] and hence the observations agree perfectly with what the first few terms of the perturbation series predict.

---

[1] Actually a simple barrier penetration calculation of the type carried out in Chapter 5 shows that the time scale is more like 10^{1000} lifetimes of the universe, for fairly reasonable fields!

Problems

1. Calculate $C_{nk}^{(2)}$ and use this to obtain an expression for $E_n^{(3)}$.

2. Consider the hydrogen atom, and assume that the proton, instead of being a point-source of the Coulomb field, is a uniformly charged sphere of radius R, so that the Coulomb potential is now modified to

$$V(r) = -\frac{3e^2}{2R^3}\left(R^2 - \frac{1}{3}r^2\right) \qquad r < R(\ll a_0)$$

$$= -\frac{e^2}{r} \qquad r > R$$

Calculate the energy shift for the $n = 1, l = 0$ state, and for the $n = 2$ states, caused by this modification, using the wave functions given in (12-30).

3. Calculate the energy shift in the ground state of the one-dimensional harmonic oscillator, when the perturbation

$$V = \lambda x^4$$

is added to

$$H = \frac{p^2}{2m} + \tfrac{1}{2}m\omega^2x^2$$

4. The bottom of an infinite well is changed to have the shape

$$V(x) = \epsilon \sin\frac{\pi x}{b} \qquad 0 \le x \le b$$

Calculate the energy shifts for all the excited states to first order in ϵ. Note that the well originally had $V(x) = 0$ for $0 \le x \le b$, with $V = \infty$ elsewhere.

5. Prove the sum rule (Thomas-Reiche-Kuhn sum rule)

$$\sum_n (E_n - E_a)|\langle n|x|a\rangle|^2 = \frac{\hbar^2}{2m}$$

[Hints: (a) Write the commutation relation $[p, x] = \hbar/i$ in the form

$$\sum_n \left\{\langle a|p|n\rangle\langle n|x|a\rangle - \langle a|x|n\rangle\langle n|p|a\rangle\right\} = \frac{\hbar}{i}\langle a|a\rangle = \frac{\hbar}{i}$$

(b) Use the fact that

$$\langle a|p|n\rangle = \left\langle a\left|m\frac{dx}{dt}\right|n\right\rangle = m\frac{i}{\hbar}\langle a|[H, x]|n\rangle$$

in working out the problem.]

6. Check the sum rule in Problem 5 for the one-dimensional harmonic oscillator, with "a" taken in the ground state.

7. Work out the first-order Stark effect in the $n = 3$ state of the hydrogen atom. Do not bother to work out all the integrals, but you should construct the correct linear combinations of states. Can you give a qualitative explanation of the pattern of energy shifts?

8. Consider an electron in a one-dimensional harmonic oscillator potential $m\omega^2 x^2/2$, placed in an electric field pointing in the x direction. Calculate the first- and second-order energy shifts. Compare your answer with the exact value of the energy shift that can be obtained in this case.

9. Consider a two-dimensional harmonic oscillator described by the Hamiltonian

$$H = \frac{1}{2m} (p_x^2 + p_y^2) + \tfrac{1}{2}m\omega^2(x^2 + y^2)$$

Generalize the approach of Chapter 7 to obtain solutions of this problem in terms of raising operators acting on the ground state. Calculate the energy shifts due to the perturbation

$$V = 2\lambda xy$$

in the ground state, and in the degenerate first excited states, using first-order perturbation theory. Can you interpret your result very simply? Solve the problem exactly, and compare it with a second-order perturbation calculation of the shift of the ground state.

10. Consider a Hamiltonian of the form

$$H = \begin{pmatrix} E_0 & 0 \\ 0 & -E_0 \end{pmatrix} + \lambda \begin{pmatrix} \alpha & U \\ U^* & \beta \end{pmatrix}$$

(a) Calculate the energy shift to first and second order in λ. Compare your results with the exact eigenvalues.

(b) Suppose U^* is replaced by $V \neq U^*$. Show that the eigenstates of the non-Hermitian new H corresponding to different eigenvalues are no longer orthogonal. (For this part of the problem you may take $\alpha = \beta = 0$ to simplify the work.)

11. Consider a particle of charge q and its antiparticle (charge $-q$ and same mass) interacting via a Coulomb potential. By how much are the low-lying energy levels shifted if the potential energy is changed by the addition of a term $V_1 = Kr$?

[*Hint:* Use (12-30).]

12. The Hamiltonian for an electron in a hydrogen atom subject to a constant magnetic field **B**, is, with the neglect of spin,

$$H = \frac{\mathbf{p}^2}{2m} - \frac{e^2}{r} + \left(\frac{e}{2mc}\right) \mathbf{L} \cdot \mathbf{B}$$

where **L** is the angular momentum operator. In the absence of the magnetic field, there will be a single line in the transition from an ($n = 4, l = 3$) state to an ($n = 3, l = 2$) state. What will be the effect of the magnetic field on that line? Sketch the new spectrum and the possible transitions, constrained by the selection rules $\Delta l_z = \pm 1, 0$. How many lines will there be? What will be the effect of constant electric field **E** parallel to **B**?

References

There are many examples of the application of first-order perturbation theory in the textbook literature, and the references listed at the end of this book may serve as a source of further examples. For a discussion of the second-order calculation of the Stark effect, see

S. Borowitz, *Fundamentals of Quantum Mechanics*, W. A. Benjamin, New York, 1967.

Chapter 17

THE REAL HYDROGEN ATOM

The discussion of hydrogenlike atoms in Chapter 12 was based on the Hamiltonian

$$H_0 = \frac{\mathbf{p}^2}{2\mu} - \frac{Ze^2}{r} \tag{17-1}$$

In a more realistic treatment, several corrections must be taken into account. We separate out the treatment of the proton motion, so that we do not clutter up our discussion of relativistic effects and spin effects with simple kinematic matters involving the motion of the proton. In the original treatment in Chapter 12, we had

$$K = \frac{\mathbf{p}_e^2}{2m_e} + \frac{\mathbf{p}_p^2}{2M_p} = \frac{1}{2}\,\mathbf{p}^2\left(\frac{1}{m_e} + \frac{1}{M_p}\right) \equiv \frac{\mathbf{p}^2}{2\mu}$$

with $\mathbf{p} = \mathbf{p}_e = -\mathbf{p}_p$ in the center-of-mass frame. The reduced mass μ differs from the electron mass by very little in the hydrogen atom

$$\frac{\mu}{m_e} \approx 1 - \frac{m_e}{M_p} \approx 1 - 5.4 \times 10^{-4}$$

In discussing the corrections to the spectrum derived in Chapter 12, we shall treat the proton as if it were infinitely massive, and return to the effect of its motion in a separate section.

RELATIVISTIC KINETIC ENERGY EFFECTS

The relativistic expression for the kinetic energy of the electron is

$$K = \sqrt{(\mathbf{p}c)^2 + (m_e c^2)^2} - m_e c^2 \approx \frac{\mathbf{p}^2}{2m_e} - \frac{1}{8}\frac{(\mathbf{p}^2)^2}{m_e^3 c^2} + \cdots \tag{17-2}$$

The second term

$$H_1 = -\frac{1}{8}\frac{(\mathbf{p}^2)^2}{m_e^3 c^2} \tag{17-3}$$

will be treated by perturbation theory. We can estimate its fractional effect on the energy eigenvalue, since

$$\frac{\langle H_1 \rangle}{\langle H_0 \rangle} \approx \frac{\mathbf{p}^2}{m_e^2 c^2} \approx (Z\alpha)^2 = (0.53 \times 10^{-4})Z^2 \tag{17-4}$$

Thus the relativistic effect is smaller than the reduced mass effect by an order of magnitude. The reduced mass corrections to the relativistic term are very tiny for hydrogen, and will be discussed later.

SPIN-ORBIT COUPLING

The existence of the electron spin gives rise to another correction that is of the same order of magnitude. It may be qualitatively understood as follows: if the electron were at rest relative to the proton (we are discussing this on a classical level), it would only see an electric field due to the proton charge. This is the Coulomb potential term that appears in H_0. Because the electron is moving, there are additional effects. In the electron rest frame, the proton is moving, so that there is a current present, and the electron "sees" a magnetic field. If the relative motion were rectilinear, the magnetic field, as seen by the electron, would be $\mathbf{v} \times \mathbf{E}/c$. This magnetic field interacts with the spin of the electron, or more precisely, with the magnetic moment of the electron.

Given that the electron has a magnetic moment $\mathbf{M}$ given in (14-60) as

$$\mathbf{M} = -\frac{eg}{2m_e c}\mathbf{S}$$

the expected additional term would be of the form

$$\begin{aligned}
-\mathbf{M} \cdot \mathbf{B} = \frac{eg}{2m_e c}\mathbf{S} \cdot \mathbf{B} &= -\frac{e}{m_e c^2}\mathbf{S} \cdot \mathbf{v} \times \mathbf{E} \\
&= \frac{e}{m_e^2 c^2}\mathbf{S} \cdot \mathbf{p} \times \boldsymbol{\nabla}\phi(r) \\
&= \frac{e}{m_e^2 c^2}\mathbf{S} \cdot \mathbf{p} \times \mathbf{r}\frac{1}{r}\frac{d\phi(r)}{dr} \\
&= -\frac{e}{m_e^2 c^2}\mathbf{S} \cdot \mathbf{L}\frac{1}{r}\frac{d\phi(r)}{dr}
\end{aligned} \tag{17-5}$$

where $\phi(r)$ is the potential due to the nuclear charge and we have taken $g = 2$. Actually this is not correct. It turns out that relativistic effects associated with the fact that the electron does not move in a straight line (the Thomas precession effect) reduce the preceding by a factor of 2. Thus the correct perturbation is

$$H_2 = -\frac{1}{2m_e^2 c^2} \mathbf{S} \cdot \mathbf{L} \frac{1}{r} \frac{d[e\phi(r)]}{dr} \tag{17-6}$$

Let us now use first-order perturbation theory to calculate the effects of H_1 and H_2 on the spectrum of hydrogenlike atoms. We can rewrite H_1 in the form

$$H_1 = -\frac{1}{8} \frac{(\mathbf{p}^2)^2}{m_e^3 c^2} = -\frac{1}{2m_e c^2} \left(\frac{\mathbf{p}^2}{2m} \right)^2$$

$$= -\frac{1}{2m_e c^2} \left(H_0 + \frac{Ze^2}{r} \right) \left(H_0 + \frac{Ze^2}{r} \right) \tag{17-7}$$

if we neglect reduced mass effects in H_1. Hence

$$\langle \phi_{nlm} | H_1 | \phi_{nlm} \rangle = -\frac{1}{2m_e c^2} \left\langle \phi_{nlm} \left| \left(H_0 + \frac{Ze^2}{r} \right) \left(H_0 + \frac{Ze^2}{r} \right) \right| \phi_{nlm} \right\rangle$$

$$= -\frac{1}{2m_e c^2} \left[E_n^2 + 2E_n Ze^2 \left\langle \frac{1}{r} \right\rangle_{nl} + (Ze^2)^2 \left\langle \frac{1}{r^2} \right\rangle_{nl} \right]$$

$$= -\frac{1}{2m_e c^2} \left\{ \left[\frac{m_e c^2 (Z\alpha)^2}{2n^2} \right]^2 - 2Ze^2 \frac{m_e c^2 (Z\alpha)^2}{2n^2} \left(\frac{Z}{a_0 n^2} \right) \right. \tag{17-8}$$

$$\left. + (Ze^2)^2 \frac{Z^2}{a_0^2 n^3 (l + 1/2)} \right\}$$

$$= -\frac{1}{2} m_e c^2 (Z\alpha)^2 \left[\frac{(Z\alpha)^2}{n^3 (l + 1/2)} - \frac{3(Z\alpha)^2}{4n^4} \right]$$

In calculating the preceding, we have used expressions for

$$\left\langle \frac{1}{r} \right\rangle_{nl} \equiv \left\langle \phi_{nlm} \left| \frac{1}{r} \right| \phi_{nlm} \right\rangle \qquad \text{and} \qquad \left\langle \frac{1}{r^2} \right\rangle_{nl} \equiv \left\langle \phi_{nlm} \left| \frac{1}{r^2} \right| \phi_{nlm} \right\rangle$$

from (12-36). The spin of the electron does not enter into this energy shift, since H_1 does not depend on the spin; H_2 does depend on the spin, and for our unperturbed wave functions we must take two-component wave functions, since what we want to calculate is the expectation value of

$$-\frac{1}{2m^2 c^2} \mathbf{S} \cdot \mathbf{L} \frac{1}{r} \frac{ed\phi(r)}{dr} = \frac{Ze^2}{2m_e^2 c^2} \mathbf{S} \cdot \mathbf{L} \frac{1}{r^3} \tag{17-9}$$

Here, again, we have an example of degenerate perturbation theory. For a given n and l, there are $2(2l + 1)$ degenerate eigenstates of H_0, with the additional factor of 2 coming from the two spin states. Thus the calculation of the energy shift involves a diagonalization of a submatrix, as in (16-23). We can save ourselves a great deal of labor by noting that

$$\mathbf{S} + \mathbf{L} = \mathbf{J} \tag{17-10}$$

implies that

$$\mathbf{S}^2 + 2\mathbf{S} \cdot \mathbf{L} + \mathbf{L}^2 = \mathbf{J}^2$$

that is,

$$\mathbf{S} \cdot \mathbf{L} = \tfrac{1}{2} (\mathbf{J}^2 - \mathbf{L}^2 - \mathbf{S}^2) \tag{17-11}$$

Thus if we combine the degenerate eigenfunctions into linear combinations that are eigenfunctions of $\mathbf{J}^2$ (they already are eigenfunctions of $J_z = L_z + S_z$), then these linear combinations will diagonalize H_2. The appropriate linear combinations were obtained in Chapter 15, (15-38) and (15-39). With these linear combinations we have

$$
\begin{aligned}
\mathbf{S} \cdot \mathbf{L} \psi_{\substack{j=l+(1/2)\\mj=m+(1/2)}} &= \frac{1}{2} \left(\mathbf{J}^2 - \mathbf{L}^2 - \mathbf{S}^2 \right) \psi_{\substack{j=l+(1/2)\\mj=m+(1/2)}} \\
&= \frac{1}{2} \hbar^2 \left[\left(l + \frac{1}{2} \right)\left(l + \frac{3}{2} \right) - l(l+1) - \frac{3}{4} \right] \psi_{\substack{j=l+(1/2)\\mj=m+(1/2)}} \\
&= \frac{1}{2} \hbar^2 l \psi_{\substack{j=l+(1/2)\\mj=m+(1/2)}}
\end{aligned}
\tag{17-12}
$$

and

$$
\begin{aligned}
\mathbf{S} \cdot \mathbf{L} \psi_{\substack{j=l-(1/2)\\mj=m+(1/2)}} &= \frac{1}{2} \hbar^2 \left[\left(l - \frac{1}{2} \right)\left(l + \frac{1}{2} \right) - l(l+1) - \frac{3}{4} \right] \psi_{\substack{j=l-(1/2)\\mj=m+(1/2)}} \\
&= -\frac{1}{2} \hbar^2 (l+1) \, \psi_{\substack{j=l-(1/2)\\mj=m+(1/2)}}
\end{aligned}
\tag{17-13}
$$

For a given l value there are $[2(l + 1/2) + 1] + [2(l - 1/2) + 1]$ states. What has happened is that the degenerate states have merely been rearranged, but the two groups that they have been split into behave differently under the action of H_2. If we call the linear combinations ϕ_{jm_jl}, then

$$
\begin{aligned}
\langle \phi_{jm_jl} | H_2 | \phi_{jm_jl} \rangle &= \frac{Ze^2}{2m_e^2c^2} \frac{\hbar^2}{2} \left\{ \begin{matrix} l \\ -l-1 \end{matrix} \right\} \\
&\times \int_0^\infty dr \, r^2 [R_{nl}(r)]^2 \frac{1}{r^3}
\end{aligned}
\tag{17-14}
$$

for $j = l \pm 1/2$, respectively.

It is possible to calculate $\langle 1/r^3 \rangle_{nl}$. The result is

$$\left\langle \frac{1}{r^3} \right\rangle_{nl} = \frac{Z^3}{a_0^3} \frac{1}{n^3 l(l+1/2)(l+1)} \tag{17-15}$$

which is valid for $l \neq 0$. Note that here

$$a_0 = \frac{\hbar}{\mu c \alpha}$$

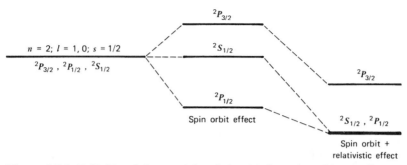

Figure 17-1. Splitting of the $n = 2$ levels by (1) the spin-orbit coupling (which leaves the S state unaffected) and (2) the relativistic effect. The final degeneracy of the $^2S_{1/2}$ and $^2P_{1/2}$ states is actually lifted by quantum electrodynamic effects. The tiny upward shift of the $^2S_{1/2}$ state is called the Lamb shift.

since we are dealing with eigenstates of H_0 in which the reduced mass of the electron is used. It follows that the energy shift is

$$\Delta E = \frac{1}{4}\, m_e c^2 (Z\alpha)^4 \, \frac{\left\{ \begin{matrix} l \\ -l - 1 \end{matrix} \right\}}{n^3 l(l + 1/2)(l + 1)} \tag{17-16}$$

valid for $l \neq 0$. When the effects of H_1 and H_2 are combined, we find, after some algebra, that

$$\Delta E = -1/2 m_e c^2 (Z\alpha)^4 \frac{1}{n^3} \left[\frac{1}{j + 1/2} - \frac{3}{4n} \right] \tag{17-17}$$

for both values of $l = j \pm 1/2$. It is necessary to work with the relativistic Dirac equation to show that the result is also correct when $l = 0$, even though the product in (17-14) is not well defined.

The splitting is depicted graphically in Fig. 17-1. A very interesting result is that the corrections add up in a manner that leaves the $^2P_{1/2}$ and the $^2S_{1/2}$ states degenerate. A more careful discussion, using the relativistic Dirac equation, does not alter this result. In 1947, a very delicate microwave absorption experiment carried out by Lamb and Retherford showed that there was, indeed, a tiny splitting of the two levels. The magnitude of the splitting, of order $m_e c^2 (Z\alpha)^4 \, \alpha \log \alpha$ could be explained by the additional interaction of the electron with its own electromagnetic field, that is, as a self-energy effect. These matters are outside of the scope of this book.

THE ANOMALOUS ZEEMAN EFFECT

Let us now turn to the discussion of the behavior of hydrogenlike atoms in an external magnetic field, that is, to the *anomalous Zeeman effect*. There is, of course, nothing anomalous about the effect; it is just that the Zeeman effect that could be explained classically was exhibited only by atoms in states in which the total electronic spin was zero. For the other states, for which there was no classical explanation (since that involves spin), the Zeeman splitting pattern was different, and therefore "anomalous."

For the unperturbed hamiltonian we take the usual H_0 together with the spin-orbit term. The reason for doing this is that the external perturbation may be small compared with the effect of what we called H_2. Thus

$$H_0 = \frac{p^2}{2\mu} - \frac{Ze^2}{r} + \frac{1}{2m^2c^2} \frac{Ze^2}{r^3} \mathbf{L} \cdot \mathbf{S} \tag{17-18}$$

The perturbation now reads

$$H_1 = \frac{e}{2mc} (\mathbf{L} + 2\mathbf{S}) \cdot \mathbf{B} \tag{17-19}$$

The first term is, in effect, the interaction of the magnetic dipole moment arising from the circulating charge, and the second term is the contribution of the intrinsic dipole moment of an object with spin

$$\mathbf{M} = -\frac{eg}{2m_e c} \mathbf{S} \tag{17-20}$$

with $g = 2$.

The choice of H_0 dictates that we calculate the expectation value of the perturbation in eigenstates of $\mathbf{J}^2$ and J_z (15-38) and (15-39). If we choose the z-axis as given by the direction of $\mathbf{B}$, then we need to calculate

$$\left\langle \phi_{jm_j l} \left| \frac{eB}{2m_e c} (L_z + 2S_z) \right| \phi_{jm_j l} \right\rangle = \left\langle \phi_{jm_j l} \left| \frac{eB}{2m_e c} (J_z + S_z) \right| \phi_{jm_j l} \right\rangle$$

$$= \frac{eB}{2m_e c} (\hbar m_j + \langle \phi_{jm_j l} | S_z | \phi_{jm_j l} \rangle) \tag{17-21}$$

To calculate the last matrix element, we carry out the calculation explicitly, using the eigenfunctions given in (15-38) and (15-39). Thus for $j = l + 1/2$ we have

$$\left\langle \sqrt{\frac{l+m+1}{2l+1}} Y_{lm}\chi_+ + \sqrt{\frac{l-m}{2l+1}} Y_{l,m+1}\chi_- \left| S_z \right| \sqrt{\frac{l+m+1}{2l+1}} Y_{lm}\chi_+ \right.$$

$$\left. + \sqrt{\frac{l-m}{2l+1}} Y_{l,m+1}\chi_- \right\rangle = \frac{\hbar}{2} \left(\frac{l+m+1}{2l+1} - \frac{l-m}{2l+1} \right) \tag{17-22}$$

$$= \frac{\hbar}{2} \frac{2m+1}{2l+1} = \frac{\hbar m_j}{2l+1}$$

and for $j = l - 1/2$, we have

$$\left\langle \sqrt{\frac{l-m}{2l+1}} Y_{lm}\chi_+ - \sqrt{\frac{l+m+1}{2l+1}} Y_{l,m+1}\chi_- \left| S_z \right| \sqrt{\frac{l-m}{2l+1}} Y_{lm}\chi_+ \right.$$

$$\left. - \sqrt{\frac{l+m+1}{2l+1}} Y_{l,m+1}\chi_- \right\rangle = \frac{\hbar}{2} \left(\frac{l-m}{2l+1} - \frac{l+m+1}{2l+1} \right) \tag{17-23}$$

$$= -\frac{\hbar}{2} \frac{2m+1}{2l+1} = -\frac{\hbar m_j}{2l+1}$$

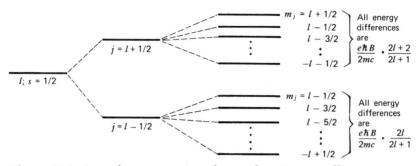

Figure 17-2. General representation of anomalous Zeeman effect.

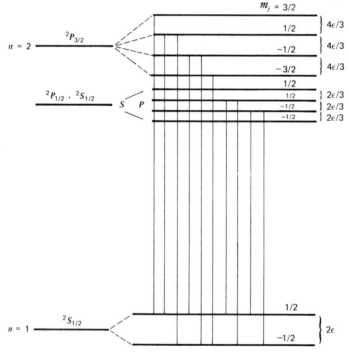

Figure 17-3. Zeeman effect in hydrogen, ϵ represents the energy $e\hbar B/2mc$. The transitions for which $l = 1$, $\Delta m = 1, 0, -1$ are drawn in the figure. The location of the unperturbed states is given by Fig. 17-1.

In both cases we used the fact that $m_j = m + 1/2$ in the preceding. Inserting the results in (17-22) and (17-23) into (17-21) we find

$$\Delta E = \frac{e\hbar B}{2m_e c} m_j \left(1 \pm \frac{1}{2l + 1} \right) \qquad j = l \pm \tfrac{1}{2} \qquad (17\text{-}24)$$

The splitting is depicted in Fig. 17-2. The selection rule[1] for the transitions is still

$$\Delta m_j = \pm 1,0 \qquad (17\text{-}25)$$

[1]The derivation of this selection rule (and others) will be discussed in Chapter 21.

but since the splitting between the lines is not the same for every multiplet, we do not get just the three lines that we obtained for the normal Zeeman effect in Chapter 13. For example, for $n = 2$, the $^2P_{3/2}$ state splits into four lines, with the splitting two times as large as that of the two states in the $^2P_{1/2}$ lines (Fig. 17-3). If the external field is very strong, so that the spin-orbit coupling can be neglected, we may use the ordinary hydrogenic wave functions simply multiplied by spinors, that is, eigenstates of $\mathbf{L}^2$, L_z, $\mathbf{S}^2$, and S_z. If we call the eigenvalues of L_z and S_z, m_l and m_s, respectively, then the expectation value of H_1 in (17-19), with $\mathbf{B}$ pointing in the z direction, is

$$\langle H_1 \rangle = \frac{e\hbar B}{2mc} (m_l + 2m_s) \tag{17-26}$$

Thus the $n = 2, l = 1$ states are split into five levels, corresponding to the values of $m_l = 1, 0, -1; m_s = 1/2, -1/2$.

HYPERFINE STRUCTURE

In addition to the *fine structure* of the levels caused by the spin-orbit coupling, there is a very tiny *hyperfine splitting*, which is really a permanent Zeeman effect due to the magnetic field generated by the magnetic dipole moment of the nucleus. If the spin of the nucleus is $\mathbf{I}$, then the magnetic dipole moment operator is

$$\mathbf{M} = \frac{Zeg_N}{2M_N c} \mathbf{I} \tag{17-27}$$

where Ze is the charge of the nucleus, M_N its mass, and g_N its gyromagnetic ratio. The vector potential due to a point dipole is, from electromagnetic theory

$$\mathbf{A(r)} = -\frac{1}{4\pi} (\mathbf{M} \times \nabla) \frac{1}{r} \tag{17-28}$$

so that the magnetic field is

$$\mathbf{B} = \nabla \times \mathbf{A} = -\frac{\mathbf{M}}{4\pi} \nabla^2 \frac{1}{r} + \frac{1}{4\pi} \nabla(\mathbf{M} \cdot \nabla) \frac{1}{r} \tag{17-29}$$

This may be written as

$$\begin{aligned} B_i &= -M_i \frac{1}{4\pi} \nabla^2 \frac{1}{r} + \frac{1}{4\pi} M_j \frac{\partial^2}{\partial x_i \partial x_j} \frac{1}{r} \\ &= M_i \, \delta(r) + \frac{1}{4\pi} M_j \frac{\partial^2}{\partial x_i \partial x_j} \frac{1}{r} \end{aligned} \tag{17-30}$$

Here we have used[2]

$$\nabla^2 \frac{1}{r} = -4\pi \, \delta(r) \tag{17-31}$$

---

[2]Only the radial part of ∇^2 is relevant. To show this, we prove that $(1/r^2)(d/dr) |r^2(d/dr)](1/r) = 0$ for $r \neq 0$, and that $\nabla^2(1/r)$ integrated over a small sphere of radius ϵ, gives a result -4π independent of ϵ.

The perturbation is

$$H_{hf} = -\mathbf{M}_e \cdot \mathbf{B} = -M_{ei}M_i\,\delta(\mathbf{r}) + \frac{1}{4\pi}M_{ei}M_j\frac{\partial^2}{\partial x_i \partial x_j}\frac{1}{r} \tag{17-32}$$

A simple dimensional analysis indicates that the perturbation is of the form $\mathbf{M}_e \cdot \mathbf{M}/a_0^3$, that is,

$$\langle H_{hf}\rangle \approx \frac{Ze^2 g_N}{m_e M_N c^2}\,\hbar^2\left(\frac{Z\alpha m_e c}{\hbar}\right)^3 \approx g_N(Z\alpha)^4\,m_e c^2(m_e/M_N)$$

which is a factor m_e/M_N smaller than the typical spin-orbit splittings.

We shall be interested in the splitting of the $l = 0$ states of the hydrogen atom, so that we need to calculate

$$\int d^3r\,|\phi_{n0}(\mathbf{r})|^2\left(-M_{ei}M_i\,\delta(\mathbf{r}) + \frac{1}{4\pi}M_{ei}M_j\frac{\partial^2}{\partial x_i \partial x_j}\frac{1}{r}\right)$$

Because of the spherical symmetry of $|\phi_{n0}(\mathbf{r})|^2$,

$$\begin{aligned}\int d^3r\,|\phi_{n0}(\mathbf{r})|^2\frac{\partial^2}{\partial x_i \partial x_j}\frac{1}{r} &= \frac{1}{3}\delta_{ij}\int d^3r\,|\phi_{n0}(\mathbf{r})|^2\,\nabla^2\frac{1}{r}\\ &= \frac{4\pi}{3}\delta_{ij}\int d^3r\,|\phi_{n0}(\mathbf{r})|^2\,\delta(\mathbf{r})\end{aligned} \tag{17-33}$$

We thus end up with

$$\langle\phi_{n0}|H_{hf}|\phi_{n0}\rangle = -\frac{2}{3}\mathbf{M}_e \cdot \mathbf{M}\,|\phi_{n0}(0)|^2 \tag{17-34}$$

Since[3]

$$|\phi_{n0}(0)|^2 = R_{n0}(0)^2 = \frac{4}{n^3}\left(\frac{Z\alpha m_e c}{\hbar}\right)^2 \tag{17-35}$$

we get

$$\langle H\rangle = \frac{4}{3}g_N\frac{m_e}{M_N}(Z\alpha)^4\,mc_e c^2\frac{1}{n^3}\left(\frac{\mathbf{S}\cdot\mathbf{I}}{\hbar^2}\right) \tag{17-37}$$

If we take the total spin of the electron and nucleus to be $\mathbf{F}$,

$$\mathbf{F} = \mathbf{S} + \mathbf{I} \tag{17-38}$$

then

$$\begin{aligned}\frac{\mathbf{S}\cdot\mathbf{I}}{\hbar^2} &= \frac{\mathbf{F}^2 - \mathbf{S}^2 - \mathbf{I}^2}{2\hbar^2} = \frac{[F(F+1) - 3/4 - I(I+1)]}{2}\\ &= \frac{1}{2}\begin{cases} I & F = I + \tfrac{1}{2}\\ -I - 1 & F = I - \tfrac{1}{2}\end{cases}\end{aligned} \tag{17-39}$$

---

[3]See, for example, Bethe and Salpeter, in the references at the end of the chapter.

For hydrogen, $g_N = g_P \cong 5.56$, and the energy difference between the excited state, characterized by $F = 1$ and the ground state of $F = 0$ is

$$\Delta E = \frac{4}{3}\,(5.56)\,\frac{1}{1840}\,\frac{1}{(137)^4}\,(m_e c^2)$$

The wavelength of the radiation corresponding to the transition between the $F = 1$ and $F = 0$ states is

$$\lambda \simeq 21.1 \text{ cm} \tag{17-40}$$

and the frequency[4] is

$$\nu = \frac{c}{\lambda} \simeq 1420 \text{ MHz} \tag{17-41}$$

The radiation arising from this transition plays an important role in astronomy. In a gas of neutral atoms, the $F = 1$ state cannot be excited by ordinary radiation, because of a selection rule that strongly suppresses transitions in which there is no change in orbital angular momentum. Both the $F = 1$ and the $F = 0$ states have zero angular momentum. On the other hand, there are other mechanisms that can cause transitions. The $F = 1$ state can, for example, be excited by collisions, and the return to the $F = 0$ ground state can be detected. From an analysis of the intensity of the 21-cm radiation received, astronomers have learned a great deal about the density distribution of neutral hydrogen in interstellar space, as well as the motion and the temperature of the gas clouds containing the hydrogen. The average number of neutral hydrogen atoms appears to be about 1 cm^{-3} in the galactic plane near the sun, and the temperature is of the order of 100 K.

COMMENTS ON REDUCED MASS EFFECTS

We ignored reduced mass effects in the perturbation calculations because of the small magnitude of m_e/M_p. In the past thirty years it has been possible to study spectra of unstable atoms such as atoms in which a short-lived μ^- (a heavy electron of mass $m_\mu = 205.8\ m_e$) forms a bound state with a proton, or positronium in which the e^- forms a bound state with the positron e^+, a particle identical to the electron, except for the sign of its charge (and magnetic moment). In these cases the reduced mass effects are more important. We briefly discuss how these affect the discussion of the relativistic effects. We shall do this in the language of the hydrogen atom.

If we take into account the relativistic kinetic energy of the proton as well as of the electron, then the corrections to the nonrelativistic kinetic energy can be written as

$$H_1 = -\frac{1}{8}\frac{(\mathbf{p}^2)^2}{c^2}\left(\frac{1}{m_e^3} + \frac{1}{M_p^3}\right)$$

---

[4]This frequency is one of the most accurately measured quantities in physics; $\nu_{\exp} = 1420405751.800 \pm 0.028$ Hz. The number involves the distribution of magnetization in the proton, but there is no theory yet that can deal with a number of this accuracy.

which, after a little algebra, can be written as

$$H_1 = \frac{1}{8} \frac{(\mathbf{p}^2)^2}{\mu m_e^2 c^2} \left(1 - \frac{m_e}{M_p} + \left(\frac{m_e}{M_p}\right)^2\right) \tag{17-42}$$

In the spin-orbit term $\mathbf{M}$ is unaltered, but $\mathbf{v} = \mathbf{p}/\mu$.

In our discussion of the anomalous Zeeman effect, the perturbation now reads

$$H_3 = \left(\frac{e}{2\mu c} \mathbf{L} + \frac{eg}{2m_e c} \mathbf{S}\right) \cdot \mathbf{B} \tag{17-43}$$

and (17-21) becomes

$$\begin{aligned}
&\left\langle \phi_{jml} \left| \frac{eB}{2\mu c} L_z + \frac{egB}{2m_e c} S_z \right| \phi_{jml} \right\rangle \\
&= B \left\langle \phi_{jml} \left| \frac{e}{2\mu c} J_z + \left(\frac{eg}{2m_e c} - \frac{e}{2\mu c}\right) S_z \right| \phi_{jml} \right\rangle \\
&= \frac{eB}{2\mu c} \left[\hbar m_j + \left(g \frac{\mu}{m_e} - 1\right) \langle \phi_{jml} |S_z| \phi_{jml} \rangle\right]
\end{aligned} \tag{17-44}$$

Reduced mass effects can be ignored in our discussion of hyperfine structure for hydrogen, since the whole effect is a factor of m_e/M_N smaller than the atomic effects under consideration.

Numerically, the reduced mass effects are small. The table below[5] lists the energy shifts in millielectronvolts including relativistic and spin-orbit terms:

Level	Energy Shift with $\mu = m_e$	Energy Shift Including Red Mass Effects
$1\,S_{1/2}$	-0.18113	-0.18074
$2\,S_{1/2}$	-0.05660	-0.05648
$2\,P_{1/2}$	-0.05660	-0.05651
$2\,P_{3/2}$	-0.01132	-0.01128

The reduced mass effects on the energy shifts are of the order of 4×10^{-7} eV. Quantum electrodynamic effects (Lamb shift) are of the order of 4×10^{-6} eV. Thus for hydrogen there is no point in keeping the reduced mass effects, unless one is interested in very high precision calculations. On the other hand, for a system like positronium, the $e^- - e^+$ bound state, reduced mass effects are crucial, since the two particles have the same mass. There the hyperfine splitting is not suppressed relative to the spin-orbit or relativistic effect.

Problems

1. If the general form of a spin-orbit coupling for a particle of mass m and spin $\mathbf{S}$ moving in a potential $V(r)$ is

[5]The table was kindly provided by Professor J. S. Tenn.

$$H_{SO} = \frac{1}{2m^2c^2} \mathbf{S} \cdot \mathbf{L} \frac{1}{r} \frac{dV(r)}{dr}$$

what is the effect of that coupling on the spectrum of a three-dimensional harmonic oscillator?

2. Consider the $n = 2$ states in the real hydrogen atom. What is the spectrum in the absence of a magnetic field? How is that spectrum changed when the atom is placed in a magnetic field of 25,000 gauss? (Ignore hyperfine structure).

3. Show that

$$\nabla^2 \frac{1}{r} = -4\pi\delta(\mathbf{r})$$

Use the procedure outlined in the footnote to Eq. (17-31).

4. Consider hydrogen gas in its ground state. What is the effect of a magnetic field on the hyperfine structure? Calculate the spectrum for $B = 1$ gauss and for $B = 10^4$ gauss.
   Hint: To solve this, set up the eigenvalue problem for the interaction $A\mathbf{S}\cdot\mathbf{I}/\hbar^2 + aS_z/\hbar + bI_z/\hbar$ with $a = \hbar B/m_ec$, $b = -eg_N\hbar B/M_Nc$ and $A = g_N(4m_e/3M_N)\alpha^4 m_ec^2$. You will need to diagonalize a 2×2 matrix.

5. Consider a harmonic oscillator in three dimensions. If the relativistic expression for the kinetic energy is used, what is the shift in the ground-state energy?

6. The deuteron consists of a proton (charge $+e$) and a neutron (charge 0) in a state of total spin 1 and total angular momentum $J = 1$. The g-factors for the proton and neutron are

$$g_P = 2(2.7896)$$

$$g_N = 2(-1.9103)$$

   (a) What are the possible orbital angular momentum states for this system? If it is known that the state is primarily $^3S_1$, what admixture is allowed given that parity is conserved?
   (b) Write an expression for the interaction of the deuteron with an external magnetic field and calculate the Zeeman splitting. Show that if the interaction with the magnetic field is written in the form

$$V = -\mu_{\text{eff}} \cdot \mathbf{B}$$

   then the effective magnetic moment of the deuteron is the sum of the proton and neutron magnetic moments, and any deviation from that result is due to an admixture of non-S state to the wave function.

7. Consider positronium, a hydrogenlike atom consisting of an electron and a positron (same mass, opposite charge). Calculate (a) the ground-state energy, and that for the $n = 2$ states; (b) the relativistic kinetic energy effect and the spin-orbit coupling; (c) the hyperfine splitting of the ground state. Compare your results with those for the hydrogen atom and explain major differences.

References

The most detailed discussion of the physics of hydrogenlike atoms may be found in

H. A. Bethe and E. E. Salpeter, *Quantum Mechanics of One- and Two-Electron Atoms*, Springer-Verlag, Berlin/New York, 1957.

The Thomas precession is discussed in

R. M. Eisberg, *Fundamentals of Modern Physics*, John Wiley & Sons, New York, 1961.

Chapter 18

THE HELIUM ATOM

THE HELIUM ATOM WITHOUT ELECTRON–ELECTRON REPULSION

The helium atom consists of a nucleus of charge $Z = 2$ and two electrons, which we label 1 and 2. Each electron is attracted to the nucleus, and the two electrons repel each other. We assume, and this will turn out to be correct, that no forces, other than the electromagnetic ones (Coulomb to a very good approximation), are necessary to describe the dynamics of the helium atom with the help of quantum mechanics.

If the nucleus is placed at the origin, and if the electron coordinates are labeled $\mathbf{r}_1$ and $\mathbf{r}_2$, then the Hamiltonian for the atom is (Fig. 18-1)

$$H = \frac{1}{2m}\,\mathbf{p}_1^2 + \frac{1}{2m}\,\mathbf{p}_2^2 - \frac{Ze^2}{r_1} - \frac{Ze^2}{r_2} + \frac{e^2}{|\mathbf{r}_1 - \mathbf{r}_2|} \tag{18-1}$$

Here m is the electron mass. We shall ignore the small effects connected with the motion of the nucleus,[1] relativistic effects, spin-orbin effects, and the effect of the

---

[1]The reduced mass effect takes a somewhat different form since one is trying to convert a three-particle problem into an effective two-particle problem. This is worked out in D. Park, *Introduction to the Quantum Theory*, McGraw-Hill, New York, Third Edition (1992).

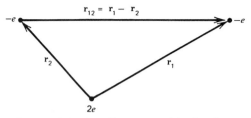

Figure 18-1. Coordinates used in the description of the helium atom.

current caused by the motion of one electron, upon the other electron. The Hamiltonian just given can be written as

$$H = H^{(1)} + H^{(2)} + V \tag{18-2}$$

with

$$H^{(i)} = \frac{1}{2m}\, \mathbf{p}_i^2 - \frac{Ze^2}{r_i} \tag{18-3}$$

and

$$V = \frac{e^2}{|\mathbf{r}_1 - \mathbf{r}_2|} \tag{18-4}$$

We shall work with the nuclear charge Z and set $Z = 2$ later. Our work on the hydrogen atom provides us with a complete set of eigenfunctions for $H^{(1)}$ and $H^{(2)}$. Thus, if we were to ignore V in the total Hamiltonian, we would have a solution to the eigenvalue problem for the two-electron system. The eigenfunctions would be

$$u(\mathbf{r}_1, \mathbf{r}_2) = \phi_{n_1 l_1 m_1}(\mathbf{r}_1)\, \phi_{n_2 l_2 m_2}(\mathbf{r}_2) \tag{18-5}$$

for the equation

$$[H^{(1)} + H^{(2)}]\, u(\mathbf{r}_1, \mathbf{r}_2) = E u(\mathbf{r}_1, \mathbf{r}_2) \tag{18-6}$$

and the energy would be given by (Fig. 18.2a)

$$E = E_{n_1} + E_{n_2} \tag{18-7}$$

where $E_n = -(mc^2/2)(Z\alpha)^2/n^2$. Thus in the idealized model in which the two electrons ignore each other, the lowest energy is

$$E = -2E_1 = -mc^2(2\alpha)^2 = -108.8 \text{ eV} \tag{18-8}$$

Note that this is $2 \times Z^2 = 8$ times the hydrogen energy of -13.6 eV.

The first excited state is one in which one electron is in its ground state, $n = 1$, and the second electron is raised to the first excited $n = 2$ state. Then

$$E = E_1 + E_2 = -68.0 \text{ eV} \tag{18-9}$$

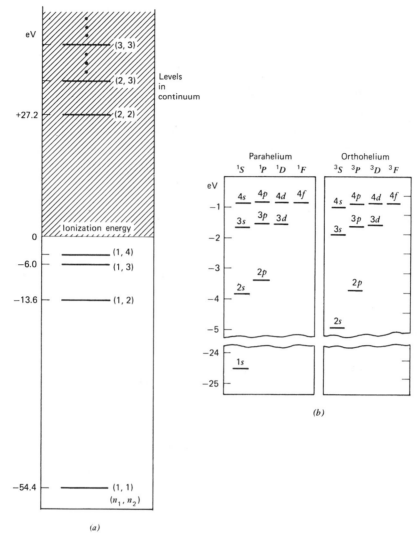

Figure 18-2. (*a*) The spectrum of helium as it would look in the absence of the electron–electron interaction. The zero energy point is chosen at the ionization energy. (*b*) The actual spectrum of helium for the singlet (parahelium) and triplet (orthohelium) states. The level labeling has a suppressed (1*s*), so that the level (2*p*) is approximately described by the (1*s*)(2*p*) orbital.

The ionization energy, that is, the energy required to remove one electron from the ground state to infinity is

$$E_{\text{ioniz}} = (E_1 + E_\infty) - 2E_1 = 54.4 \text{ eV} \qquad (18\text{-}10)$$

and, interestingly enough, the onset of the continuum lies *lower* than the excited state for which both electrons are in the $n = 2$ state. The energy of the latter state is

$$E = 2E_2 = -27.2 \text{ eV} \qquad (18\text{-}11)$$

and it brings up a new phenomenon: the existence of a discrete state in the continuum for the Hamiltonian $H^{(1)} + H^{(2)}$. We shall briefly discuss the implications of this at the end of the chapter.

Effects of the Exclusion Principle

Since the two electrons are *identical fermions* we must make the total wave function antisymmetric under the interchange of space and spin coordinates of the electrons. Thus a proper description of the ground state of this idealized model is

$$u_0(\mathbf{r}_1, \mathbf{r}_2) = \phi_{100}(\mathbf{r}_1)\ \phi_{100}(\mathbf{r}_2)\ X_{\text{singlet}} \tag{18-12}$$

The spatial part of the wave function is necessarily symmetric, and that is why the state must be a spin singlet state

$$X_{\text{singlet}} = \frac{1}{\sqrt{2}}\ (\chi_+^{(1)}\chi_-^{(2)} - \chi_-^{(1)}\chi_+^{(2)}) \tag{18-13}$$

For the first excited state, we have two possibilities, which, for $V = 0$, are degenerate in energy. These are

$$u_1^{(s)} = \frac{1}{\sqrt{2}}\ [\phi_{100}(\mathbf{r}_1)\ \phi_{2lm}(\mathbf{r}_2) + \phi_{2lm}(\mathbf{r}_1)\ \phi_{100}(\mathbf{r}_2)]\ X_{\text{singlet}} \tag{18-14}$$

and the space-antisymmetric, spin symmetric

$$u_1^{(t)} = \frac{1}{\sqrt{2}}\ [\phi_{100}(\mathbf{r}_1)\ \phi_{2lm}(\mathbf{r}_2) - \phi_{2lm}(\mathbf{r}_1)\ \phi_{100}(\mathbf{r}_2)]\ X_{\text{triplet}} \tag{18-15}$$

where

$$X_{\text{triplet}} = \begin{cases} \chi_+^{(1)}\chi_+^{(2)} \\ \dfrac{1}{\sqrt{2}}\ (\chi_+^{(1)}\chi_-^{(2)} + \chi_-^{(1)}\chi_+^{(2)}) \\ \chi_-^{(1)}\chi_-^{(2)} \end{cases} \tag{18-16}$$

is orthogonal to X_{singlet}.

EFFECT OF ELECTRON–ELECTRON REPULSION

The presence of V, the electron–electron Coulomb interaction may, in first approximation, be treated as a perturbation. Let us first compute the energy shift of the ground state to first order in V. We have

$$\Delta E = \int d^3r_1 d^3r_2\ u_0^*(\mathbf{r}_1, \mathbf{r}_2)\ \frac{e^2}{|\mathbf{r}_1 - \mathbf{r}_2|}\ u_0(\mathbf{r}_1, \mathbf{r}_2) \tag{18-17}$$

Since the perturbation does not involve the spin, we need only consider

$$\Delta E = \int d^3 r_1 d^3 r_2 |\phi_{100}(\mathbf{r}_1)|^2 \frac{e^2}{|\mathbf{r}_1 - \mathbf{r}_2|} |\phi_{100}(\mathbf{r}_2)|^2 \tag{18-18}$$

The integral has a simple physical interpretation. Since $|\phi_{100}(\mathbf{r}_1)|^2$ is the probability density of finding electron 1 at $\mathbf{r}_1$, we may interpret $e|\phi_{100}(\mathbf{r}_1)|^2$ as the charge density due to electron 1. Hence

$$U(\mathbf{r}_2) = -\int d^3 r_1 \frac{e|\phi_{100}(\mathbf{r}_1)|^2}{|\mathbf{r}_1 - \mathbf{r}_2|} \tag{18-19}$$

is the potential at $\mathbf{r}_2$ due to the charge distribution of electron 1, and

$$\Delta E = -\int d^3 r_2 \, e|\phi_{100}(\mathbf{r}_2)|^2 \, U(\mathbf{r}_2) \tag{18-20}$$

is therefore the electrostatic energy of interaction of electron 2 with that potential. The integral can be carried out. With $\phi_{100} = (2/\sqrt{4\pi})(Z/a_0)^{3/2} e^{-Zr/a_0}$ we have

$$\Delta E = \left[\frac{1}{\pi}(Z/a_0)^3\right]^2 e^2 \int_0^\infty r_1^2 \, dr_1 \, e^{-2Zr_1/a_0} \int_0^\infty r_2^2 \, dr_2 \, e^{-2Zr_2/a_0}$$

$$\int d\Omega_1 \int d\Omega_2 \frac{1}{|\mathbf{r}_1 - \mathbf{r}_2|} \tag{18-21}$$

In writing this, we used the separation

$$\int d^3 r = \int_0^\infty r^2 \, dr \, d\Omega$$

and isolated the only term that depends on the angles between $\mathbf{r}_1$ and $\mathbf{r}_2$. We have

$$\frac{1}{|\mathbf{r}_1 - \mathbf{r}_2|} = \frac{1}{(r_1^2 + r_2^2 - 2r_1 r_2 \cos \theta)^{1/2}} \tag{18-22}$$

where θ is the angle between $\mathbf{r}_1$ and $\mathbf{r}_2$. We may proceed in one of two ways.

(a) Most directly, we choose the direction of $\mathbf{r}_1$ as z-axis for the $d\Omega_2$ integration, and get

$$\int d\Omega_2 \frac{1}{|\mathbf{r}_1 - \mathbf{r}_2|} = \int_0^{2\pi} d\phi \int_{-1}^1 d(\cos \theta) \frac{1}{(r_1^2 + r_2^2 - 2r_1 r_2 \cos \theta)^{1/2}}$$

$$= -2\pi \frac{1}{r_1 r_2} \left[(r_1^2 + r_2^2 - 2r_1 r_2 \cos \theta)^{1/2}\right]_{\cos \theta = -1}^{\cos \theta = +1} \tag{18-23}$$

$$= \frac{2\pi}{r_1 r_2}(r_1 + r_2 - |r_1 - r_2|)$$

The integration over $d\Omega_1$ is trivial, since nothing depends on that angle, so that

$$\int d\Omega_1 = 4\pi \tag{18-24}$$

and we are left with

$$8e^2 \left(\frac{Z}{a_0}\right)^6 \int_0^\infty r_1 dr_1 \, e^{-2Zr_1/a_0} \int_0^\infty r_2 \, dr_2 \, e^{-2Zr_2/a_0} \times (r_1 + r_2 - |r_1 - r_2|) \tag{18-25}$$

(b) A very useful expansion, necessary when there is additional angular dependence in the numerator, is the following. For $r_1 > r_2$,

$$(r_1^2 + r_2^2 - 2r_1r_2 \cos \theta)^{-1/2} = r_1^{-1} \left(1 + \frac{r_2^2}{r_1^2} - 2\frac{r_2}{r_1} \cos \theta\right)^{-1/2}$$

$$= \frac{1}{r_1} \sum_{L=0}^\infty \left(\frac{r_2}{r_1}\right)^L P_L(\cos \theta) \tag{18-26}$$

with the roles of r_1 and r_2 reversed when $r_2 > r_1$. Thus

$$\int d\Omega_1 \int d\Omega_2 \frac{1}{|r_1 - r_2|} = \int d\Omega_1 \int d\Omega_2 \sum_{L=0}^\infty \frac{r_<^L}{r_>^{L+1}} P_L (\cos \theta) \tag{18-27}$$

where $r_>$ ($r_<$) is the larger (smaller) of r_1 and r_2. We can now proceed as before, using the fact that

$$\frac{1}{2} \int_{-1}^1 d(\cos \theta) \, P_L(\cos \theta) = \delta_{L0} \tag{18-28}$$

as a special case of

$$\frac{1}{2} \int_{-1}^1 d(\cos \theta) \, P_L(\cos \theta) \, P_{L'}(\cos \theta) = \frac{\delta_{LL'}}{2L + 1} \tag{18-29}$$

In any case, (18-25) becomes

$$\Delta E = 8e^2(Z/a_0)^6 \int_0^\infty r_1 \, dr_1 \, e^{-2Zr_1/a_0} \left\{ 2 \int_0^{r_1} r_2^2 \, dr_2 \, e^{-2Zr_2/a_0} \right.$$

$$\left. + 2r_1 \int_{r_1}^\infty r_2 \, dr_2 \, e^{-2Zr_2/a_0} \right\} \tag{18-30}$$

The integrals are straightforward and yield the answer

$$\Delta E = \frac{5}{8} \frac{Ze^2}{a_0} = \frac{5}{4} Z \left(\frac{1}{2} mc^2 \alpha^2\right) \tag{18-31}$$

This is a positive contribution, since it arises from a repulsive force, and its magnitude, for $Z = 2$ is 34 eV. When this is added to the zero-order result of -108.8 eV we obtain, to first order

$$E \simeq -74.8 \text{ eV} \tag{18-32}$$

When this is compared with

$$E_{\text{exp}} = -78.975 \text{ eV} \tag{18-33}$$

a sizable discrepancy is seen. Physically, we can attribute this discrepancy to the fact that in our calculation we took no account of "screening," that is, the effect that the presence of one electron tends to decrease the net charge "seen" by the other electron. Very roughly, if one argues that, for example, electron 1 is half the time "between" electron 2 and the nucleus, then half the time electron 2 sees a charge Z and half the time it sees a charge $Z - 1$, that is, effectively, in the expression

$$E + \Delta E = -\frac{1}{2} mc^2 \alpha^2 \left(2Z^2 - \frac{5}{4} Z \right) \tag{18-34}$$

$(Z - 1/2)$ should be substituted for Z. This does improve agreement, but the crude argument advanced is not sufficient justification for the choice of 50 percent for the probability of effective screening. We will return to this subject later in this chapter, when we discuss the Rayleigh-Ritz variational principle for the ground state energy.

EXCLUSION PRINCIPLE AND EXCHANGE INTERACTION

We next consider the first excited state of helium. It will be sufficient to calculate the energy shift with the singlet and triplet $m = 0$ states listed in (18-14) and (18-15), since the shift is caused by a perturbation that commutes with L_z. For such a perturbation, the shift must be independent of the m-value. Again, because of the spin-independence of the perturbing potential, V, we have

$$
\begin{aligned}
\Delta E_1^{(s,t)} &= \frac{1}{2} e^2 \int d^3r_1 \int d^3r_2 [\phi_{100}(\mathbf{r}_1) \, \phi_{210}(\mathbf{r}_2) \pm \phi_{210}(\mathbf{r}_1) \, \phi_{100}(\mathbf{r}_2)]^* \\
&\quad \times \frac{1}{|\mathbf{r}_1 - \mathbf{r}_2|} [\phi_{100}(\mathbf{r}_1) \, \phi_{210}(\mathbf{r}_2) \pm \phi_{210}(\mathbf{r}_1) \, \phi_{100}(\mathbf{r}_2)] \\
&= e^2 \int d^3r_1 \int d^3r_2 |\phi_{100}(\mathbf{r}_1)|^2 |\phi_{210}(\mathbf{r}_2)|^2 \frac{1}{|\mathbf{r}_1 - \mathbf{r}_2|} \\
&\quad \pm e^2 \int d^3r_1 \int d^3r_2 \, \phi_{100}^*(\mathbf{r}_1) \, \phi_{210}^*(\mathbf{r}_2) \frac{1}{|\mathbf{r}_1 - \mathbf{r}_2|} \phi_{210}(\mathbf{r}_1) \, \phi_{100}(\mathbf{r}_2)
\end{aligned} \tag{18-35}
$$

In obtaining this simplified form, we made use of the symmetry of V under $\mathbf{r}_1 \leftrightarrow \mathbf{r}_2$.

The energy shift is seen to consist of two terms: the first has the familiar form of an electrostatic interaction between two "electron clouds" distributed according to the wave functions of the two electrons. This term is just a simple generalization of the term that we found for the ground-state energy shift. The second term has no classical interpretation. Its origin lies in the Pauli principle, and its sign depends on whether the state has spin 0 or 1. Thus, because of this *exchange* contribution, the singlet and triplet terms are no longer degenerate. Although we considered $n = 2$ here, we have quite generally

$$\Delta E_{n,l}^{(t)} = J_{nl} - K_{nl}$$
$$\Delta E_{n,l}^{(s)} = J_{nl} + K_{nl}$$
(18-36)

The integrals can be evaluated in closed form [it is here that (18-27) becomes useful], but we shall not do this here. The integral J_{nl} is manifestly positive, and it turns out that this is also the case for K_{nl}. For $l = n - 1$ this is obvious: the wave functions appearing in (18-35) have no nodes in that case. That the triplet state should have a lower energy than the singlet state, that is, that

$$J_{nl} - K_{nl} < J_{nl} + K_{nl}$$

or equivalently

$$K_{nl} > 0$$
(18-37)

can be argued on qualitative grounds. For the triplet state the spatial wave function is antisymmetric, so that the electrons are somewhat constrained to stay away from each other. This tends to reduce the screening effect, so that each electron "sees" more of the nuclear charge, and it also tends to make the repulsion between the electrons less effective than for the spatially symmetric singlet state. An interesting aspect of this result is that, although the perturbing potential $e^2/|\mathbf{r}_1 - \mathbf{r}_2|$ does not depend on the spins of the electrons, the symmetry of the wave function does make the potential act as if it were spin-dependent. We may write (18-36) in a form that exhibits this. Let the spins of the two electrons be $\mathbf{s}_1$ and $\mathbf{s}_2$. Then the total spin $\mathbf{S} = \mathbf{s}_1 + \mathbf{s}_2$, and

$$\mathbf{S}^2 = \mathbf{s}_1^2 + \mathbf{s}_2^2 + 2\mathbf{s}_1 \cdot \mathbf{s}_2$$
(18-38)

If we act with this on triplet and singlet states (18-16) and (18-13) that are also eigenstates of $\mathbf{s}_1^2$ and $\mathbf{s}_2^2$, we get

$$S(S + 1)\hbar^2 = \frac{3}{4}\hbar^2 + \frac{3}{4}\hbar^2 + 2\mathbf{s}_1 \cdot \mathbf{s}_2$$

that is,

$$2\mathbf{s}_1 \cdot \mathbf{s}_2/\hbar^2 = S(S + 1) - \frac{3}{2} = \begin{cases} \dfrac{1}{2} & \text{triplet} \\[2mm] -\dfrac{3}{2} & \text{singlet} \end{cases}$$
(18-39)

We may thus write, in terms of the $\boldsymbol{\sigma}$'s related to the spins by $\mathbf{s}_i = (1/2)\hbar\boldsymbol{\sigma}_i$,

$$\Delta E_{n,l} = J_{n,l} - \frac{1}{2}(1 + \boldsymbol{\sigma}_1 \cdot \boldsymbol{\sigma}_2)K_{nl} \qquad (18\text{-}40)$$

We shall see this phenomenon again when we discuss the H_2 molecule. Usually spin-dependent forces between atoms are quite weak. As illustrated in the example of spin-orbit coupling, the spin-dependent forces tend to arise from relativistic corrections to the static forces. In the spin-orbit example, these forces are down by a factor of α^2, which is just $(v/c)^2$. Such forces could not be strong enough to keep the electron spins aligned in a ferromagnet, except at unrealistically low temperatures. The spin dependence due to exchange is much stronger than that: the force is of the same order of magnitude as the electrostatic force, and, as first observed by Heisenberg, it is responsible for the phenomenon of ferromagnetism.

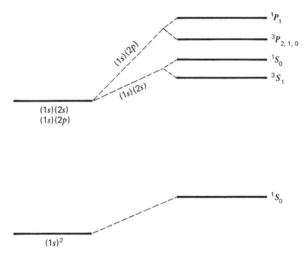

Figure 18-3. Schematic sketch of splitting of the first excited states of helium.

The spectrum of the first few excited states of helium is shown in Fig. 18-3. The notation used for the unperturbed states is that of *orbitals*, that is, the quantum numbers of the unperturbed electrons. Thus both electrons in the ground state are in $n = 1, l = 0$ states, and we write this as (1s, 1s), or more briefly (1s)^2. It should be understood that when we write (1s)(2p), as for the first excited state, this does not mean that one electron is in one state, and the other electron in the other, since we must write totally antisymmetric wave functions for the electrons. Another way of labeling the state is by the $^{2S+1}L_J$ notation, which we use for the perturbed states in the figure. We see that the singlet states lie above the triplet states in a given multiplet. This follows from the symmetry (cf. our argument that $K_{nl} > 0$) and is a special example of one of *Hund's Rules: Other things being equal, the states of highest spin will have the lowest energy.*[2]

If we excite helium from the ground state by shining ultraviolet light on it, we find that the *selection rule* $\Delta L = 1$, which we will derive later, implies an excitation

[2]Hund's rules are discussed in more detail in Chapter 19.

to the P states. Furthermore, there is a selection rule $\Delta S = 0$, that is, only transitions singlet $\rightarrow$ singlet and triplet $\rightarrow$ triplet dominate.[3] Hence the state most strongly excited from the ground state is the $^1P_1$ state. The other levels may also become occupied through other mechanisms, for example, collisional excitation. Once occupied, the radiative transitions to the ground state are very improbable. The ^3P state, which may be populated when atoms in the $^1P_1$ state undergo collisions with other atoms in the gas, can only decay to the $^3S_1$ state, and that state is *metastable*, since it cannot decay to the ground state easily. The fact that there are no transitions, to good approximation, between triplet states and singlet states, led, at one time, to the belief that there existed two kinds of helium, ortho-helium (triplet) and para-helium (singlet).

The spectrum of helium that we saw in Fig. 18-2*b* shows that the excited states $(1s)(nl)$ have energies that do not differ very much from those of the hydrogen atom levels. Thus the binding energy of one electron in the atom is 24.6 eV (total binding energy minus binding energy of singly ionized helium = 79.0 − 54.4 = 24.6 eV), whereas the energy that would be liberated if one electron were to be removed from the 2s state is of the order of 4 − 5 eV, which is comparable to the energy 3.4 eV (= $13.6/n^2$ eV) for hydrogen. The reason for this effect is that the "outer" electron sees only a unit positive charge, since the "inner" electron in the (1s) orbital tends to shield the nucleus, leaving a net effective charge $\approx Z - 1$. This is not the case for the ground state, since both electrons have access to the nucleus. Thus the ground state lies quite a bit deeper than the hydrogen ground state.

In our discussion of the first-order calculation of the ground-state energy, we noted that there was a discrepancy of about 4 eV from the experimental value. Rather than attempt an estimate of the second-order result, which would be very tedious, we turn to an entirely different method of calculating the ground-state energy—the *Ritz variational method*.

THE VARIATIONAL PRINCIPLE

Consider a Hamiltonian H, and an arbitrary square integrable function Ψ, which we choose to be normalized to unity, so that

$$\langle \Psi | \Psi \rangle = 1 \tag{18-41}$$

This function Ψ can be expanded in a complete set of eigenstates of H, denoted by ψ_n,

$$H\psi_n = E_n\psi_n \tag{18-42}$$

The expression reads

$$\Psi = \sum_n C_n\psi_n \tag{18-43}$$

---

[3]Selection rules will be discussed in Chapter 21.

Now

$$
\begin{aligned}
\langle \Psi | H | \Psi \rangle &= \sum_n \sum_m C_n^* \langle \psi_n | H | \psi_m \rangle \, C_m \\
&= \sum_n \sum_m C_n^* C_m E_m \langle \psi_n | \psi_m \rangle \\
&= \sum_n |C_n|^2 E_n \\
&\geq E_0 \sum_n |C_n|^2
\end{aligned}
\tag{18-44}
$$

Since (18-41) implies that

$$
\sum_n |C_n|^2 = 1
\tag{18-45}
$$

we obtain the result that

$$
E_0 \leq \langle \Psi | H | \Psi \rangle
\tag{18-46}
$$

We can use this result to calculate an upper bound on E_0. This can be done by choosing a Ψ that depends on a number of parameters $(\alpha_1, \alpha_2, \ldots)$, calculating $\langle \Psi | H | \Psi \rangle$, and minimizing this with respect to the parameters.

We illustrate the utility of this procedure by calculating the ground-state energy of helium with a Ψ chosen to be a product of hydrogenlike wave functions in the $(1s)$ orbitals, but corresponding to an arbitrary charge Z^*. We take

$$
\Psi(\mathbf{r}_1, \mathbf{r}_2) = \psi_{100}(\mathbf{r}_1) \, \psi_{100}(\mathbf{r}_2)
\tag{18-47}
$$

where

$$
\left(\frac{\mathbf{p}^2}{2m} - \frac{Z^* e^2}{r} \right) \psi_{100}(\mathbf{r}) = \epsilon \psi_{100}(\mathbf{r})
\tag{18-48}
$$

with $\epsilon = -(1/2) \, mc^2 (Z^* \alpha)^2$. We now need

$$
\begin{aligned}
\int d^3 r_1 \int d^3 r_2 \, \psi_{100}^*(\mathbf{r}_1) \, \psi_{100}^*(\mathbf{r}_2) &\left(\frac{\mathbf{p}_1^2}{2m} + \frac{\mathbf{p}_2^2}{2m} - \frac{Ze^2}{r_1} - \frac{Ze^2}{r_2} \right. \\
&\left. + \frac{e^2}{|\mathbf{r}_1 - \mathbf{r}_2|} \right) \psi_{100}(\mathbf{r}_1) \, \psi_{100}(\mathbf{r}_2)
\end{aligned}
\tag{18-49}
$$

We have

$$
\begin{aligned}
&\int d^3 r_1 \int d^3 r_2 \, \psi_{100}^*(\mathbf{r}_1) \, \psi_{100}^*(\mathbf{r}_2) \left(\frac{\mathbf{p}_1^2}{2m} - \frac{Ze^2}{r_1} \right) \psi_{100}(\mathbf{r}_1) \, \psi_{100}(\mathbf{r}_2) \\
&= \int d^3 r_1 \, \psi_{100}^*(\mathbf{r}_1) \left(\frac{\mathbf{p}_1^2}{2m} - \frac{Z^* e^2}{r_1} + \frac{(Z^* - Z) e^2}{r_1} \right) \psi_{100}(\mathbf{r}_1) \\
&= \epsilon + (Z^* - Z) e^2 \int d^3 r_1 |\psi_{100}(\mathbf{r}_1)|^2 \frac{1}{r_1} \\
&= \epsilon + (Z^* - Z) e^2 \frac{Z^*}{a_0} \\
&= \epsilon + Z^*(Z^* - Z) \, mc^2 \alpha^2
\end{aligned}
\tag{18-50}
$$

An identical factor comes from the Hamiltonian for electron 2, and the expectation value of the electron–electron repulsion has already been calculated in (18-31), except that we must substitute Z^* for Z there. Adding up the terms, we get

$$\langle \Psi | H | \Psi \rangle = -\frac{1}{2} mc^2 \alpha^2 \left(2Z^{*2} + 4Z^*(Z - Z^*) - \frac{5}{4} Z^* \right)$$
$$= -\frac{1}{2} mc^2 \alpha^2 \left(4ZZ^* - 2Z^{*2} - \frac{5}{4} Z^* \right) \tag{18-51}$$

Minimizing this with respect to Z^* yields

$$Z^* = Z - \frac{5}{16} \tag{18-52}$$

which is an improvement on the guess we made earlier $(Z - 1/2)$. We thus obtain

$$E_0 \leq -\frac{1}{2} mc^2 \alpha^2 \left[2 \left(Z - \frac{5}{16} \right)^2 \right] = -77.38 \text{ eV} \tag{18-53}$$

when we substitue $Z = 2$. This is much better than the first-order perturbation result.

The variational calculation can be done with more complicated trial wave functions. Pekeris[4] used a 1075-term wave function and minimized $\langle \Psi | H | \Psi \rangle$ on a computer. The resulting bound agrees, within experimental errors, with what is measured. It is, of course, true that such a complicated wave function does not have a form that is as easily interpretable as (18-47), with its partial screening effects. It does, however, provide strong support for the correctness of quantum mechanics, and for the assumption that only electromagnetic forces are required to explain the structure of atoms.

Autoionization

In conclusion, we briefly return to our observation that there exist eigenvalues of $H^{(1)} + H^{(2)}$ that lie above the ionization threshold and that are nevertheless discrete. The states labeled by the orbitals $(2s)^2$ or $(2s)(2p)$, for example, lie well above the ionization energy. This has some dramatic physical consequences. Consider, for example, the $(2s)(2p)$ state. If the electrons form a spin singlet state, then this will be a $^1P_1$ state, and it can be excited from the ground state by the absorption of radiation, since the selection rules $\Delta l = 1$ and $\Delta S = 0$ are not being violated. This state, once excited, need not decay back to the ground state $(^1S_0)$ or to another state allowed by the selection rules (a $^1D_2$ state, say), because it can go into another *channel*: it can decay into an electron and singly ionized helium, He^+, with the electron energy determined by energy conservation. This process is described as *autoionization*.

The $(2s)(2p)$ state in the continuum will show up very clearly in the scattering of electrons by He^+ ions. When the electron energy is such that the *compound state* can be formed, a very dramatic peak will occur in the scattering rate. Similarly,

---

[4]This is discussed in Bethe and Jackiw, in the references at the end of the chapter.

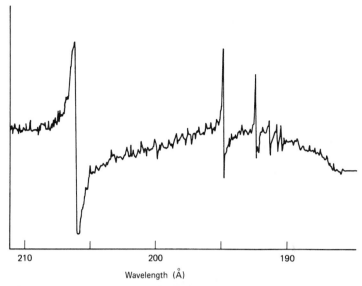

Wavelength (Å)

Figure 18-4. Resonance in the helium absorption spectrum above the continuum threshold; the first peak occurs at the energy corresponding to the location of the $(2s)(2p)$ level. (From R. P. Madden and K. Codling, *Phys. Rev. Lett.* **10**, 516 (1963), by permission.)

in the absorption of radiation by helium, in the vicinity of the energy of the compound state $(e^- - He^+)$, a sharp peak is seen in the absorption (Fig. 18-4). There is absorption at other energies, too, since the process

$$\text{Radiation} + He \rightarrow e^- + He^+$$

can occur, but the absorption at energies away from the compound state energy will vary very smoothly with energy. We can describe the state in still another way by calling it a *resonant state*. Since it decays into its constituents $e^- + He^+$, it does not exist forever. Hence, by the uncertainty relation, $\Delta E \gtrsim \hbar/\Delta t$, it appears that its energy is not precisely defined, which seems to contradict the fact that the $(2s)(2p)$ state does have a well-defined energy. It turns out that if the coupling of the discrete state to the continuum state is taken into account, the state ceases to be discrete, and its energy could lie anywhere in a narrow range about the energy as calculated without the coupling. We shall return to this topic in Chapter 23 and in Special Topics Section 4, "Lifetimes, Line Widths, and Resonances."

Problems

1. Consider the helium atom in the approximation in which the electron–electron interaction is neglected. What is the lowest ortho-helium (spin 1) state? What is its degeneracy in this approximation? Write down the expression of the splitting due to electron–electron repulsion in first-order perturbation theory, and estimate its magnitude.

2. Calculate the lowest order energy shift $\Delta E_{2,l}^{(t)}$ ($l = 0, 1$).

3. Consider the lowest state of ortho-helium. What is its magnetic moment? That is, calculate the interaction with an external magnetic field.

4. Consider

$$\text{"}E\text{"} = \langle \Psi | H | \Psi \rangle$$

with an arbitrary trial wave function Ψ. Show that if Ψ differs from the correct ground-state wave function ψ_0 by terms of order ϵ, then "E" differs from the ground-state energy by terms of order ϵ^2.

[*Note:* Do not forget the normalization condition $\langle \Psi | \Psi \rangle = 1$.]

5. Use the variational principle to estimate the ground-state energy of the three-dimensional harmonic oscillator, using the trial wave function

$$\Psi = N e^{-\alpha r}$$

6. Consider the binding of a proton and a neutron (both with $mc^2 = 938$ MeV, approximately) by means of a potential

$$V(r) = V_0 \frac{e^{-r/r_0}}{r/r_0}$$

with the system in an $\ell = 0$ state. The range of the potential is given by r_0. Use the following procedure to calculate the depth of the potential required to give the binding energy E_B. (a) Calculate an approximate value of the binding energy using the variational principle. (b) In the expression that connects the approximate value with r_0 and the depth of the potential, insert the experimental value of E_B. Do your numerical evaluation using $r_0 = 2.8 \times 10^{-13}$ cm and $E_B = -2.23$ MeV. (Do not forget the reduced mass.)

7. Consider a finite-dimensional matrix H_{ij}. Show that the condition for minimizing

$$\langle \Psi | H | \Psi \rangle = \sum_{i,j=1}^{n} a_i^* H_{ij} a_j$$

subject to the condition

$$\langle \Psi | \Psi \rangle = \sum_{i=1}^{n} a_i^* a_i = 1$$

yields the eigenvalues of the matrix H.

(*Hint:* Use the method of Lagrange multipliers.)

8. Use the variational principle to show that a one-dimensional attractive potential will always have a bound state.

(*Hint:* Evaluate $\langle \Psi | H | \Psi \rangle$ with a convenient trial function, for example, $N e^{-\beta^2 x^2}$ and show that this expectation value can always be made negative.)

9. Use the data of Fig. 18-4 to compute the location of the $(2s)(2p)$ level above the ground state of helium and compute the velocity of the electron emitted in autoionization, if the He^+ ion is in its lowest state at the end. What will it be if the He^+ ion is in its first excited state?

10. Consider a wave function $\psi(\alpha_1, \alpha_2, \ldots \alpha_n)$ for which only the dependence on some parameters is exhibited. The wave function is normalized

$$\langle \psi(\alpha_1, \alpha_2, \ldots \alpha_n) | \psi(\alpha_1, \alpha_2, \ldots \alpha_n) \rangle = 1$$

and the dependence on the parameters is so chosen that

$$\mathcal{E} = \langle \psi(\alpha_1, \ldots) | H | \psi(\alpha_1, \ldots) \rangle$$

is a minimum. Show that the parameters are determined by the set of equations

$$\left\langle \psi(\alpha_1, \ldots) | H \left| \frac{\partial \psi}{\partial \alpha_i} \right\rangle - \mu \left\langle \psi(\alpha_1, \ldots) \left| \frac{\partial \psi}{\partial \alpha_i} \right\rangle = 0 \qquad i = 1, 2, \ldots, n$$

where μ is a Lagrange multiplier. Let H depend on a parameter λ (e.g., the nuclear charge or some distance, say the internuclear distance in a molecule). Then the α_i will depend on that parameter. Prove that

$$\frac{d\mathcal{E}}{d\lambda} = \left\langle \psi(\alpha_1, \ldots) \left| \frac{\partial H}{\partial \lambda} \right| \psi(\alpha_1, \ldots) \right\rangle$$

This is known as the Feynman-Hellmann theorem and is very useful in molecular physics calculations.

11. Use the variational principle to estimate the ground-state energy for the anharmonic oscillator

$$H = \frac{p^2}{2m} + \lambda x^4$$

Compare your result with the exact result

$$E_0 = 1.060 \lambda^{1/3} \left(\frac{\hbar^2}{2m} \right)^{2/3}$$

(*Hint:* use a gaussian trial function.)

12. Given the fact that in the radial part of the Hamiltonian of the hydrogen atom the potential is given by

$$V_{\text{eff}} = -\frac{Ze^2}{r} + \frac{l(l+1)\hbar^2}{2\mu r^2}$$

and that the eigenvalues are given by

$$E(n_r, l) = -\frac{1}{2} \mu c^2 \frac{(Z\alpha)^2}{(n_r + l + 1)^2}$$

use the Feynman-Hellman theorem to calculate $\langle 1/r \rangle_{nl}$ and $\langle 1/r^2 \rangle_{nl}$ by a judicious choice of the parameter λ.

13. Use the exact result quoted in Problem 11 together with the Feynman-Hellmann theorem to calculate $\langle p^2 \rangle$ and $\langle x^4 \rangle$ for the ground state of the anharmonic oscillator.

14. According to the Ritz variational principle, the expectation of the Hamiltonian H in an arbitrary normalized state ψ obeys

$$\langle \psi | H | \psi \rangle > E_0$$

where E_0 is the lowest eigenvalue of H. Suppose H is an $N \times N$ Hermitian matrix, whose elements are H_{ij} ($i, j = 1, 2, 3, \ldots, N$) and E_0 is its lowest eigenvalue. Prove, using judicious choices of ψ, that E_0 is smaller than any one of the diagonal elements H_{ii} of the matrix H.

15. Consider two identical spin 1/2 particles in a harmonic oscillator potential, so that the Hamiltonian takes the form

$$H = \frac{p_1^2}{2m} + \frac{p_2^2}{2m} + \frac{1}{4} m\omega^2(\mathbf{r}_1 - \mathbf{r}_2)^2$$

Assume that the two-particle system has zero center-of-mass momentum, and that the two particles are in $l = 0$ states.

(a) Write down the ground-state wave function, including the spin state.

(b) Write down the first excited states in the singlet and in the triplet spin states.

(c) Assume that there is a short-range interaction between the particles, which in the $l = 0$ state can be approximated by $C\,[\delta(r)/r^2]$. Calculate the effect of this perturbation on the states obtained in (b).

References

A very nice discussion of the spectrum of helium may be found in

H. A. Bethe and R. W. Jackiw, *Intermediate Quantum Mechanics*, W. A. Benjamin, New York, 1968.

Chapter 19

THE STRUCTURE OF ATOMS

THE HARTREE APPROXIMATION

The energy eigenvalue problem for an atom with Z electrons has the form

$$\left(\sum_{i=1}^{Z} \frac{\mathbf{p}_i^2}{2m} - \frac{Ze^2}{r_i} + \sum_{i>j} \frac{e^2}{|\mathbf{r}_1 - \mathbf{r}_j|} \right) \psi(\mathbf{r}_1, \mathbf{r}_2, \ldots, r_Z) = E\psi(\mathbf{r}_1, \mathbf{r}_2, \ldots, r_Z) \quad (19\text{-}1)$$

and is a partial differential equation in $3Z$ dimensions. For light atoms it is possible to solve such an equation on a computer, but such solutions are only meaningful to the expert. We shall base our discussion of atomic structure on a different approach. As in the example of helium ($Z = 2$), it is both practical and enlightening to treat the problem as one involving Z independent electrons in a single potential, and to consider the electron–electron interaction later. Perturbation theory turned out to be adequate for $Z = 2$, but as the number of electrons increases, the shielding effects, not taken into account by first-order perturbation theory, become more and more important. The variational principle discussed at the end of Chapter 18 had the virtue of maintaining the single-particle picture, while at the same time yielding single-particle functions that incorporate the screening corrections.

To apply the variational principle, let us assume that the trial wave function is of the form

$$\psi(\mathbf{r}_1, \mathbf{r}_2, \ldots, \mathbf{r}_Z) = \phi_1(\mathbf{r}_1)\, \phi_2(\mathbf{r}_2) \cdots \phi_Z(\mathbf{r}_Z) \quad (19\text{-}2)$$

Each of the functions is normalized to unity. If we calculate the expectation value
of H in this state, we obtain

$$
\langle H \rangle = \sum_{i=1}^{Z} \int d^3 \mathbf{r}_i \, \phi_i^*(\mathbf{r}_i) \left(-\frac{\hbar^2}{2m} \nabla_i^2 - \frac{Ze^2}{r_i} \right) \phi_i(\mathbf{r}_i)
$$
$$
+ e^2 \sum_{i>j} \sum_{j} \int\!\!\int d^3 \mathbf{r}_i \, d^3 \mathbf{r}_j \, \frac{|\phi_i(\mathbf{r}_i)|^2 |\phi_j(\mathbf{r}_j)|^2}{|\mathbf{r}_i - \mathbf{r}_j|}
$$
(19-3)

The procedure of the variational principle is to pick the $\phi_i(\mathbf{r}_i)$ such that $\langle H \rangle$ is a
minimum. If we were to choose the $\phi_i(\mathbf{r}_j)$ to be hydrogenlike wave functions, with
a different Z_i for each electron (and with each electron in a different quantum state
to satisfy the Pauli exclusion principle), we would get a set of equations analogous
to (18-51) and (18-52). A more general approach is that due to Hartree. If the $\phi_i(\mathbf{r}_i)$
were the single-particle wave functions that minimized $\langle H \rangle$, then an alteration in
these functions by an infinitesimal amount

$$
\phi_i(\mathbf{r}_i) \rightarrow \phi_i(\mathbf{r}_i) + \lambda f_i(\mathbf{r}_i)
$$
(19-4)

should only change $\langle H \rangle$ by a term of order λ^2. The alterations must be such that

$$
\int d^3 \mathbf{r}_i |\phi_i(\mathbf{r}_i) + \lambda f_i(\mathbf{r}_i)|^2 = 1
$$
(19-5)

that is, to first order in λ,

$$
\int d^3 \mathbf{r}_i [\phi_i^*(\mathbf{r}_i) \, f_i(\mathbf{r}_i) + \phi_i(\mathbf{r}_i) \, f_i^*(\mathbf{r}_i)] = 0
$$
(19-6)

Let us compute the terms linear in λ that arise when (19-4) is substituted into
(19-3). Term by term, we have

$$
\sum_i \int d^3 r_i \left[\phi_i^*(\mathbf{r}_i) \left(-\frac{\hbar^2}{2m} \nabla_i^2 \right) \lambda f_i(\mathbf{r}_i) + \lambda f_i^*(\mathbf{r}_i) \left(-\frac{\hbar^2}{2m} \nabla_i^2 \right) \phi_i(\mathbf{r}_i) \right]
$$
$$
= \lambda \sum_i \int d^3 \mathbf{r}_i \left\{ f_i(\mathbf{r}_i) \left[-\frac{\hbar^2}{2m} \nabla_i^2 \, \phi_i^*(\mathbf{r}_i) \right] + f_i^*(\mathbf{r}_i) \left[-\frac{\hbar^2}{2m} \nabla_i^2 \phi_i(\mathbf{r}_i) \right] \right\}
$$
(19-7)

To obtain this we have integrated by parts two times, and used the fact that $f_i(\mathbf{r}_i)$
must vanish at infinity in order to be an acceptable variation of a square integrable
function. Next we have

$$
-\lambda \sum_i \int d^3 \mathbf{r}_i \left[f_i^*(\mathbf{r}_i) \frac{Ze^2}{r_i} \phi_i(\mathbf{r}_i) + \phi_i^*(\mathbf{r}_i) \frac{Ze^2}{r_i} f_i(\mathbf{r}_i) \right]
$$
(19-8)

and finally

$$
\lambda e^2 \sum_{i>j} \sum_{j} \int d^3 \mathbf{r}_i \int d^3 \mathbf{r}_j \, \frac{1}{|\mathbf{r}_i - \mathbf{r}_j|} \{ [f_i^*(\mathbf{r}_i) \, \phi_i(\mathbf{r}_i) + f_i(\mathbf{r}_i) \, \phi_i^*(\mathbf{r}_i)] |\phi_j(\mathbf{r}_j)|^2
$$
$$
+ [f_j^*(\mathbf{r}_j) \, \phi_j(\mathbf{r}_j) + f_j(\mathbf{r}_j) \, \phi_j^*(\mathbf{r}_j)] |\phi_i(\mathbf{r}_i)|^2 \}
$$
(19-9)

We cannot just set the sum of these three terms equal to zero because the $f_i(\mathbf{r}_i)$ are constrained by (19-6). The proper way to account for the constraint is by the use of Lagrange multipliers, that is, we multiply each of the constraining relations (19-6) by a constant (the "multiplier") and add the sum to our three terms. The total can then be set equal to zero, since the constraints on the $f_i(\mathbf{r}_i)$ are now taken care of. With a certain amount of notational foresight we label the multipliers $-\epsilon_i$, and thus get

$$\sum_i \int d^3\mathbf{r}_i \left\{ f_i^*(\mathbf{r}_i) \left[-\frac{\hbar^2}{2m} \nabla_i^2 \, \phi_i(\mathbf{r}_i) \right] - f_i^*(\mathbf{r}_i) \frac{Ze^2}{r_i} \phi_i(\mathbf{r}_i) \right\}$$

$$+ e^2 \sum_{i \neq j} \sum_j \iint d^3\mathbf{r}_i \, d^3\mathbf{r}_j f_i^*(\mathbf{r}_i) \frac{|\phi_j(\mathbf{r}_j)|^2}{|\mathbf{r}_i - \mathbf{r}_j|} \phi_i(\mathbf{r}_i) \qquad (19\text{-}10)$$

$$- \epsilon_i \int d^3 r_i f_i^*(\mathbf{r}_i) \, \phi_i(\mathbf{r}_i) + (\text{complex conjugate term}) = 0$$

In deriving the second line, first we converted the double sum $\sum_{i>j} \sum_j$ into $(1/2) \sum_{i \neq j} \sum_j$, which is unrestricted except for the requirement that $i \neq j$, and then used the fact that the integrand in (19-9) is symmetric in i and j. Now $f_i(\mathbf{r}_i)$ is completely unrestricted, so that we may treat $f_i(\mathbf{r}_i)$ and $f_i^*(\mathbf{r}_i)$ as completely independent (each one has a real and an imaginary part). Furthermore, other than being square integrable, they are completely arbitrary, so that for (19-10) to hold, the coefficients of $f_i(\mathbf{r}_i)$ and $f_i^*(\mathbf{r}_i)$ must separately vanish *at each point* $\mathbf{r}_i$, since we are allowed to make local variations in the functions $f_i(\mathbf{r}_i)$ and $f_i^*(\mathbf{r}_i)$. We are thus led to the condition that

$$\left[-\frac{\hbar^2}{2m} \nabla_i^2 - \frac{Ze^2}{r_i} + e^2 \sum_{j \neq i} \int d^3\mathbf{r}_j \frac{|\phi_j(\mathbf{r}_j)|^2}{|\mathbf{r}_i - \mathbf{r}_j|} \right] \phi_i(\mathbf{r}_i) = \epsilon_i \phi_i(\mathbf{r}_i) \qquad (19\text{-}11)$$

and the complex conjugate relation.

This equation has a straightforward interpretation: it is an energy eigenvalue equation for electron "i" located at $\mathbf{r}_i$, moving in a potential

$$V_i(\mathbf{r}_i) = -\frac{Ze^2}{r_i} + e^2 \sum_{j \neq i} \int d^3\mathbf{r}_j \frac{|\phi_j(\mathbf{r}_j)|^2}{|\mathbf{r}_i - \mathbf{r}_j|} \qquad (19\text{-}12)$$

that consists of an attractive Coulomb potential due to a nucleus of charge Z, and a repulsive contribution due to the charge density of all the other electrons. We do not, of course, know the charge densities

$$\rho_j(\mathbf{r}_j) = -e|\phi_j(\mathbf{r}_j)|^2 \qquad (19\text{-}13)$$

of all the other electrons, so that we must search for a *self-consistent* set of $\phi_i(\mathbf{r}_i)$, in the sense that their insertion in the potential leads to eigenfunctions that reproduce themselves. The equation (19-11) is a rather complicated integral equation, but it is at least an equation in three dimensions (we can replace the variable $\mathbf{r}_i$ by $\mathbf{r}$), and that makes numerical work much easier. An even greater simplification occurs when $V_i(\mathbf{r})$ is replaced by its angular average

$$V_i(r) = \int \frac{d\Omega}{4\pi} V_i(\mathbf{r}) \qquad (19\text{-}14)$$

for then the self-consistent potential becomes central, and the self-consistent solutions can be decomposed into angular and radial functions, that is, they will be
functions that can be labeled by n_i, l_i, m_i, σ_i, with the last label referring to the spin
state ($s_{iz} = \pm 1/2$).

The trial wave function (19-2) does not take into account the exclusion principle.
The latter plays an important role, since if all the electrons could be in the same
quantum state, the energy would be minimum with all the electrons in the $n = 1$,
$l = 0$ "orbital." Atoms do not have such a simple structure. To take the exclusion
principle into account, we add to the *Ansatz* represented by (19-2) the rule: *every
electron must be in a different state*, if the spin states are included in the labeling. A
more sophisticated way of doing this automatically is to replace (19-2) by a trial
wave function that is a *Slater determinant* [cf. (8-60)]. The resulting equations differ
from (19-11) by the addition of an exchange term. The new Hartree-Fock equations
have eigenvalues that turn out to differ by 10–20 percent from those obtained
using Hartree equations supplemented by the condition arising from the exclusion
principle. It is a little easier to talk about the physics of atomic structure in terms
of the Hartree picture, so we will not discuss the Hartree-Fock equations.

The potential (19-14) no longer has the $1/r$ form, and thus the degeneracy of
all states with a given n and $l \leq n - 1$ is no longer present. We may expect,
however, that for low Z at least, the splitting for different l values for a given n
will be smaller than the splitting between different n-values, so that electrons
placed in the orbitals $1s$, $2s$, $2p$, $3s$, $3p$, $3d$, $4s$, $4p$, $4d$, $4f$, ... will be successively
less strongly bound.[1] Screening effects will accentuate this: whereas s orbitals do
overlap the small r region significantly, and thus feel the full nuclear attraction,
the p-, d-, ... orbitals are forced out by the centrifugal barrier, and feel less than
the full attraction. This effect is so strong that the energy of the $3d$ electrons is very
close to that of the $4s$ electrons, so that the anticipated ordering is sometimes
disturbed. The same is true for the $4d$ and $5s$ electrons, the $4f$ and $6s$ electrons and
so on. The dominance of the l-dependence over the n-dependence becomes more
important as we go to larger Z values, as we shall see in our discussion of the
periodic table.

The number of electrons that can be placed in orbitals with a given (n, l) is
$2(2l + 1)$, since there are two spin states for given m-value. When all these
$2(2l + 1)$ states are filled, we speak of the *closing of a shell*. The charge density for
a closed shell has the form

$$-e \sum_{m=-l}^{l} |R_{nl}(r)|^2 |Y_{lm}(\theta, \phi)|^2 \tag{19-15}$$

and this is spherically symmetric because of the property of spherical harmonics
that

$$\sum_{m=-l}^{l} |Y_{lm}(\theta, \phi)|^2 = \frac{2l + 1}{4\pi} \tag{19-16}$$

---

[1]The notation is the same as that used for hydrogen. A more sensible notation, used by nuclear
shell-structure physicists, would be to replace the n by $n - l$, which is just an index representing the
ordering of a given l-state. Thus instead of starting with $3d$ states, for example, it might be more sensible
to have the lowest d state called the $1d$ state, and so on. We shall nevertheless continue to use the
conventional notation, even though the n-value does not have much to do with the ordering of levels
for large Z atoms.

THE BUILDING-UP PRINCIPLE

In this section we discuss the building up of atoms by the addition of more and more electrons to the appropriate nucleus, whose only role, to good approximation, is to provide the positive charge Ze.

HYDROGEN ($Z = 1$) There is only one electron, and the ground-state configuration is ($1s$). The ionization energy is 13.6 eV, and the amount of energy needed to excite the first state above the ground state is 10.2 eV. The radius of the atom is 0.5 Å, and its spectroscopic description is $^2S_{1/2}$.

HELIUM ($Z = 2$) The lowest two-electron state, as we saw in Chapter 18, is one in which both electrons are in the ($1s$) orbital. We denote this configuration by ($1s$)^2. In spectroscopic notation, the ground state is an $l = 0$ spin singlet state, $^1S_0$ because the exchange effect favors it. The total binding energy is 79 eV. After one electron is removed, the remaining electron is in a ($1s$) orbit about a $Z = 2$ nucleus. Thus its binding energy is $13.6Z^2$ eV $= 54.4$ eV. Thus the energy required to remove the first electron, the *ionization energy*, is $79.0 - 54.4 = 24.6$ eV. A rough estimate of the energy of the first excited state, with configuration ($1s$)($2s$), is $-13.6Z^2 - 13.6(Z - 1)^2/n^2 \approx -58$ eV for $Z = 2$ and $n = 2$. This expression takes into account shielding in the second term. Thus the excitation energy is 79 eV $-$ 58 eV ≈ 21 eV.[2] In any reaction with another substance, about 20 eV is thus required for a rearrangement of the electrons, and thus helium is chemically very inactive. This property is shared by all atoms whose electrons form closed shells, but the energy required is particularly large for helium.

LITHIUM ($Z = 3$) The exclusion principle forbids a ($1s$)^3 configuration, and the lowest energy electron configuration is ($1s$)^2($2s$). We are thus adding an electron to a closed shell, and since the shell is in a $^1S_0$ state, the spectroscopic description of the ground state is $^2S_{1/2}$, just as for hydrogen. If the screening were perfect, we should expect a binding energy of -3.4 eV (since $n = 2$). The screening is not perfect, especially since the outer *valence electron* being in an s-state, its wave function has a reasonable overlap with nucleus at $r = 0$. We can estimate the effective Z from the measured ionization energy of 5.4 eV, and it is $Z^* = 1.3$. It takes very little energy to excite the lithium atom. The six ($2p$) electronic states lie just a little above the ($2s$) state, and these ($2p$) states, when occupied, make the atom chemically active (see our more extended discussion of carbon). Lithium, like other elements that have one electron outside a closed shell, is a very active element.

BERYLLIUM ($Z = 4$) The natural place for the fourth electron to go is into the second space in the ($2s$) orbital, so that the configuration is ($1s$)^2($2s$)^2. We again have a closed shell and the spectroscopic description is $^1S_0$. As far as the energy is concerned, the situation is very much like that of helium. If the screening were

---

[2]This is a crude estimate that ignores the electron–electron repulsion and exchange effects. The difference between the 21 eV and the 24.6 eV is the 4–5 eV that will be released when the excited atom decays to its ground state. (See Fig. 18-2b.)

perfect, we might expect a binding energy like that of helium, since the inner electrons reduce the effective Z to something like $Z = 2$. Since $n = 2$, we would expect an ionization energy of $24.6/n^2 = 6.2$ eV. The shielding situation is somewhat like that for lithium, and if we make a guess that, as in lithium, the binding energy is increased by about 50 percent, we get approximately 9 eV. The experimental value is 9.3 eV. Although the shell is closed, the excitation of one of the electrons to a $(2p)$ orbital does not cost much energy. Thus, in the presence of another element a rearrangement of electrons may yield enough energy to break up the closed shell. We therefore expect beryllium not to be as inert as helium. It is generally true that atoms in which the outer electrons have their spins "paired up" into singlet states are less reactive.

BORON ($Z = 5$) After the closing of the shell, the fifth electron can either go into a $(3s)$ orbital or into a $(2p)$ orbital. The latter is lower in energy, and it is the $(2p)$ shell that begins to fill up, starting with boron. The configuration is $(1s)^2(2s)^2(2p)$, and the spectroscopic description of the state is $^2P_{1/2}$. This deserves comment: If we add spin $1/2$ to an $l = 1$ orbital state, we may have $J = 3/2$ or $1/2$. These states are split by a spin-orbit interaction

$$\frac{1}{2m^2c^2} \mathbf{L} \cdot \mathbf{S} \frac{1}{r} \frac{dV(r)}{dr} = \frac{1}{4m^2c^2} [J(J + 1) - L(L + 1) - S(S + 1)] \frac{1}{r}\frac{dV}{dr} \quad (19\text{-}17)$$

and the form of this leads to the higher J value having a higher energy, because the expectation value of $(1/r)(dV/dr)$ is still positive, even though it is no longer equal to the value given in (17-16). This conclusion depends on the degree to which the shell is filled, as specifically given by *Hund's Rules*. These will be discussed below. The ionization energy is 8.3 eV. This meets the expectation that the value should be somewhat lower than that for beryllium, since the $2p$ state energy is somewhat higher than that of the $2s$ orbital.

CARBON ($Z = 6$) The configuration for carbon is $(1s)^2(2s)^2(2p)^2$. The second electron could be in the same p-state as the first electron, with the two of them making an up–down spin pair. It is, however, advantageous for the second to stay out of the way of the first electron, thus lowering the repulsion between the electrons. It can do so because the possible $l = 1$ states Y_{11}, Y_{10}, Y_{1-1} allow for the linear combinations $\sin \theta \cos \phi$, $\sin \theta \sin \phi$, and $\cos \theta$, which are aligned along the x-, y-, and z-axes, respectively. When two electrons go into orthogonally aligned arms, the overlap is minimized and the repulsion is reduced. The electrons are in different spatial states, so that their spins do not have to be antiparallel. One might expect carbon to be divalent. This is not so, because of the subtleties that arise from close-lying energy levels. It costs very little energy to promote one of the $(2s)$ electrons into the third unoccupied $l = 1$ state. The configuration $(1s)^2(2s)(2p)^3$ has four "unpaired" electrons, and the gain in energy from the formation of four bonds with other atoms more than makes up for the energy needed to promote the $(2s)$ electron. The reduction in the repulsion leads to a somewhat larger ionization energy than that for boron, 11.3 eV. The spectroscopic description of the ground state is $^3P_0$. We can have a total spin of 0 or 1 for the two $2p$ electrons, and, since we are adding two $l = 1$ states, the total orbital angular momentum can be 0, 1, or 2. Of the various states, $^1S_0$, $^3P_{2,1,0}$, and $^1D_2$, the state of higher spin has the lower energy (cf. our discussion of helium) and by another of *Hund's rules*, the $^3P_0$ state has the lowest energy.

NITROGEN ($Z = 7$) Here the configuration is $(1s)^2(2s)^2(2p)^3$, sometimes described as $(2p)^3$ for brevity (the closed shells and subshells are omitted). The three electrons can all be in nonoverlapping p-states, and thus we expect the increase in ionization energy to be the same as the increase from boron to carbon. This is in agreement with the measured value of 14.5 eV.

OXYGEN ($Z = 8$) The configuration may be abbreviated to $(2p)^4$, and the shell is more than half full. Since there are four electrons, it appears as if the determination of the ground-state spectroscopic state would be very difficult. We can, however, look at the shell in another way. We know that when two more electrons are added to make a $(2p)^6$ configuration, then the shell is filled, and the the total state has $L = S = 0$. We can thus think of oxygen as having a closed $2p$ shell with two *holes* in it. These holes are just like antielectrons, and we can look at two-hole configurations. These will be the same as two-electron configurations, since the holes also have spin $1/2$. Thus, as with carbon, the possible states consistent with the antisymmetry of the two-fermion (two-hole) wave function are ^1S, ^3P, ^1D, and the four electrons must be in the same states, since they, together with the two-hole system, give $S = 0$, $L = 0$. The highest spin is $S = 1$, and thus we must have a ^3P state. Hund's rule, which will be discussed in the next section, yields the $^3P_2$ state. When the fourth electron is added to the nitrogen configuration, it must go into an orbital with an m-value already occupied. Thus two of the electron wave functions overlap, and this raises the energy because of the repulsion. It is therefore not surprising that the ionization energy drops to the value of 13.6 eV.

FLUORINE ($Z = 9$) Here the configuration is $(2p)^5$. The monotonic increase in the ionization energy resumes, with the experimental value 17.4 eV. Fluorine is chemically very active, because it can "accept" an electron to form a closed shell $(2p)^6$, which is very stable. Since the addition of a single electron with $s = 1/2$ and $l = 1$ yields a $^1S_0$ state, the shell with the hole in it must have $s = 1/2$ and $l = 1$. It is therefore a ^2P state, and by Hund's rule, as we shall see, the state is $^2P_{3/2}$.

NEON ($Z = 10$) With $Z = 10$ the $(2p)$ shell is closed, and all electrons are paired off. The ionization energy is 21.6 eV, continuing the monotonic trend. Here, as in helium, the first available state that an electron can be excited into has a higher n value, and thus it takes quite a lot of energy to perturb the atom. Neon, like helium, is an inert gas.

At this point, the addition of another electron requires putting it in an orbit with a higher n value ($n = 3$), and thus neon marks the end of a *period* in the periodic table, as did helium. In neon, as in helium, the first available state into which an electron can be excited has a higher n-value, so that it takes quite a lot of energy to perturb the atom. Neon shares with helium the property of being an *inert gas*.

The next period again has eight elements in it. First the $(3s)$ shell is filled, with sodium ($Z = 11$) and magnesium ($Z = 12$), and then the $3p$ shell, which includes, in order, aluminum ($Z = 13$), silicon ($Z = 14$), phosphorus ($Z = 15$), sulphur ($Z = 16$), chlorine ($Z = 17$) and, closing the shell, argon ($Z = 18$). These elements are chemically very much like the series: lithium, . . . , neon, and the spectroscopic descriptions of the ground states are the same. The only difference is that, since $n = 3$, the ionization energies are somewhat smaller, as can be seen from the periodic table at the end of the chapter.

It might appear a little strange that the period ends with argon, since the $(3d)$

shell, accommodating ten elements, remains to be filled. The fact is that the self-consistent potential is not of the $1/r$ form, and the intrashell splitting here is sufficiently large that the $(4s)$ state lies lower than the $(3d)$ state, though not by much. Hence a competition develops, and in the next period we have $(4s)$, $(4s)^2$, $(4s)^2(3d)$, $(4s)^2(3d)^2$, $(4s)^2(3d)^3$, $(4s)(3d)^5$, $(4s)^2(3d)^5$, $(4s)^2(3d)^6$, $(4s)^2(3d)^7$, $(4s)^2(3d)^8$, $(4s)(3d)^{10}$, $(4s)^2(3d)^{10}$ and then the $4p$ shell gets filled until the period ends with kyrpton $(Z = 36)$. The chemical properties of elements at the beginning and end of this period are similar to those of elements at the beginning and end of other periods. Thus potassium, with the single $(4s)$ electron, is an alkali metal, like sodium with its single $(3s)$ electron outside a closed shell. Bromine, with the configuration $(4s)^2(3d)^{10}(4p)^5$, has a single hole in a p-shell and thus is chemically like chlorine and fluorine. The series of elements in which the $(3d)$ states are being filled all have rather similar chemical properties. The reason for this again has to do with the details of the self-consistent potential. It turns out that the radii of these orbits[3] are somewhat smaller than those of the $(4s)$ electrons, so that when the $(4s)^2$ shell is filled, these electrons tend to shield the $(3d)$ electrons, no matter how many there are, from outside influences. The same effect occurs when the $(4f)$ shell is being filled, just after the $(6s)$ shell has been filled. The elements here are called the *rare earths*.

SPECTROSCOPIC DESCRIPTION OF GROUND STATES

In our discussion of the light atoms we often gave the spectroscopic description of the ground states, for example, $^3P_2$ for oxygen, $^2P_{3/2}$ for fluorine, and so on. The knowledge of S, L, and J for the ground states are important, because selection rules allow us to determine these quantities for the excited states of atoms. We referred to *Hund's rules* in the determination of these, and these rules are the subject of this section.

What determines the ground-state quantum numbers is an interplay of spin-orbit coupling and the exchange effect discussed in connection with helium in Chapter 18. For the lighter atoms $(Z < 40)$, for which the motion of the electrons is nonrelativistic, the electron–electron repulsion effects are more important than the spin-orbit coupling. This means that it is a fairly good approximation to view L and S as separately good quantum numbers: we add up all the spins to form an S, and all the orbital angular momenta of the electrons to form an L, and these are then coupled to obtain a total J. For heavier atoms it is a better approximation to first couple the spin and orbital angular momentum to form a total angular momentum for that electron, and then to couple all of the J's together. The former case is described as Russell-Saunders coupling, the latter as j-j coupling. For Russell-Saunders coupling F. Hund summarized the results of various calculations by a set of rules that give the overall quantum numbers of the lowest states. The rules are:

1. The state with largest S lies lowest.

2. For a given value of S, the state with maximum L lies lowest.

---

[3]It is understood that this is just a way of talking about the peaking tendencies of the charge distribution.

3. For a given L and S, if the incomplete shell is not more than half-filled, the lowest state has the minimum value of $J = |L - S|$; if the shell is more than half-filled, the state of lowest energy has $J = L + S$.

In applying these rules we must be careful not to violate the Pauli principle.

The first of these rules is easy to understand: the largest S state is symmetric in all the spins (since it contains the state $S_z = S_{max}$ for which all the spins are parallel), and thus the spatial wave function is antisymmetric, which minimizes the electron overlap, and thus the expectation value of the repulsive potentials.

The second rule emerges qualitatively from the fact that the higher the L-value, the more lobes the wave function has, as shown in Fig. 12-3. This allows the electrons to stay away from one another, and reduce the effect of the Coulomb repulsion.

The third rule follows from the form of the spin-orbit coupling. Since the expectation value of $[1dV(r)/r \, dr]$ is positive, the perturbation due to the spin-orbit coupling splits the degenerate J states (for a given L and S) and it is clear from (19-17) that the lowest value of J will give the lowest lying state. Once we get to the point of having a shell that is more than half filled, it is clearer to look at the atom as consisting of a filled shell with a number of holes, as we discussed in our description of oxygen. These holes act as if they had positive charge, and for the spin-orbit interaction of the holes, the sign of $[1dV(r)/r \, dr]$ is reversed. Thus the multiplet is *inverted* and it is the largest value of J that gives the lowest lying state.

Let us illustrate the application of the Hund rules to some atoms, and the need to keep track of the Pauli principle. We shall consider the quantum numbers of carbon $(2p)^2$, oxygen $(2p)^4$, and manganese $(3d)^5$. In the first two cases we have p-states, so that we can draw a set of "shelves" corresponding to $L_z = 1, 0, -1$. The electrons are placed, as far as is possible, on different shelves, to minimize the repulsion. For carbon we place them in $L_z = 1$ and $L_z = 0$ states. By Hund's first rule, the spins will be parallel (strictly speaking they will be in a triplet state) (Fig. 19-1). Thus we have $S_z = 1$ for the largest possible value, and we get a triplet state. The largest possible value of L_z gives the L value, which is 1. The third rule thus gives $J = |L - S| = 0$, and we have a $^3P_0$ state. For the $(2p)^4$ case, we fill all three shelves with one electron each, and then put the last electron in the $L_z = 1$ state, for example. The Pauli principle demands that the two electrons in the $L_z = 1$ state form a singlet. Thus only the other two electrons are relevant, and since $S_z = 1, S = 1$. The maximum value of $L_z = [2 + 0 + (-1)] = 1$, so that $L = 1$. Now, however, we have more than a half-filled shell, so that $J = L + S = 2$, and oxygen has a $^3P_2$ ground state.

For manganese the shelves have $L_z = 2, 1, 0, -1, -2$, as shown in Fig. 19-2. There are five electrons, and thus each of the spaces is filled by one. With the spins parallel we get $S_z = 5/2$. This implies that $S = 5/2$. The total value of $L_z = 0$, and thus we have an S-state. This means that the ground state is an $^6S_{5/2}$ state.

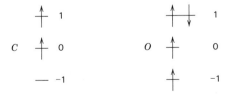

Figure 19-1.

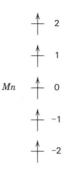

Figure 19-2.

Limitations of space prevent us from a more detailed discussion of the periodic table. A few additional comments are, however, in order.

(a) There is nothing in atomic structure that limits the number of elements. The reason that atoms with $Z \gtrsim 100$ do not occur naturally is that heavy *nuclei* undergo spontaneous fission. If new, superheavy (meta)stable nuclei are ever discovered, there will presumably exist corresponding atoms, and it is expected that their structure will conform to the prediction of the building-up approach outlined in this chapter.

(b) The ionization energies all lie in the vicinity of 5–15 eV. The reason is that in spite of the increasing number of electrons, the outermost electrons "see" a charge that lies in the range $Z = 1$–2. In addition, because of the departures from a point charge distribution, the dependence of the energy is no longer of the $1/n^2$ form. Consequently, the wave functions of the outermost electrons do not extend much further than that of the electron in the hydrogen atom. Atoms are more or less the same size!

(c) We went to a great deal of trouble to specify the S, L, and J quantum numbers of the ground states of the various elements. The reason for doing this is that in spectroscopy, the quantum numbers are of particular interest because of the selection rules

$$\Delta S = 0$$

$$\Delta L = \pm 1 \tag{19-18}$$

$$\Delta J = 0, \pm 1 \qquad (\text{no } 0 - 0)$$

that will be derived later, and that may then be used to determine the quantum numbers of the excited states. The spectroscopy of atoms, once we get beyond hydrogen and helium, is very complicated. Consider, as a relatively simple example, the first few states of carbon, which are formed from different configurations of the two electrons that lie outside the closed shell in the $(2p)^2$ orbitals. As already pointed out, the possible states are $^1S_0$, $^3P_{2,1,0}$, and $^1D_2$. The $^3P_0$ state lies lowest, but the other states are still there. The first excited states may be described by the orbitals $(2p)(3s)$. Here $S = 0$ or 1, but $L = 1$ only. Since the n-values are different, the exclusion principle does not restrict the states in any way, and all of the states $^1P_1$, $^3P_{2,1,0}$ are possible, while the excited states that arise from the orbitals $(2p)(3p)$ can have $S = 0$, 1 and $L = 2$, 1, 0, leading to all the states $^1D_2$, $^1P_1$, $^1S_0$, $^3D_{3,2,1}$, $^3P_{2,1,0}$, and $^3S_1$. Even with the restrictions provided by the selection rules, there are numerous transitions. Needless to say, the ordering of these levels represents a delicate balance between various competing effects, and the prediction of the more complex spectra is very difficult. That task is not really of interest to us, since the main point that we want to make is that quantum mechanics provides a qualitative, and quantitative, detailed explanation of the chemical properties of atoms and of their spectra, without assuming an interaction other than the electromagnetic interaction between charged particles. We shall have occasion to return to the topic of spectra.

Periodic Table

Z	Element	Configuration	Term[1]	Ionization Potential eV	Radius[2] in Å
1	H	$(1s)$	$^2S_{1/2}$	13.6	0.53
2	He	$(1s)^2$	$^1S_0$	24.6	0.29
3	Li	$(He)(2s)$	$^2S_{1/2}$	5.4	1.59
4	Be	$(He)(2s)^2$	$^1S_0$	9.3	1.04
5	B	$(He)(2s)^2(2p)$	$^2P_{1/2}$	8.3	0.78
6	C	$(He)(2s)^2(2p)^2$	$^3P_0$	11.3	0.62
7	N	$(He)(2s)^2(2p)^3$	$^4S_{3/2}$	14.5	0.52
8	O	$(He)(2s)^2(2p)^4$	$^3P_2$	13.6	0.45
9	F	$(He)(2s)^2(2p)^5$	$^2P_{3/2}$	17.4	0.40
10	Ne	$(He)(2s)^2(2p)^6$	$^1S_0$	21.6	0.35
11	Na	$(Ne)(3s)$	$^2S_{1/2}$	5.1	1.71
12	Mg	$(Ne)(3s)^2$	$^1S_0$	7.6	1.28
13	Al	$(Ne)(3s)^2(3p)$	$^2P_{1/2}$	6.0	1.31
14	Si	$(Ne)(3s)^2(3p)^2$	$^3P_0$	8.1	1.07
15	P	$(Ne)(3s)^2(3p)^3$	$^4S_{3/2}$	11.0	0.92
16	S	$(Ne)(3s)^2(3p)^4$	$^3P_2$	10.4	0.81
17	Cl	$(Ne)(3s)^2(3p)^5$	$^2P_{3/2}$	13.0	0.73
18	Ar	$(Ne)(3s)^2(3p)^6$	$^1S_0$	15.8	0.66
19	K	$(Ar)(4s)$	$^2S_{1/2}$	4.3	2.16
20	Ca	$(Ar)(4s)^2$	$^1S_0$	6.1	1.69
21	Sc	$(Ar)(4s)^2(3d)$	$^2D_{3/2}$	6.5	1.57
22	Ti	$(Ar)(4s)^2(3d)^2$	$^3F_2$	6.8	1.48
23	V	$(Ar)(4s)^2(3d)^3$	$^4F_{3/2}$	6.7	1.40
24	Cr	$(Ar)(4s)(3d)^5$	$^7S_3$	6.7	1.45
25	Mn	$(Ar)(4s)^2(3d)^5$	$^6S_{3/2}$	7.4	1.28
26	Fe	$(Ar)(4s)^2(3d)^6$	$^5D_4$	7.9	1.23
27	Co	$(Ar)(4s)^2(3d)^7$	$^4F_{9/2}$	7.8	1.18
28	Ni	$(Ar)(4s)^2(3d)^8$	$^3F_4$	7.6	1.14
29	Cu	$(Ar)(4s)(3d)^{10}$	$^2S_{1/2}$	7.7	1.19
30	Zn	$(Ar)(4s)^2(3d)^{10}$	$^1S_0$	9.4	1.07
31	Ga	$(Ar)(4s)^2(3d)^{10}(4p)$	$^2P_{1/2}$	6.0	1.25
32	Ge	$(Ar)(4s)^2(3d)^{10}(4p)^2$	$^3P_0$	8.1	1.09
33	As	$(Ar)(4s)^2(3d)^{10}(4p)^3$	$^4S_{3/2}$	10.0	1.00
34	Se	$(Ar)(4s)^2(3d)^{10}(4p)^4$	$^3P_2$	9.8	0.92
35	Br	$(Ar)(4s)^2(3d)^{10}(4p)^5$	$^2P_{3/2}$	11.8	0.85
36	Kr	$(Ar)(4s)^2(3d)^{10}(4p)^6$	$^1S_0$	14.0	0.80

Periodic Table   (*continued*)

Z	Element	Configuration	Term[1]	Ionization Potential eV	Radius[2] in Å
37	Rb	$(Kr)(5s)$	$^2S_{1/2}$	4.2	2.29
38	Sr	$(Kr)(5s)^2$	$^1S_0$	5.7	1.84
39	Y	$(Kr)(5s)^2(4d)$	$^2D_{3/2}$	6.6	1.69
40	Zr	$(Kr)(5s)^2(4d)^2$	$^3F_2$	7.0	1.59
41	Nb	$(Kr)(5s)(4d)^4$	$^6D_{1/2}$	6.8	1.59
42	Mo	$(Kr)(5s)(4d)^5$	$^7S_3$	7.2	1.52
43	Tc	$(Kr)(5s)^2(4d)^5$	$^6S_{5/2}$	Not known	1.39
44	Ru	$(Kr)(5s)(4d)^7$	$^5F_5$	7.5	1.41
45	Rh	$(Kr)(5s)(4d)^8$	$^4F_{9/2}$	7.7	1.36
46	Pd	$(Kr)(4d)^{10}$	$^1S_0$	8.3	0.57
47	Ag	$(Kr)(5s)(4d)^{10}$	$^2S_{1/2}$	7.6	1.29
48	Cd	$(Kr)(5s)^2(4d)^{10}$	$^1S_0$	9.0	1.18
49	In	$(Kr)(5s)^2(4d)^{10}(5p)$	$^2P_{1/2}$	5.8	1.38
50	Sn	$(Kr)(5s)^2(4d)^{10}(5p)^2$	$^3P_0$	7.3	1.24
51	Sb	$(Kr)(5s)^2(4d)^{10}(5p)^3$	$^4S_{3/2}$	8.6	1.19
52	Te	$(Kr)(5s)^2(4d)^{10}(5p)^4$	$^3P_2$	9.0	1.11
53	I	$(Kr)(5s)^2(4d)^{10}(5p)^5$	$^2P_{3/2}$	10.4	1.04
54	Xe	$(Kr)(5s)^2(4d)^{10}(5p)^6$	$^1S_0$	12.1	0.99
55	Cs	$(Xe)(6s)$	$^2S_{1/2}$	3.9	2.52
56	Ba	$(Xe)(6s)^2$	$^1S_0$	5.2	2.06
57	La	$(Xe)(6s)^2(5d)$	$^2D_{3/2}$	5.6	1.92
58	Ce	$(Xe)(6s)^2(4f)(5d)$	$^3H_5$	6.9	1.98
59	Pr	$(Xe)(6s)^2(4f)^3$	$^4I_{9/2}$	5.8	1.94
60	Nd	$(Xe)(6s)^2(4f)^4$	$^5I_4$	6.3	1.92
61	Pm	$(Xe)(6s)^2(4f)^5$	$^6H_{5/2}$	Not known	1.88
62	Sm	$(Xe)(6s)^2(4f)^6$	$^7F_0$	5.6	1.84
63	Eu	$(Xe)(6s)^2(4f)^7$	$^8S_{7/2}$	5.7	1.83
64	Gd	$(Xe)(6s)^2(4f)^7(5d)$	$^9D_2$	6.2	1.71
65	Tb	$(Xe)(6s)^2(4f)^9$	$^6H_{15/2}$	6.7	1.78
66	Dy	$(Xe)(6s)^2(4f)^{10}$	$^5I_8$	6.8	1.75
67	He	$(Xe)(6s)^2(4f)^{11}$	$^4I_{15/2}$	Not known	1.73
68	Er	$(Xe)(6s)^2(4f)^{12}$	$^3H_6$	Not known	1.70
69	Tm	$(Xe)(6s)^2(4f)^{13}$	$^2F_{7/2}$	Not known	1.68
70	Yb	$(Xe)(6s)^2(4f)^{14}$	$^1S_0$	6.2	1.66
71	Lu	$(Xe)(6s)^2(4f)^{14}(5d)$	$^2D_{3/2}$	5.0	1.55
72	Hf	$(Xe)(6s)^2(4f)^{14}(5d)^2$	$^3F_2$	5.5	1.48
73	Ta	$(Xe)(6s)^2(4f)^{14}(5d)^3$	$^4F_{3/2}$	7.9	1.41
74	W	$(Xe)(6s)^2(4f)^{14}(5d)^4$	$^5D_0$	8.0	1.36
75	Re	$(Xe)(6s)^2(4f)^{14}(5d)^5$	$^6S_{5/2}$	7.9	1.31
76	Os	$(Xe)(6s)^2(4f)^{14}(5d)^6$	$^5D_4$	8.7	1.27
77	Ir	$(Xe)(6s)^2(4f)^{14}(5d)^7$	$^4F_{9/2}$	9.2	1.23
78	Pt	$(Xe)(6s)(4f)^{14}(5d)^9$	$^3D_3$	9.0	1.22
79	Au	$(Xe)(6s)(4f)^{14}(5d)^{10}$	$^2S_{1/2}$	9.2	1.19
80	Hg	$(Xe)(6s)^2(4f)^{14}(5d)^{10}$	$^1S_0$	10.4	1.13
81	Tl	$(Xe)(6s)^2(4f)^{14}(5d)^{10}(6p)$	$^2P_{1/2}$	6.1	1.32
82	Pb	$(Xe)(6s)^2(4f)^{14}(5d)^{10}(6p)^2$	$^3P_0$	7.4	1.22
83	Bi	$(Xe)(6s)^2(4f)^{14}(5d)^{10}(6p)^3$	$^4S_{3/2}$	7.3	1.30
84	Po	$(Xe)(6s)^2(4f)^{14}(5d)^{10}(6p)^4$	$^3P_2$	8.4	1.21
85	At	$(Xe)(6s)^2(4f)^{14}(5d)^{10}(6p)^5$	$^2P_{3/2}$	Not known	1.15
86	Rn	$(Xe)(6s)^2(4f)^{14}(5d)^{10}(6p)^6$	$^1S_0$	10.7	1.09

For Z = 57 through Z = 71 the bracket is labelled **Lanthanides (Rare Earths)**.

Periodic Table (*continued*)

Z		Element	Configuration	Term[1]	Ionization Potential eV	Radius[2] in Å
87		Fr	$(Rn)(7s)$		Not known	2.48
88		Ra	$(Rn)(7s)^2$	$^1S_0$	5.3	2.04
89		Ac	$(Rn)(7s)^2(6d)$	$^2D_{3/2}$	6.9	1.90
90		Th	$(Rn)(7s)^2(6d)^2$	$^3F_2$		
91		Pa	$(Rn)(7s)^2(5f)^2(6d)$	$^4K_{11/2}$		
92		U	$(Rn)(7s)^2(5f)^3(6d)$	$^5L_6$		
93		Np	$(Rn)(7s)^2(5f)^4(6d)$	$^6L_{11/2}$		
94		Pu	$(Rn)(7s)^2(5f)^6$	$^7F_0$		
95		Am	$(Rn)(7s)^2(5f)^7$	$^8S_{7/2}$		
96	Actinides	Cm	$(Rn)(7s)^2(5f)^7(6d)$	$^9D_2$		
97		Bk	$(Rn)(7s)^2(5f)^9$	$^6H_{15/2}$		
98		Cf	$(Rn)(7s)^2(5f)^{10}$	$^5I_8$		
99		Es	$(Rn)(7s)^2(5f)^{11}$	$^4I_{15/2}$		
100		Fm	$(Rn)(7s)^2(5f)^{12}$	$^3H_6$		
101		Md	$(Rn)(7s)^2(5f)^{13}$	$^2F_{7/2}$		
102		No	$(Rn)(7s)^2(5f)^{14}$	$^1S_0$		

[1]Term designation is equivalent to spectroscopic description.
[2]Radius is defined by the peak of the calculated charge density of the outermost orbital.

Problems

1. List the spectroscopic states (in the form $^{2S+1}L_J$) that can arise from combining

$$S = 1/2, L = 3$$
$$S = 2, L = 1$$
$$S_1 = 1/2, S_2 = 1, L = 4$$
$$S_1 = 1, S_2 = 1, L = 3$$
$$S_1 = 1/2, S_2 = 1/2, L = 2$$

Which states are excluded, among the two-spin questions, if the particles are identical?

2. Consider the following states

$$^1D, \, ^2P, \, ^4F, \, ^3G, \, ^2D, \, ^3H$$

What are the possible J values associated with each?

3. Consider the states $^1D, \, ^3P, \, ^3S, \, ^5G, \, ^5P, \, ^5S$. Given that in each one the state consists of two identical particles in their largest possible spin state, which of the states are disallowed by the exclusion principle?

4. Use Hund's rules to find the spectroscopic description of the ground states of the following atoms:

$$N(Z = 7), K(Z = 19), Sc(Z = 21), Co(Z = 27).$$

Figure out the electronic configurations as far as you are able to.

5. Use Hund's rules to check the (S, L, J) quantum numbers of the elements with $Z = 14, 15, 24, 30, 34$.

6. Define Z_{eff} of the valence electrons by the ionization potential, using

$$\text{Ioniz. pot.} = \frac{13.6 Z_{eff}^2}{n^2} \text{ eV}$$

List Z_{eff} for $Z = 1$–40 and discuss any patterns you see.

7. Consider $Z = 11$. What would you expect the L, S, J values for the first excited state to be? What are the possible values of these quantum numbers? Estimate the excitation energy, using estimates of the centrifugal barrier and Z_{eff} estimated as in Problem 6.

(*Hint:* Take note of the fact that it takes a lot of energy to break up the $Z = 10$ closed shell.)

8. Plot the ionization potentials given in the periodic table against Z. Observe the peaks indicating the shell structure of the atoms.

References

An excellent introductory treatment of atomic structure can be found in

G. Herzberg, *Atomic Spectra and Atomic Structure*, Dover, New York, 1944.

An advanced treatment that is definitive is

I. I. Sobelman, *Introduction to the Theory of Atomic Spectra*, Pergamon Press, New York, 1972. This is a very advanced book.

Chapter 20

MOLECULES

Just as atoms are aggregates of electrons and a single nucleus, so molecules are aggregates of electrons and several nuclei. Molecules in their lowest energy state are stable, that is, it takes a certain amount of energy to dissociate them into their components. Since dissociation of molecules into atoms is the most common occurrence when enough energy is transferred to the system, we may call molecules bound states of atoms, although we shall see that this description hides much of what makes up the structure of molecules. The purpose of this chapter, is to show that quantum mechanics is successful in describing the properties and behavior of molecules.

The simplest molecules are those that involve two nuclei, the diatomic molecules. Even they are more complex systems than atoms because, after the center of mass is fixed in space, the nuclei are still free to move. This leads to an increase in the number of degrees of freedom. Thus for the simplest of all molecules, the H_2^+ molecule, consisting of two protons and one electron, there are still six degrees of freedom left, three for the electron and three for the relative motion of the two protons. As with atoms, a frontal attack on the problem of the dynamics of molecules, that is, a numerical solution of the Schrödinger equation in many dimensions is possible. For our purposes cruder but more physical approaches will be more enlightening.

Insight into the dynamics of molecules can be obtained from use of the fact that nuclei are a great deal more massive than electrons ($M/m_e \gg 10^3$) and thus their motion is a great deal slower. One may view the motion of the electrons as if the nuclei were fixed in space. The motion of the nuclei, on the other hand, is in an average field due to the electrons. For a given set of nuclear coordinates, there will be a Hamiltonian for the electrons. The lowest eigenvalue of that Hamiltonian will depend on these coordinates, and its minimum value will determine the positions of the nuclei. This picture must be modified a little for nuclei that are not infinitely massive, since they can also move. Their motion depends on the electrons, but they only "see" an average charge distribution due to the rapid motion of the electrons, and to first approximation, they move in a harmonic potential about the locations determined by the minimum in the energy of the electrons.

THE H_2^+ MOLECULE

We begin by considering the simplest molecule of all, the H_2^+ ion, which consists of two protons and one electron. In the center-of-mass system of the two nuclei, the nuclei are located at $\mathbf{R}/2$ and $-\mathbf{R}/2$, and the nuclear kinetic energy is described by $-(\hbar^2/2M^*)\nabla_\mathbf{R}^2$, where M^* is the reduced mass of the two protons ($M^* = M/2$). The electron coordinate is $\mathbf{r}$. The energy eigenvalue equation is

$$\left(-\frac{\hbar^2}{2M^*}\,\nabla_\mathbf{R}^2 - \frac{\hbar^2}{2m}\,\nabla_\mathbf{r}^2 - \frac{e^2}{|\mathbf{r} - \mathbf{R}/2|} - \frac{e^2}{|\mathbf{r} + \mathbf{R}/2|} + \frac{e^2}{R}\right)\Psi(\mathbf{r},\,\mathbf{R}) = E\,\Psi(\mathbf{r},\,\mathbf{R})$$

$$(20\text{-}1)$$

The first term represents the kinetic energy of the protons and the second is the kinetic energy of the electron. The next two terms represent the attraction between the electron and the two protons located at $\pm\mathbf{R}/2$, and the last term represents the repulsion between the two protons, separated by a distance $R = |\mathbf{R}|$. The qualitative features of the potential with $\mathbf{R}$ held fixed and the proton kinetic energy absent are shown in Fig. 20-1. For R very large, the electron will be bound to one of the protons, and the energy of the system is -13.6 eV, the energy of a single

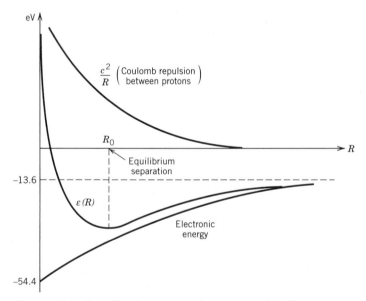

Figure 20-1. Contributions to "nuclear potential." The Coulomb repulsion and the electronic energy combine to give a curve with a minimum at R_0.

hydrogen atom. When $R \rightarrow 0$, and we leave out the proton–proton repulsion, the electron will be bound to a $Z = 2$ nucleus, and the binding energy will be $-13.6Z^2$ eV $= -54.4$ eV. The electronic energy, as a function of R interpolates smoothly between these values. When the energy of repulsion e^2/R is added to this, the curve for $E(R)$ results. This curve dips below the level of the binding energy of an electron to a single proton, showing that the system of two protons and one electron has a lower energy than a proton and a hydrogen atom separately. Such a dip does not always exist, so that some atoms do not form molecules as we shall soon see.

To find $E(R)$ we must solve the electron energy eigenvalue equation

$$\left(-\frac{\hbar^2}{2m} \nabla_r^2 - \frac{e^2}{|\mathbf{r} - \mathbf{R}/2|} - \frac{e^2}{|\mathbf{r} + \mathbf{R}/2|} + \frac{e^2}{R} \right) u(\mathbf{r}, \mathbf{R}) = E(R) \, u(\mathbf{r}, \mathbf{R}) \quad (20\text{-}2)$$

This can actually be solved in elliptical coordinates, but we will get more insight from using the *variational principle*, with trial wave functions that reflect some physical intuition about the system.

Molecular Orbitals

A reasonable trial wave function is a linear combination of $u_{100}(r_1, \mathbf{R})$ and $u_{100}(r_2, \mathbf{R})$, the ground-state wave functions of electrons bound to one or the other of the protons, so that $r_1 = |\mathbf{r} - \mathbf{R}/2|$ and $r_2 = |\mathbf{r} + \mathbf{R}/2|$. Since the two nuclei are identical, the Hamiltonian is symmetric about reflections in the plane bisecting the line joining them, that is, the Hamiltonian is invariant under the changes $\mathbf{r} \rightarrow -\mathbf{r}$ and $\mathbf{R} \rightarrow -\mathbf{R}$. Thus we can take linear combinations that are also eigenfunctions of parity. These are the even and odd combinations of

$$u_{100}(r_1, \mathbf{R}) \equiv \psi_1(\mathbf{r}, \mathbf{R}) = \left(\frac{1}{\pi a_0^3} \right)^{1/2} e^{-|\mathbf{r} - \mathbf{R}/2|/a_0} \quad (20\text{-}3)$$

and

$$u_{100}(r_2, \mathbf{R}) \equiv \psi_2(\mathbf{r}, \mathbf{R}) = \left(\frac{1}{\pi a_0^3} \right)^{1/2} e^{-|\mathbf{r} + \mathbf{R}/2|/a_0} \quad (20\text{-}4)$$

Our trial wave functions associate the electron with each of the protons, and they are called *molecular orbitals (MO)*.

We take[1]

$$\psi_g(\mathbf{r}, \mathbf{R}) = C_+(R)[\psi_1(\mathbf{r}, \mathbf{R}) + \psi_2(\mathbf{r}, \mathbf{R})]$$
$$\psi_u(\mathbf{r}, \mathbf{R}) = C_-(R)[\psi_1(\mathbf{r}, \mathbf{R}) - \psi_2(\mathbf{r}, \mathbf{R})] \quad (20\text{-}5)$$

The normalization factors are given by

$$\frac{1}{C_\pm^2} = \langle \psi_1 \pm \psi_2 | \psi_1 \pm \psi_2 \rangle$$
$$= 2 \pm 2 \int d^3r \, \psi_1(\mathbf{r}, \mathbf{R}) \, \psi_2(\mathbf{r}, \mathbf{R}) \quad (20\text{-}6)$$

---

[1]The labeling is historical: "g" stands for "gerade," which means even in German, and "u" for "ungerade," odd.

The integral appearing in the preceding equation is called the *overlap integral*, and it can be calculated. The calculation of

$$S(R) = \int d^3r \, \psi_1(\mathbf{r}, \mathbf{R}) \, \psi_2(\mathbf{r}, \mathbf{R})$$

$$= \frac{1}{\pi a_0^3} \int d^3r \, e^{-|\mathbf{r} - \mathbf{R}/2|/a_0} \, e^{-|\mathbf{r} + \mathbf{R}/2|a_0} \tag{20-7}$$

$$= \frac{1}{\pi a_0^3} \int d^3r' \, e^{-|\mathbf{r}' - \mathbf{R}|/a_0} \, e^{-r'/a_0}$$

is straightforward, though tedious. The result is

$$S(R) = \left(1 + \frac{R}{a_0} + \frac{R^2}{3a_0^2}\right) e^{-R/a_0} \tag{20-8}$$

The expectation value of H_0 in the two states is

$$\langle H \rangle_{g,u} = \frac{1}{2[1 \pm S(R)]} \langle \psi_1 \pm \psi_2 | H_0 | \psi_1 \pm \psi_2 \rangle$$

$$= \frac{1}{2[1 \pm S(R)]} \{\langle \psi_1 | H_0 | \psi_1 \rangle + \langle \psi_2 | H_0 | \psi_2 \rangle \pm \langle \psi_1 | H_0 | \psi_2 \rangle \pm \langle \psi_2 | H_0 | \psi_1 \rangle\} \tag{20-9}$$

$$= \frac{\langle \psi_1 | H_0 | \psi_1 \rangle \pm \langle \psi_1 | H_0 | \psi_2 \rangle}{1 \pm S(R)}$$

where use has been made of the symmetry under $\mathbf{R} \to -\mathbf{R}$. The two terms in the numerator can be calculated:

$$\langle \psi_1 | H_0 | \psi_1 \rangle = \int d^3r \psi_1^*(\mathbf{r}, \mathbf{R}) \left(\frac{p_e^2}{2m} - \frac{e^2}{|\mathbf{r} - \mathbf{R}/2|} - \frac{e^2}{|\mathbf{r} + \mathbf{R}/2|} + \frac{e^2}{R}\right)$$

$$\times \, \psi_1(\mathbf{r}, \mathbf{R}) \tag{20-10}$$

$$= E_1 + \frac{e^2}{R} - e^2 \int d^3r \, \frac{|\psi_1(\mathbf{r}, \mathbf{R})|^2}{|\mathbf{r} + \mathbf{R}/2|}$$

The first term is just the energy of a single hydrogen atom $E_1 = -13.6$ eV; the second term is the proton–proton repulsion, and the third term is the electrostatic potential energy due to the electron charge distribution about one proton being attracted to the other proton. The last integral can be evaluated, so that finally

$$\langle \psi_1 | H_0 | \psi_1 \rangle = E_1 + \frac{e^2}{R} \left(1 + \frac{R}{a_0}\right) e^{-2R/a_0} \tag{20-11}$$

Similarly, we find

$$\langle \psi_1 | H_0 | \psi_2 \rangle = \int d^3r_1 \, \psi_1^*(\mathbf{r}, \mathbf{R}) \left(E_1 + \frac{e^2}{R} - \frac{e^2}{|\mathbf{r} + \mathbf{R}/2|}\right) \psi_2(\mathbf{r}, \mathbf{R})$$

$$= \left(E_1 + \frac{e^2}{R}\right) S(R) - e^2 \int d^3r \, \frac{\psi_1^*(\mathbf{r}, \mathbf{R}) \, \psi_2(\mathbf{r}, \mathbf{R})}{|\mathbf{r} + \mathbf{R}/2|} \tag{20-12}$$

Here the last term is the exchange integral, which can also be evaluated, yielding

$$e^2 \int d^3r \, \frac{\psi_1^*(\mathbf{r}, \mathbf{R}) \, \psi_2(\mathbf{r}, \mathbf{R})}{|\mathbf{r} + \mathbf{R}/2|} = \frac{e^2}{a_0} \left(1 + \frac{R}{a_0}\right) e^{-R/a_0} \tag{20-13}$$

When all of this is put together we get, using $R/a_0 = y$ and $e^2/a_0 = -2E_1$,

$$\langle H \rangle_{g,u} = E_1 \frac{1 - (2/y)(1 + y)e^{-2y} \pm \{(1 - 2/y)(1 + y + y^2/3)e^{-y} - 2(1 + y)e^{-y}\}}{1 \pm (1 + y - y^2/3)e^{-y}}$$

$$\tag{20-14}$$

Figure 20-2 shows the energies as functions of $R = a_0 y$.

The exact solution, which according to the variational principle must lie below the curves obtained, differs little from the minimum. In our approximation, we see that the even solution yields binding, while the odd one does not. The difference between the even and the odd solutions is that in the former, the electron has a high probability of being located between the two protons, where the attractive contribution is maximized; for the odd solution, which has a node midway between the protons, the electron tends to be excluded from that region.

The experimental separation between the protons is 1.06 Å, and the binding energy is -2.8 eV. The calculations based on (20-14) lead to a separation of 1.3 Å and a binding energy of -1.76 eV. Thus our wave function is not as compact as it should be. The reason is that when R is small, the wave function should approach that of a He$^+$ ion, which (20-5) does not. One could improve the calculation by introducing an effective charge for the proton and minimizing $\langle H_0 \rangle_g$ with respect to that parameter in addition to R, as in our illustration involving the helium atom. Since we are more interested in a qualitative understanding of the problem than in improving the variational calculation, we do not pursue this idea.

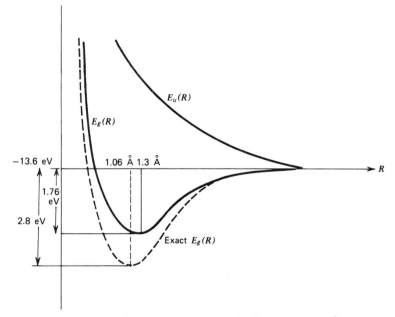

Figure 20-2. Results of variational calculation for H$_2^+$.

The orbitals that we have considered do not depend on the azimuthal angle about the axis of the molecule. Since the Hamiltonian is invariant under rotations about the axis, we can classify the solutions by the angular momentum component along the axis. If we choose $\mathbf{R}$ to define the z-axis, our eigenstates will be simultaneous eigenstates of L_z The solutions will, in general, have the dependence $e^{im\phi}$ with $m = 0, \pm 1, \pm 2, \ldots$. These are labeled $\sigma, \pi, \delta, \ldots$ in analog with $S, P, D, \ldots$. There is also the labeling "g" and "u" that is applicable to all diatomic molecules for which the atoms are the same (*homonuclear* molecules). Thus in our example the ground state could be labeled $1s\sigma_g$, and the antisymmetric state could be labeled $1s\sigma_u^*$, the asterisk indicating that the state is unbound. Excited states of the H_2^+ molecule can be formed with higher orbitals.

THE H_2 MOLECULE

We next discuss the H_2 molecule in some detail, because there are two electrons (in contrast to the H_2^+ molecule), and the exclusion principle and electron spin considerations make their first appearance. As in the case of the H_2^+ molecule we treat the nuclei as fixed.

The nuclei (protons) will be labeled A, B and the two electrons "1" and "2" (Fig. 20-3). The Hamiltonian has the form

$$H = H_1 + H_2 + \frac{e^2}{r_{12}} + \frac{e^2}{R_{AB}} \tag{20-15}$$

where

$$H_i = \frac{\mathbf{p}_i^2}{2m} - \frac{e^2}{r_{Ai}} - \frac{e^2}{r_{Bi}} \qquad (i = 1, 2) \tag{20-16}$$

depends only on the coordinates of the electron i relative to the nuclei. We will again compute an upper bound to $E(R_{AB})$ by constructing the expectation value of H with a trial wave function. Since

$$\tilde{H}_i = H_i + \frac{e^2}{R_{AB}} \tag{20-17}$$

are just hamiltonians for the H_2^+ molecule (Eq. 20-3) it is suggestive to take as our trial wave function a product of two $1s\sigma_g$ functions (Eq. 20-5) for the H_2^+ molecule:

$$\psi_g(\mathbf{r}_1, \mathbf{r}_2) = \frac{1}{2[1 + S(R_{AB})]} \, [\psi_A(\mathbf{r}_1) + \psi_B(\mathbf{r}_1)][\psi_A(\mathbf{r}_2) + \psi_B(\mathbf{r}_2)]X_{\text{singlet}} \tag{20-18}$$

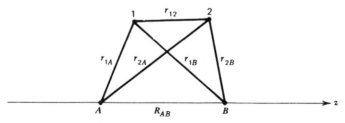

Figure 20-3. Coordinate labels in the discussion of the H_2 molecule.

The electron spin state is a singlet, since the spatial part of the wave function is taken to be symmetric. In this trial wave function, *each electron is associated with both protons,* that is, the trial wave function is said to be a product of *molecular orbitals.* The description in terms of molecular orbitals is sometimes called the *MO* method.

The calculation of $\langle \psi_g | H | \psi_g \rangle$ yields

$$\left\langle \psi_g \left| \left(\tilde{H}_1 - \frac{e^2}{R_{AB}} \right) + \left(\tilde{H}_2 - \frac{e^2}{R_{AB}} \right) + \frac{e^2}{r_{12}} + \frac{e^2}{R_{AB}} \right| \psi_g \right\rangle$$

$$= E(R_{AB}) + E(R_{AB}) + \left\langle \psi_g \left| \frac{e^2}{r_{12}} \right| \psi_g \right\rangle - \frac{e^2}{R_{AB}} \qquad (20\text{-}19)$$

$$= 2E(R_{AB}) - \frac{e^2}{R_{AB}} + \left\langle \psi_g \left| \frac{e^2}{r_{12}} \right| \psi_g \right\rangle$$

where $E(R_{AB})$ is the energy of the H_2^+ molecule calculated in (20-14). The first-order electron–electron repulsion contribution can also be calculated, and when the total energy so computed is minimized with respect to the separation R_{AB}, it is found that the binding energy and internuclear separation are given by

$$E_b = -2.68 \text{ eV}$$
$$R = \quad 0.85 \text{ Å} \qquad (20\text{-}20)$$

The experimental values are

$$E_b = -4.75 \text{ eV}$$
$$R = \quad 0.74 \text{ Å} \qquad (20\text{-}21)$$

Evidently the approximation is not a very good one. We noted in our discussion of the H_2^+ molecule that the trial wave functions (the *MO's*) are inaccurate for small proton–proton separations, and the fact that the *MO's* are too spread out in space shows up in the numbers above. The trial wave function also has some undesirable features for large R_{AB}. The product in (20-18) may be rewritten in the form

$$[\psi_A(\mathbf{r}_1) + \psi_B(\mathbf{r}_1)][(\psi_A(\mathbf{r}_2) + \psi_B(\mathbf{r}_2)]$$
$$= [\psi_A(\mathbf{r}_1)\,\psi_A(\mathbf{r}_2) + \psi_B(\mathbf{r}_1)\,\psi_B(\mathbf{r}_2)] + [\psi_A(\mathbf{r}_1)\,\psi_B(\mathbf{r}_2) + \psi_A(\mathbf{r}_2)\,\psi_B(\mathbf{r}_1)] \qquad (20\text{-}22)$$

The first term is called an "ionic" term, since it describes both electrons bound to one proton or the other. The second term, the "covalent" term, is a description in terms of linear combinations of atomic orbitals (LCAO). Our trial wave function thus implies, since the two terms enter with equal weight, that for large R_{AB} the molecule is as likely to dissociate into the ions H^+ and H^-, as it is into two hydrogen atoms, and this is patently false.

The Valence Bond Method

The last difficulty can be avoided with the use of the *Valence Bond* (also called Heitler-London) method, in which linear combinations of atomic orbitals are used. The singlet wave function used as a trial wave function in the variational principle is taken to be

$$\psi(\mathbf{r}_1, \mathbf{r}_2) = \left\{ \frac{1}{2[1 + S^2(R_{AB})]} \right\}^{1/2} [\psi_A(\mathbf{r}_1)\,\psi_B(\mathbf{r}_2) + \psi_A(\mathbf{r}_2)\,\psi_B(\mathbf{r}_1)]\, X_{\text{singlet}} \quad (20\text{-}23)$$

where, as before, the $\psi_A(\mathbf{r}_i)$ are hydrogenic wave functions for the *i*th electron about proton A. We could, in principle, add a triplet term to our variational trial wave function. However, a triplet wave function must be spatially antisymmetric and yields low probability for the electrons being located in the region between the protons. We saw in our discussion of the H_2^+ molecule that just this configuration led to the lowest energy. Although it is not immediately obvious that the attraction is still largest in this configuration when there are *two* electrons that repel each other in the system, it is in fact so. The results of a variational calculation with the *VB* trial wave function is

$$E_b = -3.14 \text{ eV}$$
$$R = \quad 0.87 \text{ Å} \qquad\qquad (20\text{-}24)$$

This is not a significant improvement over the *MO* results, for the simple reason that the inadequacy of the trial wave functions for small R_{AB} carries more weight. There should be no question about the quantitative successes of quantum mechanics in molecular physics. More sophisticated trial wave functions have to be used; for example, a 50-term trial wave function yields complete agreement with observations for the H_2 molecule, but it does not, as the *MO* and *VB* functions do, give us something of a qualitative feeling of what goes on between the atoms. In what follows, we will explore the relevance of these approaches to a qualitative understanding of some aspects of chemistry.

The expectation value of H for the H_2 molecule in the *VB* approach has the following schematic form

$$\langle \psi | H | \psi \rangle = \frac{1}{2(1 + S^2)} \langle \psi_{A1}\psi_{B2} + \psi_{A2}\psi_{B1} | H | \psi_{A1}\psi_{B2} + \psi_{A2}\psi_{B1} \rangle$$

$$= \frac{1}{1 + S^2} \left\langle \psi_{A1}\psi_{B2} \left| \left(T_1 + T_2 - \frac{e^2}{r_{A1}} - \frac{e^2}{r_{A2}} - \frac{e^2}{r_{B1}} - \frac{e^2}{r_{B2}} + \frac{e^2}{r_{12}} \right. \right. \right. \qquad (20\text{-}25)$$

$$\left. \left. \left. + \frac{e^2}{R_{AB}} \right) \right| \psi_{A1}\psi_{B2} + \psi_{A2}\psi_{B1} \right\rangle$$

where T_i is the kinetic energy of the *i*th electron, and since

$$\left(T_1 - \frac{e^2}{r_{A1}} \right) \psi_{A1} = E_1 \psi_{A1}$$

and so forth, this can be simplified to

$$\frac{1}{1+S^2}\left(\left\langle\psi_{A1}\psi_{B2}\left|2E_1-\frac{e^2}{r_{B1}}-\frac{e^2}{r_{A2}}+\frac{e^2}{r_{12}}+\frac{e^2}{R_{AB}}\right|\psi_{A1}\psi_{B2}\right\rangle\right.$$

$$\left.+\left\langle\psi_{A1}\psi_{B2}\left|2E_1-\frac{e^2}{r_{B2}}-\frac{e^2}{r_{A1}}+\frac{e^2}{r_{12}}+\frac{e^2}{R_{AB}}\right|\psi_{A2}\psi_{B1}\right\rangle\right)$$

$$=\frac{1}{1+S^2}\left\{\left(2E_1+\frac{e^2}{R_{AB}}\right)(1+S^2)-2e^2\left\langle\psi_{A1}\left|\frac{1}{r_{B1}}\right|\psi_{A1}\right\rangle\right.\qquad(20\text{-}26)$$

$$-2e^2S\left\langle\psi_{A1}\left|\frac{1}{r_{A1}}\right|\psi_{B1}\right\rangle+e^2\iint\frac{|\psi_{A1}|^2|\psi_{B2}|^2}{r_{12}}$$

$$\left.+e^2\iint\frac{\psi_{A1}^*\psi_{B1}\psi_{B2}^*\psi_{A2}}{r_{12}}\right\}$$

In obtaining this, liberal use has been made of symmetry. The terms that can make this expression more negative are

$$\left\langle\psi_{A1}\left|\frac{1}{r_{B1}}\right|\psi_{A1}\right\rangle\qquad\text{and}\qquad\frac{S}{1+S^2}\left\langle\psi_{A1}\left|\frac{1}{r_{A1}}\right|\psi_{B1}\right\rangle$$

The former is just the attraction of the electron cloud about one proton to the other proton; the second is the overlap of the two electrons (weighted with $1/r_{A1}$). If this can be large, there will be binding. The two electrons can only overlap significantly, however, if their spins are antiparallel; this is a consequence of the exclusion principle. The region of overlap is between the two nuclei, and there the attraction to the nuclei generally overcomes the electrostatic repulsion between the electrons.

In the *MO* picture, too, it is an overlap term—the last term in (20-12)—that is crucial to bonding, and again, bonding occurs because the electron charge distribution is large between the nuclei. Thus, although here the orbitals belong to the whole molecule rather than to individual atoms, the physical reason for bonding is the same.

We note that in general there may be several bound states of the nuclei, corresponding to different electronic configurations. For example, if in (20-23) we take the $\psi(\mathbf{r}_2)$ wave function to be a u_{200} eigenfunction, while the $\psi(\mathbf{r}_1)$ remains a u_{100} eigenfunction, the overlap may be such as to provide a second, more weakly bound state of the protons. We are not going to pursue this, except to point out the important fact that the $E(R)$ is *different for each electronic state*.

THE IMPORTANCE OF UNPAIRED VALENCE ELECTRONS

We will discuss some molecules in terms of these two approaches to the description of the electronic charge distribution. An important simplification occurs because we really do not need to take all electrons into account. In the construction of orbitals, be it valence or molecular, only the outermost electrons, not in closed shells, that is, the so-called *valence electrons* have a chance to contribute to the bonding. The inner electrons, being closer to the nucleus, are less affected by the

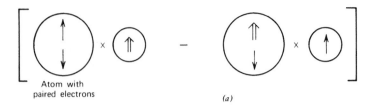

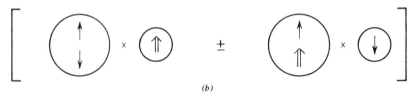

Figure 20-4. Illustration of why paired electrons do not give rise to bonding.
(*a*) If parallel electrons exchange, wave function is spatially antisymmetric.
(*b*) If antiparallel electrons exchange, one term in the wave function has electrons in the same spin state, which may require promotion to a higher energy
orbit.

presence of another atom in the vicinity.[2] Furthermore, not all valence electrons
contribute equally: if two electrons are in a spin 0 state—we call them *paired elec-*
trons—they will *not give rise to bonding*. To see why this is so, consider what happens when an atom with a single valence electron is brought near an atom with
two paired electrons. There are two cases to be considered (Fig. 20-4).

(a) If the two electrons that are parallel exchange [i.e., are put into a form such as
(20-23) with a $\pm$ sign between the terms], then they must be in a triplet state,
and hence the spatial wave function of this pair must be antisymmetric. This
reduces the overlap, and it turns out that the exchange integral gives a repulsive contribution to the energy.

(b) When the two electrons that are antiparallel exchange, then one atom finds
itself some of the time with two electrons in the same spin state. The original
atomic state will frequently no longer be a possible one, and one of the electrons will have to be promoted into another atomic orbital. Sometimes this
may cost very little energy, but usually this is not the case, and again bonding
is not achieved. *Chemical activity depends on the presence of unpaired outer elec-*
trons. An example of this is the nonexistence of the H-He molecule. In He we
have two electrons in the 1s state; promotion of one of them into a 2s state
costs a lot of energy. It is for this reason that the atoms for which the outer
shells are closed are *inert*. Not all unpaired electrons are of equal significance.
As noted before, the unpaired *d*- and *f*-electrons in the transition elements

---

[2]It may happen in atoms that even the valence electrons are rather close to the nucleus. This is the
case for the rare earths. A consequence of the fact that the outer electrons in 5*d* and 4*f* shells lie close
in is that the rare earths are chemically less active than the transition metals ($Z \simeq 20\text{–}30$).

tend to be close to the nucleus, and hence inactive. Thus, mainly s- and p-electrons in the outer shells contribute to chemical activity. The pairing effect is also responsible for what is called the "saturation of chemical binding forces": once two unpaired electrons from different atoms form a singlet state (and cause bonding), they become paired; an electron from a third atom must find an unpaired electron elsewhere, that is, participate in a different bond. Another consequence is that molecules have spin 0 in most cases.

OVERVIEW OF SOME SIMPLE MOLECULES

Let us next go through a process analogous to the building up of the electronic shells in atoms. In Fig. 20-5 we show pictures of atomic orbitals, in particular the $s(Y_{00})$ orbital and the p-orbitals. For the latter the linear combinations $p_x(Y_{11} + Y_{1,-1})$ and $p_y(Y_{11} - Y_{1,-1})$ are plotted in addition to $p_z(Y_{10})$. The corresponding d-orbitals, $d_{xy}, d_{xz}, d_{yz}, d_{zz}$, and $d_{xx} - d_{yy}$ are not shown, because the d-electrons will play no role in our discussion. Figure 20-6 represents what happens when atomic orbitals are brought together and exchange occurs. Thus, two 1s atomic orbitals may combine into a spatially symmetric MO (hence with spin 0) or into a spatially antisymmetric MO, which is antibonding since the wave function between the nuclei is small. Similarly, the formation of bonding and antibonding

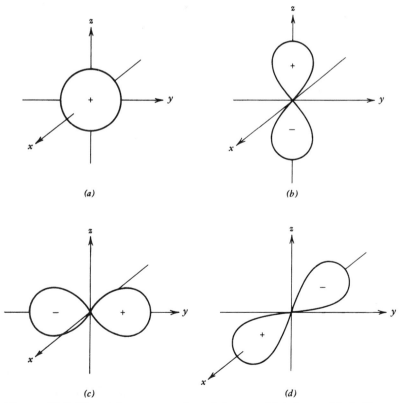

Figure 20-5. Pictorial representation of shapes of (a) the s-orbital, (b) the $p_z(Y_{10})$, (c) the $p_y(Y_{11} - Y_{1,-1})$, and (d) the $p_x(Y_{11} + Y_{1,-1})$ orbitals. The signs refer to the signs of the wave function in the given region.

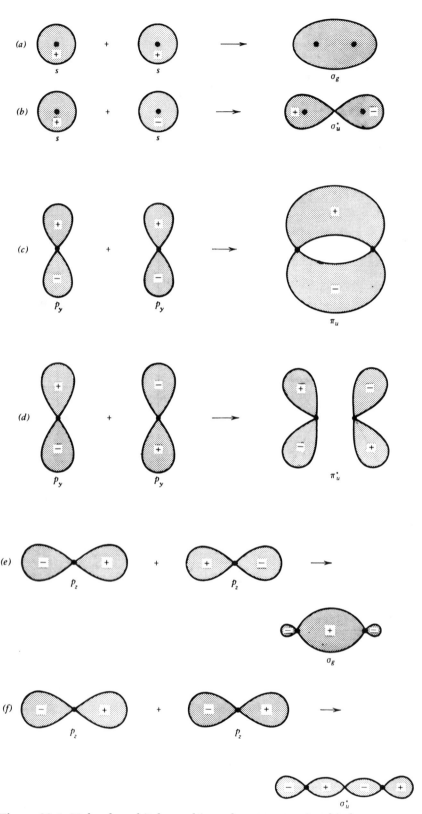

Figure 20-6. Molecular orbitals resulting when two atomic orbitals are brought together. (*a*) Two *s*-orbitals combine to form the spatially symmetric MO σ_g that gives rise to bonding. (*b*) Two *s*-orbitals combine to form the spatially antisymmetric antibonding MO σ_u^*. (*c*) and (*d*) show bonding and antibonding with p_y atomic orbitals. (*e*) and (*f*) show bonding and antibonding with p_z orbitals. The *z*-axis is along the line connecting the nuclei, which are represented by black dots.

MO's with *p*-orbitals is illustrated in the figure. Note that (1) the parity "g" or "u" can be read off from the figures, since these indicate the signs of the wave functions; the distributions that change sign upon reflection in the *x*-*y* plane, here represented by a vertical line, are odd; (2) since the p_x- and p_y-orbitals have $m_l = \pm 1$, the molecular orbital formed from them is a π-orbital. It should be stressed that in the figure we are *not* just bringing two charge distributions together, but are trying to suggest the probability amplitude that results when wave functions are combined, that is, the *MO*'s such as $\psi_{1s}(\mathbf{r}_A) \pm \psi_{1s}(\mathbf{r}_B)$ and $\psi_{2p_y}(\mathbf{r}_A) \pm \psi_{2p_y}(\mathbf{r}_B)$ for R_{AB} large and for R_{AB} small.

We can use the *MO*'s to discuss the properties of a few diatomic homonuclear molecules:

H₂ This molecule was discussed in some detail. We merely repeat that the two electrons can go into a $1s\sigma_g$ *MO*, and since this orbital has a lower energy than the separated $1s$ atomic orbitals, there is stability.

He₂ Of the four electrons, only two can go into a bonding $1s\sigma_g$ orbital; the other two must form an antibonding $1s\sigma_u^*$ orbital. The net energy is greater than that of the separated He atoms, *so that no molecule is formed*. In terms of the valence bond picture, both atoms have paired electrons and the conclusion is the same. In general, electrons in bonding orbitals and in antibonding orbitals tend to cancel each other out. Since there are two electrons involved in a full bond, we can speak of a *bond number*, given by

$$\begin{pmatrix} \text{Bond} \\ \text{number} \end{pmatrix} = \frac{1}{2} \left[\begin{pmatrix} \text{electrons in} \\ \text{bonding orbitals} \end{pmatrix} - \begin{pmatrix} \text{electrons in} \\ \text{antibonding orbitals} \end{pmatrix} \right]$$

This number vanishes for He₂.

Li₂ The atomic structure of Li is $(1s)^2(2s)$. Thus, the $2s$ electrons are unpaired, and they can form $2s\sigma_g$ bonding orbitals. We thus expect the molecule to exist, but because of the $n = 2$ value of the orbital, we would expect the binding to be significantly smaller than for the H₂ molecule.

Be₂ Here the atomic structure is $(1s)^2(2s)^2$; there are no unpaired electrons, and hence we expect no molecule to exist. This is indeed so.

B₂ The atomic structure indicates that there is an unpaired $2p$ electron in each atom. It can be in any one of the states $2p_x$, $2p_y$, and $2p_z$. They may combine either into a $2p\pi_u$ or into a $2p\sigma_g$ *MO*. The former has a lower energy, so that here the ground state is a triplet. This is in agreement with Hund's rule: The state with highest multiplicity has the lowest energy.

The reason why $2p\sigma_g$ has a higher energy is that there exist $2s\sigma_g$ orbitals. Whenever there are states that have the same quantum numbers, "mixing" occurs, and states that are almost degenerate tend to repel each other. The state that is largely $2p\sigma_g$ is pushed up. We begin to see the appearance of complications similar to the ones that appeared in our discussion of atomic structure!

C₂ The atomic structure is $(1s)^2(2s)^2(2p)^2$, that is, each atom has two unpaired electrons. Since each electron can be in any one of three p states, two bonding *MO*'s can be formed. The *MO* description turns out to be $(2p\sigma_g)(2p\pi_u)$.

N$_2$ Here the situation is very similar to that of C$_2$ except that three bonding MO's can be formed. The MO description turns out to be $(2p\sigma_g)(2p\pi_u)^2$.

O$_2$ Here things get a little more interesting, because the atomic structure is $(1s)^2(2s)^2(2p)^4$, that is, there are four valence electrons. In terms of molecular orbitals, three bonds, as in N$_2$, can be formed, but this leaves two electrons that cannot possibly form a bonding orbital. What is the least harmful antibonding orbital? The two electrons should avoid each other as much as possible, and this can be done by means of a triplet state, with the electrons in orthogonal orbitals, for example, one in a p_x, the other in a p_y state, with the two spatially antisymmetrized. In this case the spin of O$_2$ is 1, an exception to the strong tendency toward zero spin that was mentioned earlier.

In the valence bond picture, two of the four valence electrons in oxygen must be paired, so that two bonds will exist orthogonal to each other, as p_x is to p_z, for example. One can see the effect of this directionality in a molecule like H$_2$O. Each H uses up one bond, and we would expect the shape of the molecule to be an **L** with 90° between the equal length arms. Actually, the two hydrogen nuclei repel each other, and one might expect the angle to be a little larger than 90°. Experimentally it is around 105°! It is the directionality of the p-orbitals that explains the shape of simple molecules.

Even more than in the case of atoms, we have barely indicated the range of possibilities in the structure of molecules.

THE ROTATION OF MOLECULES

We still consider molecules in the static approximation. The rigid structure, in which the electron distribution holds the nuclei fixed, can rotate. For example, the mass distribution of the H$_2$ molecule is like a dumbbell, with two point masses separated by the distance R_{AB}. This system has two rotational degrees of freedom: if the internuclear separation defines the z-axis, there are rotations about the x- and the y-axes. There is no angular momentum about the z-axis, that is, $L_z = 0$. Typically

$$E_{rot} = \frac{L_x^2 + L_y^2}{2\mathcal{I}} = \frac{\mathbf{L}^2 - L_z^2}{2\mathcal{I}} = \frac{L(L+1)\hbar^2}{2\mathcal{I}} \tag{20-27}$$

where $\mathcal{I}$ is the moment of inertia of the molecule. For the homopolar molecules, $\mathcal{I} = MR^2/2$. Since $R \approx 2a_0$

$$E_{rot} \approx (m/M)E_{elec} \tag{20-28}$$

that is, the rotational splittings are about three orders of magnitude smaller than the electronic splittings, which are typically a few electronvolts. Thus the wavelengths of the radiation emitted in transitions between rotational levels are of the order of 10^7 Å ≈ 1 mm.

We do not go into further details of the rotational spectra in molecules, except to note that for homopolar molecules the Pauli principle plays a significant role. Consider, for example, the H$_2$ molecule for which the two nuclei are identical and each has spin 1/2. Thus the total wave function must be antisymmetric under the interchange of the two nuclei. The two protons in this example may be in the antisymmetric spin singlet ($S = 0$) state, in which case the rotational state must be described by a symmetric function, so that the angular momentum is even. If the two protons are in the symmetric spin triplet ($S = 1$) state, the angular momentum of rotation must be odd. In a gas, collisions among the H_2 molecules will randomize the distribution of spin states, and assuming that they have equal probability, the number of molecules in a given spin state will be proportional to the degeneracy ($2S + 1$). Thus there will be three times as many odd L molecules as there are even L H_2 molecules in the gas. This will manifest itself in the intensity of the spectral lines associated with the transitions between rotational levels. More generally, if each nucleus has spin I, then the spin states $2I, 2I - 2, 2I - 4, \ldots$, and the spin states $2I - 1, 2I - 3, \ldots$, will have opposite symmetry. If, for example, I is an integer, then the first series of spin states will be associated with even orbital angular momentum, since the nuclei are bosons in this case. Their total number is

$$\sum_{k=0}^{I} [2(2I - 2k) + 1] = (4I + 1)(I + 1) - \frac{4I(I + 1)}{2}$$
$$= (2I + 1)(I + 1) \tag{20-29}$$

whereas the remaining

$$(2I + 1)^2 - (2I + 1)(I + 1) = (2I + 1)I \tag{20-30}$$

states will be associated with odd orbital angular momentum. Thus for integral I, the ratio of even L to odd L intensities for a given L is $(I + 1)/I$. For fermions that ratio is inverted.

The energies of the rotational states are

$$E_L = \frac{\hbar^2 L(L + 1)}{2\mathcal{I}} \tag{20-31}$$

where $\mathcal{I}$ is the moment of inertia of the homonuclear molecule under consideration. Transitions between adjacent L values (to conform with the selection rule $\Delta L = \pm 1$, still to be derived) yield radiation with frequencies

$$\omega(L + 1 \rightarrow L) = \frac{\hbar}{2\mathcal{I}} [(L + 1)(L + 2) - L(L + 1)]$$
$$= \frac{\hbar}{\mathcal{I}} (L + 1) \tag{20-32}$$

VIBRATIONS OF THE NUCLEI IN MOLECULES

Up to now we treated the nuclei in molecules as fixed. This fixed distribution of positive charges gives rise to an electron distribution that depends on the location of the nuclei. The electrons move much more rapidly than the nuclei, $v_e/v_N \approx M/m$, and any motion of the nuclei is one that the electron distribution can adjust to *adiabatically*. This means that the electron energy $E(R)$ is not expected to change, even though the nuclei slowly change their location about the minimum of the $E(R)$ curve. In particular, a slow change in the location of the nuclei will not change the electronic state. The frequency associated with a change ΔR in the nuclear position is of the order of $\nu \approx v_N/\Delta R$, while the frequency associated with a change in the electronic energy is of the order of $\nu_e \approx \alpha c/a_0$. Since ΔR is of the same order of magnitude as a_0 and $v_N/\alpha c \approx 10^{-4}$, the time-dependence of the nuclear motion will not be able to induce transitions in the electronic state. A more detailed understanding of these remarks will emerge from the study of Chapter 21.

In the adiabatic approximation, we may view $E(R)$ as a fixed potential in which the nuclei move. In particular, motion away from the R_0 lying at the minimum of the $E(R)$ curve will be simple harmonic:

$$E(R) \approx E(R_0) + \frac{1}{2} (R - R_0)^2 \left(\frac{\partial^2 E(R)}{\partial R^2}\right)_0 \tag{20-33}$$

since $\partial E(R)/\partial R = 0$ at $R = R_0$. Thus for small displacements, the nuclei move in a harmonic oscillator, with angular frequency

$$\omega = \sqrt{\frac{1}{M}\left(\frac{\partial^2 E}{\partial R^2}\right)_0} \tag{20-34}$$

For homopolar molecules this motion is one-dimensional, along the line joining the two nuclei, and the energy eigenvalues for this motion are

$$E_{\text{vib}} = \left(n_v + \frac{1}{2}\right)\hbar\omega \tag{20-35}$$

with $n_v = 0, 1, 2, 3, \ldots$. We can estimate this energy by looking at Fig. 20-1. In the vicinity of the minimum, $E(R)$ is the sum of e^2/R and the slowly changing electronic energy, which in the vicinity of the minimum can be taken to be of the form $\kappa(R = R_0)$. Thus the second derivative comes primarily from the e^2/R term, and thus

$$\left(\frac{\partial^2 E(R)}{\partial R^2}\right)_0 \approx \frac{2e^2}{R_0^3} \tag{20-36}$$

Since $R_0 \approx 2a_0$, we get

$$\hbar\omega \approx \sqrt{\frac{\hbar^2 e^2}{4Ma_0^3}} \approx \sqrt{\frac{m}{M}} \tag{20-37}$$

This means that the wavelength of the radiation emitted in transitions between vibrational levels are of the order of 10^5 Å.

Molecular spectra exhibit a hierarchy of levels. There are the electronic levels, separated by energies in the electronvolt range. Upon each electronic energy level there is a series of vibrational levels, separated by energies of the order of tens of millielectronvolts, and associated with each electronic level there is a set of rotational energy levels. Figure 20-7 shows the complex structure of the molecular energy spectrum.

In our discussion so far, we have treated these levels as quite separate, but in

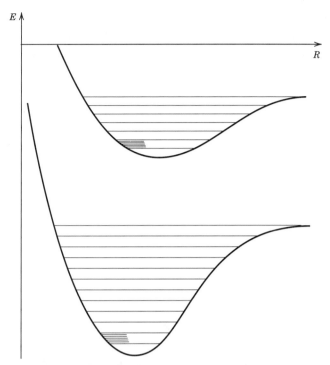

Figure 20-7. Vibrational levels superimposed on two
electronic levels of a diatomic molecule. The rotational
levels associated with the lowest vibrational level are
also sketched in. None of this is to scale.

reality there is a weak coupling between all of them. For example, the rotational
motion gives rise to a centrifugal distortion of the molecule (analogous to the
equatorial bulge in the earth's rotation) which actually has an effect on the rotational energy levels, since R_0 changes (see Problem 4). The form of $E(R)$ is different
for each electronic level, and thus the pattern of spectral lines (simplified only by
the selection rules $\Delta L = 1$ and $\Delta n_v = 1$) is complex indeed. We shall not get further
into the improvements beyond the adiabatic approximation used here. The *Born-
Oppenheimer* procedure gives a systematic way of improving the approximations,
but this is beyond the scope of this book.

Problems

1. In HCl a number of absorption lines with wave numbers (in cm^{-1}) 83.03, 103.73,
   124.30, 145.03, 165.51, 185.86 have been observed. Are these vibrational or rotational transitions? If the former, what is the characteristic frequency? If the
   latter, what J values do they correspond to, and what is the moment of inertia
   of HCl? In that case, estimate the separation between the nuclei. (In radiation
   the quantum numbers change by one unit.)

2. What is the ratio of the number of HCl molecules in a state with $J = 10$ to the
   number in a state with $J = 0$, if the gas of molecules is at a temperature of
   300 K?

3. The frequency of vibration of the CO molecule in its lowest state is $\nu_0 = 2 \times 10^{13}$ Hz. What is the wavelength of the radiation emitted in the lowest vibrational excitation? What is the probability that the first vibrational state is excited, relative to the probability that CO is in its vibrational ground state, when the temperature is 300 K?

4. Consider the vibrational and rotational energy of a molecule in the approximation

$$E_J(R) = \frac{1}{2} m\omega^2 (R - R_0)^2 + \frac{J(J + 1)\, \hbar^2}{2mR^2}$$

Find the position where the energy is a minimum. If the moment of inertia of the molecule is calculated using the new internuclear separation, show that the rotational energy can be written in the form

$$E_J = AJ(J + 1) + B[J(J + 1)]^2 + \cdots$$

Determine the coefficients A and B (the latter is the effect of centrifugal distortion).

References

This chapter owes much to the brief discussion of molecules in G. Baym, *Lectures on Quantum Mechanics*, W. A. Benjamin, New York, 1969. The interested reader should consult the following books for more information:

M. Karplus and R. N. Porter, *Atoms and Molecules*, W. A. Benjamin, New York, 1970.
M. W. Hanna, *Quantum Mechanics in Chemistry*, W. A. Benjamin, New York, 1969.
U. Fano and L. Fano, *Physics of Atoms and Molecules*, Chicago University Press, Chicago, 1972.
G. W. King, *Spectroscopy and Molecular Structure*, Holt, Rinehart & Winston, New York, 1964.

There are, of course, hundreds of books on quantum chemistry, molecular structure, and molecular spectroscopy, and the sampling listed here is just the one familiar to the author. For a better reference list, the reader should consult any physical chemist.

Chapter 21

THE RADIATION
OF ATOMS

In the study of spectra, that is, the study of transitions between atomic levels accompanied by the emission or absorption of radiation, one is interested in the interaction between atoms and the electromagnetic field. Since the radiation field oscillates, it is time dependent. It is therefore necessary to study the effect of time-dependent perturbations.

TIME-DEPENDENT PERTURBATION THEORY

The problem is, given the complete set of solutions to

$$H_0 \phi_n = E_n^0 \phi_n \tag{21-1}$$

to solve for $\psi(t)$, which obeys the equation

$$i\hbar \frac{\partial \psi(t)}{\partial t} = [H_0 + \lambda V(t)]\, \psi(t) \tag{21-2}$$

The standard procedure is to expand $\psi(t)$ in a complete set of states:

$$\psi(t) = \sum_n c_n(t)\, e^{-iE_n^0 t/\hbar}\, \phi_n \tag{21-3}$$

The time-dependence associated with the ϕ_n is explicitly inserted in the expansion, so that if $V(t) = 0$, the $c_n(t)$ would be constants. The expansion coefficients $c_n(t)$

satisfy a set of equations that may be obtained by substituting (21-3) into the time-dependent Schrödinger equation (21-2). We get

$$\sum_n \left[i\hbar \frac{dc_n(t)}{dt} + E_n^0 c_n(t) \right] e^{-iE_n^0 t/\hbar} \; \phi_n = H\psi(t) = \sum_n [E_n^0 + \lambda V(t)] \; c_n(t) \; e^{-iE_n^0 t/\hbar} \; \phi_n$$

that is,

$$i\hbar \sum_n \frac{dc_n(t)}{dt} e^{-iE_n^0 t/\hbar} \; \phi_n = \lambda \sum_n V(t) c_n(t) \; e^{-iE_n^0 t/\hbar} \; \phi_n \qquad (21\text{-}4)$$

Taking the scalar product with ϕ_m and using the orthonormality of the ϕ_m,

$$\langle \phi_m | \phi_n \rangle = \delta_{mn} \qquad (21\text{-}5)$$

yields, after the factor $e^{-iE_m^0 t/\hbar}$ is divided out, the set of equations

$$i\hbar \frac{dc_m(t)}{dt} = \lambda \sum_n c_n(t) \; e^{i(E_m^0 - E_n^0)t/\hbar} \; \langle \phi_m | V(t) | \phi_n \rangle \qquad (21\text{-}6)$$

We shall solve these to first order in the parameter λ. As an initial condition at $t = 0$ we take the system to be in a particular state ϕ_k, so that $\psi(0) = \phi_k$, that is,

$$c_n(0) = \delta_{nk} \qquad (21\text{-}7)$$

Sometimes the initial state is specified in the distant past. In that case (21-7) reads

$$\operatorname*{Lim}_{t_0 \to -\infty} c_n(t_0) = \delta_{nk}$$

Since departures from these values at later times will depend on λ, we may, for a first-order calculation, substitute (21-7) into the right side of (21-6). This yields the differential equation (for $m \neq k$)

$$i\hbar \frac{dc_m(t)}{dt} = \lambda \; e^{i(E_m^0 - E_k^0)t/\hbar} \; \langle \phi_m | V(t) | \phi_k \rangle \qquad (21\text{-}8)$$

which is easily solved

$$c_m(t) = \frac{\lambda}{i\hbar} \int_0^t dt' \; e^{i(E_m^0 - E_k^0)t'/\hbar} \; \langle \phi_m | V(t') | \phi_k \rangle \qquad (21\text{-}9)$$

The probability that at a later time t, the state $\psi(t)$ is an eigenstate of H_0 with energy E_n^0, that is, that it is ϕ_n, is, according to the expansion postulate

$$P_n(t) = |\langle \phi_n | \psi(t) \rangle|^2 = |c_n(t)|^2 \qquad (21\text{-}10)$$

This general result can only be made more specific if $V(t)$ is known.

Example A hydrogen atom in the ground state is subject to an electric field that is turned on and off so that

$$\mathbf{E}(t) = \mathbf{E}_0 \, e^{-t^2/\tau^2}$$

What is the probability that the hydrogen atom ends up in the $n = 2$, $l = 1$, $m = 0$ state after a long time ($t \gg \tau$)?

We see from (16-32) that the perturbation is

$$\lambda V(t) = eE_0 z \, e^{-t^2/\tau^2}$$

Also, since the time $t \gg \tau$, we can take the upper and lower limits on the time-integral in (21-9) to be ∞ and $-\infty$, respectively. Thus

$$c_{210}(\infty) = \frac{eE_0}{i\hbar} \langle \phi_{210}|z|\phi_{100}\rangle \int_{-\infty}^{\infty} dt' \, e^{i(E_{210}-E_{100})t'/\hbar} \, e^{-t'^2/\tau^2}$$

$$= \frac{eE_0}{i\hbar} \langle \phi_{210}|z|\phi_{100}\rangle \, \tau\sqrt{\pi} \, e^{-\omega^2\tau^2/4}$$

where $\omega = (E_{210} - E_{100})/\hbar$ is equal to the angular frequency of a photon emitted in the transition $(2, 1, 0) \to (1, 0, 0)$. The probability is given by the absolute square of this

$$P = \frac{e^2 E_0^2 \tau^2 \pi}{\hbar^2} |\langle \phi_{210}|z|\phi_{100}\rangle|^2 \, e^{-\omega^2\tau^2/2}$$

Note that for $\tau \to \infty$, $P \to 0$. When the electric field turns on very slowly, then the transition probability goes to zero, that is, the atom adjusts *adiabatically* to the presence of the electric field, without being "jolted" into making a transition.

HARMONIC TIME-VARIATION OF THE POTENTIAL

In a large number of examples, the potential has the time dependence given by

$$V(t) = Ve^{-i\omega t} + V^\dagger e^{i\omega t} \tag{21-11}$$

where V and V^+ are operators that do not have an explicit time dependence. In this case,

$$c_m(t) = \frac{\lambda}{i\hbar} \int_0^t dt' \, e^{i\omega_{mk}t'} \, [e^{-i\omega t'} \langle \phi_m|V|\phi_k\rangle + e^{i\omega t'} \langle \phi_m|V^\dagger|\phi_k\rangle] \tag{21-12}$$

where $\omega_{mk} = (E_m^0 - E_k^0)/\hbar$. We will in most cases be interested in $t \to \infty$. We work out the time integrals

$$\int_0^t dt' \, e^{i(\omega_{mk}-\omega)t'} = \frac{e^{i(\omega_{mk}-\omega)t} - 1}{i(\omega_{mk} - \omega)} = e^{i(\omega_{mk}-\omega)t/2} \frac{\sin(\omega_{mk} - \omega)t/2}{(\omega_{mk} - \omega)/2} \tag{21-13}$$

The second integral can be obtained by changing $\omega \to -\omega$. We next need

$$
|e^{i(\omega_{mk} - \omega)t/2} \frac{\sin(\omega_{mk} - \omega)t/2}{(\omega_{mk} - \omega)/2} \langle \phi_m | V | \phi_k \rangle
$$
$$
+ e^{i(\omega_{mk} + \omega)t/2} \frac{\sin(\omega_{mk} + \omega)t/2}{(\omega_{mk} + \omega)/2} \langle \phi_m | V^+ | \phi_k \rangle |^2 \tag{21-14}
$$

The form of this expression suggests that only when $\omega = \pm \omega_{mk}$ will the terms be important. We have three terms: one of them is

$$
\left(\frac{\sin(\omega_{mk} - \omega)t/2}{(\omega_{mk} - \omega)/2} \right)^2 |\langle \phi_m | V | \phi_k \rangle|^2 \tag{21-15}
$$

the second one is indentical to this, with the change $\omega \to -\omega$; and the third term has a time dependence of the form

$$
e^{i\omega t} \frac{\sin(\omega_{mk} - \omega)t/2}{(\omega_{mk} - \omega)/2} \frac{\sin(\omega_{mk} + \omega)t/2}{(\omega_{mk} + \omega)/2} + \text{complex conjugate} \tag{21-16}
$$

It can be shown, using a mathematical theorem called the Riemann-Lesbegue lemma, that as $t \to \infty$, $t^n \sin \alpha t \to 0$ faster than any power of t, and similarly for $t^n \cos \alpha t$. When this is applied to the cross terms, one finds that there is one term left, independent of t (see Problem 9), but this is negligible compared with the terms such as appear in (21-15), which grow linearly with t, as we shall demonstrate.

These terms have the form

$$
F(t) = \frac{4}{\Delta^2} \sin^2 \frac{t\Delta}{2} \tag{21-17}
$$

with

$$
\Delta = \frac{E_m^0 - E_k^0 \mp \hbar\omega}{\hbar} \tag{21-18}
$$

Figure 21-1 shows the behavior of this function. For large t it becomes strongly peaked at $\Delta = 0$, and away from $\Delta = 0$ it oscillates very rapidly. This is the kind of behavior that we associate with a delta function. In fact, if $f(\Delta)$ is a smooth function of Δ, then, for t large

$$
\int_{-\infty}^{\infty} f(\Delta) \frac{4}{\Delta^2} \sin^2 \frac{t\Delta}{2} d\Delta \approx f(0) \int_{-\infty}^{\infty} d\Delta \frac{4}{\Delta^2} \sin^2 \frac{t\Delta}{2}
$$
$$
= 2tf(0) \int_{-\infty}^{\infty} dy \frac{1}{y^2} \sin^2 y = 2\pi t f(0) \tag{21-19}
$$

that is, for t large

$$
\frac{4}{\Delta^2} \sin^2 \frac{t\Delta}{2} \to 2\pi t \, \delta(\Delta) = 2\pi \hbar t \delta(E_m^0 - E_k^0 \mp \hbar\omega) \tag{21-20}
$$

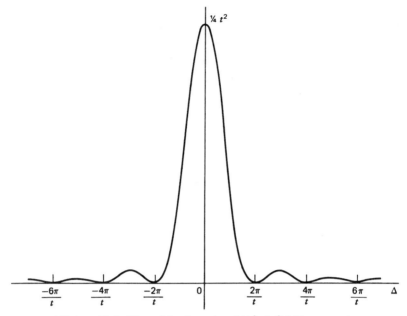

Figure 21-1. Plot of the function $1/\Delta^2 \sin^2 t\Delta/2$ versus Δ.

Thus the transition probability in (21-14) grows linearly with time, and for long times, the interference term, which is independent of t, becomes increasingly irrelevant. We thus get for the *transition probability per unit time*

$$\Gamma_{k \to m} = \frac{2\pi}{\hbar} |\langle\phi_m|V|\phi_k\rangle|^2 \; \delta(E_m^0 - E_k^0 - \hbar\omega)$$
$$+ \frac{2\pi}{\hbar} |\langle\phi_m|V^\dagger|\phi_k\rangle|^2 \; \delta(E_m^0 - E_k^0 + \hbar\omega)$$

(21-21)

Since $E_m^0 - E_k^0$ is fixed, only one of the two terms can contribute for a fixed ω. The delta functions ensure that if $E_k^0 > E_m^0$, that is, if the atom makes a transition from a higher energy state to a lower energy state, then only the second term contributes, and it does so only if $\hbar\omega = E_k^0 - E_m^0$. If the relevant potential term in (21-11) were of the form

$$\lambda V(t) = \int_0^\infty d\omega' \; V(\omega')e^{-i\omega't} + \int_0^\infty d\omega' \; V^\dagger(\omega')e^{i\omega't}$$

then the delta function would pick out just the contribution in the second term for which $\hbar\omega'$ is equal to $\Delta E = E_k^0 - E_m^0$. Equation (6-67) with $A = H$, shows that

$$\left\langle\frac{dH}{dt}\right\rangle = \left\langle\frac{\partial H}{\partial t}\right\rangle$$

(21-22)

so that with a time-dependent potential, energy is not a constant of the motion. Here we see more explicitly how much energy a time-dependent potential can absorb or emit.

THE COUPLING OF ATOMS
TO THE ELECTROMAGNETIC FIELD

The Hamiltonian describing the interaction of an electron in a static potential ($V(r)$) with an electromagnetic field described by the vector potential $\mathbf{A}(\mathbf{r}, t)$ is given by

$$H = \frac{[\mathbf{p} + (e/c)\,\mathbf{A}(\mathbf{r}, t)]^2}{2m} + V(r) \tag{21-23}$$

as we saw in Chapter 13. Thus, if we write

$$H_0 = \frac{\mathbf{p}^2}{2m} + V(r) \tag{21-24}$$

we find that

$$\lambda V(t) = \frac{e}{mc}\,\mathbf{A}(\mathbf{r}, t) \cdot \mathbf{p} \tag{21-25}$$

In obtaining the last expression we have specified the gauge so that

$$\boldsymbol{\nabla} \cdot \mathbf{A}(\mathbf{r}, t) = 0 \tag{21-26}$$

Under these circumstances, $\mathbf{p} \cdot \mathbf{A} = \mathbf{A} \cdot \mathbf{p}$, and we have dropped the term quadratic in $\mathbf{A}(\mathbf{r}, t)$. If we treat e, the electron charge as the parameter of smallness λ, then the $\mathbf{A}^2$ term is a second-order term. We will see that the $\mathbf{A}^2$ term will contribute to the scattering of light by an atom and to the transition with the emission of two photons, but it will not contribute to the transition accompanied by the emission (or absorption) of a single photon. The probability for a transition involving two photons involves a factor $(e^2)^2$, whereas the one-photon transition probability is proportional to e^2. Recalling that the appropriate dimensionless number involving e^2 is $\alpha = e^2/\hbar c \cong 1/137$, we are justified in concentrating on transitions that are accompanied by the emission of a single photon.

To give a real justification of the association of each $\mathbf{A}(\mathbf{r}, t)$ with the emission or absorption of a single photon—so that higher powers of $\mathbf{A}(\mathbf{r}, t)$ imply the presence of more photons—one must treat the electromagnetic field quantum mechanically, that is, treat the fields at each point $\mathbf{r}$ as operators. This is fundamentally not terribly complicated, but it is outside the scope of this book. The reader will have to take the following assertions on faith.

If we write

$$\mathbf{A}(\mathbf{r}, t) = \mathbf{A}_0^*(\mathbf{r})\,e^{i\omega t} + \mathbf{A}_0(\mathbf{r})\,e^{-i\omega t} \tag{21-27}$$

then in the *emission of a photon*, only the first term, with the time dependence $e^{i\omega t}$, is to be included in $\lambda V(t)$, whereas in the *absorption of a photon*, only the second term, with the time dependence $e^{-i\omega t}$, appears. This is a consequence of the general association of $\mathbf{A}_0^*(\mathbf{r})$ with the creation of a photon and $\mathbf{A}_0(\mathbf{r})$ with the annihilation of a photon, and the time dependence is just what one would expect from the harmonic oscillator (7-67). The resemblance to the harmonic oscillator problem is not accidental, since in the quantization of the electromagnetic field, what is done is a normal mode decomposition, according to which one finds that the field is really a collection of simple harmonic oscillators; these are then quantized. The "occupation number" n that labels the harmonic oscillator state vector may be associated with the number of photons, hence $\mathbf{A}_0^*$ raises the photon number by unity and $\mathbf{A}_0$ lowers the photon number by unity.

The more quantitative description of $\mathbf{A}_0^*(\mathbf{r})$ and $\mathbf{A}_0(\mathbf{r})$ need not, fortunately, involve the full machinery of quantum electrodynamics. We may use correspondence principle arguments to find these quantities, and then just state the quantum mechanical modifications. Away from the sources, the electromagnetic field has a very simple spatial behavior. If we look back at (13-14) and substitute (21-27), we find that

$$-\nabla^2 \mathbf{A}_0(\mathbf{r}) - \frac{\omega^2}{c^2} \mathbf{A}_0(\mathbf{r}) = 0 \tag{21-28}$$

whose solution is

$$\mathbf{A}_0(\mathbf{r}) = \mathbf{A}_0\, e^{i\mathbf{k}\cdot\mathbf{r}} \tag{21-29}$$

with

$$\mathbf{k}^2 = \frac{\omega^2}{c^2} \tag{21-30}$$

The choice of gauge (21-26) implies that

$$\mathbf{k} \cdot \mathbf{A}_0 = 0 \tag{21-31}$$

The electric and magnetic fields corresponding to this vector potential are

$$\mathbf{E} = -\frac{1}{c}\frac{\partial \mathbf{A}}{\partial t} = \frac{i\omega}{c}\, \mathbf{A}_0\, e^{i(\mathbf{k}\cdot\mathbf{r}-\omega t)} + \text{complex conjugate}$$

$$\mathbf{B} = \nabla \times \mathbf{A} = i\mathbf{k} \times \mathbf{A}_0\, e^{i(\mathbf{k}\cdot\mathbf{r}-\omega t)} + \text{complex conjugate} \tag{21-32}$$

Now the energy density of the electromagnetic field is given by

$$\frac{1}{8\pi}(\mathbf{E}^2 + \mathbf{B}^2) = \frac{1}{8\pi}\left[2\frac{\omega^2}{c^2}\, \mathbf{A}_0 \cdot \mathbf{A}_0^* + 2(\mathbf{k} \times \mathbf{A}_0) \right. $$
$$\left. \cdot\, (\mathbf{k} \times \mathbf{A}_0^*) + \text{oscillating terms} \right] \tag{21-33}$$

If we average over time, so that the oscillating terms drop out, and make use of the fact that with (21-31)

$$(\mathbf{k} \times \mathbf{A}_0) \cdot (\mathbf{k} \times \mathbf{A}_0^*) = k^2 \mathbf{A}_0 \cdot \mathbf{A}_0^* \tag{21-34}$$

and that $k^2 = \omega^2/c^2$, we get

$$\frac{1}{8\pi}(\mathbf{E}^2 + \mathbf{B}^2) = \frac{\omega^2}{2\pi c^2}\, \mathbf{A}_0 \cdot \mathbf{A}_0^* \tag{21-35}$$

If the system is enclosed in a box of volume, V, then the total energy in the electromagnetic field is

$$\int d^3r \, \frac{1}{8\pi} \, (E^2 + B^2) = \frac{\omega^2 V}{2\pi c^2} \, |A_0|^2 \tag{21-36}$$

If this is to be carried by N photons, each with energy $\hbar\omega$, we have

$$\frac{\omega^2 V}{2\pi c^2} \, |A_0|^2 = N\hbar\omega \tag{21-37}$$

The direction of A_0 is determined by the polarization of the electric field, and will be denoted by the unit vector ε. It must satisfy

$$\begin{aligned}\varepsilon \cdot \varepsilon &= 1 \\ \varepsilon \cdot k &= 0\end{aligned} \tag{21-38}$$

We therefore obtain

$$A(r, t) = \left(\frac{2\pi c^2 N\hbar}{\omega V}\right)^{1/2} \varepsilon \, e^{i(k \cdot r - \omega t)} \tag{21-39}$$

The quantum electrodynamic modification is the following: For the absorption of a light quantum by a charged particle from an initial state that already has N photons of angular frequency ω,

$$A(r, t) = \left(\frac{2\pi c^2 N\hbar}{\omega V}\right)^{1/2} \varepsilon \, e^{i(k \cdot r - \omega t)} \tag{21-40}$$

For the emission of a light quantum by a charged particle into a final state that has $N + 1$ quanta, that is, from an initial state with N quanta of frequency ω,

$$A(r, t) = \left[\frac{2\pi c^2 (N + 1) \, \hbar}{\omega V}\right]^{1/2} \varepsilon \, e^{-i(k \cdot r - \omega t)} \tag{21-41}$$

Hence for the emission of a single photon of frequency ω from a state that has no photons, we have, according to (21-25),

$$\lambda V(t) = \frac{e}{mc} \left(\frac{2\pi c^2 \hbar}{\omega V}\right)^{1/2} \varepsilon \cdot p \, e^{-i(k \cdot r - \omega t)} \tag{21-42}$$

Let us now consider the transition probability per unit time in (21-21) and assume that $E_k^0 > E_m^0$, so that the transition corresponds to the emission of a photon of energy $\hbar\omega$. The transition rate is

$$\Gamma_{k \to m} = 2\pi\hbar \, \frac{2\pi e^2}{m^2 \hbar \omega V} \, |\langle \phi_m | e^{-ik \cdot r} \, \varepsilon \cdot p | \phi_k \rangle|^2 \, \delta(E_k^0 - E_m^0 - \hbar\omega) \tag{21-43}$$

As things stand, the reader undoubtedly feels swindled. First of all the manipulations leading to (21-21) certainly are not straightforward. They involve vague notions such as "*t* large," which cannot be taken too seriously, since a transition probability that grows linearly with time must sooner or later exceed unity. Second, they lead to a nonsensical formula, according to which a perfectly reasonable quantity, like the transition rate, is proportional to a delta function. A more satisfactory discussion can be found in the Special Topics section on lifetimes, linewidths, and resonances. At this point we merely note that (21-43) is correct, if properly used.

For this, we note that $\Gamma_{k \to m}$ is really the *transition probability per unit time for the atom making a transition from the state ϕ_k to the state ϕ_m, accompanied by the emission of a photon of energy $\hbar\omega$.* The delta function, unappealing as it is, does tell us that energy must be conserved, that is,

$$\hbar\omega = E_k^0 - E_m^0 \tag{21-44}$$

The delta function is actually integrated over, if we take into account that the photon energy $\hbar\omega$ does not uniquely specify the photon state. The photon will in general be detected in some momentum interval $(\mathbf{k}, \mathbf{k} + \Delta\mathbf{k})$ in the vicinity of $|\mathbf{k}| = \omega/c$, and the transition rate that is measured is really

$$R_{k \to m} = \sum_{\Delta\mathbf{k}} \Gamma_{k \to m} \tag{21-45}$$

summed over all the possible photon states in that interval. Note that the various final states in the interval $\Delta\mathbf{k}$ are in principle distinguishable, so that it is the probabilities that are summed. We will see that the sum (21-45) is well defined, since, in effect, it involves the integral of a delta function and a smooth function. The summation is in the next section.

PHASE SPACE

We will now calculate the number of photon states in the momentum interval $(\mathbf{k}, \mathbf{k} + \Delta\mathbf{k})$, that is, the density of photon states. For the purpose at hand, we write the vector potential $\mathbf{A}(\mathbf{r}, t)$ in the form

$$\mathbf{A}(\mathbf{r}, t) = \frac{1}{\sqrt{V}} \mathbf{a}\, e^{i(\mathbf{k}\cdot\mathbf{r} - \omega t)} + \text{complex conjugate} \tag{21-46}$$

where V is the volume of the enclosure in which the calculation is done. This "box" is just a convenience to save us the trouble of working with wave packets for the free particles (the photons, here—cf. Chapter 4). Its shape and the conditions at the boundary may be chosen at will, but it must be large. At the end, we will take $V \to \infty$. We will find it convenient to take the box to be a cube of side L, and to impose periodic boundary conditions, that is,

$$\mathbf{A}(x + L, y, z, t) = \mathbf{A}(x, y, z, t) \tag{21-47}$$

and so on. This implies, just as in the solution of a particle in a one-dimensional box, that the wave numbers, that is, the momenta, are quantized. The form (21-46) requires that

$$e^{ik_x L} = e^{ik_y L} = e^{ik_z L} = 1 \tag{21-48}$$

that is, that the wave numbers be of the form

$$k_x = \frac{2\pi}{L} n_x \qquad k_y = \frac{2\pi}{L} n_y \qquad k_z = \frac{2\pi}{L} n_z \tag{21-49}$$

where n_x, n_y, and n_z are integers. We also have

$$\Delta \mathbf{k} = \Delta k_x \, \Delta k_y \, \Delta k_z = \left(\frac{2\pi}{L}\right)^3 \Delta n_x \, \Delta n_y \, \Delta n_z \tag{21-50}$$

and

$$\omega = |\mathbf{k}|c = \frac{2\pi c}{L} (n_x^2 + n_y^2 + n_z^2)^{1/2} \tag{21-51}$$

When we carry out a sum like that in (21-45), we sum over all values of (n_x, n_y, n_z) in the range specified by (21-50) consistent with the constraint of the delta function. Thus

$$R_{k\to m} = \sum_{\Delta \mathbf{k}} \Gamma_{k\to m} \tag{21-52}$$

In this chapter we shall consider a very large volume V, so that the states become very dense, and the sum in (21-52) can be converted into an integral. In that case

$$R_{k\to m} = \int d^3\mathbf{n} \, \Gamma_{k\to m} = \int \frac{L^3 d^3\mathbf{k}}{(2\pi)^3} \Gamma_{k\to m}$$
$$= \int \frac{V d^2\mathbf{p}}{(2\pi\hbar^3)} \Gamma_{k\to m} \tag{21-53}$$

In the last line we make use of the relation $\mathbf{p} = \hbar\mathbf{k}$.

The integration is over the volume in momentum space defined by the experimental arrangement. If we write

$$d^3\mathbf{p} = d\Omega_\mathbf{p} \, p^2 dp = d\Omega_\mathbf{p} \left(\frac{\omega}{c}\right)^2 d\left(\frac{\omega}{c}\right) \hbar^3 \tag{21-54}$$

where $d\Omega_\mathbf{p}$ is the solid angle differential, we find that the energy conserving delta function is integrated over and the result is

$$R_{k\to m} = \int \frac{4\pi^2 e^2}{m^2 \omega V} |\langle \phi_m | e^{-i\mathbf{k}\cdot\mathbf{r}} \, \boldsymbol{\varepsilon} \cdot \mathbf{p} | \phi_k \rangle|^2 \, d\Omega_\mathbf{p} \frac{V}{(2\pi\hbar)^3}$$
$$\times \hbar^3 \frac{\omega^3}{c^3} \frac{d(\hbar\omega)}{\hbar} \delta(E_k^0 - E_m^0 - \hbar\omega) \tag{21-55}$$
$$= \int d\Omega_\mathbf{p} \frac{\alpha}{2\pi} \omega_{km} \left| \frac{1}{mc} \langle \phi_m | e^{-i\mathbf{k}\cdot\mathbf{r}} \, \boldsymbol{\varepsilon} \cdot \mathbf{p} | \omega_k \rangle \right|^2$$

where

$$\omega_{km} = \frac{E_k^0 - E_m^0}{\hbar} \tag{21-56}$$

If the experimental apparatus does not discriminate between the polarization states of the photon, the rate calculation must include a sum over those two independent final states. Furthermore, the sum should also include all the final states of the atom if these are degenerate. This will be discussed in a later section.

The phase space

$$d^3\mathbf{n} = \frac{V\,d^3\mathbf{p}}{(2\pi\hbar)^3} \tag{21-57}$$

is not restricted to photons. An electron that is free is described by the plane wave function $1/\sqrt{V}\,e^{i\mathbf{p}\cdot\mathbf{r}/\hbar}$ and it will have the same density of states. The only difference is that the relation between energy (which appears in the delta function) and momentum is $E = \mathbf{p}^2/2m$ [or, relativistically, $E = (\mathbf{p}^2 c^2 + m^2 c^4)^{1/2}$] instead of $E = pc$.

If we have several free particles in the final state, the density of states is the product

$$\prod_k \frac{V\,d^3\mathbf{p}_k}{(2\pi\hbar)^3} \tag{21-58}$$

The expression (21-53) combined with (21-21) then generalizes to

$$R_{i\to j} = \frac{2\pi}{\hbar} \int_{\substack{\text{independent} \\ \text{momenta}}} \prod_k \frac{V\,d^3\mathbf{p}_k}{(2\pi\hbar)^3} |M_{fi}|^2\,\delta\left(E_f^0 + \sum_k E_k - E_i^0\right) \tag{21-59}$$

where M_{fi} is the matrix element of the perturbation between the initial and final states of the unperturbed system. The delta function again expresses energy conservation, that is, the energy carried off by the free particles is equal to the energy change in the system, and *the integration is over independent momenta*. Thus, if a system decays into three particles, there are only two independent momenta, since the third one is determined by momentum conservation. Note, however, that the product of factors in (21-58) is over *all* the particles in the final state, that is, it involves V^n if there are n particles in the final state. Equivalently we could write (21-59) as an integral over *all* momenta, with a delta function that includes a statement of momentum conservation. The reason that such a delta function did not appear in our derivation is that we treated the nucleus of the atom as fixed in space, which is justified since the atom is so much more massive than the electrons. Under these circumstances, the atom momentum is not a dynamical variable. At any rate, the general result

$$R_{i\to f} = \frac{2\pi}{\hbar} \int \prod_k \frac{V\,d^3\mathbf{p}_k}{(2\pi\hbar)^3}$$
$$\times |M_{fi}|^2\,\delta\left(E_i^0 - E_f^0 - \sum E_k\right)\delta\left(\mathbf{p}_i - \mathbf{p}_f - \sum \mathbf{p}_k\right) \tag{21-60}$$

which could also be abbreviated by

$$R_{i\to f} = \frac{2\pi}{\hbar}|M_{fi}|^2\,\rho(E) \tag{21-61}$$

with $\rho(E)$ called the density of states, is a fundamental result, and has been named the *Golden Rule* by Fermi.

Note that the volume of the box always drops out. For n free particles in the final state, there is a V^n from the density of states (phase space) and a $1/\sqrt{V}$ for each free particle in the matrix element, coming from their wave function

$$\prod_k \frac{e^{i\mathbf{p}_k \cdot \mathbf{r}/\hbar}}{\sqrt{V}} \tag{21-62}$$

There are n of these factors, and thus the V dependence of the square of the matrix element just cancels the V^n from the phase space. We will have further occasion to use the Golden Rule, but at this point we turn to the evaluation of the matrix element for the radiative transition.

THE MATRIX ELEMENT AND SELECTION RULES

Our next task is to calculate

$$\langle \phi_m | e^{-i\mathbf{k}\cdot\mathbf{r}} \, \varepsilon \cdot \mathbf{p} | \phi_k \rangle \tag{21-63}$$

We begin by estimating its magnitude. For a typical atomic transition

$$\varepsilon \cdot \mathbf{p} \sim |\mathbf{p}| \sim Zmc\alpha \tag{21-64}$$

We also need to estimate the exponent, since it is an oscillating factor and could change the result significantly. With

$$r \sim \frac{\hbar}{mcZ\alpha} \tag{21-65}$$

and

$$|k| \sim \frac{\hbar\omega}{\hbar c} \sim \frac{\frac{1}{2}mc^2(Z\alpha)^2}{\hbar c} \sim \frac{mc}{2\hbar}(Z\alpha)^2 \tag{21-66}$$

we have

$$kr \sim \tfrac{1}{2}Z\alpha \tag{21-67}$$

Hence, for $Z\alpha \ll 1$, the order of magnitude of the matrix element is indeed $Zmc\alpha$, thus

$$R_{k\to m} \sim 2\alpha\omega(Z\alpha)^2 \sim \alpha(Z\alpha)^2 \frac{mc^2(Z\alpha)^2}{\hbar}$$

$$\sim \alpha(Z\alpha)^4 \frac{mc^2}{\hbar} \sim 2 \times 10^{10} \, Z^4 \, \sec^{-1} \tag{21-68}$$

It simplifies matters that in the expansion

$$e^{-i\mathbf{k}\cdot\mathbf{r}} = \sum_{n=0}^{\infty} \frac{(-i)^n}{n!} (\mathbf{k}\cdot\mathbf{r})^n \tag{21-69}$$

the successive terms are estimated to decrease as $Z\alpha$. Thus, to order $Z\alpha$,

$$\langle\phi_m|e^{-i\mathbf{k}\cdot\mathbf{r}} \; \varepsilon\cdot\mathbf{p}|\phi_k\rangle \simeq \langle\phi_m|\varepsilon\cdot\mathbf{p}|\phi_k\rangle \tag{21-70}$$

We can write this as

$$\begin{aligned}
\varepsilon\cdot\langle\phi_m|\mathbf{p}|\phi_k\rangle &= m\varepsilon\cdot\langle\phi_m|d\mathbf{r}/dt|\phi_k\rangle \\
&= \frac{im}{\hbar}\varepsilon\cdot\langle\phi_m|[H,\mathbf{r}]|\phi_k\rangle \\
&= im\frac{(E_m^0 - E_k^0)}{\hbar}\varepsilon\cdot\langle\phi_m|\mathbf{r}|\phi_k\rangle \\
&= im\omega\;\varepsilon\cdot\langle\phi_m|\mathbf{r}|\phi_k\rangle
\end{aligned} \tag{21-71}$$

Thus we are interested in calculating the matrix element of the operator $\mathbf{r}$, and that is one reason for calling the approximation (21-70) the electric *dipole approximation*.

It is worth pointing out that in the dipole approximation, the perturbation in (21-25),

$$\lambda(V(t) = \frac{e}{mc}\,\mathbf{A}(\mathbf{r},\,t)\cdot\mathbf{p}$$

can, with the help of (21-32) and (21-71) be rewritten in the form

$$\lambda V(t) = e\mathbf{E}(\mathbf{r},\,t)\cdot\mathbf{r} \tag{21-72}$$

This is, in fact, the potential energy of the interaction of a dipole of dipole moment $\mathbf{d} = -e\mathbf{r}$ in an electric field $\mathbf{E}$.

If the initial state ϕ_k is a hydrogenlike state characterized by the "initial" quantum numbers n_i, l_i, and m_i, and the state ϕ_m, the final state, by the quantum numbers n_f, l_f, and m_f, then what needs to be evaluated is

$$\begin{aligned}
\langle\phi_m|\varepsilon\cdot\mathbf{r}|\phi_k\rangle &= \int_0^{\infty} r^2\,dr\int d\Omega R_{n_f l_f}^*(r)\;Y_{l_f m_f}^*(\theta,\,\phi)\;\varepsilon\cdot\mathbf{r}R_{n_i l_i}(r)\;Y_{l_i m_i}(\theta,\,\phi) \\
&= \int_0^{\infty} r^2\,dr R_{n_f l_f}^*(r)\;rR_{n_i l_i}(r) \\
&\quad\times\int d\Omega Y_{l_f l_f}^*(\theta,\,\phi)\;\varepsilon\cdot\hat{\mathbf{r}}Y_{l_i m_i}(\theta,\,\phi)
\end{aligned} \tag{21-73}$$

The radial integral will be discussed for a special case in the next section. Here we concentrate on the angular integral. We have

$$\varepsilon\cdot\hat{\mathbf{r}} = \varepsilon_x\sin\theta\cos\phi + \varepsilon_y\sin\theta\sin\phi + \varepsilon_z\cos\theta$$

and making use of

$$\sqrt{\frac{3}{4\pi}}\, Y_{1,0}(\theta,\, \phi) = \cos\theta \qquad \sqrt{\frac{3}{8\pi}}\, Y_{1,\pm 1}(\theta,\, \phi) = \mp\sin\theta\, e^{\pm i\phi} \qquad (21\text{-}74)$$

a little algebra yields

$$\varepsilon\cdot\hat{\mathbf{r}} = \sqrt{\frac{4\pi}{3}}\left(\varepsilon_z Y_{1,0} + \frac{-\varepsilon_x + i\varepsilon_y}{\sqrt{2}}\, Y_{1,1} + \frac{\varepsilon_x + i\varepsilon_y}{\sqrt{2}}\, Y_{1,-1}\right) \qquad (21\text{-}75)$$

Thus the angular integral in (21-73) involves

$$\int d\Omega Y^*_{l_f m_f}(\theta,\, \phi)\, Y_{1,m}(\theta,\, \phi)\, Y_{l_i m_i}(\theta,\, \phi) \qquad (21\text{-}76)$$

Let us first consider the azimuthal integration. It yields

$$\int_0^{2\pi} d\phi\, e^{-im_f\phi}\, e^{im\phi}\, e^{im_i\phi} = 2\pi\, \delta_{m=m_f-m_i} \qquad (21\text{-}77)$$

We thus get the first *selection rule*

$$m_f - m_i = m = 1,\, 0,\, -1 \qquad (21\text{-}78)$$

This was the selection rule that was mentioned in our discussion of the Zeeman effect. Specifically, if we define the z-axis to lie along the photon momentum direction $\mathbf{k}$, then the condition (21-38) implies that $\varepsilon_z = 0$ and hence $m = \pm 1$ only appears, so that

$$m_f - m_i = \pm 1 \qquad (21\text{-}79)$$

As a special case, we note that if the final state is the ground state, with $l_f = m_f = 0$, then $m = -m_i$. For example, if $m_i = 1$, then $m = -1$ and hence the polarization vector for the radiation is $(\varepsilon_x + i\varepsilon_y)/\sqrt{2}$. The implication is that if the atom in the initial state is polarized along the z-axis with $m_i = 1$, then in a decay to a state with zero angular momentum, the conservation of the z-component of angular momentum demands that the photon carry this off. The photon must therefore have its spin aligned along the positive z-axis, that is, it must have positive helicity (helicity $= +1$), or, equivalently, it must be left-circularly polarized. This is just what the term $(\varepsilon_x + i\varepsilon_y)/\sqrt{2}$ indicates.

The θ integration gives rise to another selection rule. Consider first the special case that $l_f = 0$. Since $Y_{0,0} = 1/\sqrt{4\pi}$, the angular integration (21-76) involves

$$\frac{1}{\sqrt{4\pi}}\int d\Omega Y_{1,m}(\theta,\, \phi)\, Y_{l_i m_i}(\theta,\, \phi) = \frac{1}{\sqrt{4\pi}}\, \delta_{l_i 1}\, \delta_{m_i,-m} \qquad (21\text{-}80)$$

which implies that *the initial state must have* $l_i = 1$. In hydrogen, the dominant transitions to the ground state will be $np \to 1s$.

More generally, when l_i and l_f do not vanish, we still get a selection rule. The derivation, beyond the scope of the mathematical knowledge about special functions assumed in this book, makes use of the *addition theorem* for spherical harmonics, which reads

$$Y_{l_1 m_1}(\theta, \phi) \, Y_{l_2 m_2}(\theta, \phi) = \sum_{L=|l_1-l_2|}^{l_1+l_2} C(L, m_1 + m_2; l_1, l_2, m_1, m_2) \, Y_{L, m_1+m_2}(\theta, \phi)$$

(21-81)

The coefficients $C(L, m_1 + m_2; l_1, l_2, m_1, m_2)$ are the same Clebsch–Gordan coefficients that appear in (15-46). The possible angular momenta on the right side are just those that could be obtained from the addition of the angular momenta l_1 and l_2. Substitution into (21-76) yields

$$\int d\Omega Y^*_{l_f m_f}(\theta, \phi) \sum_{L=|l_i-1|}^{l_i+1} C(L, m + m_i; 1, l_i, m, m_i) \, Y_{L, m+m_i}(\theta, \phi) = 0$$

unless

$$l_f = l_i + 1, \, l_i, \, |l_i - 1|$$

(21-82)

This is the general form of the *electric dipole radiation selection rule*

$$\Delta l = 1, 0, -1$$

(21-83)

with the observation, obvious from (21-80) that there are *no zero-zero transitions.* There is a further constraint that comes from parity conservation. Since $\mathbf{r}$ is odd under reflections, there is an additional selection rule for the electric dipole transitions:

$$\text{The atomic state must change} \atop \text{parity}$$

(21-84)

Since parity is given by $(-1)^l$, this implies that the l-value must actually change. Thus, for example $3p \to 2p$ transitions are not allowed to order $Z\alpha$.

To the extent that the only perturbation is the coupling

$$\frac{e}{mc} \mathbf{p} \cdot \mathbf{A}(\mathbf{r}, t)$$

(21-85)

there is no spin dependence in it, and hence the spins cannot flip in the transition. This leads to the additional selection rule

$$\Delta S = 0$$

(21-86)

mentioned earlier in connection with the spectrum of helium.

The selection rules stated above are not absolute. The conservation laws of angular momentum and parity (for electromagnetic processes) are absolute, but (21-83) is only approximately true. Transitions between states that involve a change of l larger than 1 cannot take place through the electric dipole mechanism. They can still take place, provided there is a nonvanishing matrix element

$$\langle \phi_f | e^{-i\mathbf{k}\cdot\mathbf{r}} \, \boldsymbol{\varepsilon} \cdot \mathbf{p} | \phi_i \rangle \tag{21-87}$$

For $\Delta l = 2$, the first power of $\mathbf{k} \cdot \mathbf{r}$ will give a nonvanishing contribution. We may write

$$
\begin{aligned}
\mathbf{k} \cdot \mathbf{r}\boldsymbol{\varepsilon} \cdot \mathbf{p} &= \tfrac{1}{2}(\boldsymbol{\varepsilon} \cdot \mathbf{p}\mathbf{k} \cdot \mathbf{r} + \boldsymbol{\varepsilon} \cdot \mathbf{r}\mathbf{p} \cdot \mathbf{k}) + \tfrac{1}{2}(\boldsymbol{\varepsilon} \cdot \mathbf{p}\mathbf{k} \cdot \mathbf{r} - \boldsymbol{\varepsilon} \cdot \mathbf{r}\mathbf{p} \cdot \mathbf{k}) \\
&= \tfrac{1}{2}(\boldsymbol{\varepsilon} \cdot \mathbf{p}\mathbf{k} \cdot \mathbf{r} + \boldsymbol{\varepsilon} \cdot \mathbf{r}\mathbf{p} \cdot \mathbf{k}) + \tfrac{1}{2}(\mathbf{k} \times \boldsymbol{\varepsilon}) \cdot (\mathbf{r} \times \mathbf{p})
\end{aligned}
\tag{21-88}
$$

The first of these terms is called an electric quadrupole term, and the second is clearly related to an $\mathbf{L} \cdot \mathbf{B}$ term, and is called a magnetic dipole term. For these transitions, whose matrix element we estimated to be $Z\alpha$ times smaller than the leading term, we will have $\Delta l = 2$ and $\Delta l = 0$, respectively. Since the operators in (21-88) are even, there will be no parity change between the atomic states. Transitions between $3d \rightarrow 1s$, for example, cannot go via the electric dipole mechanism, but can go via the electric quadrupole mechanism. Actually, it turns out to be much more probable that the $3d$ state decays first into a $2p$ state, and the latter then undergoes the favored $2p \rightarrow 1s$ transition, that is, two photons are emitted in succession.

The spin selection rule $\Delta S = 0$ too is not sacred. In addition to the coupling (21-85) there is the coupling discussed in connection with the anomalous Zeeman effect

$$\lambda V(t) = \frac{ge}{2mc} \, \mathbf{S} \cdot \mathbf{B}(\mathbf{r}, t) \tag{21-89}$$

The matrix element for the $\Delta S \neq 0$ transition-inducing term can be estimated. We compare it with the electric dipole matrix element

$$\frac{(eg/2mc)\,\hbar|\mathbf{k} \times \boldsymbol{\varepsilon}|}{(2e/mc)|\mathbf{p} \cdot \boldsymbol{\varepsilon}|} \simeq \frac{\hbar|\mathbf{k}|}{|\mathbf{p}|} \simeq \frac{\hbar\omega}{|\mathbf{p}|c} \simeq \frac{mc^2(Z\alpha)^2}{mc^2(Z\alpha)} \simeq Z\alpha \tag{21-90}$$

and see that it is suppressed, just like the magnetic dipole matrix element, which it strongly resembles in form. As an example of a situation where the coupling (21-89) plays an important role, we consider the nuclear process of photodisintegration of the deuteron

$$\gamma + d \rightarrow n + p \tag{21-91}$$

The deuteron, to a very good approximation, is a $^3S_1$ state. An electric dipole transition must involve the final $(n - p)$ system in a ^3P state since $\Delta l = 1$ and $\Delta S = 0$. It turns out, however, that just above threshold for the reaction, the two nucleons are unlikely to be in a relative P-state. In general, particles will be in a relative angular momentum L state with any appreciable probability only if

$$|\mathbf{p}|a \gtrsim \hbar L \tag{21-92}$$

where $\mathbf{p}$ is the relative momentum and a are the dimensions of the system. For the deuteron for γ's below 10 MeV in energy, the $(n - p)$ system is unlikely to be in a P-state. The additional coupling

$$-\frac{e}{2Mc}(g_p\mathbf{s}_p + g_n\mathbf{s}_n) \cdot \mathbf{B} \tag{21-93}$$

can, however, lead to a transition between the $^3S_1$ state, and the unbound $^1S_0$ state. The interaction may be rewritten in the form

$$-\frac{e}{2Mc}\left[\frac{1}{2}(g_p + g_n)(\mathbf{s}_p + \mathbf{s}_n) + \frac{1}{2}(g_p - g_n)(\mathbf{s}_p - \mathbf{s}_n)\right] \cdot \mathbf{B} \qquad (21\text{-}94)$$

The first term is symmetric under the $n \leftrightarrow p$ exchange, and hence cannot contribute to a transition between a symmetric and an antisymmetric spin state. The second term does, however, contribute. The coefficients are actually quite large, since $g_p \cong 5.56$ and $g_n = -3.81$.

There is one selection rule that is sacred, and that is the one forbidding zero-zero transitions (referring to *total* angular momentum $j = 0$) in one-photon processes. A general way of arguing the absoluteness of this selection rule is the following: The matrix element, a scalar quantity, must involve the photon polarization linearly, and must therefore be of the form $\boldsymbol{\varepsilon} \cdot \mathbf{V}$, where $\mathbf{V}$ is some vector that enters into the problem. If the initial and final states are $j = 0$ states, that is, have no directionality associated with them, then the only vector is $\mathbf{k}$, the photon momentum. However $\boldsymbol{\varepsilon} \cdot \mathbf{k} = 0$, so that there is no way of constructing a matrix element. It must therefore not exist.[1]

THE $2p \rightarrow 1s$ TRANSITION

Let us now specialize to the transition $2p \rightarrow 1s$ in (21-73). We need to evaluate the radial integral

$$\int_0^\infty dr\, r^3 R_{10}^*(r)\, R_{21}(r)$$

$$= \int_0^\infty dr\, r^3 \left[2\left(\frac{Z}{a_0}\right)^{3/2} e^{-Zr/a_0}\right]\left[\frac{1}{\sqrt{24}}\left(\frac{Z}{a_0}\right)^{5/2} r\, e^{-Zr/2a_0}\right]$$

$$= \frac{1}{\sqrt{6}}\left(\frac{Z}{a_0}\right)^4 \int_0^\infty dr\, r^4\, e^{-3Zr/2a_0} \qquad (21\text{-}95)$$

$$= \frac{1}{\sqrt{6}}\left(\frac{Z}{a_0}\right)^4 \left(\frac{2a_0}{3Z}\right)^5 \int_0^\infty dx\, x^4\, e^{-x} = \frac{24}{\sqrt{6}}\left(\frac{2}{3}\right)^5 Z^{-1}a_0$$

The angular integral is

$$\int d\Omega\, Y_{0,0}^*\, \boldsymbol{\varepsilon} \cdot \hat{\mathbf{r}}\, Y_{1,m} = \frac{1}{\sqrt{4\pi}}\int d\Omega\, \sqrt{\frac{4\pi}{3}}\left(\varepsilon_z Y_{1,0} + \frac{-\varepsilon_x + i\varepsilon_y}{\sqrt{2}} Y_{1,1}\right.$$

$$\left. + \frac{\varepsilon_x + i\varepsilon_y}{\sqrt{2}} Y_{1,-1}\right) Y_{1,m} \qquad (21\text{-}96)$$

$$= \frac{1}{\sqrt{3}}\left(\varepsilon_z\, \delta_{m,0} + \frac{-\varepsilon_x + i\varepsilon_y}{\sqrt{2}}\, \delta_{m,-1} + \frac{\varepsilon_x + i\varepsilon_y}{\sqrt{2}}\, \delta_{m,1}\right)$$

---

[1] The relation $\boldsymbol{\varepsilon} \cdot \mathbf{k} = 0$ is independent of the choice of gauge, and is a statement about the transversality of the electromagnetic field. Such arguments "by enumeration" are frequently used in elementary particle physics, where the interaction is not really known.

Now the absolute square of the product of (21-95) and (21-96) is

$$96 \left(\frac{2}{3}\right)^{10} \left(\frac{a_0}{Z}\right)^2 \frac{1}{3} \left[\delta_{m0} \varepsilon_z^{\,2} + \frac{1}{2} (\delta_{m,1} + \delta_{m,-1})(\varepsilon_x^2 + \varepsilon_y^2) \right] \qquad (21\text{-}97)$$

so that the transition rate is for a given m-value of the excited atom,

$$
\begin{aligned}
R_{2p \to 1s} = \int d\Omega_{\mathbf{p}} \left(\frac{\alpha}{2\pi}\right) \frac{\omega}{m^2 c^2} m^2 \omega^2 \frac{2^{15}}{3^{10}} \left(\frac{a_0}{Z}\right)^2 \\
\times \left[\delta_{m,0} \varepsilon_z^2 + \frac{1}{2} (\delta_{m,1} + \delta_{m,-1})(\varepsilon_x^2 + \varepsilon_y^2) \right]
\end{aligned}
\qquad (21\text{-}98)
$$

where

$$
\begin{aligned}
\omega &= \frac{1}{\hbar} \left[\frac{1}{2} mc^2 (Z\alpha)^2 \left(1 - \frac{1}{4}\right) \right] \\
&= \frac{3}{8} \frac{mc^2}{\hbar} (Z\alpha)^2
\end{aligned}
\qquad (21\text{-}99)
$$

is the frequency of the radiation emitted in the transition.

The angular integration in (21-98) is over the photon directions, and this is not trivial, since ε is constrained to be perpendicular to the photon momentum direction. The integration is very simple if the initial p-state is unaligned, that is, it occurs in the three possible m-states ($m = 1, 0, -1$) with equal probability. The rate is then

$$R_{2p \to 1s} = \frac{1}{3} \sum_{m=-1}^{1} R_{2p \to 1s}(m) \qquad (21\text{-}100)$$

Since

$$\sum_{m=-1}^{1} [\delta_{m0} \varepsilon_z^2 + \tfrac{1}{2} (\delta_{m1} + \delta_{m,-1})(\varepsilon_x^2 + \varepsilon_y^2)] = \varepsilon_x^2 + \varepsilon_y^2 + \varepsilon_z^2 = 1 \quad (21\text{-}101)$$

the integrand becomes independent of the photon direction. This result should also be multiplied by a factor of 2. The reason is that there are two possible polarization states for the photon, and we are detecting both of them. A more careful way of writing (21-55) would have been

$$\int d\Omega_{\mathbf{p}} \frac{\alpha}{2\pi} \frac{\omega_{km}}{m^2 c^2} \sum_{\lambda=1}^{2} |\langle \phi_m| e^{-i\mathbf{k}\cdot\mathbf{r}} \, \varepsilon^{(\lambda)} \cdot \mathbf{p}|\phi_k\rangle|^2 \qquad (21\text{-}102)$$

with λ denoting the polarizations. The two polarization states are orthogonal, so that we have

$$\varepsilon^{(\lambda)} \cdot \varepsilon^{(\lambda')} = \delta_{\lambda\lambda'} \qquad (21\text{-}103)$$

When all of this is put together, we get

$$R_{2p \to 1s} = 2 \cdot 4\pi \, \frac{\alpha}{2\pi} \frac{1}{c^2} \left(\frac{3}{8} \frac{mc^2}{\hbar} Z^2 \alpha^2 \right)^3 \frac{2^{15}}{3^{10}} \left(\frac{\hbar}{mcZ\alpha} \right)^2 \frac{1}{3}$$

$$= \frac{2^8}{3^8} \frac{mc^2}{\hbar} \, \alpha (Z\alpha)^4 \cong 0.6 \times 10^9 \, Z^4 \; \text{sec}^{-1} \tag{21-104}$$

This differs by a factor of about 25 from the estimate made in (21-68). Thus detailed factors in the matrix elements are important and guesses cannot replace a calculation. Nevertheless, dimensional considerations and a proper counting of powers of α do give us an order of magnitude guidance to how large physical quantities in atomic physics are.

The expression for the rate

$$R_{fi} = \frac{d\Omega_{\mathbf{p}}}{2\pi} \frac{e^2}{\hbar c} \frac{\omega^3}{c^2} \sum_{\lambda=1}^{2} |\langle f|\mathbf{r}|i \rangle \cdot \boldsymbol{\varepsilon}^{(\lambda)}|^2 \tag{21-105}$$

may be translated into a formula for the intensity of radiation by multiplying it by the energy of the light quantum $\hbar\omega$. Thus

$$I_{fi} = d\Omega_{\mathbf{p}} \frac{e^2}{2\pi c^3} \omega^4 \sum_{\lambda=1}^{2} |\langle f|\mathbf{r}|i \rangle \cdot \boldsymbol{\varepsilon}^{(\lambda)}|^2 \tag{21-106}$$

This, however, is just the *classical* formula for the intensity of light emitted by an oscillating dipole, of dipole moment

$$\mathbf{d} = e\langle f|\mathbf{r}|i \rangle \, e^{-i\omega t} \tag{21-107}$$

which provides another illustration of the correspondence principle.

SPIN AND INTENSITY RULES

The inclusion of spin does not change things very much. It is true that the initial states and the final states can each be in an "up" or "down" spin state, but since the interaction in atomic transitions is spin independent, only "up" → "up" and "down" → "down" transitions are allowed. Hence the transition rates will not only be independent of m_l (as we saw in the last section) but also of m_s, and hence, m_j. With the inclusion of spin-orbit coupling, there will be small (on the scale of the $2p - 1s$ energy difference) level splittings. For example, the $n = 1$ and $n = 2$ level structure is changed, as shown in Fig. 21-2. The spectral line corresponding

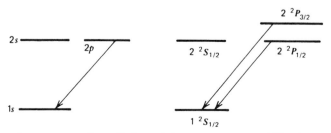

Figure 21-2. The splitting of the $2p - 1s$ spectral line by spin-orbit coupling.

to the transition $2p \rightarrow 1s$ is split into two lines, $2^2P_{3/2} \rightarrow 1^2S_{1/2}$ and $2^2P_{1/2} \rightarrow 1^2S_{1/2}$. For the split states, the radial integral and the phase space are almost unchanged, and hence *the ratio of the intensity of the two lines can be determined from the angular parts of the integral alone, that is, purely from angular momentum considerations.*

The following table lists the wave functions for the states in question.

J	m_j	Odd Parity $l = 1$	Even Parity $l = 0$
3/2	3/2	$Y_{11}\chi_+$	—
3/2	1/2	$\sqrt{2/3}\,Y_{10}\chi_+ + \sqrt{1/3}\,Y_{11}\chi_-$	—
3/2	−1/2	$\sqrt{1/3}\,Y_{1,-1}\chi_+ + \sqrt{2/3}\,Y_{10}\chi_-$	—
3/2	−3/2	$Y_{1,-1}\chi_-$	—
1/2	1/2	$\sqrt{1/3}\,Y_{10}\chi_+ - \sqrt{2/3}\,Y_{11}\chi_-$	$Y_{00}\chi_+$
1/2	−1/2	$\sqrt{2/3}\,Y_{1,-1}\chi_+ - \sqrt{1/3}\,Y_{10}\chi_-$	$Y_{00}\chi_-$

In the squares of the matrix elements, the radial parts are common to all of them. Thus, in considering the rates for $P_{3/2} \rightarrow S_{1/2}$ we must add the squares of the transition matrix elements for $m_j = 3/2 \rightarrow m_j = 1/2$, $m_j = 3/2 \rightarrow m_j = -1/2, \ldots, m_j = -3/2 \rightarrow m_j = -1/2$, while the rate for $P_{1/2} \rightarrow S_{1/2}$ involves the sum of the squares of the matrix elements for $m_j = 1/2 \rightarrow m_j = 1/2, \ldots, m_j = -1/2 \rightarrow m_j = -1/2$. This can be done directly by techniques that are quite sophisticated and beyond the scope of this book. One can, however, work out these quantities in detail, using the fact that the spin wave functions are orthonormal.

$P_{3/2} \rightarrow S_{1/2}$		
$m_j = 3/2 \rightarrow m_j = 1/2$	$|\langle Y_{11}|\mathbf{r}\cdot\varepsilon|Y_{00}\rangle|^2 = C$	
$3/2 \rightarrow -1/2$	0 $\quad$ since $\chi_+^*\chi_- = 0$	
$1/2 \rightarrow \;\;1/2$	$|\langle\sqrt{2/3}\,Y_{10}|\mathbf{r}\cdot\varepsilon|Y_{00}\rangle|^2 = 0$	$(\Delta m = 0)$
$1/2 \rightarrow -1/2$	$|\langle\sqrt{1/3}\,Y_{11}|\mathbf{r}\cdot\varepsilon|Y_{00}\rangle|^2 = C/3$	
$-1/2 \rightarrow \;\;1/2$	$|\langle\sqrt{1/3}\,Y_{1,-1}|\mathbf{r}\cdot\varepsilon|Y_{00}\rangle|^2 = C/3$	
$-1/2 \rightarrow -1/2$	$|\langle\sqrt{2/3}\,Y_{10}|\mathbf{r}\cdot\varepsilon|Y_{00}\rangle|^2 = 0$	$(\Delta m = 0)$
$-3/2 \rightarrow \;\;1/2$	0	
$-3/2 \rightarrow -1/2$	$|\langle Y_{1,-1}|\mathbf{r}\cdot\varepsilon|Y_{00}\rangle|^2 = C$	

If we sum the terms we get

$$\sum R = \frac{8C}{3} \tag{21-108}$$

Similarly

$P_{1/2} \rightarrow S_{1/2}$	
$m_j = 1/2 \rightarrow m_j = 1/2$	$|\langle\sqrt{1/3}\,Y_{10}|\varepsilon\cdot\mathbf{r}|Y_{00}\rangle|^2 = 0$
$1/2 \rightarrow -1/2$	$|\langle-\sqrt{2/3}\,Y_{11}|\varepsilon\cdot\mathbf{r}|Y_{00}\rangle|^2 = 2C/3$
$-1/2 \rightarrow \;\;1/2$	$|\langle\sqrt{2/3}\,Y_{1,-1}|\varepsilon\cdot\mathbf{r}|Y_{00}\rangle|^2 = 2C/3$
$-1/2 \rightarrow -1/2$	$|\langle-\sqrt{1/3}\,Y_{10}|\varepsilon\cdot\mathbf{r}|Y_{00}\rangle|^2 = 0$

Again

$$\sum R = \frac{4C}{3} \tag{21-109}$$

Thus the ratio of the intensities is

$$\frac{R(P_{3/2} \to S_{1/2})}{R(P_{1/2} \to S_{1/2})} = \frac{8C/3}{4C/3} = 2 \tag{21-110}$$

The reason for *summing* over all the initial states is that when the atom is excited, all the p-levels are equally occupied, since their energy difference is so tiny compared to the $2p - 1s$ energy difference. We also sum over all the final states if we perform an experiment that does not discriminate between them, as is the case for a spectroscopic measurement. In our calculation of the $2p \to 1s$ transition rate, we *averaged* over the initial m-states. There we were concerned with the problem of asking: "If we have N atoms in the $2p$ states, how many will decay per second?" The averaging came about because of the fact that under most circumstances, when N atoms are excited, about $N/3$ go into each one of the $m = 1, 0, -1$ states. Here, the fact that there are more levels in the $P_{3/2}$ state than there are in the $P_{1/2}$ state is relevant. There will be altogether six levels, (four with $j = 3/2$ and two with $j = 1/2$), and there will be on the average $N/6$ atoms in each of the states. The fact that there are more atoms in the $j = 3/2$ subset of levels just means that more decay, and that therefore the intensity will be larger.

LIFETIME AND LINEWIDTH

The number $R(i \to f)$ that we learned to calculate in this chapter represents the probability for the transition $i \to f$, divided by the time during which the perturbation has acted. This time must be long compared to $\hbar/(E_m^0 - E_k^0 + \hbar\omega)$ in order that the transition probability be proportional to t, but it clearly cannot be too long. If we ask for the probability that the initial state remain intact, we get

$$P_i(t) = 1 - \left[\sum_{f \neq i} R(i \to f) \right] t \tag{21-111}$$

where the sum is over all final states that are accessible. This clearly has no meaning for long enough times, since probabilities are positive. It turns out that if the calculation of the time development of the system is done more carefully,[2] then it can be shown that the right side of (21-111) just represents an approximation (to lowest order in the perturbation) to the correct expression—again only true for long times—that

$$P_i(t) = e^{-t\sum_{f \neq i} R(i \to f)} \tag{21-112}$$

One may thus speak of a *lifetime* of the initial state

$$\tau = \frac{1}{R} = \frac{1}{\sum_{f \neq i} R(i \to f)} \tag{21-113}$$

[2]This is done in the Special Topics section, "Lifetimes, Line Widths, and Resonances."

The total transition rate R is the sum of partial transition rates into the possible *channels* f. In the example that was discussed in detail, the $2p \rightarrow 1s$ transition in hydrogenlike atoms, no other channels are available, so that the lifetime of the $2p$ state is

$$\tau = 1.6 \times 10^{-9} \, Z^{-4} \, \sec \tag{21-114}$$

This is in excellent agreement with experiment. Let us compare this (we take $Z = 1$) with the time it takes the electron to "go once around the nucleus." The velocity is αc, and the distance is of the order 3×10^{-8} cm, so that the characteristic time is of the order of 1.4×10^{-16} sec. In terms of this time, the $2p$ state is very long lived.

Since the $2p$ state has a finite lifetime, it should, by the uncertainty principle, have an uncertainty in the energy, of magnitude

$$\Delta E \sim \frac{\hbar}{\tau} \tag{21-115}$$

The way in which this manifests itself is that the intensity of the line, as a function of frequency, is not completely sharp at the value $\omega_0 = (E_{2p} - E_{1s})/\hbar$, but it has a distribution of the form

$$I(\omega) \propto \frac{(R/2)^2}{(\omega - \omega_0)^2 + R^2/4} \tag{21-116}$$

Note that in the limit that $R \rightarrow 0$, that is, in the limit that perturbation theory is strictly applicable, we get, as a consequence of the formula

$$\lim_{\epsilon \rightarrow 0} \frac{\epsilon}{(\omega - \omega_0)^2 + \epsilon^2} = \pi \, \delta \, (\omega - \omega_0) \tag{21-117}$$

the line shape represented by the energy-conservation delta function. The width of the line (21-116) is R, and this is a measure of the uncertainty in the energy. This line shape is called a Lorentzian line shape.

Problems

1. A hydrogen atom is placed in an electric field $\mathbf{E}(t)$ that is uniform and has the time dependence

$$\mathbf{E}(t) = 0 \qquad t < 0$$
$$= \mathbf{E}_0 \, e^{-\gamma t} \qquad t > 0$$

What is the probability that as $t \rightarrow \infty$, the hydrogen atom, if initially in the ground state, makes a transition to the $2p$ state?

2. Calculate an expression for $c_n(t)$ to second order in λ. Write $c_n(t) = \delta_{nk} + \lambda c_n^{(1)}(t) + \lambda^2 c_n^{(2)}(t) + \cdots$. Show that the first-order equation for $c_n^{(1)}(t)$ is unaltered, and $c_n^{(2)}(t)$ obeys the equation

$$C_m^{(2)}(t) = \frac{1}{i\hbar\lambda} \int_0^t dt' \sum_n c_n^{(1)}(t')\, e^{i(E_m^0 - E_n^0)t'/\hbar} \langle \phi_m | V(t') | \phi_n \rangle$$

Substitute the solution $c_n^{(1)}(t')$ and simplify your expression as far as possible.

3. Consider a harmonic oscillator described by

$$H = \frac{1}{2m} p_x^2 + \tfrac{1}{2} m\omega^2(t)\, x^2$$

where

$$\omega(t) = \omega_0 + \delta\omega \cos ft$$

and $\delta\omega \ll \omega_0$. Calculate the probability that a transition occurs from the ground state, as a function of time, given that the system is in the ground state at $t = 0$. Use perturbation theory. Use the fact that for $n \neq 0$,

$$\langle n | x^2 | 0 \rangle = \hbar/\sqrt{2}m\omega \qquad \text{for } n = 2$$
$$= 0 \qquad \text{otherwise.}$$

4. Suppose a particle of rest mass M decays into two particles of rest mass m_1 and m_2, respectively. Use the relativistic relation between energy and momentum to compute the density of states ρ that appears in (21-61).

[Hint: There is only one independent momentum, say $\mathbf{p}$, and what is needed is

$$\int \frac{d^3\mathbf{p}}{(2\pi\hbar)^6}\, \delta\left(E_{\text{initial}} - \sum_{\substack{\text{final} \\ \text{states}}} E \right)$$

5. Consider the preceding calculation when the decay is of the form

$$A \to B + C + D$$

with particles C and D massless.

(Hint: There are now two independent momenta.)

6. In this problem the *adiabatic theorem* is to be illustrated. The theorem states that if the Hamiltonian is changed very slowly from H_0 to H, then a system in a given eigenstate of H_0 goes over into the corresponding eigenstate of H, but does not make any transitions. To be specific, consider the ground state, so that

$$H_0\phi_0 = E_0\phi_0$$

Let $V(t) = f(t)V$, where $f(t)$ is a slowly varying function, as shown in the graph.

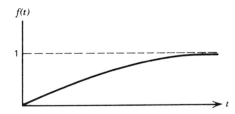

If the ground state of $H = H_0 + V$ is ω_0, the theorem states that

$$|\langle \omega_0 | \psi(t) \rangle| \to 1$$

The steps to be carried out are the following:

(a) Show that

$$\frac{1}{i\hbar} \int_0^t dt' \, e^{i(E_m^0 - E_0^0)t'/\hbar} \, f(t') \rightarrow \frac{e^{i(E_m^0 - E_0^0)t/\hbar}}{E_m^0 - E_0^0}$$

for times t such that $f(t) = 1$. Use the fact that

$$\frac{df(t')}{dt'} \ll \frac{E_m^0 - E_0^0}{\hbar} f(t')$$

Either construct an example of a function $f(t)$ or use integration by parts, that is, write

$$e^{i\omega t'} = \frac{1}{i\omega} \frac{d}{dt'} e^{i\omega t'}$$

in the preceding.

(b) Calculate $\psi(t)$ using (21-3) and (21-9). Compare this with the formula (16-19), which here reads

$$w_0 = \phi_0 + \sum_{m \neq 0} \frac{\langle \phi_m | V | \phi_0 \rangle}{E_0^0 - E_m^0} \phi_m$$

and thus show that

$$|\langle w_0 | \psi(t) \rangle| \rightarrow 1$$

7. Work out the $2p \rightarrow 1s$ transition rate for the three-dimensional oscillator, following the steps carried out in this chapter.

8. Nuclei sometimes decay from excited states to the ground state by *internal conversion*, a process in which one of the $1s$ electrons is emitted instead of a photon. Let the initial and final nuclear wave functions be

$$\phi_I(\mathbf{r}_1, \mathbf{r}_2, \ldots, \mathbf{r}_A) \quad \text{and} \quad \phi_F(\mathbf{r}_1, \mathbf{r}_2, \ldots, \mathbf{r}_A)$$

where $\mathbf{r}_i$ ($i = 1, 2, \ldots, Z$) describe the protons and $\mathbf{r}_{Z+1} \ldots \mathbf{r}_A$ the neutrons. The perturbation giving rise to the transition is just the nucleus–electron interaction

$$V = -\sum_{i=1}^{Z} \frac{e^2}{|\mathbf{r} - \mathbf{r}_i|}$$

where $\mathbf{r}$ is the electron coordinate. Thus the matrix element is given by

$$-\int d^3r \int d^3r_1 \cdots d^3r_A \phi_F^* \frac{e^{-i\mathbf{p}\cdot\mathbf{r}/\hbar}}{\sqrt{V}} \sum_{i=1}^{Z} \frac{e^2}{|\mathbf{r} - \mathbf{r}_i|} \phi_I \psi_{100}(\mathbf{r})$$

(a) What is the magnitude of $\mathbf{p}$, the free electron momentum?

(b) Calculate the rate for the process for a dipole transition in terms of

$$\mathbf{d} = \sum \int d^3r_1 \cdots d^3r_A \phi_F^* \mathbf{r}_i \phi_I$$

by making use of the expansion

$$\frac{1}{|\mathbf{r} - \mathbf{r}_i|} \approx \frac{1}{r} + \frac{\mathbf{r} \cdot \mathbf{r}_i}{r^3}$$

9. Show that as $t \to \infty$

$$e^{i\omega t} \sin (\omega_0 - \omega)t \sin (\omega_0 + \omega)t \to -\tfrac{1}{2}$$

(*Hint:* use $\lim_{t \to \infty} \sin At = \lim_{t \to \infty} \cos At = 0$.)

References

The addition theorem that leads to the more general derivation of selection rules is discussed in all of the more advanced textbooks listed at the end of this volume, and also in

M. E. Rose, *Elementary Theory of Angular Momentum*, John Wiley & Sons, New York, 1957.

The radial integrals for the more general case are discussed in

H. A. Bethe and R. W. Jackiw, *Intermediate Quantum Mechanics*, W. A. Benjamin, New York, 1968.

H. A. Bethe and E. E. Salpeter, *Quantum Mechanics of One- and Two-Electron Atoms*, Springer-Verlag, Berlin/New York, 1957.

E. U. Condon and G. H. Shortley, *The Theory of Atomic Spectra*, Cambridge University Press, Cambridge, England, 1959.

SELECTED TOPICS IN RADIATION THEORY

In Chapter 21 [see (21-40) and (21-41)] we pointed out that the vector potential to be used for the *absorption* of a light quantum from an initial state of n photons of angular frequency ω and polarization λ was

$$\mathbf{A}(\mathbf{r}, t) = \sqrt{\frac{2\pi c^2 \hbar}{\omega V}} \sqrt{n} \; \varepsilon^{(\lambda)} \, e^{i(\mathbf{k}\cdot\mathbf{r} - \omega t)} \tag{22-1}$$

while that for the *emission* of a light quantum into an initial state of n photons of angular frequency ω and polarization λ was

$$\mathbf{A}(\mathbf{r}, t) = \sqrt{\frac{2\pi c^2 \hbar}{\omega V}} \sqrt{n + 1} \; \varepsilon^{(\lambda)} \, e^{-i(\mathbf{k}\cdot\mathbf{r} - \omega t)} \tag{22-2}$$

where we have modified our equations to take into account the polarization labeling. Strictly speaking the number of photons also depends on the momentum and polarization of the photon, so that we really should write

$$\mathbf{A}(\mathbf{r}, t) = \sqrt{\frac{2\pi c^2 \hbar}{\omega V}} \sqrt{n_\lambda(\mathbf{k})} \; \varepsilon^{(\lambda)} \, e^{i(\mathbf{k}\cdot\mathbf{r} - \omega t)} \tag{22-3}$$

for *absorption* and

$$\mathbf{A}(\mathbf{r}, t) = \sqrt{\frac{2\pi c^2 \hbar}{\omega V}} \sqrt{n_\lambda(\mathbf{k}) + 1} \; \varepsilon^{(\lambda)} \, e^{-i(\mathbf{k}\cdot\mathbf{r} - \omega t)} \tag{22-4}$$

for *emission*. Physically these factors imply that the presence of photons of a particular frequency *enhances* the probability of emission of another photon of that same frequency. The photons are said to stimulate or induce the emission of radiation.

The presence of these n-dependent factors emerges quite naturally when one goes through the process of "quantizing" the electromagnetic field, that is, treating the electric and magnetic fields as dynamical variables, like $p(t)$ and $x(t)$ in one-dimensional motion of a particle. This is beyond the scope of this book. Instead, we discuss the remarkable derivation of the $\sqrt{n}$ and $\sqrt{n+1}$ factors by Einstein carried out in 1917 before quantum mechanics was discovered.

THE EINSTEIN A AND B COEFFICIENTS

Einstein used the Planck theory, the Bohr theory, and statistical mechanics and argued as follows:

Consider a gas of molecules interacting with radiation in a cavity at temperature T. According to the Bohr theory the molecules can be in a variety of stationary states. In equilibrium, the ratio of the number of molecules in a state m to the number of molecules in a state n is given by

$$\frac{N_m}{N_n} = \frac{g_m}{g_n} \frac{e^{-E_m/kT}}{e^{-E_n/kT}} = \frac{g_m}{g_n} e^{-(E_m - E_n)/kT} \tag{22-5}$$

where the g_m represent the degeneracy, that is the number of levels with energy E_m. (From quantum mechanics we now know that $g_m = 2J_m + 1$, where J_m is the total angular momentum of the state with energy E_m, but this is not needed in what follows.) Now consider a pair of levels E_1 and E_0 with $E_1 > E_0$ (Fig. 22-1). The rate of transitions from the lower energy level to the higher energy level, R_{01}, must be proportional to the number N_0 of molecules with energy E_0 and to the intensity of the radiation in the cavity with the frequency of the absorbed light, ν. Thus

$$R_{01} = N_0 u(\nu, T) B_{01} \tag{22-6}$$

B_{01} is the coefficient that describes the absorption of radiation by the molecules with energy E_0. It is called a coefficient of *induced absorption* because the process is induced by the radiation present. For the transition rate from the higher state to the lower state, Einstein uses the Bohr postulate that spontaneous emission can occur with a rate independent of the radiation present, but there must also exist *induced emission*. The number of molecules in the state 1 is N_1, and thus

$$R_{10} = N_1(u(\nu, T) B_{10} + A_{10}) \tag{22-7}$$

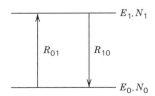

Figure 22-1. Schematics of two-level system.

B_{10} describes the induced emission and A_{10} the spontaneous emission. In equilibrium there must be as many transition $1 \to 0$ as $0 \to 1$, so that $R_{10} = R_{01}$, from which it follows that

$$N_0 u(\nu, T)B_{01} = N_1(u(\nu, T)B_{10} + A_{10}) \tag{22-8}$$

When this is combined with (22-6), we get

$$\frac{u(\nu, T)B_{10} + A_{10}}{u(\nu, T)B_{01}} = \frac{N_0}{N_1} = \frac{g_0}{g_1} e^{-(E_0 - E_1)/kT}$$

which we can rewrite in the form

$$g_1 A_{10} = u(\nu, T)(g_0 B_{01} e^{(E_1 - E_0)/kT} - g_1 B_{10}) \tag{22-9}$$

A number of consequences can be drawn from this formula.

(a) As $T \to \infty$ for fixed $E_1 - E_0$, $e^{(E_1 - E_0)/kT} \to 1$ and from the Rayleigh-Jeans law (see Chapter 1), $u(\nu, T) \to (8\pi\nu^2/c^3)kT$. Since the left side of the equation is independent of T, it follows that

$$g_0 B_{01} = g_1 B_{10} \tag{22-10}$$

This means that the rate per molecule of induced emission is equal to the rate per molecule of induced absorption.

(b) When this result is inserted into (22-9) we obtain

$$g_1 A_{10} = u(\nu, T)(g_1 B_{10})(e^{(E_1 - E_0)/kT} - 1)$$

or, equivalently

$$u(\nu, T) = \frac{A_{10}/B_{10}}{e^{(E_1 - E_0)/kT} - 1} \tag{22-11}$$

The left-hand side obeys the Wien law (1-4), so that

$$\nu^3 g\left(\frac{\nu}{T}\right) = \frac{A_{10}/B_{10}}{e^{(E_1 - E_0)/kT} - 1} \tag{22-12}$$

Since the left-hand side of the equation is a universal function, and A_{10}/B_{10} cannot depend on the temperature, it follows that A_{10}/B_{10} must be proportional to ν^3, and $(E_1 - E_0)$ must be proportional to ν. Thus $E_1 - E_0 = h\nu$, where h is a constant, and consequently

$$u(\nu, T) = \frac{A_{10}/B_{10}}{\nu^3} \frac{\nu^3}{e^{h\nu/kT} - 1} \tag{22-13}$$

When this is finally compared with the Planck formula, we see that

$$\frac{A_{10}}{B_{10}} = \frac{8\pi h \nu^3}{c^3} \tag{22-14}$$

results. One needs quantum mechanics to calculate A_{10}, and this was done in Chapter 21.

The rate of emission *per molecule* R_{10}/N_1 may be written in the form

$$R_{10}/N_1 = u(\nu, T)B_{10} + A_{10} = A_{10}\left(1 + \frac{B_{10}}{A_{10}} u(\nu, T)\right)$$

$$= A_{10}\left(1 + \frac{1}{e^{h\nu/kT} - 1}\right) \tag{22-15}$$

However, the average number of photons per unit volume in a black-body cavity at temperature T is given by

$$\langle n \rangle = \frac{\sum\limits_{n} n e^{-nh\nu/kT}}{\sum\limits_{n} e^{-nh\nu/kT}} = \left. \frac{d/dx \sum_n e^{-nx}}{\sum_n e^{-nx}} \right|_{x=h\nu/kT}$$

$$= \frac{1}{e^{h\nu/kT} - 1} \tag{22-16}$$

so that the emission rate per molecule may be written as

$$\frac{R_{10}}{N_1} = A_{10}(1 + \langle n(\nu, T) \rangle) \tag{22-17}$$

Thus the emission rate per molecule is proportional to $(1 + \langle n \rangle)$, where n is the average number of photons present. Similarly R_{01}/N_0 is proportional to $\langle n \rangle$, the average number of photons present. Although the absorption rate and the emission rate are proportional to $\langle n(\nu, T) \rangle$ and $\langle n(\nu, T) + 1 \rangle$, respectively, where $\langle n(\nu, T) \rangle$ is the average number of photons of the appropriate frequency in blackbody radiation, these factors do not depend on the fact that the radiation has a particular frequency distribution. Each absorption or emission event involves one photon only, and thus the factors in the *amplitudes* are $\sqrt{n(\nu)}$ and $\sqrt{n(\nu) + 1}$ respectively.

It is quite amazing how much Einstein was able to learn by combining powerful statistical reasoning and a primitive knowledge of quantum effects.

LASERS

Induced emission finds its most dramatic technological application in the production of coherent, monochromatic, highly directional electromagnetic radiation by means of *light amplification through stimulated emission* in a device called the *laser*.[1] The basic components of a laser are the following:

1. A lasing medium in which there are at least two energy levels, separated by an energy gap, such that the atoms in the upper level can make a transition that is stimulated by the presence of photons of the right frequency.

---

[1]The phenomenon holds for radiation of all frequencies and was indeed first studied in the microwave region. The device was called the Maser.

2. Some mechanism for repopulating the upper level for repeated operation.
3. A suitable cavity in which the stimulating photons can be contained and which also contains the lasing medium.

Conditions for the Operation of a Laser

Consider a material in which we focus on two energy levels with energies E_1 and E_0, with $E_1 > E_0$. The rate of change of $n(v)$, the number of photons with frequency v such that $hv = E_1 - E_0$, can be computed using the rate of increase because of stimulated and spontaneous emission by the N_1 atoms in the state E_1, the rate of decrease because of stimulated absorption by the N_0 atoms in the state E_0, and the rate of loss of photons due to leakage from the cavity, which is proportional to $n(v)$. The equation is

$$\frac{dn(v)}{dt} = N_1(u(v)B_{10} + A_{10}) - N_0 u(v)B_{01} - \frac{n(v)}{\tau_0} \tag{22-18}$$

τ_0 has the dimensions of time, and the cavity must be designed such that τ_0 is large compared with the traversal time of the photons in the cavity. There is a relation between the photon energy density and the number of photons. The formula

$$u(v) = \frac{8\pi v^2}{c^3} hvn(v) \tag{22-19}$$

which holds for blackbody radiation, is more general, since it describes the density as the product of the number of modes per unit frequency interval, the energy at the frequency, and the number of photons with that energy. Thus

$$\frac{dn(v)}{dt} = n(v)\left[\left(N_1 - \frac{g_1}{g_0}N_0\right)A_{10} - \frac{1}{\tau_0}\right] + N_1 A_{10} \tag{22-20}$$

Thus the number of photons will decrease with time, unless

$$N_1 - \frac{g_1}{g_0}N_0 > \frac{1}{A_{10}\tau} \tag{22-21}$$

Since in thermal equilibrium

$$N_1 - \frac{g_1}{g_0}N_0 = N_1\left(1 - \frac{N_0/g_0}{N_1/g_1}\right) = N_1(1 - e^{-(E_0-E_1)/kT})$$
$$= N_1(1 - e^{hv/kT}) < 0 \tag{22-22}$$

we see that a laser must operate in a nonequilibrium mode. Equation (22-21) shows that we must create an excess population of atoms in the level E_1, that is, we must set up a *population inversion*. One way of doing this is described below.

Optical Pumping

A way of setting up a population inversion is to use a material in which the process involves transitions between three levels. Figure 22-2 shows a three level system.

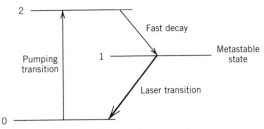

Figure 22-2. Schematic picture of pumping transition, followed by fast decay to metastable state, from which laser transition occurs.

The equations describe the rate of change of N_2, N_1, N_0, in the presence of a light beam of energy density u_p with frequency such that it *pumps* atoms from the ground state "0" to the higher of the two excited states, "2." The contributions to the change in N_2 are: spontaneous and stimulated decay to the states "1" and "0" which decrease the occupation number of level "2" and stimulated excitation from the states "1" and "0," which increases N_2:

$$\frac{dN_2}{dt} = -N_2 A_{21} - N_2 A_{20} - N_2 B_{21} u(\nu_{12}) - N_2 B_{20} u_p(\nu_{02})$$
$$+ N_1 B_{12} u(\nu_{12}) + N_0 B_{02} u_p(\nu_{02}) \tag{22-23}$$

Similarly

$$\frac{dN_1}{dt} = -N_1 A_{10} - N_1 B_{12} u(\nu_{12}) - N_1 u(\nu_{01}) B_{10} + N_2 A_{21} + N_2 B_{21} u(\nu_{12}) \tag{22-24}$$

In the steady state both time derivatives vanish.

Suppose that the material is such that $R_{21} \gg R_{10}$. In that case there will be no buildup of atoms in the state "2" and there will be no buildup of the density of radiation at ν_{12}. The steady-state equations, with $u(\nu_{12}) = 0$, can easily be shown to yield

$$\frac{N_1}{N_0} = \frac{R_{21}}{R_{10}} \frac{B_{02} u_p}{A_{21} + A_{20} + B_{02} u_p} \tag{22-25}$$

For a large pumping energy density u_p the second factor is of order unity, and $N_1 \gg N_0$, which describes a population inversion.

Physically, the atoms are pumped to the higher excited state "2" and decay rapidly to the *metastable* state "1" in which a population inversion is created, and laser transitions to the ground state take place. The three-level system is used in the ruby laser. There are, of course, many other types of laser, and the preceding discussion is only meant to show one way in which a population inversion can be created. Thus different mechanisms of pumping and different lasing materials allow for lasers that operate in different parts of the electromagnetic spectrum. It is possible to use materials in which the laser radiation terminates on a variety of final (closely spaced) vibrational levels, which allows one to construct *tunable* lasers.

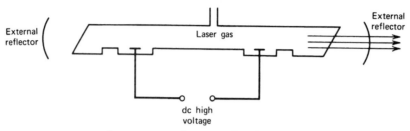

Figure 22-3. Schematic sketch of laser.

The Cavity

The production of a collimated, narrow beam requires a cylindrical structure for the cavity, with slightly transparent mirrors at the ends. The mirrors are there to keep the photons inside the cavity to build up the energy density $u(\nu_{10})$. Recall that the rate of change of photon number is

$$\frac{dn(\nu)}{dt} = N_1(u(\nu)B_{10} + A_{10}) - N_0 u(\nu)B_{01} - \frac{n(\nu)}{\tau_0}$$

with $N_1 \gg N_0$. We can estimate τ_0, the "lifetime of photons in the cavity" as follows: Assume that the radiation only travels back and forth in the cylindrical cavity of length L. If n^* is the refractive index of the medium, then the traversal time is n^*L/c. If the reflection coefficient of the mirror is r (≈ 0.99), then after k traversals, the intensity of the light is reduced to r^k, so that, with $r = 1 - \epsilon$

$$I_k/I_0 = (1 - \epsilon)^k \approx e^{-k\epsilon} \tag{22-26}$$

The intensity is reduced to $1/e$ of its original value after $k = 1/\epsilon = 1/(1 - r)$ traversals. Thus the lifetime of the radiation in the cavity is

$$\tau_0 \approx kn^*L/c = n^*L/c(1 - r) \tag{22-27}$$

The reflection coefficient can be brought closer to unity if the cavity is terminated with Brewster angle windows (which almost perfectly contain one state of polarization of the radiation), beyond which identical spherical mirrors, separated by a distance equal to their common radius of curvature are placed (Fig. 22-3). Note that the linewidth of the radiation is

$$\Delta\nu = 1/\tau_0 = c(1 - r)n^*L \tag{22-28}$$

which is very small compared with $c/2n^*L$, the frequency separation of two adjacent modes in a cavity of dimension L. Thus a nearly monochromatic beam is produced.

THE COOLING OF ATOMS

In this section we describe one application of lasers that is directly related to the study of atoms, namely their use to slow down atoms and in this way to reduce the *Doppler broadening* of spectral lines beyond the form described in (21-116).

Spectral lines can be broadened in several ways; atoms are generally not observed in isolation. In a gas of atoms, there will generally be collisions, and the time between collisions τ_c, if shorter than the mean life of the state under consideration, determines the width of the spectral line, because τ_c is effectively the lifetime of the state, and $\hbar/\tau_c$ is then larger than the natural width of the line. Collision broadening can be reduced by reducing the density (equivalently, the pressure) of the gas of atoms under consideration. There is also *Doppler broadening*. When an atom is moving with speed v, then the frequency of the radiation emitted by the atom is shifted by $\Delta\omega = (v/c)\omega$. If for v we take $v_{rms} = \sqrt{3kT/M_{atom}}$, the root-mean-square velocity of atoms in a gas at temperature T, then for hydrogen, for example, $\Delta\omega/\omega \approx 0.3 \times 10^{-6} \sqrt{T}$. This is much larger than the natural value of $\Delta\omega/\omega$, which for the $2p \to 1s$ line is $\approx 3 \times 10^{-8}$. Thus to overcome the Doppler broadening, the atoms need to be *cooled*. This is done by placing the atom in a laser beam. Consider an atom moving in the positive z direction with a velocity v. Let the energy difference between the two levels for which a transition occurs be $\hbar\omega_0$, and let the frequency of the (monochromatic) laser beam be ω. We shall take ω a little different from ω_0, and take the *detuning parameter* $\delta = \omega - \omega_0 < 0$. If the laser beam propagates along the z-axis in the positive z direction, then the atom "sees" a red-shifted beam, since the origin of the beam appears to be moving away. Thus the frequency as seen by the atom is $\omega(1 - v/c)$. Thus the amount of detuning of the laser beam away from the resonant frequency is $\omega(1 - v/c) - \omega_0 = \delta - \omega v/c$. Since $\delta < 0$, the magnitude of this is larger than the magnitude of δ, and the probability of absorption of the photons is decreased, since the frequency of the light absorbed is farther out on the resonance curve. On the other hand, with another laser beam propagating in the negative z direction, the detuning is $\omega(1 + v/c) - \omega_0 = \delta + \omega v/c$, so that the frequency of the light absorbed is closer to the peak of the resonance (with $\omega v/c = |\delta|$ one would be at resonance). Thus the probability of absorbing a photon is larger. This means that the atom experiences a net force in the negative z direction. The atom re-emits the photon, because it decays, but the emission has no preferred direction associated with it, and on the average, it is spherically symmetric.[2] Thus there is a net loss of momentum for the atom along the z direction. If there are three pairs of lasers, aligned along the x-, y-, and z-axes and pairwise of equal intensity, then all of the degrees of freedom are covered.

To make this more quantitative, we calculate the radiation pressure force exerted on the atom by the beam. We assume that there are only two levels that are relevant in the atom. This is a good approximation when the frequency of the oscillating electric field is close to the frequency associated with the energy of excitation of the state above the ground state. The rate at which momentum of magnitude $\hbar\omega/c$ is absorbed was calculated to be

$$
\begin{aligned}
R &= \frac{2\pi}{\hbar} |\langle 1|e\mathbf{r} \cdot \mathbf{E}|0\rangle|^2 \, \delta(E_1 - E_0 - \hbar\omega) \\
&= \frac{2\pi}{\hbar^2} |\langle 1|e\mathbf{r} \cdot \mathbf{E}|0\rangle|^2 \, \delta(\omega_0 - \omega)
\end{aligned}
\tag{22-29}
$$

[2]We assume that the intensity of the laser beam is not very large, so that the excitation and the subsequent decay are events that are well separated. Later in this chapter we shall see situations in which the system oscillates rapidly between the ground state and the excited state. In that case, the emission is coherent with the excitation, and directional information is not lost.

where the dipole moment operator er gives rise to the transition. We now take into account that the delta function needs to be modified to take into account the linewidth of the excited state $|1\rangle$, so that we make the replacement (for a justification of this, see the Special Topics section 4).

$$\pi\delta(\omega_0 - \omega) \rightarrow \frac{R/2}{(\omega_0 - \omega)^2 + R^2/4} \tag{22-30}$$

where R is the spontaneous decay rate. Thus the radiation pressure force is

$$F = \frac{\hbar\omega}{c} \frac{2}{\hbar^2} |\langle 1|er \cdot E|0\rangle|^2 \frac{R/2}{(\omega_0 - \omega)^2 + R^2/4} \tag{22-31}$$

We introduce the dimensionless quantity

$$I = \frac{|\langle 1|er \cdot E|0\rangle|^2}{(\hbar R/2)^2} \tag{22-32}$$

in terms of which

$$F = \frac{\hbar\omega}{c} IR \frac{R^2/4}{(\omega_0 - \omega)^2 + R^2/4} \tag{22-33}$$

For a weak field, that is, for I small, the force is very small. For a very large field, that is for large I, the induced decays overwhelm the spontaneous ones, and under those circumstances the atom oscillates rapidly between the ground state and the excited state.[3] In particular, the emitted photons are coherent with the absorbed photons, so that the atom does not absorb any net momentum. It turns out that the optimum laser intensity is such that $I \approx 1$.

For an atom moving in the same direction as the beam, there is the additional Doppler shift detuning, so that

$$F = \frac{\hbar\omega}{c} IR \left[\frac{R^2/4}{(\omega - \omega_0 - \omega v/c)^2 + R^2/4} \right] \tag{22-34}$$

If now a standing wave beam is considered, or, equivalently we add a laser beam with the same frequency ω traveling in the negative z direction, we get a force in the opposite direction, with the frequency blue-shifted to $\omega(1 + v/c)$. Thus the net force on the beam is

$$F_{net} = \frac{\hbar\omega IR}{c} \left[\frac{R^2/4}{(\omega - \omega_0 - \omega v/c)^2 + R^2/4} - \frac{R^2/4}{(\omega - \omega_0 + \omega v/c)^2 + R^2/4} \right]$$

To the lowest order in v/c, one gets

$$\begin{aligned} F_{net} &= \hbar\omega IR/c^2 \frac{R^2/4}{(\omega - \omega_0)^2 + R^2/4} \frac{4\omega(\omega - \omega_0)}{(\omega - \omega_0)^2 + R^2/4} v \\ &= \frac{\hbar\omega IR}{c^2} \frac{R^2/4}{\delta^2 + R^2/4} \frac{4\omega\delta}{\delta^2 + R^2/4} v \end{aligned} \tag{22-35}$$

[3]See the next section.

Since $\delta < 0$, that is, the laser frequency is chosen to be a little below the resonance peak ω_0, the force is in a direction opposite to that of v, that is, it is a *frictional force*, of the form $F_{net} = -\beta v$. The force depends on the amount of detuning, and it is largest when $dF/d\delta = 0$, that is, when $|\delta| = \omega_0 - \omega = R/2\sqrt{3}$. The atomic motion in all three dimensions must be slowed down, and for this purpose three pairs of lasers are used, to create an environment commonly called *optical molasses*.

The maximum frictional force for $I \approx 1$ is therefore

$$F \approx -\sqrt{\frac{27}{4}} \frac{\hbar \omega_0^2}{c^2} v \qquad (22\text{-}36)$$

The atoms are also subject to a random force due to the random encounters with the photons that make up the laser beam. They thus behave like particles in a fluid undergoing Brownian motion. The details of the cooling process are beyond the scope of our discussion, but the prediction of the semiclassical argument just presented is that the atoms will be cooled down to a temperature given by

$$T = \frac{\hbar \omega}{kc}$$

where k is Boltzmann's constant. This is generally in the range of 2×10^{-4} K. A deeper examination of the details of the process, which takes into account the degeneracy of the excited levels and the polarization of the photons, shows that one should expect to be able to cool atoms to a temperature of order 4×10^{-5} K, which agrees with what has been found experimentally.[4] In fact, recent improvements in the technology of laser cooling of atoms have yielded atomic temperatures of 7×10^{-7} K. When atoms are cooled to such low temperatures, measurements of the natural linewidth become possible, and this has had a large impact on the testing of quantum electrodynamics predictions.

TWO-LEVEL ATOM IN A MONOCHROMATIC ELECTRIC FIELD

Let us denote the two states of the atom under consideration as eigenstates of a Hamiltonian H_0, such that

$$H_0|\phi_1\rangle = E_1|\phi_1\rangle$$
$$H_0|\phi_0\rangle = E_0|\phi_0\rangle \qquad (22\text{-}37)$$

We choose $E_1 > E_0$ and we shall find it useful to use the notation

$$\omega_d = (E_1 - E_0)\hbar \qquad (22\text{-}38)$$

and make the replacement $|\phi_1\rangle = |1\rangle$, $|\phi_0\rangle = |0\rangle$. Let us now consider the two-level system in the presence of an electric field. We shall assume that the electric field is strong enough that we cannot limit ourselves to first and second order effects

---

[4]This is discussed in nontechnical detail by C. N. Cohen-Tannoudji and W. D. Phillips in *Phys. Today*, 43(10), 33 (1990). This article also contains many references.

in **E**, as was done in the discussion of the Stark effect in Chapter 16. We will thus not use perturbation theory. The Hamiltonian now takes the form

$$H = H_0 + V(t) \tag{22-39}$$

where, as before

$$V(t) = \frac{e}{mc} \mathbf{A}(\mathbf{r}, t) \cdot \mathbf{p} \tag{22-40}$$

Here **p** is the momentum operator for the electron, and thus

$$\mathbf{p} = \frac{im}{\hbar} [H_0, \mathbf{r}] \tag{22-41}$$

We will also make the dipole approximation, so that

$$\mathbf{A}(\mathbf{r}, t) = \mathbf{A}_0 \, e^{-i\omega t} + \mathbf{A}_0^* e^{i\omega t} \tag{22-42}$$

that is, the vector potential is constant over the dimensions of the atom. We write this in terms of the electric field. It follows from

$$\mathbf{E} = -\frac{1}{c} \frac{\partial \mathbf{A}}{\partial t} \equiv \mathbf{E}_0 e^{-i\omega t} + \mathbf{E}_0^* e^{i\omega t}$$

that $\mathbf{E}_0 = (i\omega/c)\mathbf{A}_0$, $\mathbf{E}_0^* = (-i\omega/c)\mathbf{A}_0^*$, so that

$$V(t) = \frac{e}{\hbar\omega} (\mathbf{E}_0 e^{-i\omega t} - \mathbf{E}_0^* e^{i\omega t}) \cdot [H_0, \mathbf{r}] \tag{22-43}$$

Thus any matrix element of the operator $V(t)$ will take the form

$$\langle 0|V(t)|1\rangle = \frac{e}{\hbar\omega} (E_0 - E_1)\{\langle 0|\mathbf{E}_0 \cdot \mathbf{r}|1\rangle e^{-i\omega t} - \langle 0|\mathbf{E}_0^* \cdot \mathbf{r}|1\rangle e^{i\omega t}\} \tag{22-44}$$

Consider now a solution of the time-dependent Schrodinger equation $|\psi(t)\rangle$. We again use the expansion theorem, as in Chapter 21, to write

$$|\psi(t)\rangle = C_0(t)e^{-iE_0 t/\hbar}|0\rangle + C_1(t)e^{-iE_1 t/\hbar}|1\rangle \tag{22-45}$$

Thus

$$i\hbar \frac{d}{dt} |\psi(t)\rangle = (H_0 + V(t))|\psi(t)\rangle$$

with the help of (22-45) leads to

$$i\hbar \frac{d}{dt} C_0(t)e^{-iE_0 t/\hbar}|0\rangle + i\hbar \frac{d}{dt} C_1(t)e^{-iE_1 t/\hbar}|1\rangle$$
$$= V(t)C_0(t)e^{-iE_0 t/\hbar}|0\rangle + V(t)C_1(t)e^{-iE_1 t/\hbar}|1\rangle$$

If we take matrix elements of this equation, by multiplying by $\langle 0|$ and $\langle 1|$ respectively, and if we take into account that

$$\langle 0|\mathbf{r}|0\rangle = \langle 1|\mathbf{r}|1\rangle = 0 \tag{22-46}$$

we get

$$i\hbar \frac{d}{dt} C_0(t) = C_1(t) \, e^{-i\omega_d t} \, \langle 0|V(t)|1\rangle \tag{22-47}$$

and

$$i\hbar \frac{d}{dt} C_1(t) = C_0(t) \, e^{i\omega_d t} \, \langle 1|V(t)|0\rangle \tag{22-48}$$

In more detail the first of these equations reads

$$i\hbar \frac{d}{dt} C_0(t) = \frac{\omega_d}{\omega} C_1(t)\{-\langle 0|e\mathbf{E}_0 \cdot \mathbf{r}|1\rangle \, e^{-i(\omega + \omega_d)t} + \langle 0|e\mathbf{E}_0^* \cdot \mathbf{r}|1\rangle \, e^{-i(\omega_d - \omega)t}\}$$

We shall be interested in physical situations in which ω is chosen to be close to or equal to ω_d. The term involving $e^{-i(\omega_d + \omega)t}$ will oscillate very rapidly, and contribute nothing when averaged over time. As in our discussion of paramagnetic resonance in Chapter 14, we drop such terms. This is often called the *rotating wave approximation*. When this is done, we are left with

$$i\hbar \frac{d}{dt} C_0(t) = \frac{\omega_d}{\omega} C_1(t) \, \langle 0|e\mathbf{E}_0^* \cdot \mathbf{r}|1\rangle \, e^{-i(\omega_d - \omega)t}$$
$$\equiv \hbar\gamma C_1(t) e^{-i(\omega_d - \omega)t} \tag{22-49}$$

In terms of the *detuning* parameter

$$\delta = \omega - \omega_d \tag{22-50}$$

this reads

$$i\hbar \frac{d}{dt} C_0(t) = \hbar\gamma C_1 \, e^{i\delta t} \tag{22-51}$$

In the same approximation

$$i\hbar \frac{d}{dt} C_1(t) = \frac{\omega_d}{\omega} C_0(t) \, \langle 1|e\mathbf{E}_0 \cdot \mathbf{r})|0\rangle \, e^{i(\omega_d - \omega)t}$$
$$= \hbar\gamma C_0(t) \, e^{-i\delta t} \tag{22-52}$$

where

$$\gamma = \frac{\omega_d}{\hbar\omega} \langle 0|e\mathbf{E}_0^* \cdot \mathbf{r}|1\rangle \tag{22-53}$$

can be chosen to be real. Differentiating (22-49) with respect to time yields

$$\frac{d^2}{dt^2} C_0(t) = \gamma \delta C_1 e^{i\delta t} - i\gamma \frac{d}{dt} C_1(t) e^{i\delta t}$$

$$= i\delta \frac{d}{dt} C_0(t) - \gamma^2 C_0(t) \qquad (22\text{-}54)$$

A trial solution of the form

$$C_i(t) = e^{-i\Omega t} \qquad (22\text{-}55)$$

leads to

$$\Omega^2 + \delta\Omega - \gamma^2 = 0 \qquad (22\text{-}56)$$

that is,

$$\Omega = \Omega_{\pm} = -\frac{1}{2} \delta \pm \sqrt{\frac{1}{4} \delta^2 + \gamma^2} \qquad (22\text{-}57)$$

Thus a general solution is

$$C_0(t) = e^{i\delta t/2}(A \cos \sqrt{\delta^2/4 + \gamma^2}t + B \sin\sqrt{\delta^2/4 + \gamma^2}t) \qquad (22\text{-}58)$$

and

$$C_1(t) = \frac{1}{\gamma} e^{-i\delta t} \frac{d}{dt} C_0(t)$$

$$= -e^{-i\delta t/2} \left\{ \frac{\delta}{2\gamma} (A \cos \sqrt{\delta^2/4 + \gamma^2}\, t + B \sin \sqrt{\delta^2/4 + \gamma^2}\, t) \right. \qquad (22\text{-}59)$$

$$\left. - i \frac{\sqrt{\delta^2/4 + \gamma^2}}{\gamma} (A \sin \sqrt{\delta^2/4 + \gamma^2}\, t - B \cos \sqrt{\delta^2/4 + \gamma^2}\, t) \right\}$$

If the system is in the $|0\rangle$ state at time $t = 0$, $C_0(0) = 1$ and $C_1(0) = 0$ imply that $A = 1$ and $(\delta/2)A + i\sqrt{\delta^2/4 + \gamma^2} B = 0$, i.e., $B = -i\delta/2\sqrt{\delta^2/4 + \gamma^2}$. Thus at a later time

$$|C_0(t)|^2 = \cos^2 \sqrt{\delta^2/4 + \gamma^2}\, t + \frac{\delta^2}{\delta^2 + 4\gamma^2} \sin^2\sqrt{\delta^2/4 + \gamma^2}\, t$$

$$= 1 - \frac{4\gamma^2}{\delta^2 + 4\gamma^2} \sin^2 \sqrt{\delta^2/4 + \gamma^2}\, t \qquad (22\text{-}60)$$

In the case of perfect tuning, $\delta = 0$, we get from the above

$$|C_0(t)|^2 = \cos^2 \gamma t \qquad (22\text{-}61)$$

The system continues to oscillate between the two states with frequency γ, and, on the average, it spends half the time in the upper state, and half the time in the lower state. The frequency at $\delta = 0$, that is, when $\omega = \omega_d$, is

$$\gamma = \frac{1}{\hbar} \langle 0|e\mathbf{E}_0^* \cdot \mathbf{r}|1\rangle \tag{22-62}$$

and is called the *Rabi frequency*. We note that $\mathbf{E}_0^*$ is proportional to $\mathbf{A}_0^*$, and thus when the lower state $|0\rangle$ also has n photons of frequency ω present, then, according to (22-2), $\mathbf{E}_0^*$ is proportional to $\sqrt{n+1}$. We may thus write

$$\gamma = \sqrt{n+1}\, \gamma_0 \tag{22-63}$$

It is fairly straightforward to show that if the initial state is $|1\rangle$, and it has n photons of frequency ω present, then the frequency has the form $\gamma = \sqrt{n}\, \gamma_0$.

The study of single atoms in a cavity in which there is only one mode of an oscillating electric field present has been made possible through the invention and construction of radio-frequency traps by H. Dehmelt and collaborators. The oscillations predicted by equation (22-61) have been experimentally established.

Observation of Quantum Jumps

The invention of traps in which single ions could be studied allows for a variety of new techniques to study atoms. One ingenious idea was first proposed by H. Dehmelt and observed by several experimental groups in the past decade. The principle of the experiment is the following: Consider a three-level system, consisting of the ground state $|0\rangle$ and two excited states $|1\rangle$ and $|2\rangle$. Transitions between the states $|0\rangle$ and $|1\rangle$ are allowed, while those between $|2\rangle$ and the ground state are forbidden (but not absolutely forbidden), so that the the state $|2\rangle$ is metastable. A strong light source (laser) tuned to the angular frequency $\omega_{10} = (E_1 - E_0)/\hbar$ is now directed at the atom, as is a weak light source tuned to the frequency $\omega_{20} = (E_2 - E_0)/\hbar$. The number of transitions between the ground state and the allowed excited state is very large. Effectively the strong laser field excites the electron to the state $|1\rangle$ at a very rapid rate, and the electron decays to the ground state also at a very rapid rate. Thus a continuous signal of light emitted by the atom is observed. This is just a manifestation of the Rabi oscillations discussed in the last section. Very occasionally the electron is excited by the weak laser to the state $|2\rangle$. Since the electron now finds itself in a metastable state, the time until it decays to the ground state may be as long as seconds. *During that time there is no fluorescence*—that is—the atom is dark. When the electron finally does decay to the ground state, it is immediately excited by the strong laser field to the allowed excited state, and rapidly decays, giving rise to a resumption of the fluorescent radiation. Effectively the fluorescence monitors the quantum jumps between the ground state and the metastable state (Fig. 22-4).

A simple analysis of the process proceeds as follows: Let A_{10}, B_{10} and A_{20}, B_{20} be the Einstein coefficients for the transitions connecting $(0, 1)$ and $(0, 2)$ respectively. The conditions on the transitions imply that $A_{10} \gg A_{20}$. If the energy densities of the laser beams are U_1 and U_2, respectively, then the rate equation for the probability that the atom is in the excited state $|1\rangle$ is, with the assumption of nondegeneracy (so the g_i are 1)

$$\frac{dP_1}{dt} = -P_1(A_{10} + B_{10}U_1) + B_{10}U_1P_0 \tag{22-64}$$

This describes the loss of probability due to spontaneous and stimulated emission, and its buildup due to stimulated absorption from the ground state, the latter

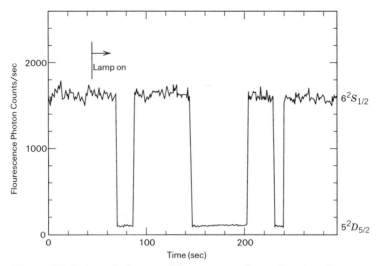

Figure 22-4. A typical trace of fluorescence from showing the quantum jumps. The atom is definitely known to be in the shelf level during the low fluorescence periods. (From W. Nagourney, J. Sandberg, and H. Dehmelt, *Phys. Rev. Lett.* **56**, 2797 (1986), by permission.)

proportional to the probability P_0 that the electron is in the ground state. Similarly the rate equation for the probability that the atom is in the metastable state $|2\rangle$ is

$$\frac{dP_2}{dt} = -(A_{20} + B_{20}U_2)P_2 + B_{20}U_2P_0 \tag{22-65}$$

The probabilities add up to 1, so that $P_0 + P_1 + P_2 = 1$. If the laser coupling the ground state to the excited state $|1\rangle$ is intense, then $U_1 \to \infty$ and $P_0 = P_1$. In terms of the probability P_+ that the metastable state is excited ($P_+ = P_2$) and the probability that it is *not* excited $P_- = 1 - P_2 = P_0 + P_1$, these equations lead to

$$\frac{dP_+}{dt} = -R_-P_+ + R_+P_- \tag{22-66}$$

where

$$R_+ = \frac{1}{2} B_{20}U_2$$
$$R_- = A_{20} + B_{20}U_2 \tag{22-67}$$

The equation

$$\frac{dP_-}{dt} = R_-P_+ - R_+P_- \tag{22-68}$$

follows automatically from the fact that $P_+ + P_- = 1$. We can view these equations as representing a two-level system in which the upward transition rate is R_+ and the downward transition rate is R_-.

Experimentally interesting quantities are the probabilities that in the time interval from t to $t + T$ no transitions have occurred, and that at the end of the interval the electron ends either in the excited state, P_{0+} or in the ground state P_{0-}. A little reflection shows that once the experiment starts and the laser beam intensities are independent of time, these probabilities only depend on the length of the time interval T. We can write rate equations for these probabilities, and these are

$$\frac{dP_{0+}}{dT} = -R_- P_{0+} \qquad (22\text{-}69)$$

$$\frac{dP_{0-}}{dT} = -R_+ P_{0-} \qquad (22\text{-}70)$$

The "initial conditions" for these equations require that we know $P_{0\pm}(T = 0)$. Suppose that the intensity of fluorescence turns off at some time t. We choose that time to denote $T = 0$, so that $P_{0+}(T = 0) = 1$. The solution to the equation with this initial condition is

$$P_{0+}(T) = e^{-R_- T} \qquad (22\text{-}71)$$

This is the probability that the signal is still "off" after a time T. Similarly one can show that if the fluorescence has just turned on, then the probability that it is still on after a time interval T is

$$P_{0-}(T) = e^{-R_+ T} \qquad (22\text{-}72)$$

A statistical analysis of the distribution of on and off times as shown in Fig. 22-5 can be used to measure A_2. For very long-lived states, direct measurements of A_2 are very difficult, since the rate at which photons are emitted is very small, and they can be emitted in all directions, so counting them is a very slow process.

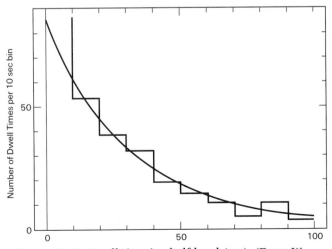

Figure 22-5. Dwell time in shelf level (sec). (From W. Nagourney, J. Sandberg, and H. Dehmelt, *Phys. Rev. Lett.* **56,** 2797 (1986), by permission.)

It is interesting to note that quantum mechanics usually involves the study of a large ensemble of identical systems. In this case, one is studying a single atom and the ensemble is replaced by rebuilding over and over again that single member of the ensemble with identical initial conditions. This is generally not possible when radiation is emitted into a continuum of states, but it is possible in the special case of a single mode electric field.

THE MÖSSBAUER EFFECT

An atom (or any other quantum system) can act as a very accurate clock, since its transitions are signaled by radiation of a very well-defined frequency. If the only limitation were the natural linewidth, great accuracy could be obtained.

Unfortunately, as stated in our discussion of the cooling of atoms, the motion of atoms leads to the broadening of lines by the Doppler effect. One might think that the use of a liquid or solid source would eliminate this effect, but then broadening caused by the effect of neighboring atoms is just as harmful. Let us consider nuclear transitions. A nucleus such as $_{77}\text{Ir}^{191}$ emits a γ-ray of energy of the order of 100 keV, with a lifetime of 10^{-10} s. This corresponds to

$$\frac{\Delta\omega}{\omega} = \frac{\Delta E}{E} = \frac{\hbar/\tau}{E} \cong \frac{10^{-27}/10^{-10}}{10^5 \times 1.6 \times 10^{-12}} \simeq 0.6 \times 10^{-10} \qquad (22\text{-}73)$$

There will, unfortunately, be a recoil shift of the line. The γ-ray carries off momentum $\hbar\omega/c$, and the nucleus, to conserve momentum, must recoil with the same momentum. This gives rise to a recoil energy

$$\Delta E = \frac{P_{\text{recoil}}^2}{2M} = \frac{1}{2M}\left(\frac{\hbar\omega}{c}\right)^2 \qquad (22\text{-}74)$$

and thus a decrease in the energy radiated. The fractional change in frequency is

$$\frac{\Delta E}{\hbar\omega} \simeq \frac{\hbar\omega}{2Mc^2} \simeq \frac{10^{-1}(\text{MeV})}{2 \times 940 \times 191(\text{MeV})} \simeq 3 \times 10^{-7} \qquad (22\text{-}75)$$

The observation of radiation of this energy cannot be carried out with the conventional, extremely accurate spectroscopic methods, but must utilize a detector that is extremely "well tuned" to the radiation. This is best done by using the same material (e.g., $_{77}\text{Ir}^{191}$) as an absorber. The absorption will be very much enhanced at the "resonant" frequency at which the radiation is emitted, but here, too, there will be a recoil shift. The overall shift is thus $\Delta\omega/\omega \simeq 6 \times 10^{-7}$. Thus, the "fine tuning" does not work, since the line is shifted by far more than the width, which is of the order of $10^{-10}\,\omega$. One could try to compensate for the recoil by moving the emitter with the recoil velocity. This is given by

$$\frac{v}{c} = \frac{P_{\text{recoil}}}{Mc} = \frac{\hbar\omega/c}{Mc} = 2\,\frac{\hbar\omega}{2Mc^2} \simeq 6 \times 10^{-7} \qquad (22\text{-}76)$$

that is, $v = 1.7 \times 10^4$ cm/sec. This presents technical difficulties, but it has been achieved with an ultracentrifuge.

A major breakthrough came with the discovery by Mössbauer in 1958 that under certain conditions there is a high probability of *recoilless emission*. The emission is not recoilless, of course, but the recoil is not taken up by the nucleus, but instead by a large part of the crystal that the nucleus is imbedded in. Since the mass of the nucleus is 10^{22} times smaller than that of the crystal, the recoil energy is completely negligible. To get some intuition about what is happening, let us consider the nucleus as moving in a harmonic oscillator well, with characteristic frequency ω_0. The energy levels of the oscillator are

$$E_n = \hbar\omega_0 \left(n_x + n_y + n_z + \frac{3}{2} \right) \tag{22-77}$$

The harmonic well is just an approximate description of the crystalline forces that are responsible for the properties of the lattice. If the forces that tie the nucleus to its neighbors are strong—if the "springs" are stiff—then ω_0 is large; if the "springs" are soft, then ω_0 is small. In terms of level spacing, a "stiff spring" has widely separated levels, that is, a low density of states, whereas a "soft spring" has a high density of states. Let us now consider the matrix element for a transition from a nuclear state described by $\Psi_i (\mathbf{r}_1, \mathbf{r}_2, \ldots \mathbf{r}_N)$ to a nuclear state described by $\Psi_f(\mathbf{r}_1, \mathbf{r}_2, \ldots, \mathbf{r}_N)$, and we take the interaction to be

$$- \frac{e}{Mc} \sum_{\text{protons}} \mathbf{p}_k \cdot \mathbf{A}_k(\mathbf{r}_k, t) \tag{22-78}$$

The matrix element then is proportional to

$$- \frac{e}{Mc} \int \cdots \int d^3\mathbf{r}_1, \ldots, d^3\mathbf{r}_N \Psi_f^*(\mathbf{r}_1, \ldots, \mathbf{r}_N) \sum_k \boldsymbol{\varepsilon} \cdot \mathbf{p}_k \, e^{-i\mathbf{k}\cdot\mathbf{r}_k} \, \Psi_i(\mathbf{r}_1, \ldots, \mathbf{r}_N) \tag{22-79}$$

If we introduce the center-of-mass coordinate $\mathbf{R} = (1/N) \sum_i \mathbf{r}_i$ then (a) the interaction term takes the form

$$- \frac{e}{Mc} e^{-i\mathbf{k}\cdot\mathbf{R}} \sum_{\text{protons}} \boldsymbol{\varepsilon} \cdot \mathbf{p}_k \, e^{-i\mathbf{k}\cdot\boldsymbol{\rho}_k} \tag{22-80}$$

where $\boldsymbol{\rho}_i = \mathbf{r}_i - \mathbf{R}$, and (b) the nuclear wave function decomposes into a product describing the internal motion and the motion of the nuclear center of mass in the harmonic potential

$$\Psi(\mathbf{r}_1, \ldots, \mathbf{r}_N) = \psi_{n_x n_y n_z}(\mathbf{R}) \, \phi(\boldsymbol{\rho}_1, \ldots, \boldsymbol{\rho}_{N-1}) \tag{22-81}$$

Thus the matrix element (22-79) becomes

$$- \frac{e}{Mc} \int d^3\mathbf{R} \psi_{nf}^*(\mathbf{R}) \, e^{-i\mathbf{k}\cdot\mathbf{R}} \psi_{ni}(\mathbf{R})$$

$$\times \int d^3\boldsymbol{\rho}_1, \ldots, d^3\boldsymbol{\rho}_{N-1} \phi_f^*(\boldsymbol{\rho}_1, \ldots, \boldsymbol{\rho}_{N-1}) \sum_{\text{protons}} \boldsymbol{\varepsilon} \cdot \mathbf{p}_k \, e^{-i\mathbf{k}\cdot\boldsymbol{\rho}_k} \, \phi_i(\boldsymbol{\rho}_1, \ldots, \boldsymbol{\rho}_{N-1}) \tag{22-82}$$

We may write this in the form

$$M = M_{\text{internal}} \int d^3R \psi_{nf}^*(\mathbf{R}) \, e^{-i\mathbf{k}\cdot\mathbf{R}} \psi_0(\mathbf{R}) \tag{22-83}$$

where we have set $n_i = 0$, since the initial state is in the ground state of the lattice. The probability that the radiative transition leaves the nucleus in the lattice ground state is

$$P_0(k) = \frac{|M_{\text{int}}|^2 \left| \int d^3R \psi_0^*(\mathbf{R}) \, e^{-i\mathbf{k}\cdot\mathbf{R}} \psi_0(\mathbf{R}) \right|^2}{|M_{\text{int}}|^2 \sum_{nf} \left| \int d^3R \psi_{nf}^*(\mathbf{R}) \, e^{-i\mathbf{k}\cdot\mathbf{R}} \psi_0(\mathbf{R}) \right|^2}$$

$$= \left| \int d^3R \psi_0^*(\mathbf{R}) \, e^{-i\mathbf{k}\cdot\mathbf{R}} \psi_0(\mathbf{R}) \right|^2 \tag{22-84}$$

In the last step we replaced the sum in the denominator by unity, using completeness.[5] To calculate this, we use the normalized ground-state wave function of the oscillator. We found in Chapter 7 that the one-dimensional ground-state wave function is

$$\psi_0(x) = \left(\frac{m\omega_0}{\pi\hbar} \right)^{1/4} e^{-m\omega_0 x^2/2\hbar}$$

Hence, for three dimensions we have

$$\psi_0(\mathbf{R}) = \psi_0(x) \, \psi_0(y) \, \psi_0(z) = \left(\frac{m\omega_0}{\pi\hbar} \right)^{3/4} e^{-m\omega_0 \mathbf{R}^2/2\hbar} \tag{22-85}$$

We thus calculate

$$\left| \left(\frac{M_N\omega_0}{\pi\hbar} \right)^{3/2} \int d^3R \, e^{-M_N\omega_0 \mathbf{R}^2/\hbar} \, e^{-i\mathbf{k}\cdot\mathbf{R}} \right|^2$$

where M_N is the mass of the nucleus. We get

$$P_0 = \left(\frac{M_N\omega_0}{\pi\hbar} \right)^3 \left| \int d^3R \, e^{-(M_N\omega_0/\hbar[\mathbf{R}+i\mathbf{k}(\hbar/2M_N\omega_0)]^2} \, e^{-k^2\hbar/4M_N\omega_0} \right|^2$$

$$= e^{-\hbar^2 k^2/2M_N\hbar\omega_0} \tag{22-86}$$

$$= e^{-\text{recoil energy/level spacing}}$$

[5]The formal proof is quickest. We have

$$\sum_{nf} |\langle n_f | e^{i\mathbf{k}\cdot\mathbf{R}} | 0 \rangle|^2 = \sum_{nf} \langle 0 | e^{-i\mathbf{k}\cdot\mathbf{R}} | n_f \rangle \langle n_f | e^{i\mathbf{k}\cdot\mathbf{R}} | 0 \rangle$$

Using

$$1 = \sum |n_f\rangle\langle n_f|$$

one gets

$$\langle 0 | e^{-i\mathbf{k}\cdot\mathbf{R}} \, e^{i\mathbf{k}\cdot\mathbf{R}} | 0 \rangle = 1$$

since $P_{\text{recoil}} = \hbar k$ and $\hbar\omega_0$ is the level spacing in the lattice. Thus, if the level spacing is large, that is, we have a stiff spring, recoilless emission becomes more probable. The model of the lattice that was used here, that of each nucleus moving in its own harmonic potential, is the Einstein model of a lattice, and the frequency ω_0 is the so-called *Debye frequency*, so that we should really replace ω_0 by ω_D, which is related to the Debye temperature T_D by

$$\hbar\omega_D = kT_D \tag{22-87}$$

A more accurate treatment of the lattice using the Debye model for its description merely changes the exponent by a factor of 3/2.

It is not quite correct to say that the whole crystal recoils; instead, in a time τ equal to the lifetime of the transition (1.4×10^{-7} sec for Fe^{57}), only a region of the crystal of magnitude

$$L = v_s\tau$$

where v_s is the velocity of propagation of a lattice disturbance, (i.e., the velocity of sound) absorbs the recoil. Now a reasonable estimate of v_s is given by

$$v_s \simeq \frac{a\omega_D}{2\pi}$$

where a is the lattice spacing. Thus

$$\frac{L}{a} \simeq \frac{\omega_D\tau}{2\pi}$$

and with $\omega_D \simeq 10^{13}$ sec^{-1}, the number of nuclei absorbing the recoil, $\sim(L/a)^3$ is still enormous.

The preceding estimates, combined with the uncertainty relation, may be used to show that it is not possible to determine whether it is a single nucleus that "really" recoils. To measure the recoil energy $\hbar^2k^2/2M_N$ takes a time of the order of

$$\Delta t \gg \frac{\hbar}{(\hbar^2k^2/2M_N)}$$

The condition for the Mössbauer effect to occur is that

$$\frac{\hbar^2k^2}{2M_N} < \hbar\omega_D$$

Hence

$$\Delta t \gg \frac{1}{\omega_D}$$

During that time the disturbance will have traveled a distance

$$d \simeq v_s \, \Delta t \sim \frac{a \omega_D}{2\pi} \, \Delta t \gg \frac{a}{2\pi}$$

that is, over a distance covering many nuclei.

The question arises of how did we manage to get away from the problem of recoil and momentum conservation by talking about the energy states of the nucleus in the crystal lattice? Where does it say that the crystal absorbed the momentum? The quantum mechanical answer is that, if we want to talk about momentum, we should work in a momentum representation. This, however, is complicated, since it is difficult to describe the crystal forces in that representation. What one must do is to decompose the crystal motion (the crystal is just a lot of oscillators with nearest neighbor "springs") into normal modes and quantize these. The quanta of the lattice motion, analogs of photons, are the *phonons*. Recoilless emission then means a transition in which phonons are not emitted. The resulting formula is very similar to (22-86). Under these circumstances, the recoil broadening is infinitesimal compared to the natural linewidth. There is still Doppler broadening because of the thermal motion, but this can be handled by cooling the emitter and absorber.

Recoilless emitters provide us with a superb clock, and research utilizing the Mössbauer effect has been done in many fields, such as solid-state physics and chemistry. We will mention just one application, the terrestrial measurement of the gravitational red shift. It follows from the Equivalence Principle that a photon will have its frequency shifted by

$$\frac{\Delta \omega}{\omega} = \frac{gx}{c^2} \tag{22-88}$$

if it falls through a height x. This can be compensated by a recoil of velocity v, where

$$v^2 = 2gx \tag{22-89}$$

(If the photon and the absorber were to fall freely together, there would be resonant absorption.) If the absorber or the source is allowed to oscillate rapidly—one uses a transducer—and the absorption curve is correlated with the oscillations, it is possible to check the gravitational shift. Since the velocity, for a separation $x = 20$ m, is of the order of ~ 20 m/sec, the experiment is feasible, and was carried out by several groups. Within the errors, the effect is confirmed. For example, for Fe^{57} the predicted shift is $\Delta \omega / \omega = 4.92 \times 10^{-15}$, and the experimental shift found by Pound and Rebka is $(5.13 \pm 0.51) \times 10^{-15}$. A similar experiment in which the energy shift of the γ-ray emitted by Fe^{57} accelerated on a rapidly rotating turntable was measured again yielded results in agreement with the Equivalence Principle.

References

A good discussion of the quantum theory of light, with applications to lasers, can be found in R. Loudon, *The Quantum Theory of Light* (2nd edition), Clarendon Press, Oxford, 1986. The Mössbauer effect is discussed in H. Lipkin *Quantum Mechanics–New Approaches to Selected Topics*, North-Holland, Amsterdam, 1973.

$C\ h\ a\ p\ t\ e\ r$ **23**

COLLISION THEORY

Atomic and molecular structure was largely explored through spectroscopy. When it comes to trying to understand nuclear forces and the laws that govern the interactions of elementary particles, the only technique available is that of scattering a variety of particles by a variety of targets. In some sense, spectroscopy is also a form of "scattering." The atom in the ground state is excited by some projectile (it may be electrons in a discharge tube or collisions with other target particles, as in heating up of the gas), and then an outgoing photon is observed, with the atom going into the ground state again, or possibly another excited state. We do not usually describe these processes as "collision processes" because the atom has very well-defined energy levels, in which it stays for times that are enormously long compared to collision times,[1] so that it is possible to separate the "decay" from the excitation process. In particular, the characteristics of the decay are not sensitive to the particular mode of excitation. In nuclei and also in elementary particles, there exist levels, but frequently the lifetime is not sufficiently long to warrant a separation into excitation and decay, especially since accompanying the "resonant" scattering there is also nonresonant "background" scattering, and the disentangling of the two is sometimes complicated. In this chapter we will therefore discuss the process as a whole.

---

[1] Recall that the lifetime of a $2p$ hydrogen state is 1.6×10^{-9} sec, which is large compared to the characteristic time $a_0/\alpha c \simeq 2 \times 10^{-17}$ sec.

COLLISION CROSS SECTION

The ideal way to talk about scattering is to formulate equations that describe exactly what happens: an incident particle, described by a wave packet, approaches the target. The wave packet must be spatially large, so that it does not spread appreciably during the experiment, and it must be large compared with the target particle, but small compared with the dimensions of the laboratory, that is, it must not simultaneously overlap the target and detector. The lateral dimensions are, in fact, determined by the beam size in the accelerator. There follows an interaction with the target, and finally we see two wave packets: one continues in the forward direction, describing the unscattered part of the beam, and the other flies off at some angle and describes the scattered particles. The number of particles scattered into a given solid angle per unit time and unit incident flux is defined to be the *differential scattering cross section*. We will not follow this approach directly,[2] but will instead use some of the material developed in Chapter 10 to obtain the differential cross section. We will, however, keep the wave-packet treatment in mind as we interpret our formal results.

In our discussion of the continuum solutions of the Schrödinger equation in Chapter 10 we concluded that: (a) A solution of the Schrödinger equation in the absence of a potential is the plane wave form $e^{i\mathbf{k}\cdot\mathbf{r}}$, which describes a flux

$$\mathbf{j} = \frac{\hbar}{2im}(\psi^*\boldsymbol{\nabla}\psi - \psi\boldsymbol{\nabla}\psi^*) = \frac{\hbar k}{m} \tag{23-1}$$

If we choose $\mathbf{k}$ to define the z-axis, then the large r behavior of this solution may be written (cf. 10-72) in the form of an incoming + an outgoing spherical wave

$$e^{i\mathbf{k}\cdot\mathbf{r}} \Rightarrow \frac{i}{2k}\sum_{l=0}^{\infty}(2l+1)i^l\left[\frac{e^{-i(kr-l\pi/2)}}{r} - \frac{e^{i(kr-l\pi/2)}}{r}\right]P_l(\cos\theta) \tag{23-2}$$

(b) The conservation of particles forces us to the conclusion that the presence of a radial potential can only alter this to a function, whose asymptotic form is

$$\psi(\mathbf{r}) \Rightarrow \frac{i}{2k}\sum_{l=0}^{\infty}(2l+1)\,i^l\left[\frac{e^{-i(kr-l\pi/2)}}{r} - S_l(k)\frac{e^{i(kr-l\pi/2)}}{r}\right]P_l(\cos\theta) \tag{23-3}$$

subject to[3]

$$|S_l(k)| = 1 \tag{23-4}$$

The asymptotic form (23-3) may be rewritten, with the help of (23-2), as

$$\psi(\mathbf{r}) \Rightarrow e^{i\mathbf{k}\cdot\mathbf{r}} + \left[\sum_{l=0}^{\infty}(2l+1)\frac{S_l(k)-1}{2ik}P_l(\cos\theta)\right]\frac{e^{ikr}}{r} \tag{23-5}$$

corresponding to a plane wave + an outgoing spherical wave. Note that we are working with the effective one-particle Schrödinger equation, so that m is the

[2]This is done very nicely in R. Hobbie, *American Journal of Physics*, 30, 857 (1962), at the level of mathematics that we use in this book.
[3]See the discussion following (10-89). $S_l(k)$ is the standard notation for $e^{2i\delta_l(k)}$ defined in (10-88).

reduced mass and θ is the center of mass angle between the direction of $\mathbf{k}$ (the z-axis) and the asymptotic point $\mathbf{r}$, where, presumably the counter will be set up. When the target is much more massive than the projectile, there is no distinction between the laboratory angle and the center-of-mass angle. Note also that we could, of course, have set up a solution that has the asymptotic form of a plane wave + an incoming spherical wave since it is the first term in (23-3) that could be modified by a coefficient satisfying (23-4). However, the solution that describes the scattering is the one involving the outgoing wave. Let us calculate the flux for the asymptotic solution (23-5).

$$\mathbf{j} = \frac{\hbar}{2im} \left\{ \left[e^{i\mathbf{k}\cdot\mathbf{r}} + f(\theta)\frac{e^{ikr}}{r} \right]^* \nabla \left[e^{i\mathbf{k}\cdot\mathbf{r}} + f(\theta)\frac{e^{ikr}}{r} \right] - \text{complex conjugate} \right\} \quad (23\text{-}6)$$

where we have defined

$$f(\theta) = \sum_{l=0}^{\infty} (2l + 1)f_l(k)\, P_l(\cos\theta) \quad (23\text{-}7)$$

with

$$f_l(k) = [S_l(k) - 1]/2ik \quad (23\text{-}8)$$

Calculating the gradient gives

$$\mathbf{j} = \frac{\hbar}{2im} \left\{ \left[e^{-i\mathbf{k}\cdot\mathbf{r}} + f^*(\theta)\frac{e^{-ikr}}{r} \right] \left[i\mathbf{k}\, e^{i\mathbf{k}\cdot\mathbf{r}} + \hat{\imath}_\theta\frac{1}{r}\frac{\partial f(\theta)}{\partial\theta}\frac{e^{ikr}}{r} \right. \right.$$

$$\left. + \hat{\imath}_r f(\theta)\left(ik\frac{e^{ikr}}{r} - \frac{e^{ikr}}{r^2} \right) \right] - \text{complex conjugate} \Bigg\}$$

$$= \frac{\hbar}{2im}\left[i\mathbf{k} + i\mathbf{k}f^*(\theta)\frac{e^{-ikr(1-\cos\theta)}}{r} + ik\hat{\imath}_r f(\theta)\frac{e^{ikr(1-\cos\theta)}}{r} + ik\hat{\imath}_r |f(\theta)|^2\frac{1}{r^2} \right.$$

$$\left. - \hat{\imath}_r f(\theta)\frac{e^{ikr(1-\cos\theta)}}{r^2} + \hat{\imath}_\theta\frac{\partial f(\theta)}{\partial\theta}\frac{e^{ikr(1-\cos\theta)}}{r^2} - \text{complex conjugate} \right\}$$

We have used $\mathbf{k}\cdot\mathbf{r} = kr\cos\theta$, in the exponential factors. $\hat{\imath}_r$ is the unit vector in the $\mathbf{r}$ direction.

In carrying out this calculation we have left out the $1/r^3$ terms since these are dominated by the $1/r^2$ terms for large r. We will be interested in the flux at a detector, which is located at a distance r from the origin, where the potential that gives rise to the scattering is localized, and thus in large r. Thus the flux is

$$\mathbf{j} = \frac{\hbar k}{m} + \frac{\hbar k}{m}\hat{\imath}_r |f(\theta)|^2\frac{1}{r^2}$$

$$+ \frac{\hbar\mathbf{k}}{2m}\frac{1}{r}\left[f^*(\theta)\, e^{-ikr(1-\cos\theta)} + f(\theta\, e^{ikr(1-\cos\theta)} \right]$$

$$+ \frac{\hbar k}{2m}\frac{\hat{\imath}_r}{r}\left[f^*(\theta)\, e^{-ikr(1-\cos\theta)} + f(\theta)\, e^{ikr(1-\cos\theta)} \right] \quad (23\text{-}9)$$

$$- \frac{\hbar}{2im}\frac{\hat{\imath}_r}{r^2}\left[f(\theta)\, e^{ikr(1-\cos\theta)} - f^*(\theta)\, e^{-ikr(1-\cos\theta)} \right]$$

$$+ \frac{\hbar}{2im}\frac{\hat{\imath}_\theta}{r^2}\left[\frac{\partial f(\theta)}{\partial\theta}e^{ikr(1-\cos\theta)} - \frac{\partial f^*(\theta)}{\partial\theta}e^{-ikr(1-\cos\theta)} \right]$$

This rather involved expression simplifies considerably when we consider that $\theta \neq 0$, since one never does a scattering experiment directly in the forward direction,[4] and that in a measurement one always integrates the flux over a small but finite solid angle. Thus in the last four terms of this expression we should replace $e^{ikr(1-\cos\theta)}$ by

$$\int \sin\theta \, d\theta \, d\phi \, g(\theta, \phi) \, e^{ikr(1-\cos\theta)} \qquad (23\text{-}10)$$

where $g(\theta, \phi)$ is some sort of smooth, localized acceptance function for the counter. Now, as $r \to \infty$, we have an integral over a product of a smooth function and an extremely rapidly varying one, and this vanishes faster than any power of $1/r$. This is what is known in the mathematical literature as the Riemann-Lesbegue lemma, and it is illustrated in Problem 7. Thus, only the first two terms remain, so that

$$\mathbf{j} = \frac{\hbar\mathbf{k}}{m} + \frac{\hbar k}{m} \, \hat{\imath}_r \, \frac{|f(\theta)|^2}{r^2} \qquad (23\text{-}11)$$

In the absence of a potential, only the first term is there: it represents the incident flux. In a wave-packet treatment, $\hbar\mathbf{k}/m$ would be multiplied by a function that defines the lateral dimensions of the beam. Thus, if we ask for the *radial flux*, $\hat{\imath}_r \cdot \mathbf{j}$, then that term gives a contribution $\hbar\mathbf{k} \cdot \hat{\imath}_r/m = \hbar k \cos\theta/m$, but only within a finite region of the z-axis (see Fig. 23-1). Since the counter is put outside of that

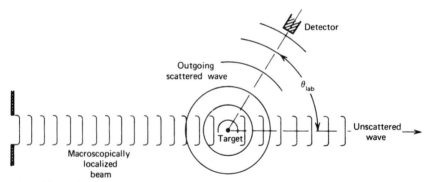

Figure 23-1. Schematic layout for scattering experiment. The scattering angle is the laboratory angle.

region, this first term does not contribute to the radial flux in the asymptotic region, so that only the second term in (23-11) contributes, and

$$\mathbf{j} \cdot \hat{\imath}_r = \frac{\hbar k}{m} \cdot \frac{|f(\theta)|^2}{r^2} \qquad (23\text{-}12)$$

Thus the number of particles crossing the area that subtends a solid angle $d\Omega$ at the origin (the target) is

$$\mathbf{j} \cdot \hat{\imath}_r \, dA = \frac{\hbar k}{m} \cdot \frac{|f(\theta)|^2}{r^2} \, r^2 d\Omega \qquad (23\text{-}13)$$

[4]How could one tell scattered from unscattered particles?

Notice that the factor of r^2 drops out. This justifies the dropping of the $1/r^3$ terms in (23-9), since these would contribute terms of order $1/kr$ to the number of particles. The differential cross section is this number, divided by the incident flux, $\hbar k/m$, that is,

$$d\sigma = |f(\theta)|^2 \, d\Omega \tag{23-14}$$

If the potential has spin dependence, there may be an azimuthal dependence, so that more generally,

$$\frac{d\sigma}{d\Omega} = |f(\theta, \phi)|^2 \tag{23-15}$$

The total cross section is given by

$$\sigma_{\text{tot}}(k) = \int d\Omega \, \frac{d\sigma}{d\Omega} \tag{23-16}$$

If we now use $f(\theta)$ as expressed in terms of $S_l(k)$, and express the latter in terms of the phase shift (cf. 10-86) to (10-89) $S_l(k) = e^{2i\delta_l(k)}$, so that

$$f(\theta) = \frac{1}{k} \sum_{l=0}^{\infty} (2l + 1) \, e^{i\delta_l(k)} \sin \delta_l(k) \, P_l(\cos \theta) \tag{23-17}$$

then

$$\sigma_{\text{tot}} = \int d\Omega \left[\frac{1}{k} \sum_l (2l + 1) \, e^{i\delta_l(k)} \sin \delta_l(k) \, P_l(\cos \theta) \right]$$
$$\left[\frac{1}{k} \sum_{l'} (2l' + 1) \, e^{-i\delta_{l'}(k)} \sin \delta_{l'}(k) \, P_{l'}(\cos \theta) \right]$$

and using

$$\int d\Omega P_l(\cos \theta) \, P_{l'}(\cos \theta) = \frac{4\pi}{2l + 1} \, \delta_{ll'} \tag{23-18}$$

we get

$$\sigma_{\text{tot}} = \frac{4\pi}{k^2} \sum_{l=0}^{\infty} (2l + 1) \sin^2 \delta_l(k) \tag{23-19}$$

It is an interesting fact that

$$\text{Im} f(0) = \frac{1}{k} \sum_{l=0}^{\infty} (2l + 1) \, \text{Im}[e^{i\delta_l(k)} \sin \delta_l(k)] \, P_l(1)$$
$$= \frac{1}{k} \sum_{l=0}^{\infty} (2l + 1) \sin^2 \delta_l(k) = \frac{k}{4\pi} \, \sigma_{\text{tot}} \tag{23-20}$$

This relation is known as the *optical theorem* and it is true even when inelastic processes can occur, as they do in nuclear and particle physics scattering processes. It is a very useful relation and in wave language it follows from the fact that the total cross section represents the removal of flux from the incident beam. Such a removal can only occur as a result of destructive interference, and the latter can only occur between the incident wave and the elastically scattered wave in the forward direction. This explains why $f(0)$ appears linearly.

This hand-waving argument does not explain why it is the imaginary part that is involved, but this can be shown to be generally true.[5]

Elastic and Inelastic Scattering

The requirement that $|S_l(k)| = 1$ followed from conservation of flux. Actually, in many scattering experiments there is *absorption* of the incident beam; the target may merely get excited, or change its state, or another particle may emerge. Under these circumstances our discussion is unchanged except that

$$S_l(k) = \eta_l(k) \, e^{2i\delta_l(k)} \tag{23-21}$$

is to be used, with

$$0 \leq \eta_l(k) \leq 1 \tag{23-22}$$

because we are dealing with absorption. The partial wave scattering amplitude is now

$$f_l(k) = \frac{S_l(k) - 1}{2ik} = \frac{\eta_l(k) \, e^{2i\delta_l(k)} - 1}{2ik} = \frac{\eta_l \sin 2\delta_l}{2k} + i \frac{1 - \eta_l \cos 2\delta_l}{2k} \tag{23-23}$$

and the total *elastic* cross section is

$$\begin{aligned}
\sigma_{e1} &= 4\pi \sum_l (2l + 1)|f_l(k)|^2 \\
&= 4\pi \sum_l (2l + 1) \frac{1 + \eta_l^2 - 2\eta_l \cos 2\delta_l}{4k^2}
\end{aligned} \tag{23-24}$$

There is also a cross section for the *inelastic* processes. Since we do not specify what the inelastic processes consist of, we can only talk about the *total inelastic cross section*, which describes the loss of flux. If we look at a particular term in (23-3), the inward radial flux carried by

$$\frac{i}{2k} \frac{e^{-ikr}}{r} P_l(\cos \theta)$$

is

$$\left(\frac{\hbar k}{m} \right) \left[\frac{4\pi}{(2k)^2} \right]$$

---

[5]See L. I. Schiff, *Progr. Theor. Phys.* (Kyoto), 11, 288 (1954).

(Recall that $Y_{l0} = P_l(\cos \theta)/\sqrt{4\pi}$). The outward radial flux is $(\hbar k/m)(|S_l(k)|^2$ $4\pi/4k^2)$, so that the net flux lost is $(\hbar k/m)(\pi/k^2)[1 - \eta_l^2(k)]$ for each l-value. Hence, dividing by the incident flux, we get

$$\sigma_{\text{inel}} = \frac{\pi}{k^2} \sum_l (2l + 1)[1 - \eta_l^2(k)] \tag{23-25}$$

Thus the total cross section is

$$\sigma_{\text{tot}} = \sigma_{\text{el}} + \sigma_{\text{inel}}$$

$$= \frac{\pi}{k^2} \sum_l (2l + 1)(1 + \eta_l^2 - 2\eta_l \cos 2\delta_l + 1 - \eta_l^2) \tag{23-26}$$

$$= \frac{2\pi}{k^2} \sum_l (2l + 1)(1 - \eta_l \cos 2\delta_l)$$

It also follows from (23-23) that

$$\text{Im} f(0) = \sum_l (2l + 1) \, \text{Im} \, f_l(k)$$

$$= \sum_l (2l + 1) \frac{1 - \eta_l \cos 2\delta_l}{2k} = \frac{k}{4\pi} \sigma_{\text{tot}} \tag{23-27}$$

so that the optical theorem is indeed satisfied.

If $\eta_l(k) = 1$, we have no absorption, and the inelastic cross section vanishes. When $\eta_l(k) = 0$ we have total absorption. Nevertheless there is still elastic scattering in that partial wave. This becomes evident in *scattering by a black disc*. The black disc is described as follows: (a) it has a well-defined edge and (b) it is totally absorbing. Since we will consider scattering for short wavelengths, that is, large k-values, condition (a) specifies that we only consider partial waves $l \le L$, where

$$L = ka \tag{23-28}$$

and a is the radius of the disc. Condition (b) specifies that $\eta_l(k) = 0$ for the relevant values of $l \le L$. Thus

$$\sigma_{\text{inel}} = \frac{\pi}{k^2} \sum_{l=0}^{L} (2l + 1) = \frac{\pi}{k^2} L^2 = \pi a^2 \tag{23-29}$$

and

$$\sigma_{\text{el}} = \frac{\pi}{k^2} \sum_{l=0}^{L} (2l + 1) = \pi a^2 \tag{23-30}$$

so that the total cross section is

$$\sigma_{\text{tot}} = \sigma_{\text{el}} + \sigma_{\text{inel}} = 2\pi a^2 \tag{23-31}$$

The result looks peculiar; on purely classical grounds we might perhaps expect that the total cross section cannot exceed the area presented by the disc; we might

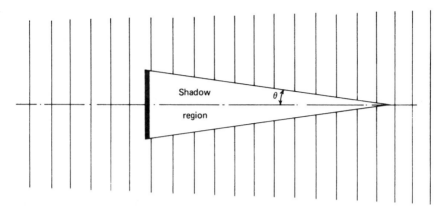

Figure 23-2. Black disc scattering and the shadow effect.

also expect to see no elastic scattering when there is total absorption. This is wrong; the absorptive disc takes flux proportional to πa^2 out of the incident beam (Fig. 23-2), and this leads to a shadow behind the disc. Far away, however, the shadow gets filled in—far enough away you cannot "see" the disc—and the only way in which this can happen is through the diffraction of some of the incident wave at the edge of the disc. The amount of incident wave that must be diffracted is the same amount as was taken out of the beam to make the shadow. Thus the elastically scattered flux must also be proportional to πa^2. The elastic scattering that accompanies absorption is called *shadow scattering* for the above reason. It is strongly peaked forward. The angle to which it is confined can be estimated from the uncertainty principle: an uncertainty in the lateral direction of magnitude a will be accompanied by an uncontrolled lateral momentum transfer of magnitude $p_\perp \sim \hbar/a$. This, however, is equal to $p\theta$, so that

$$\theta \sim \frac{\hbar}{ap} \sim \frac{1}{ak} \tag{23-32}$$

This agrees with the optical result $\theta \sim \lambda/a$. These features are observed both in nuclear scattering and in particle scattering at high energies, since the central region of nuclei and of protons is strongly absorptive, and the edges of these objects are moderately sharp. (See Fig. 23-3.)

SCATTERING AT LOW ENERGIES

The phase shift expansion (23-17) may be used to express the differential cross section in terms of the phase shifts

$$\frac{d\sigma}{d\Omega} = \frac{1}{k^2} \left| \sum_l (2l + 1)\, e^{i\delta_l(k)} \sin \delta_l(k)\, P_l(\cos \theta) \right|^2 \tag{23-33}$$

We expect, on grounds of correspondence with classical theory, that the angular momentum involved in the scattering is bounded by pa where p is the center-of-mass momentum and a is the range of the forces. Thus we expect that

$$l \lesssim \frac{pa}{\hbar} = ka \tag{23-34}$$

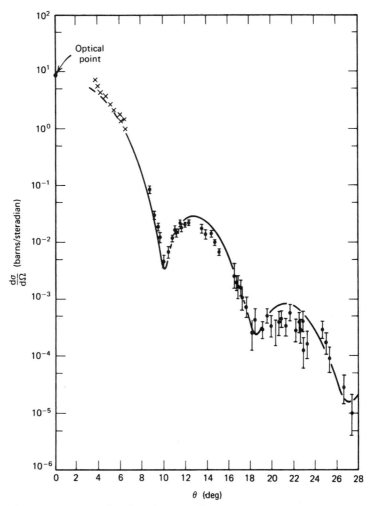

Figure 23-3. Angular distribution of 1000 MeV (1 GeV) protons scattered by ^{16}O nuclei. The angular distribution shows the dips that characterize diffraction scattering. The departures from the shape of Frauenhofer scattering in optics is due to the fact that nuclei are not sharp, nor are they totally absorbing. The curve is the result of a theoretical calculation that takes these effects into account. (From H. Palevsky et al., *Phys. Rev. Lett.* **18**, 1200 (1967), by permission.)

With the sum in (23-33) limited, one can try, by fitting the differential cross section measured at a number of angles to a form like

$$\frac{d\sigma}{d\Omega} = \sum_{n=0}^{N} A_n (\cos\,\theta)^n \tag{23-35}$$

to determine the phase shifts for a finite number of l-values. There are ambiguities, for example, the cross section is unaltered when all the phase shifts change their sign, but these can be resolved with the help of theory, continuity from low energies, and other tricks of the trade. The hope is that one can learn something about the interaction from the phase shifts, which form empirical data somewhat closer to the theory than the cross sections do.

The connection between the phase shifts $\delta_l(k)$ and the potential $V(r)$ is via the Schrödinger equation; the radial equation will have a solution that asymptotically behaves as

$$R_l(r) \sim \frac{1}{r} \sin \left[kr - \frac{l\pi}{2} + \delta_l(k) \right] \tag{23-36}$$

aside from an amplitude factor in front. Thus, given $V(r)$, a straightforward way to calculate $\delta_l(k)$ is to integrate the radial equation numerically to values of r that are far out of the range of the potential, and to examine the asymptotic behavior. This is, in fact, what one does, but this does not give us any insight into the properties of the phase shifts. To learn more about the phase shifts, we consider the square well potential. We found in Chapter 10 that

$$\tan \delta_l(k) = -\frac{C}{B} \tag{23-37}$$

where the ratio is obtained by matching the internal to the external radial wave function (10-85)

$$\kappa \frac{j_l'(\kappa a)}{j_l(\kappa a)} = k \frac{j_l'(ka) + (C/B)\, n_l'(ka)}{j_l(ka) + (C/B)\, n_l(ka)} \tag{23-38}$$

In this equation

$$\kappa^2 = \frac{2m}{\hbar^2} (E + V_0) \qquad k^2 = \frac{2mE}{\hbar^2} \tag{23-39}$$

and the ' denotes differentiation, with respect to the argument. $V_0 > 0$ for an attractive potential. Thus

$$\tan \delta_l(k) = \frac{kj_l'(ka)\, j_l(\kappa a) - \kappa j_l(ka)\, j_l'(\kappa a)}{kn_l'(ka)\, j_l(\kappa a) - \kappa n_l(ka)\, j_l'(\kappa a)} \tag{23-40}$$

This is not a particularly transparent expression, but it simplifies in some limiting cases.

(a) Consider the case that

$$ka \ll 1 \tag{23-41}$$

We do not insist that $\kappa a \ll l$. With the help of the formulas (10-66) and (10-67) we get

$$\tan \delta_l(k) \simeq \frac{2l + 1}{[1.3.5 \ldots (2l + 1)]^2}\, (ka)^{2l+1}\, \frac{lj_l(\kappa a) - \kappa a j_l'(\kappa a)}{(l + 1)\, j_l(\kappa a) + \kappa a j_l'(\kappa a)} \tag{23-42}$$

after a little algebra. One can show that for large l, this drops faster than e^{-l} even if $ka \gg 1$. The behavior

$$\tan \delta_l(k) \sim k^{2l+1} \tag{23-43}$$

for $ka \to 0$ is not restricted to the square well potential, but is true for all reasonably smooth potentials. It is a consequence of the centrifugal barrier, which keeps waves of energy far below the barrier from feeling the effect of the potential.

(b) For certain values of the energy, the denominator in (23-40) will vanish, so that at these energies the phase shift passes through $\pi/2$, or more generally through $(n + 1/2)\,\pi$. When the phase shift is $\pi/2$, then the partial wave cross section

$$\sigma_l(k) = \frac{4\pi(2l + 1)}{k^2} \sin^2 \delta_l(k) \tag{23-44}$$

has the largest possible value. One says that when $\tan \delta_l(k)$ rises rapidly to infinity and continues rising from $-\infty$, we have *resonant scattering*. To justify this terminology, and explain when resonant scattering occurs, let us consider a very deep potential, and also l large, so that

$$\kappa a \gg l \gg ka \tag{23-45}$$

We may then use (23-42) for $\tan \delta_l(k)$, and this will become infinite when

$$(l + 1)\, j_l(\kappa a) + \kappa a j_l'(\kappa a) = 0 \tag{23-46}$$

Since $\kappa a \gg l$, this condition is approximately equivalent to

$$\frac{(l + 1)}{\kappa a} \cos\left(\kappa a - \frac{l}{2}\,\pi\right) - \sin\left(\kappa a - \frac{l}{2}\,\pi\right) = 0$$

that is,

$$\tan\left(\kappa a - \frac{l}{2}\,\pi\right) \simeq \frac{l + 1}{\kappa a} \tag{23-47}$$

Since the right side is very small, the resonance condition is

$$\kappa a - \frac{l}{2}\,\pi \cong n\pi + \frac{l + 1}{\kappa a} \tag{23-48}$$

Now this is just the condition (10-76) for the existence of discrete levels in a three-dimensional box, so that resonant scattering occurs when the incident energy is just such as to match an energy level. Since $E > 0$, these levels are not really bound states. As Fig. 23-4 indicates, these are levels that would be bound states if the barrier were infinitely thick. It is not, but a particle being scattered at just the right energy still "knows" that there is a virtual level there.

In the Special Topics section 4 we discuss the scattering of a photon at an energy corresponding to a state that would be stationary in the absence of coupling to the radiation field. There we have the same situation, and we also see a resonant behavior.

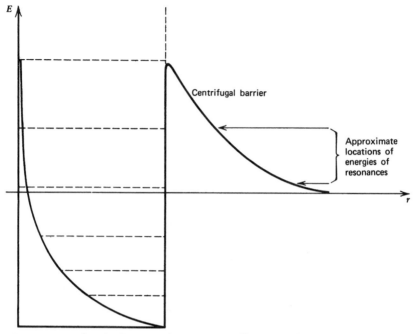

Figure 23-4. Sketch showing the square well potential with the centrifugal barrier tail. The dashed lines represent the energy levels in an infinite square well of range a, and the approximate locations of the scattering resonance energies are indicated on the right. The lower one will be much sharper than the upper one.

The Breit-Wigner Formula

As (23-42) shows, the phase shift is very tiny for ka small. Nevertheless, as ka changes and goes through the resonance, δ_l rises very rapidly, increasing by π; thus the partial wave cross section (23-44) will exhibit a very sharp peak at the resonant energy. This behavior (Fig. 23-5) is very similar to the cross section for the scattering of electrons by He^+ at the energy corresponding to the $(2s)^2$ excited state (Fig. 18-4). In the neighborhood of the resonant energy, the phase shift rises through $\pi/2$ very rapidly. We may represent this behavior by

$$\tan \delta_l \approx \frac{\gamma(ka)^{2l+1}}{E - E_{res}} \tag{23-49}$$

This leads to the partial wave cross section

$$\sigma_l = \frac{4\pi(2l + 1)}{k^2} \frac{\tan^2 \delta_l}{1 + \tan^2 \delta_l} = \frac{4\pi(2l + 1)}{k^2} \frac{[\gamma(ka)^{2l+1}]^2}{(E - E_{res})^2 + [\gamma(ka)^{2l+1}]^2} \tag{23-50}$$

This is the well-known *Breit-Wigner formula* for resonant cross sections. Again, the behavior is not a peculiarity of the square well potential, but is characteristic of

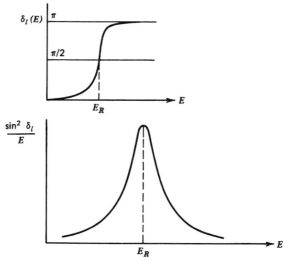

Figure 23-5. The partial wave cross section corresponding to the phase shift sketched in the upper insert.

all potentials that have a shape such that metastable states can simulate bound states above $E = 0$ in it. We just note for completeness that

$$f_l(k) = \frac{e^{2i\delta_l(k)} - 1}{2ik} = \frac{\dfrac{1 + i \tan \delta_l}{1 - i \tan \delta_l} - 1}{2ik}$$

$$= \frac{\tan \delta_l}{k(1 - i \tan \delta_l)} = \frac{\gamma(ka)^{2l+1}/k}{E - E_{\text{res}} - i\gamma(ka)^{2l+1}}$$

(23-51)

If there is nonresonant scattering that is appreciable, then the scattering amplitude is of the form

$$f_l(k) = f_l^{\text{res}}(k) + f_l^{\text{nonres}}(k)$$

(23-52)

S-Wave Scattering for Square Well

At low energies, the scattering is primarily in S-states, so that we may concentrate on $l = 0$. It is simpler to derive the phase shift directly than to work out (23-40). The solution inside the well that is regular at $r = 0$ is

$$u(r) = rR(r) = C \sin \kappa r$$

(23-53)

and this is to be matched onto

$$u(r) = \sin (kr + \delta)$$

(23-54)

the solution outside the well. The continuity of $(1/u)(du/dr)$ at $r = a$ implies that

$$\kappa \cot \kappa a = k \cot (ka + \delta)$$

that is,

$$\tan \delta = \frac{(k/\kappa) \tan \kappa a - \tan ka}{1 + (k/\kappa) \tan \kappa a \tan ka} \tag{23-55}$$

Note that if we define

$$\tan qa = \frac{k}{\kappa} \tan \kappa a$$

then

$$\tan \delta = \frac{\tan qa - \tan ka}{1 + \tan qa \tan ka} = \tan (qa - ka)$$

that is,

$$\delta = \tan^{-1} \left(\frac{k}{\kappa} \tan \kappa a \right) - ka \tag{23-56}$$

We have, following (23-39),

$$(\kappa a)^2 = (ka)^2 + \frac{2mV_0 a^2}{\hbar^2} \tag{23-57}$$

with $V_0 > 0$ for an attractive potential. Thus, at very low energies, using $\tan x \simeq x$ for $x \ll 1$, we get

$$\tan \delta \approx \delta \approx ka \left(\frac{\tan \kappa a}{\kappa a} - 1 \right) \tag{23-58}$$

When κa goes through $\pi/2$ (we imagine that we are slowly deepening the potential well), which is just the condition that the well be deep enough for a bound state to develop [cf. Eq. (5-69)], then $\tan \kappa a \to \infty$ and (23-55) shows that

$$\tan \delta = \frac{1}{\tan ka} \to \infty \tag{23-59}$$

that is, δ goes through $\pi/2$. In a sense, a bound state at zero energy is like a resonance.

As the well becomes a little deeper, we again have $\tan \delta \sim O\,(ka)$, and continuity demands that the branch is such that

$$\delta \approx ka \left(\frac{\tan \kappa a}{\kappa a} - 1 \right) \qquad \text{(no bound state)}$$

$$\delta \approx \pi + ka \left(\frac{\tan \kappa a}{\kappa a} - 1 \right) \qquad \text{(with bound state)} \tag{23-60}$$

As the potential becomes still deeper, a second bound state can appear, κa goes through $3\pi/2$, and we have $\delta \approx 2\pi + ka[(\tan \kappa a/\kappa a) - 1]$, and so on. There is a general result known as Levinson's theorem, which states

$$\delta(0) - \delta(\infty) = N_B \pi \tag{23-61}$$

where N_B is the number of bound states, and (23-60) is an example of it.

Connection Between Scattering Amplitude and Binding Energy

At very low energies the cross section only has the $l = 0$ contribution to it, and it is

$$\sigma \cong \frac{4\pi}{k^2} (ka)^2 \left(\frac{\tan \kappa a}{\kappa a} - 1\right)^2 = 4\pi a^2 \left(\frac{\tan \kappa a}{\kappa a} - 1\right)^2 \tag{23-62}$$

that is, it is a constant. There will, of course, be a correction of order $(ka)^2$ to this result. If we consider neutron–proton scattering, then we know that the potential must be such as to give the right binding energy of the deuteron. If we let

$$E = -\frac{\hbar^2 \alpha^2}{2m}$$

and

$$\kappa = \sqrt{-\alpha^2 + \frac{2mV_0}{\hbar^2}}$$

(effectively $k^2 = -\alpha^2$ for the bound-state problem), then the matching of the wave function outside the potential $u(r) = A\, e^{-\alpha r}$ to the solution inside $B \sin \kappa r$ at the boundary gives

$$\kappa \cot \kappa a = -\alpha \tag{23-63}$$

For $k \ll \kappa$, we have

$$\left(\frac{\tan \kappa a}{\kappa a}\right)_{\text{scatt}} \cong \left(\frac{\tan \kappa a}{\kappa a}\right)_{\text{deuteron}} = -\frac{1}{a\alpha} \tag{23-64}$$

Thus

$$\sigma \cong 4\pi a^2 \left(1 + \frac{1}{a\alpha}\right)^2 \cong \frac{4\pi}{\alpha^2} (1 + 2a\alpha) \tag{23-65}$$

Thus making the low energy approximation expressed by (23-64) allows us to bypass the problem of determining the potential and *then* calculating the cross section. The approximation only works when the binding energy is small. The quantity $1/\alpha$ is the distance over which the deuteron wave function spills over, and this is always much larger than the range of the potential a for a loosely bound system. It is $1/\alpha$ and not the range of the potential that determines the scattering cross section at low energies.

Spin-Dependent Scattering

In the 1930s there was great interest in the form of the neutron–proton potential, since it was hoped that this would give some fundamental clues concerning the nuclear forces in general. Rudimentary experiments at low energies were fitted with a variety of potentials. It became evident after a while that almost any reasonably shaped potential would work, provided that one chose the appropriate depth and range. It was shown in 1947 by Schwinger (and subsequently derived by Bethe in a simpler manner) that at low energies it is always a good approximation to write

$$k \cot \delta = -\frac{1}{A} + \frac{1}{2} r_0 k^2 \qquad (23\text{-}66)$$

where A is called the scattering length, and r_0 is the effective range. The cross section at threshold determines the scattering length

$$\sigma \cong 4\pi A^2 \qquad (23\text{-}67)$$

and the energy dependence determines the effective range. The relation between these parameters and the parameters describing the potential vary with the shape, but a two-parameter fit to the data is always possible. This *effective range formula* shows that if we want to probe the shape of the potential, we must go to higher energies.

The binding energy of the deuteron is 2.23 MeV. Thus, remembering that in our discussion m is the reduced mass, that is, $M_p/2$,

$$\frac{1}{\alpha} = \sqrt{\frac{\hbar^2}{2mE}} = \frac{\hbar c}{\sqrt{M_p c^2 E}} = \frac{\hbar}{M_p c} \sqrt{\frac{M_p c^2}{E}}$$

$$\cong \frac{10^{-27}}{1.6 \times 10^{-24} \times 3 \times 10^{10}} \sqrt{\frac{940}{2.23}} = 4.3 \times 10^{-13} \text{ cm}$$

so that

$$\frac{4\pi}{\alpha^2} \simeq 2.5 \times 10^{-24} \text{ cm}^2 \simeq 2.5 \text{ barns}$$

A more accurate determination leads to the prediction that the cross section at threshold is four barns. The measurement, carried out with neutrons at thermal speeds yields 21 barns!

The explanation of this disagreement came with the realization that the spin of the neutron and the proton had not been taken into account. If the potential were spin independent, then all spin states would scatter the same way, that is, it would not matter whether the spins of the particles are "up" or "down." If the potential does depend on the spin, a possible form could be

$$V(r) = V_1(r) + \boldsymbol{\sigma}_p \cdot \boldsymbol{\sigma}_n V_2(r) \qquad (23\text{-}68)$$

In this case spin is no longer a good quantum number, and the states must be classified by total angular momentum and total spin, that is, with $l = 0$, the four states divide up into a $^3S_1$ triplet of states, and a singlet $^1S_0$. These need not scatter the same way, so that there are really two phase shifts, δ_t for the triplet, and δ_s for the singlet. There are no triplet–singlet transitions, since the total angular momentum J must be the same in the initial and final states. The total cross section is weighted by the number of final states in each case (the cross section involves a *sum* over final states and is independent of the value of the z-component of the angular momentum), so that

$$\sigma = \frac{3}{4}\, \sigma_t + \frac{1}{4}\, \sigma_s \tag{23-69}$$

For spin independent forces, $\sigma = \sigma_t = \sigma_s$.

The deuteron is a $^3S_1$ state, so that the four barns are really predicted for σ_t. This implies that

$$\sigma_s = 4\sigma - 3\sigma_t = 72 \text{ barns} \tag{23-70}$$

Since we are at threshold, this implies that

$$|A_s| = \sqrt{\frac{72 \times 10^{-24}}{4\pi}} \cong 2.4 \times 10^{-12} \text{ cm} \tag{23-71}$$

The earlier result implied that

$$|A_t| = \sqrt{\frac{4 \times 10^{-24}}{4\pi}} \cong 4.7 \times 10^{-13} \text{ cm} \tag{23-72}$$

The question of the signs of A_t and A_s now arises. At threshold we have $k \cot \delta \approx k/\delta \approx -1/A$ so that $\delta_s = -A_s k$ and $\delta_t = -A_t k$. Thus, the asymptotic wave functions have the form

$$\sin\,(kr + \delta_{t,s}) \simeq \sin k(r - A_{t,s}) \simeq k(r - A_{t,s}) \tag{23-73}$$

The two possible cases are shown in Fig. 23-6. We know that for the triplet state the wave function turns over just before the edge of the well (since there is a bound state), so that it must correspond to the situation $A_t > 0$.

If A_s were positive, too, one would expect a singlet bound state, with very much weaker binding, since the internal wave function ties onto a much flatter asymptotic form. In fact, the binding energy would be 70 keV. Such a bound state was not found, suggesting that $A_s < 0$.

This choice of sign was actually confirmed by the scattering of neutrons off the H_2 molecule. As we know, the H_2 molecule can exist as ortho-H_2, with the spins in a triplet state, and para-H_2, with the two proton spins in a singlet state. For neutrons at very low energies, such that the wavelength is much larger than the proton–proton separation in the molecule, the scattering amplitude for neutron-H_2, scattering is just the sum of the amplitudes for the individual scatterings. One may show that the amplitude off para-H_2 is different from the amplitude off ortho-H_2 and these separately involve linear combinations of A_s and A_t. The fact that $\sigma_{para} \cong 3.9$ barns, while $\sigma_{ortho} \cong 125$ barns can be explained in this way. The calculation is complicated by a number of effects that must be taken into account, for example, that the effective mass of the proton in a molecule is different from that of a free proton, and that the molecules are not really at rest, but are moving

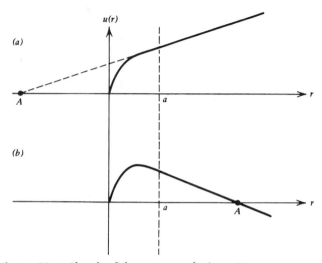

Figure 23-6. Sketch of the s-wave solution $u(r)$ near threshold. Outside the range radius $r = a$, the wave function has the form $C(r - A)$. [This is not in conflict with (23-73), which is an expansion of $\sin(kr + \delta)$. We could equally well have taken the form of $u(r)$ to be $(C/k) \sin(kr + \delta)$, since the normalization is arbitrary. It is, in fact, the interior wave function and the position of A that determine the slope of the line.] The sign of A depends on whether the interior wave function has or has not turned over cases (*b*) and (*a*), respectively. Since the wave function must turn over if there is a weakly bound state (so that it can match a slowly falling exponential) and since one does not expect the wave function inside the potential to be very sensitive to variations in E about zero, one expects that for a potential that has a bound state with E_B small, $A > 0$.

with a distribution appropriate to the (low $\sim$ 20 K) temperature. The large discrepancy between the two cross sections is not changed much by these corrections, and it can only be explained if A_s is indeed negative.

THE BORN APPROXIMATION

At higher energies many partial waves contribute to the scattering, and it is therefore preferable to avoid the angular momentum decomposition. A procedure that leads to a very useful approximation both when the potential is weak and when the energy is high is the Born approximation, in which we consider the scattering process as a transition, just like the transitions studied in Chapter 21. The difference is that here we consider the transitions

$$\text{continuum} \rightarrow \text{continuum}$$

If we work in the center-of-mass system, we have effectively a one-particle problem, and this particle makes a transition from an initial state, described by the eigenfunction

$$\psi_i(\mathbf{r}) = \frac{1}{\sqrt{V}} \, e^{i\mathbf{p}_i \cdot \mathbf{r}/\hbar} \tag{23-74}$$

to the final state, described by

$$\psi_f(\mathbf{r}) = \frac{1}{\sqrt{V}} \, e^{i\mathbf{p}_f \cdot \mathbf{r}/\hbar} \tag{23-75}$$

where $\mathbf{p}_i$ and $\mathbf{p}_f$ are the initial and final momenta, respectively. The transition rate, following the Golden Rule (21-59) is given by

$$R_{i \to f} = \frac{2\pi}{\hbar} \int \frac{V \, d^3\mathbf{p}_f}{(2\pi\hbar)^3} \, |M_{fi}|^2 \, \delta\left(\frac{p_f^2}{2m} - \frac{p_i^2}{2m}\right) \tag{23-76}$$

The delta function expresses energy conservation. If the particles that emerge have a different mass from those that enter, or if the target is excited, that delta function takes a somewhat different form. It will, however, always be of the form $\delta[(p_f^2/2m) - E]$ where E is the energy available for kinetic energy of the final particle. The matrix element M_{fi} is given by

$$\begin{aligned} M_{fi} = \langle \psi_f | V | \psi_i \rangle &= \int d^3\mathbf{r} \frac{e^{-i\mathbf{p}_f \cdot \mathbf{r}/\hbar}}{\sqrt{V}} \, V(\mathbf{r}) \, \frac{e^{i\mathbf{p}_i \cdot \mathbf{r}/\hbar}}{\sqrt{V}} \\ &= \frac{1}{V} \int d^3\mathbf{r} \, e^{-i\mathbf{\Delta} \cdot \mathbf{r}} \, V(\mathbf{r}) \end{aligned} \tag{23-77}$$

where $\mathbf{\Delta} = \frac{1}{\hbar} (\mathbf{p}_f - \mathbf{p}_i)$. We write the matrix element as

$$M_{fi} = \frac{1}{V} \, \tilde{V}(\mathbf{\Delta}) \tag{23-78}$$

The integral in (23-76) can be rewritten in the form

$$\begin{aligned} R_{i \to f} &= \frac{2\pi}{\hbar} \int d\Omega \, \frac{V p_f^2 \, dp_f}{(2\pi\hbar)^3} \frac{1}{V^2} \, |\tilde{V}(\mathbf{\Delta})|^2 \, \delta\left(\frac{p_f^2}{2m} - E\right) \\ &= \frac{2\pi}{\hbar} \frac{1}{(2\pi\hbar)^3} \frac{1}{V} \int d\Omega \, p_f m \, \frac{p_f \, dp_f}{m} \, \delta\left(\frac{p_f^2}{2m} - E\right) |\tilde{V}(\mathbf{\Delta})|^2 \\ &= \frac{1}{4\pi^2\hbar^4} \frac{1}{V} \int d\Omega \, p_f m |\tilde{V}(\mathbf{\Delta})|^2 \end{aligned} \tag{23-79}$$

To get the last line, we noted that $p_f \, dp_f / m = d(p_f^2/2m)$ and carried out the delta function integration. Thus, p_f must be evaluated at $p_f = (2mE)^{1/2}$, and we must not forget that m here is the reduced mass in the final state.

This expression has an undesirable dependence on the volume of the quantization box, but this is not really surprising. Our wave functions were normalized to one particle in the box V, so that the number of transitions should certainly go down as V increases. This difficulty arises because we are asking a question that does not correspond to an experiment. What one does is send a flux of incident particles at each other (in the center-of-mass frame; in the laboratory, one particle is stationary, of course). If we want a flux of one particle per square centimeter per second, we must multiply the preceding by V divided by the volume of a cylinder with 1 cm^2 base, and the relative velocity of the particles in the center-of-mass frame in the initial state. The number of transitions for unit flux is just the cross section. We therefore have

$$d\sigma = \frac{1}{4\pi^2\hbar^4} \frac{1}{|v_{\rm rel}|} d\Omega p_f m |\tilde{V}(\mathbf{\Delta})|^2 \tag{23-80}$$

Since in the center-of-mass frame the two incident particles are moving toward each other with equal and opposite momenta of magnitude p_i, their relative velocity is

$$|v_{\rm rel}| = \frac{p_i}{m_1} + \frac{p_i}{m_2} = p_i \left(\frac{1}{m_1} + \frac{1}{m_2}\right) = \frac{p_i}{m_{\rm red}^{(i)}} \tag{23-81}$$

if m_1 and m_2 are their masses. Thus, if the initial and final reduced masses and momenta are not the same, we have

$$\frac{d\sigma}{d\Omega} = \frac{1}{4\pi^2} \frac{p_f}{p_i} m_{\rm red}^{(f)} m_{\rm red}^{(i)} \left|\frac{1}{\hbar^2} \tilde{V}(\mathbf{\Delta})\right|^2 \tag{23-82}$$

When the initial and final particles are the same,

$$\frac{d\sigma}{d\Omega} = \frac{m_{\rm red}^2}{4\pi^2} \left|\frac{1}{\hbar^2} \tilde{V}(\mathbf{\Delta})\right|^2 \tag{23-83}$$

When one particle is a great deal more massive than the other, $m_{\rm red} \to m$, the mass of the lighter particle. When we compare the above with (23-15) we see that

$$f(\theta, \phi) = -\frac{m_{\rm red}}{2\pi\hbar^2} \tilde{V}(\mathbf{\Delta}) \tag{23-84}$$

Actually, to determine the sign, one must go through a more detailed comparison with the partial wave expansion. We will not bother to do this here.

As an illustration of the application of the Born approximation, we will calculate the cross section for the scattering of a particle of mass m and charge Z_1 by a Coulomb potential of charge Z_2. The source of the Coulomb field is taken to be infinitely massive, so that the mass in (23-83) is the mass of the incident particle. For generality (and, as we will see, for technical reasons) we take the Coulomb field to be screened, so that

$$V(\mathbf{r}) = Z_1 Z_2 e^2 \frac{e^{-r/a}}{r} \tag{23-85}$$

where a is the screening radius. We thus need to evaluate

$$\tilde{V}(\mathbf{\Delta}) = Z_1 Z_2 e^2 \int d^3\mathbf{r}\, e^{-i\mathbf{\Delta}\cdot\mathbf{r}} \frac{e^{-r/a}}{r} \tag{23-86}$$

We choose the direction of Δ as z-axis, and then get

$$
\int d^3\mathbf{r}\, e^{-i\Delta\cdot\mathbf{r}}\frac{e^{-r/a}}{r} = \int_0^{2\pi} d\phi \int_0^{\pi} \sin\theta\, d\theta \int_0^{\infty} r^2\, dr\, e^{-i\Delta r\cos\theta}\frac{e^{-r/a}}{r}
$$

$$
= 2\pi \int_0^{\infty} r dr\, e^{-r/a} \int_{-1}^{1} d(\cos\theta)\, e^{-i\Delta r\cos\theta}
$$

$$
= \frac{2\pi}{i\Delta} \int_0^{\infty} dr\, e^{-r/a}\, (e^{i\Delta r} - e^{-i\Delta r})
$$

$$
= \frac{2\pi}{i\Delta}\left(\frac{1}{(1/a)-i\Delta} - \frac{1}{(1/a)+i\Delta}\right) = \frac{4\pi}{(1/a^2)+\Delta^2}
$$

(23-87)

Now

$$
\Delta^2 = \frac{1}{\hbar^2}(\mathbf{p}_f - \mathbf{p}_i)^2 = \frac{1}{\hbar^2}(2p^2 - 2\mathbf{p}_f\cdot\mathbf{p}_i) = \frac{2p^2}{\hbar^2}(1 - \cos\theta)
$$

(23-88)

so that the cross section becomes

$$
\frac{d\sigma}{d\Omega} = \frac{m^2}{4\pi^2}\frac{1}{\hbar^4}(Z_1Z_2e^2)^2\frac{16\pi^2}{[(2p^2/\hbar^2)(1-\cos\theta)+(1/a^2)]^2}
$$

$$
= \left(\frac{2mZ_1Z_2e^2}{4p^2\sin^2(\theta/2)+(\hbar^2/a^2)}\right)^2
$$

(23-89)

$$
= \left(\frac{Z_1Z_2e^2}{4E\sin^2(\theta/2)+(\hbar^2/2ma^2)}\right)^2
$$

In the last line we replaced $p^2/2m$ by E, and we used $\frac{1}{2}(1 - \cos\theta) = \sin^2(\theta/2)$. The angle θ defined in (23-88) is the center-of-mass scattering angle. In the absence of screening ($a \to \infty$) this reduces to the well-known Rutherford formula. There is no $\hbar$ in it, and it is the same as the classical formula. Had we left out the screening factor in (23-86) we would have had an ill-defined integral. One often evaluates ambiguous integrals with the aid of such convergence factors.

The Born approximation has its limitations. For example, we found that $\tilde{V}(\Delta)$ was purely real so that $f(\theta)$ is also real in this approximation. This implies, by the optical theorem, that the cross section is zero. In fact, the Born approximation is only good when either (a) the potential is weak, so that the cross section is of second order in a small parameter; this would make the use of it consistent with the optical theorem, or (b) at high energies for potentials such that the cross section goes to zero. This is true for most smooth potentials. It is not true for real particles; there it seems that the cross sections stay constant at very high energies, and one cannot expect the Born approximation to serve as more than a guide of the behavior of the scattering amplitude.

As a final comment, we observe that if the potential V has a spin dependence, then (23-77) is trivially modified by the requirement that the initial and final states be described by their spin wave functions, in addition to the spatial wave functions. Thus, for example, if the neutron-proton potential has the form

$$
V(r) = V_1(r) + \boldsymbol{\sigma}_P \cdot \boldsymbol{\sigma}_N V_2(r)
$$

the Born approximation reads

$$M_{fi} = \frac{1}{V} \int d^3r \, e^{-i\Delta \cdot r} \, \xi_f^+ \, V(r) \, \xi_i$$

where ξ_i and ξ_f represent the initial and final spin states of the neutron-proton system.

SCATTERING OF IDENTICAL PARTICLES

When two identical particles scatter, there is no way of distinguishing a deflection of a particle through an angle θ and a deflection of $\pi - \theta$ in the center-of-mass frame, since momentum conservation demands that if one of the particles scatters through θ, the other goes in the direction $\pi - \theta$ (Fig. 23-7). Classically, too, the cross section for scattering is affected by the identity of particles, since the number of counts at a certain counter will be the sum of the counts due to the two particles. Thus

$$\sigma_{cl}(\theta) = \sigma(\theta) + \sigma(\pi - \theta) \tag{23-90}$$

In quantum mechanics there is no way of distinguishing the two final states, so that the two *amplitudes* $f(\theta)$ and $f(\pi - \theta)$ can interfere. Thus the cross section for the scattering of two identical spin zero (boson) particles, for example, α-particles, is

$$\frac{d\sigma}{d\Omega} = |f(\theta) + f(\pi - \theta)|^2 \tag{23-91}$$

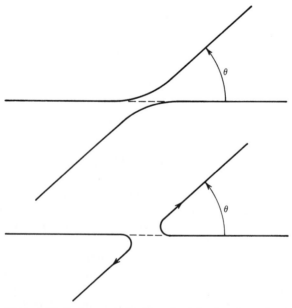

Figure 23-7. Asymptotic directions in the scattering of two identical particles through a center-of-mass angle θ.

This differs from the classical result by the interference term

$$\frac{d\sigma}{d\Omega} = |f(\theta)|^2 + |f(\pi - \theta)|^2 + [f^*(\theta)f(\pi - \theta) + f(\theta)f^*(\pi - \theta)] \quad (23\text{-}92)$$

and it leads to an enhancement at $\pi/2$, for example,

$$\left(\frac{d\sigma}{d\Omega}\right)_{\pi/2} = 4 \left| f\left(\frac{\pi}{2}\right) \right|^2 \quad (23\text{-}93)$$

compared to the result that would be obtained without interference:

$$\left(\frac{d\sigma}{d\Omega}\right)_{\pi/2} = 2 \left| f\left(\frac{\pi}{2}\right) \right|^2 \quad (23\text{-}94)$$

When the scattering of two spin 1/2 particles is considered, for example, proton–proton scattering or electron–electron scattering, then the amplitude should reflect the basic antisymmetry of the total wave function under the interchange of the two particles. If the two particles are in a spin singlet state, then the spatial wave function is symmetric, and

$$\frac{d\sigma_s}{d\Omega} = |f(\theta) + f(\pi - \theta)|^2 \quad (23\text{-}95)$$

If the two particles are in a spin triplet state, then the spatial wave function is antisymmetric, and

$$\frac{d\sigma_t}{d\Omega} = |f(\theta) - f(\pi - \theta)|^2 \quad (23\text{-}96)$$

In the scattering of two unpolarized protons, all spin states are equally likely, and thus the probability of finding the two protons in a triplet state is three times as large as finding them in a singlet state, so that

$$
\begin{aligned}
\frac{d\sigma}{d\Omega} &= \frac{3}{4}\frac{d\sigma_t}{d\Omega} + \frac{1}{4}\frac{d\sigma_s}{d\Omega} \\
&= \tfrac{3}{4}|f(\theta) - f(\pi - \theta)|^2 + \tfrac{1}{4}|f(\theta) + f(\pi - \theta)|^2 \quad (23\text{-}97) \\
&= |f(\theta)|^2 + |f(\pi - \theta)|^2 - \tfrac{1}{2}[f(\theta)f^*(\pi - \theta) + f^*(\theta)f(\pi - \theta)]
\end{aligned}
$$

For proton–proton scattering as well as for α–α scattering, the basic amplitude $f(\theta)$ is the sum of a nuclear term (if the energies are not too low) and a Coulomb term. Whether the identical particles be bosons or fermions, there is symmetry under the interchange $\theta \rightarrow \pi \rightarrow \theta$.

Scattering by Atoms on a Lattice

Symmetry considerations also play a role in the scattering of particles by a crystal lattice. If we ignore spin, so that we do not have to worry whether the electron does or does not flip its spin ("up" → "down" or vice versa), then at low energies, the scattering amplitude $f(\theta)$ is independent of angle (S-wave scattering), and the solution of the Schrödinger equation by a single atom located at the lattice point $\mathbf{a}_i$ has the asymptotic form

$$\psi(\mathbf{r}) \sim e^{i\mathbf{k}\cdot(\mathbf{r}-\mathbf{a}_i)} + f \frac{e^{ik|\mathbf{r}-\mathbf{a}_i|}}{|\mathbf{r}-\mathbf{a}_i|} \tag{23-98}$$

Now

$$\begin{aligned}
k|\mathbf{r}-\mathbf{a}_i| &= k(r^2 - 2\mathbf{r}\cdot\mathbf{a}_i + a_i^2)^{1/2} \\
&\cong kr\left(1 - \frac{2\mathbf{r}\cdot\mathbf{a}_i}{r^2}\right)^{1/2} \\
&\cong kr - k\hat{\imath}_r\cdot\mathbf{a}_i
\end{aligned} \tag{23-99}$$

and since $k\hat{\imath}_r$ is a vector of magnitude k and it points in the direction $\mathbf{r}$, the point of observation, it is the final momentum $\mathbf{k}'$. If we divide out the phase factor $e^{-i\mathbf{k}\cdot\mathbf{a}_i}$, the wave function has the asymptotic form

$$\psi \sim e^{i\mathbf{k}\cdot\mathbf{r}} + f\, e^{-i\mathbf{k}'\cdot\mathbf{a}_i}\, e^{i\mathbf{k}\cdot\mathbf{a}_i} \frac{e^{ikr}}{r} + 0\left(\frac{1}{r^2}\right) \tag{23-100}$$

so that the scattering amplitude is

$$f(\theta) = f\, e^{-i\boldsymbol{\Delta}\cdot\mathbf{a}_i} \qquad \boldsymbol{\Delta} = \mathbf{k}' - \mathbf{k} \tag{23-101}$$

The total amplitude is the sum of all individual scattering amplitudes when we have a situation in which we cannot tell which atom in the crystal did the scattering. This is indeed the case for elastic low energy scattering when recoil is not observed and spins are not measured. Thus, for the *coherent* process we have

$$\frac{d\sigma}{d\Omega} = \left| f \sum_{\text{atoms}} e^{-i\boldsymbol{\Delta}\cdot\mathbf{a}_i} \right|^2 \tag{23-102}$$

If we have a simple cubic array of lattice points, such that

$$\mathbf{a}_i = a(n_x\hat{\imath}_x + n_y\hat{\imath}_y + n_z\hat{\imath}_z) \qquad -N \le n_x,\, n_y,\, n_z \le N \tag{23-103}$$

(spacings are integral multiples of a in all directions), then

$$\sum e^{-i\boldsymbol{\Delta}\cdot\mathbf{a}_i} = \sum_{n_x=-N}^{N}\sum_{n_y=-N}^{N}\sum_{n_z=-N}^{N} e^{-ia\Delta_x n_x}\, e^{-ia\Delta_y n_y}\, e^{-ia\Delta_z n_z}$$

We use

$$\begin{aligned}
\sum_{n=-N}^{N} e^{ia n} &= e^{-iaN}(1 + e^{ia} + e^{2ia} + \cdots + e^{2iaN}) \\
&= e^{iaN}\frac{e^{ia(2N+1)} - 1}{e^{ia} - 1} = \frac{e^{ia(N+1)} - e^{-iaN}}{e^{ia} - 1} \\
&= \frac{e^{ia(N+1/2)} - e^{-ia(N+1/2)}}{e^{ia/2} - e^{-ia/2}} = \frac{\sin \alpha\,(N + \tfrac{1}{2})}{\sin \alpha/2}
\end{aligned} \tag{23-104}$$

to obtain the result

$$\frac{d\sigma}{d\Omega} = |f|^2 \frac{\sin^2 \alpha_x(N + \frac{1}{2})}{\sin^2 \alpha_x/2} \cdot \frac{\sin^2 \alpha_y(N + \frac{1}{2})}{\sin^2 \alpha_y/2} \cdot \frac{\sin^2 \alpha_z(N + \frac{1}{2})}{\sin^2 \alpha_z/2} \qquad (23\text{-}105)$$

where

$$\alpha_x = a\,\Delta_x - 2\pi\nu_x \qquad (\nu_x = \text{integer}), \text{ etc.} \qquad (23\text{-}106)$$

We can make the generalization exhibited earlier, since a change $\alpha \to \alpha - 2\pi\nu$, with ν an integer, does not change (23-105). The expression (23-105) is not very transparent. However, when N is large, each of the factors becomes very strongly peaked when $\alpha_x, \ldots$ are near zero. In fact, using

$$\frac{\sin^2 Nu}{u^2/4} \to 4\pi N\,\delta(u) \qquad (23\text{-}107)$$

a formula easily derived from (21-20) by a simple change of variables, we get

$$\frac{d\sigma}{d\Omega} = |f|^2\,(2\pi)^3\,(2N)^3\,\delta(a\boldsymbol{\Delta} - 2\pi\boldsymbol{\nu}) \qquad (23\text{-}108)$$

Now the total number of atoms is $(2N)^3$ and hence the cross section per atom is

$$\frac{d\sigma}{d\Omega} = |f|^2\,\frac{(2\pi)^3}{a^3}\,\delta\!\left(\mathbf{k'} - \mathbf{k} - \frac{2\pi\boldsymbol{\nu}}{a}\right) \qquad (23\text{-}109)$$

Thus the differential cross section is very small, except in the directions given by

$$\mathbf{k'} - \mathbf{k} = \frac{2\pi}{a}\,\boldsymbol{\nu} \qquad (23\text{-}110)$$

where it is strongly peaked. The conditions above are called the *Bragg conditions*, and the integers ν_x, ν_y, ν_z are called the *Miller indices of the Bragg planes*.

The relations just derived can be generalized to more complicated crystals. They are used to study crystal structure, using neutrons or X-rays as incident particles, or using a known crystal to study X-rays that are emitted in atomic transitions involving energetic photons.

Problems

1. Show that for a central potential $V(\mathbf{r}) = V(r)$, the matrix element M_{fi} in (23-77) can be written in the form

$$M_{fi} = \frac{1}{V}\frac{4\pi\hbar}{\Delta}\int_0^\infty r\,dr\,V(r)\,\sin r\Delta$$

Note that this is an even function of Δ, that is, a function of

$$\boldsymbol{\Delta}^2 = (\mathbf{p}_f - \mathbf{p}_i)^2/\hbar^2$$

2. Consider a potential of the form

$$V(r) = V_G \, e^{-r^2/a^2}$$

Calculate, using the Born approximation, the differential cross section $d\sigma/d\Omega$ as a function of the center-of-mass scattering angle θ. Compare your result with the differential cross section for a Yukawa potential

$$V(r) = V_0 b \, \frac{e^{-r/b}}{r}$$

[Already done in (23-85)–(23-89)]. To make the comparison, adjust the parameters in the two cases so that the two differential cross sections and their slopes are the same in the forward direction at $\Delta = 0$. It might be convenient to pick some definite numerical values for V_G, V_0, a, and b to depict this graphically. Can you give a qualitative argument explaining the large difference between the predictions for large momentum transfers?

3. Consider the potential

$$V(r) = V_0 a \, \frac{e^{-r/a}}{r}$$

If the range parameter is $a = 1.2 \text{ fm} = 1.2 \times 10^{-13}$ cm and $V_0 = 100$ MeV in magnitude, what is the total cross section for proton–proton scattering at 100-MeV center-of-mass energy, calculated in Born approximation? Ignore Coulomb scattering, but not identity of two protons.
[*Note:* It is useful to use the relation

$$\hbar^2 \, \Delta^2 = (\mathbf{p}_f - \mathbf{p}_i)^2 = 2p^2(1 - \cos\theta)$$

to write

$$d\Omega = 2\pi \, d \, (\cos\theta) = \frac{\hbar^2 \pi}{p^2} \, d(\Delta^2).]$$

4. Suppose the scattering amplitude for neutron–proton scattering is given by the form

$$f(\theta) = \xi_f^\dagger \, (A + B\boldsymbol{\sigma}_P \cdot \boldsymbol{\sigma}_N) \, \xi_i$$

where ξ_i and ξ_f are the initial and final spin states of the neutron–proton system. The possible states are

$$\xi_i = \chi_\uparrow^{(P)} \chi_\uparrow^{(N)} \qquad \xi_f = \chi_\uparrow^{(P)} \chi_\uparrow^{(N)}$$
$$\chi_\uparrow^{(P)} \chi_\downarrow^{(N)} \qquad\qquad \chi_\uparrow^{(P)} \chi_\downarrow^{(N)}$$
$$\chi_\downarrow^{(P)} \chi_\uparrow^{(N)} \qquad\qquad \chi_\downarrow^{(P)} \chi_\uparrow^{(N)}$$
$$\chi_\downarrow^{(P)} \chi_\downarrow^{(N)} \qquad\qquad \chi_\downarrow^{(P)} \chi_\downarrow^{(N)}$$

Use

$$\boldsymbol{\sigma}_P \cdot \boldsymbol{\sigma}_N = \sigma_z^{(P)}\sigma_z^{(N)} + 2(\sigma_+^{(P)}\sigma_-^{(N)} + \sigma_-^{(P)}\sigma_+^{(N)})$$

where

$$\sigma_+ = \frac{\sigma_x + i\sigma_y}{2} = \begin{pmatrix} 0 & 1 \\ 0 & 0 \end{pmatrix} \qquad \sigma_- = \frac{\sigma_x - i\sigma_y}{2} = \begin{pmatrix} 0 & 0 \\ 1 & 0 \end{pmatrix}$$

in the representation in which $\sigma_z = \begin{pmatrix} 1 & 0 \\ 0 & -1 \end{pmatrix}$ and $\chi_\uparrow = \begin{pmatrix} 1 \\ 0 \end{pmatrix}$, $\chi_\downarrow = \begin{pmatrix} 0 \\ 1 \end{pmatrix}$ to calculate all 16 scattering amplitudes. Make a table of your results and also tabulate the cross sections.

5. If any one of the spin states (e.g., initial proton, or initial neutron) is not measured, the cross section is the sum over the unmeasured spin states. Suppose both the initial and final proton spins are not measured. Write down expressions for the cross sections for the final neutron "up" and the final neutron "down," given that the initial neutron state is "up." What is the polarization P, defined by

$$P = \frac{\sigma \uparrow - \sigma \downarrow}{\sigma \uparrow + \sigma \downarrow}$$

where $\sigma \uparrow$ is the cross section with the final neutron up and so on?

6. Use the table computed in Problem 4 to calculate the cross sections for triple $\rightarrow$ triplet and singlet $\rightarrow$ singlet scattering, respectively. Show the triplet $\rightarrow$ singlet scattering vanishes. Check your results by observing that since (in units of $\hbar$)

$$\tfrac{1}{2}\boldsymbol{\sigma}_P + \tfrac{1}{2}\boldsymbol{\sigma}_N = \mathbf{S}$$

one has

$$\boldsymbol{\sigma}_P \cdot \boldsymbol{\sigma}_N = 2\mathbf{S}^2 - 3$$
$$= \begin{array}{ll} 1 & \text{when acting on triplet state} \\ -3 & \text{when acting on singlet state} \end{array}$$

Note that the amplitude is independent of m_S so that m_S must be the same in the initial and final spin states. There are three states in the triplet, all contributing an equal amount to the cross section, and only one to the singlet cross section.

   [*Caution:* In calculating amplitudes such as

$$\frac{1}{\sqrt{2}} (\chi_\uparrow^{(P)} \chi_\downarrow^{(N)}) - \chi_\downarrow^{(P)} \chi_\uparrow^{(N)})(A + B\boldsymbol{\sigma}_P \cdot \boldsymbol{\sigma}_N) \frac{1}{\sqrt{2}} (\chi_\uparrow^{(P)} \chi_\downarrow^{(N)}) - \chi_\downarrow^{(P)} \chi_\uparrow^{(N)})$$

the amplitudes are added for the four terms before squaring. Can you explain why?]

7. Consider the integral

$$I(kr) = \int_0^\pi d\theta \sin \theta \, g(\cos \theta) \, e^{-ikr\cos\theta} = \int_{-1}^1 du \, g(u)e^{-ikru}$$

where $g(\cos \theta)$ is strongly localized about $\theta = \theta_0$, and is infinitely differentiable. An example of such a function would be

$$g = e^{-\alpha^2(\cos\theta - \cos\theta_0)^2}$$

with α large. We may thus assume that $g(u)$ and all of its derivatives at $u = \pm 1$ vanish. In that case, show that $I(kr)$ vanishes faster than any power of kr as $kr \rightarrow \infty$.

(*Hint:* Write $e^{-ikru} = \frac{i}{kr} \frac{d}{du} e^{-ikru}$ and integrate by parts repeatedly.)

References

Scattering theory is discussed in all of the textbooks listed at the end of this volume. In addition there exist a number of advanced treatises that are devoted to this subject alone. Most accessible to students at this level is

N. F. Mott and H. S. W. Massey, *The Theory of Atomic Collisions* (3rd edition), Oxford University Press (Clarendon), Oxford, 1965.

More formal is

L. S. Rodberg and R. M. Thaler, *Introduction to the Quantum Theory of Scattering,* Academic Press, New York, 1967.

More advanced are

M. L. Goldberger and K. M. Watson, *Collison Theory,* John Wiley & Sons, New York, 1965.
R. Newton, *Scattering Theory of Waves and Particles,* McGraw-Hill, New York, 1966.

Chapter 24

THE ABSORPTION OF RADIATION IN MATTER

The process that is the inverse of radiative decay of atoms, namely the capture of photons accompanied by the excitation of atoms, can also take place. For photon energies exceeding the ionization energy of the atom, the electron is excited to the continuum. This is called the *photoelectric effect*, and is an important mechanism in the absorption of radiation in matter.

According to the Golden Rule (21-59), the transition rate for the process

$$\gamma + (\text{atom}) \rightarrow (\text{atom})' + e \qquad (24\text{-}1)$$

is given by

$$
\begin{aligned}
R &= \frac{2\pi}{\hbar} \int \frac{V\, d^3\mathbf{p}_e}{(2\pi\hbar)^3}\, |M_{fi}|^2\, \delta\left(\hbar\omega - E_B - \frac{p_e^2}{2m}\right) \\
&= \frac{2\pi}{\hbar} \int \frac{d\Omega V}{(2\pi\hbar)^3} \int mp_e\, d\left(\frac{p_e^2}{2m}\right) |M_{fi}|^2\, \delta\left(\hbar\omega - E_B - \frac{p_e^2}{2m}\right) \qquad (24\text{-}2) \\
&= \frac{2\pi V}{\hbar} \int d\Omega\, \frac{mp_e}{(2\pi\hbar)^3}\, |M_{fi}|^2
\end{aligned}
$$

In this expression, m is the electron mass, the delta function represents energy conservation, E_B is the magnitude of the binding energy of the electron in the atom, and in the last line, p_e is evaluated at the vanishing of the argument of the delta function.

415

The matrix element is given by

$$\frac{e}{mc} \left(\frac{2\pi\hbar c^2}{\omega V}\right)^{1/2} \int d^3r \psi_f^*(\mathbf{r})\, \boldsymbol{\varepsilon} \cdot \mathbf{p}\, e^{i\mathbf{k}\cdot\mathbf{r}}\, \psi_i(\mathbf{r}) \tag{24-3}$$

The vector potential is normalized, as in Chapter 21, to one photon in the volume V, and $\psi_i(\mathbf{r})$, $\psi_f(\mathbf{r})$ are the wave functions for the electron in the initial and final states. If we consider a hydrogenlike atom, and assume that the electron is in the ground state, we have

$$\psi_i(\mathbf{r}) = \frac{1}{\sqrt{\pi}} \left(\frac{Z}{a_0}\right)^{3/2} e^{-Zr/a_0} \tag{24-4}$$

The final-state wave function should be taken to be a solution of the Schrödinger equation with a Coulomb potential with $E > 0$. We did not discuss these solutions when we studied the hydrogen atom. They can be written in closed form but they are quite complicated, as is the integral in (24-3). If the photon energy is much larger than the ionization energy, then the residual interaction of the outgoing electron with the ion that it leaves behind becomes less important, and we may approximate $\psi_f(\mathbf{r})$ by a plane wave. Since we assume that we only have one atom in our volume, we will have only one electron in the volume, and hence the normalization is such that

$$\psi_f(\mathbf{r}) = \frac{1}{\sqrt{V}} e^{i\mathbf{p}_e\cdot\mathbf{r}/\hbar} \tag{24-5}$$

The factor V that appears in the phase space $[V\, d^3p/(2\pi\hbar)^3]$ corresponds to the same normalization, that is, the two factors are not independent. The square of the matrix element is somewhat simplified since the final state is an eigenstate of momentum, so that

$$\langle f | \boldsymbol{\varepsilon} \cdot \mathbf{p}_{op}\, e^{i\mathbf{k}\cdot\mathbf{r}} | i \rangle = \boldsymbol{\varepsilon} \cdot \mathbf{p}_e \langle f | e^{i\mathbf{k}\cdot\mathbf{r}} | i \rangle \tag{24-6}$$

Hence the square of the matrix element is

$$|M_{fi}|^2 \simeq \left(\frac{e}{mc}\right)^2 \frac{2\pi\hbar c^2}{\omega V} \cdot \frac{1}{V}\frac{1}{\pi}\left(\frac{Z}{a_0}\right)^3 (\boldsymbol{\varepsilon} \cdot \mathbf{p}_e)^2$$
$$\times \left| \int d^3r\, e^{i(\mathbf{k}-\mathbf{p}_e/\hbar)\cdot\mathbf{r}}\, e^{-Zr/a_0} \right|^2 \tag{24-7}$$

We will evaluate the integral later. At this point we note that the rate has a $1/V$ behavior due to the fact that we are dealing with a single photon in the volume V. Instead, we will consider the cross section for the photoelectric effect. To have a flux of one photon per square centimeter we must have a density of photons $1/c$ per cubic centimeter (so that a cylinder of unit area base and length c corresponding to a time interval of 1 sec contain one photon), that is, we must multiply the rate by V/c. We get, combining (24-2) and (24-7), the differential cross section

$$\frac{d\sigma}{d\Omega} = \frac{2\pi}{\hbar} \frac{mp_e}{(2\pi\hbar)^3} \left(\frac{e}{mc}\right)^2 \frac{2\pi\hbar c^2}{\omega} \frac{1}{\pi} \left(\frac{Z}{a_0}\right)^3 (\boldsymbol{\varepsilon} \cdot \mathbf{p}_e)^2$$
$$\times \frac{1}{c} \left| \int d^3\mathbf{r}\, e^{i(\mathbf{k} - \mathbf{p}e/\hbar)\cdot\mathbf{r}}\, e^{-r/a_0} \right|^2 \tag{24-8}$$

In this expression $d\Omega$ is the solid angle into which $\mathbf{p}_e$ points. The integral over all electron directions yields the total cross section σ for the photoelectric effect. If the target atoms are distributed with a density of N atoms per cubic centimeter, then in a slab of target material of area A and thickness dx, there are $NA\,dx$ target atoms. Each atom has a cross section σ for the reaction under consideration, so that the total effective area presented to the beam is $NA\,\sigma\,dx$. If there are n incident particles in the bombarding beam, then the number of particles that interact in the thickness dx of the target is given by

$$\frac{\text{interacting particles}}{\text{incident particles}} = \frac{\text{cross section}}{\text{total area}}$$

that is,

$$\frac{dn}{n} = -\frac{NA\sigma\,dx}{A} = -N\sigma\,dx \tag{24-9}$$

The minus sign indicates that particles are removed from the beam. Integration gives

$$n(x) = n_0\, e^{-N\sigma x} \tag{24-10}$$

where n_0 is the number of incident particles and $n(x)$ is the number of particles left in the beam after traversing a thickness x of the target. The quantity $\lambda = 1/N\sigma$ has the dimensions of a length and is called the *mean free path*. One sometimes speaks of the mean free path for the photoelectric effect, for pair production, and so on, even though what is measured is the cross section.

To get an idea of the magnitudes of mean free paths, note that $N = N_0\rho/A$ where $N_0 = 6.02 \times 10^{23}$ is Avogadro's number, ρ is the density in grams per cubic centimeter, and A is the atomic weight. Cross sections for molecular collisions can be estimated from the properties of gases, and they turn out to have magnitudes of the order of 10^{-16} cm^2, consistent with the fact that atomic dimensions are of the order of 10^{-8} cm.[1] Is this a reasonable guess for the photoelectric cross section? We shall soon examine the reasons why it is not. In the meantime we write the mean free path in centimeters, in a material of density ρ and atomic weight A, with the cross section expressed in units of 10^{-24} cm^2, called *barns* thus

$$\lambda = \frac{1}{N\sigma} = \frac{A}{\rho} \frac{1}{6.02 \times 10^{23}\,\sigma}$$
$$= \frac{A}{\rho} \frac{1.67}{\sigma\,(\text{barns})} \tag{24-11}$$

[1] By the same token, nuclear cross sections tend to be of the order of 10^{-24} cm^2 (barns), particle physics cross sections are of the order of 10^{-27} cm^2 (millibarns), going down to microbarns for rarer reactions, and even down to 10^{-44} cm^2 for the extremely rare neutrino reactions at low energies.

To evaluate the cross section in (24-8) we need to work out the integral

$$\int d^3\mathbf{r} \ e^{i(\hbar \mathbf{k} - \mathbf{p}e)\cdot\mathbf{r}/\hbar} \ e^{-Zr/a_0} \tag{24-12}$$

If we use the integral evaluated in (23-87), which, with a slight change of notation reads

$$\int d^3\mathbf{r} \ e^{i\Delta\cdot\mathbf{r}} \frac{e^{-\mu r}}{r} = \frac{4\pi}{\mu^2 + \Delta^2} \tag{24-13}$$

we can, by differentiating with respect to μ obtain

$$\int d^3r \ e^{-i\Delta\cdot\mathbf{r}} \ e^{-\mu r} = \frac{8\pi\mu}{(\mu^2 + \Delta^2)^2} \tag{24-14}$$

so that we can finally calculate the cross section. After some judicious combining of factors, we end up with

$$\frac{d\sigma}{d\Omega} = 32Z^5a_0^2 \left(\frac{p_ec}{\hbar\omega}\right)\left(\frac{\varepsilon\cdot\mathbf{p}_e}{mc}\right)^2 \frac{1}{(Z^2 + a_0^2\Delta^2)^4} \tag{24-15}$$

where

$$\Delta = \frac{(\hbar\mathbf{k} - \mathbf{p}_e)}{\hbar} = \frac{(\mathbf{p}_\gamma - \mathbf{p}_e)}{\hbar}$$

Since the electron and photon energies are related by

$$\hbar\omega = E_B + \frac{p_e^2}{2m} \tag{24-16}$$

we see that for energies quite a bit above the binding energies, $\hbar\omega \cong p_e^2/2m$. Hence

$$\frac{p_ec}{\hbar\omega}\left(\frac{\varepsilon\cdot\mathbf{p}_e}{mc}\right) \cong \frac{2p_e}{mc}(\varepsilon\cdot\hat{\mathbf{p}}_e)^2$$

$$\Delta^2 = \frac{1}{\hbar^2}(\mathbf{p}_\gamma - \mathbf{p}_e)^2 = \frac{1}{\hbar^2}\left[\left(\frac{\hbar\omega}{c}\right)^2 - 2\frac{\hbar\omega}{c}p_e\hat{\mathbf{p}}_\gamma\cdot\hat{\mathbf{p}}_e + p_e^2\right]$$

$$\cong \frac{1}{\hbar^2}\left[p_e^2 - \left(\frac{p_e^3}{mc}\right)\hat{\mathbf{p}}_\gamma\cdot\hat{\mathbf{p}}_e\right] \tag{24-17}$$

$$\cong \frac{p_e^2}{\hbar^2}\left(1 - \frac{v_e}{c}\hat{\mathbf{p}}_\gamma\cdot\hat{\mathbf{p}}_e\right)$$

for nonrelativistic electrons, $p_e \ll mc.$[2] We have used the notation $\mathbf{p}_\gamma = \hbar\mathbf{k}$ for the photon momentum, and as usual the ˆ denotes the unit vector. Thus

$$\frac{d\sigma}{d\Omega} = 64Z^5 a_0^2 \left(\frac{p_e}{mc}\right) \frac{(\varepsilon \cdot \hat{\mathbf{p}}_e)^2}{\left[Z^2 + \dfrac{p_e^2}{\alpha^2 m^2 c^2}\left(1 - \dfrac{v_e}{c}\,\hat{\mathbf{p}}_e \cdot \hat{\mathbf{p}}_\gamma\right)\right]^4}$$

$$= \frac{64Z^5 \alpha^8 a_0^2 \left(\dfrac{p_e}{mc}\right)(\varepsilon \cdot \hat{\mathbf{p}}_e)^2}{\left[(\alpha Z)^2 + \dfrac{p_e^2}{m^2 c^2}\left(1 - \dfrac{v_e}{c}\,\hat{\mathbf{p}}_e \cdot \hat{\mathbf{p}}_\gamma\right)\right]^4}$$

(24-18)

If we choose the photon direction to define the z-axis, and the two photon polarization directions $\varepsilon^{(1)}$, $\varepsilon^{(2)}$ to point in the x and y directions, respectively, then, writing

$$\hat{\mathbf{p}}_e = (\sin\theta\cos\phi,\ \sin\theta\sin\phi,\ \cos\theta) \tag{24-19}$$

we have $(\mathbf{p}_e \cdot \varepsilon^{(1)})^2 = \sin^2\theta\cos^2\phi$ and $(\mathbf{p}_e \cdot \varepsilon^{(2)})^2 = \sin^2\theta\sin^2\phi$ so that the *average* of the numerator over the two polarization directions (we are calculating the photoeffect cross section with unpolarized photons) is

$$\overline{(\hat{\mathbf{p}}_e \cdot \varepsilon)^2} = \tfrac{1}{2}(\sin^2\theta\sin^2\phi + \sin^2\theta\cos^2\phi) = \tfrac{1}{2}\sin^2\theta. \tag{24-20}$$

Also

$$\hat{\mathbf{p}}_e \cdot \hat{\mathbf{p}}_\gamma = \cos\theta \tag{24-21}$$

so that, writing $p_e^2/2m = E$, we get

$$\frac{d\sigma}{d\Omega} = \frac{32\sqrt{2}\,Z^5 \alpha^8 a_0^2 (E/mc^2)^{1/2}\sin^2\theta}{\left[(\alpha Z)^2 + \dfrac{2E}{mc^2}\left(1 - \dfrac{v_e}{c}\cos\theta\right)\right]^4} \tag{24-22}$$

For light elements, the condition that we imposed earlier, $\hbar\omega \gg E_B$, which is equivalent to

$$E \gg \tfrac{1}{2} mc^2(Z\alpha)^2 \tag{24-23}$$

is satisfied over a reasonably wide range of energies. If we insert (24-23) into the cross section, we find that the denominator simplifies, and we get

$$\frac{d\sigma}{d\Omega} = 2\sqrt{2}\,Z^5 \alpha^8 a_0^2 \left(\frac{E}{mc^2}\right)^{-7/2} \frac{\sin^2\theta}{\left(1 - \dfrac{v_e}{c}\cos\theta\right)^4} \tag{24-24}$$

Let us discuss various aspects of this formula.

[2]For relativistic electrons one should really use the Dirac equation to describe the process. Effects other than the photoelectric effect are more important when $E_e \simeq 1$ MeV.

(1) First, the vague guess that since atomic sizes tend to be of the order of 10^{-8} cm, the cross sections should be of order 10^{-16} cm^2 is wrong! It is true that the factor a_0^2 is of that magnitude, but it is multiplied by $(1/137)^8$, which is dimensionless, but hardly negligible! We should try to understand how one could be so wrong in order to have some guidance on what one must be careful about in making estimates. If we ignore the last angular factor, which we will discuss later, we see that we may, with the help of

$$E = \tfrac{1}{2} m v_e^2$$

write the factor in front as

$$2\sqrt{2}\, a_0^2 Z^5 \alpha^8 \left(\frac{mc^2}{E}\right)^{7/2} = 32 a_0^2 Z^5 \alpha^8 \left(\frac{c}{v_e}\right)^7$$

$$= 32 \left(\frac{a_0}{Z}\right)^2 \alpha \left(\frac{\alpha Z c}{v_e}\right)^7 \tag{24-25}$$

This is a more useful form. It shows, first of all, the presence of a single factor α, which should always be present when a single photon is emitted or absorbed. The coupling of the vector potential to a charge is proportional to the charge e, and the square of this will lead to the α. The factor $(a_0/Z)^2$ is a better measure of the area of the atom than a_0^2, since we are considering a hydrogenlike atom of charge Z. What remains is a rather high power of the ratio of the "orbital" velocity of the electron in the atom to the velocity of the outgoing free electron.

It is the ratio $(\alpha Z c/v_e)$ [rather than just (c/v_e), which is also dimensionless] that appears, because the matrix element involves the overlap between the free electron wave function and the bound electron wave function, that is, the square matrix element is related to the probability that a measurement of the momentum of the bound electron yields p_e. The functional dependence $f(\alpha Z c/v_e)$, in this case the eighth power,[3] cannot be guessed at on general qualitative grounds. For example, if the electron wave function were Gaussian [$\psi_i(\mathbf{r}) \propto e^{-r^2/a^2}$], the falloff with increasing velocity would be much faster than the eighth power. The reason why a guess is hard to make is that the momentum distribution of the electron is localized in a region of spread

$$\Delta p \sim \frac{\hbar}{a_0/Z} \sim \frac{\hbar Z}{\hbar/mc\alpha} \sim Z\alpha mc \tag{24-26}$$

and for $p_e \gg Z\alpha mc$ one is far out in the tail of the momentum distribution. This, again by the uncertainty relation, depends on the small r-distribution of the wave function, and depends sensitively on the state, in particular on the angular momentum. This does make photodisintegration in nuclear physics a very useful tool.

(2) The angular distribution of $d\sigma/d\Omega$ is given by

$$F(\theta) = \frac{\sin^2 \theta}{[1 - (v_e/c)\cos \theta]^4} \tag{24-27}$$

---

[3]There is a factor p_e in the phase space so that the matrix element squared gives an eighth power of $(\alpha c Z/v_e)$.

We note, first of all, that the cross section vanishes in the forward direction. This is a consequence of the fact that photons are transversely polarized. The matrix element is proportional to $\mathbf{p}_e \cdot \boldsymbol{\varepsilon}$, and when $\mathbf{p}_e$ is parallel to the photon momentum, this factor vanishes. The factor in the denominator has, because of the fourth power, a strong influence on the angular distribution. When v_e/c approaches unity, this becomes very dramatic, but even for moderate v_e/c there is significant peaking in the near forward direction, where the denominator is at its smallest. This corresponds to the minimum value of the momentum transfer between photon and electron $(\mathbf{p}_\gamma - \mathbf{p}_e)^2$.

More detailed calculations need to be done to cover the relativistic region. The formula just derived works well in the region of its validity.[4] At very low energies, a more accurate wave function for the outgoing electron must be used. Such a wave function will reflect the Coulomb interaction between the nucleus and the electron. There will, of course, be no photoeffect below the threshold for ionizing the least-bound electron from an outer shell. As the energy increases above the threshold, electrons from deeper shells will be photoproduced. If one plots the integrated cross section, or preferably the *mass absorption coefficient*[5], $N\sigma/\rho$, as a function of the photon wavelength, one finds the data shown on Fig. 24-1. The so-

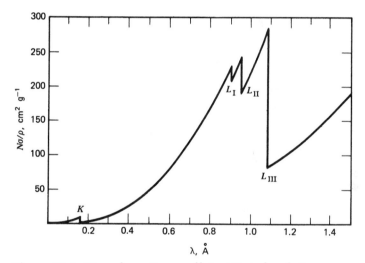

Figure 24-1. Mass absorption coefficient $N\sigma/\rho$ for platinum as a function of photon wavelength.

called K-edge corresponds to the ejection of the $n = 1$ electrons; the L-edges correspond to the various electrons in the $n = 2$ states. The edges occur at the binding energies of the various electrons. Moseley's empirical law states that they are located at

$$E = 13.6 \, \frac{(Z - \sigma_n)^2}{n^2} \, \text{eV} \tag{24-28}$$

---

[4] In calculating absorption of radiation, the result that we derived must be multiplied by 2, since there are two electrons in the ground state, except in hydrogen.

[5] This is equal to $N_0\sigma/A$ where N_0 is Avogadro's number and A is the atomic weight.

where σ_n, the "screening constants," are approximately given by $\sigma_n = 2n + 1$. This formula is just what we expect for the ns orbitals, and the screening is the effect of all the other s electrons.

At relativistic energies the cross section drops less precipitously, with an $(E/m)^{-1}$ behavior instead of $(E/m)^{-7/2}$, but by the time energies of 0.5 MeV are reached, the photoelectric effect ceases to be of any importance as far as the absorption of radiation is concerned. In the energy region of 0.5–5 MeV, say, it is the *Compton effect* that is the dominant absorptive effect.

Here free electrons scatter photons. At low frequencies the effect can be understood classically; electromagnetic radiation impinging on the electron accelerates it, and the radiation emitted by the accelerated charge is the scattered radiation. The classically calculated *Thomson* cross section is

$$\sigma_T = \frac{8\pi}{3} \left(\frac{e^2}{mc^2} \right)^2 \tag{24-29}$$

In quantum mechanics, the scattering amplitude (matrix element) must be proportional to e^2, since two photons are involved. Since the perturbation in the Hamiltonian is

$$\frac{e}{mc} \mathbf{p} \cdot \mathbf{A}(\mathbf{r}, t) + \frac{e^2}{2mc^2} \mathbf{A}^2(\mathbf{r}, t) \tag{24-30}$$

when both terms in the expansion of (21-23) are kept, we see than an e^2 contribution to the scattering amplitude can come from two sources.

1. The first source is a first-order contribution from the term $e^2\mathbf{A}^2(\mathbf{r}, t)/2mc^2$.

2. The second source is a second-order perturbation term from the coupling $e\mathbf{p} \cdot \mathbf{A}(\mathbf{r}, t)/mc$. Since we have not developed the second-order perturbation formalism, we will restrict ourselves to stating the results.

(a) At threshold, with the gauge that we have been using, $\boldsymbol{\nabla} \cdot \mathbf{A}(\mathbf{r}, t) = 0$, the whole amplitude comes from the term involving $e^2\mathbf{A}^2(\mathbf{r}, t)/2mc^2$.

(b) The matrix element in second order has the form

$$-\sum_n \frac{\langle f|e\mathbf{p} \cdot \mathbf{A}/mc|n\rangle\langle n|e\mathbf{p} \cdot \mathbf{A}/mc|i\rangle}{E_n - E_i} \tag{24-31}$$

where the "sum" over intermediate states "n" also implies integration over all the momenta, when "n" includes continuum states. It is not enough to include intermediate one-electron states corresponding to the sequence

$$\gamma_i + e_i \to e' \to \gamma_f + e_f$$

and the intermediate states containing an electron and two photons, corresponding to the process

$$e_i + \gamma_i \to \gamma_i + \gamma_f + e' \to \gamma_f + e_f$$

It turns out that it is necessary to include the possibility of the "virtual" creation of an electron–positron pair by the incident photon, followed by the annihilation of the positron by the incident electron, with the emission of the final photon, as in

$$e_i + \gamma_i \rightarrow e_i + e_f + e^{+'} \rightarrow \gamma_f + e_f$$

and the process

$$e_i + \gamma_i \rightarrow e_i + \gamma_i + \gamma_f + e_f + e^{+'} \rightarrow \gamma_f + e_f$$

The calculation leads to the *Klein-Nishina formula*

$$\sigma = 2\pi \left(\frac{e^2}{mc^2}\right)^2 \left\{\frac{1+x}{x^2}\left[\frac{2(1+x)}{1+2x} - \frac{1}{x}\log(1+2x)\right]\right.$$
$$\left. + \frac{1}{2x}\log(1+2x) - \frac{1+3x}{(1+2x)^2}\right\} \tag{24-32}$$

$$x = \frac{\hbar\omega}{mc^2}$$

which is in excellent agreement with experiment. At low frequencies this becomes

$$\sigma = \frac{8\pi}{3}\left(\frac{e^2}{mc^2}\right)^2 (1 - 2x) \tag{24-33}$$

and at high frequencies ($x \gg 1$) this reads

$$\sigma = \pi \left(\frac{e^2}{mc^2}\right)^2 \frac{1}{x}(\log 2x + \tfrac{1}{2}) \tag{24-34}$$

Thus the Compton cross section, too, drops off at high energies. At energies above a few MeV, the dominant absorptive process is *pair production*.

It is a remarkable fact that a photon at high enough energies, $\hbar\omega > 2mc^2$ can "materialize" into an electron and a positron (Fig. 24-2). The latter can be properly called an "antielectron"; it has the same mass as the electron and the same spin, but its charge and magnetic moment have the same value with *opposite sign* as those for the electron, and the nonrelativistic coupling with the electromagnetic field is obtained by the replacement of $\mathbf{p}$ by $\mathbf{p} - e\mathbf{A}(\mathbf{r}, t)/c$. Such a materialization can only occur in the presence of a third particle, a nucleus, for example, since energy and momentum conservation cannot hold for the process

$$\gamma \rightarrow e + e^+$$

To see this without going through a long kinematical calculation, consider the inverse process $e + e^+ \rightarrow \gamma$ in the center-of-mass frame. The electron and positron have equal and opposite momenta, so that the final state has energy $2(m^2c^4 +$

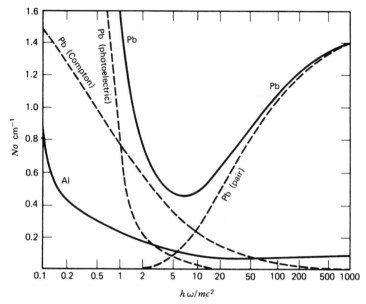

Figure 24-2. Total absorption coefficient for lead and aluminum as a function of energy, in units of the electron rest energy (0.51 MeV). The photoelectric cross section for Al is negligible on the scale depicted here.

$p^2c^2)^{1/2}$ and momentum 0. A photon of energy E must carry momentum E/c. If there is a nucleus present, it can absorb momentum and energy (for a massive nucleus this will be very small, $p^2/2M$), so that it becomes possible to balance energy and momentum.

The calculation of

$$\gamma + \text{nucleus} \rightarrow e + e^+ + \text{nucleus}$$

is beyond the scope of this book. The theory of quantum electrodynamics that is used in these calculations also shows that we can transfer particles from one side of the equation to the other, provided we change the transferred particles to their antiparticles. Thus one predicts that

$$\text{nucleus} + e^{\pm} \rightarrow \text{nucleus} + e^{\pm} + \gamma$$

should also occur, with a matrix element very closely related to that of pair production. This is in agreement with experiment, and the last process is responsible for *cosmic-ray showers*.

An incident γ ray of very high energy (it may come from the decay $\pi^\circ \rightarrow 2\gamma$, with the π° produced when a primary cosmic-ray proton hits a nucleus at the top of the atmosphere) will make a pair, with each member carrying roughly half the original energy. Each member can produce a photon, as indicated earlier,[6] and the end products can make further photons and pairs. Showers coming from extremely high energy events occurring at the top of the atmosphere can cover areas

---

[6]This process is called *Bermsstrahlung*, and can be understood classically; a charge deflected in the Coulomb field of the nucleus is accelerated, and hence radiates.

of several square miles! Less spectacular showers in counters are used to identify photons or electrons. An incident particle that is charged, but much heavier, will be deflected less, and will therefore radiate less.

Detailed calculations show that energy lost in material through these processes follows the law

$$E(x) = E_{inc} \, e^{-x/L} \tag{24-35}$$

where the "radiation length" is given by

$$L = \frac{(m^2c^2/\hbar^2) \, A}{4Z^2\alpha^3 N_0\rho \, \log \, (183/Z^{1/3})} \tag{24-36}$$

where $N_0 = 6.02 \times 10^{23}$ is Avogadro's number, m is the electron mass, A is the atomic weight, Z is the charge of the nucleus, and ρ is the density of the material in grams per cubic centimeter. The "pair production length" is given by

$$L_{pair} = \frac{9}{7} L \tag{24-37}$$

The formula is not good for very low Z. Typical values of L are

Air	330 m
Al	9.7 cm
Pb	0.53 cm

Bremsstrahlung is the dominant energy loss mechanism for electrons at high energies. At low energies ionization dominates. Lack of space keeps us from discussing this essentially classical effect.

Problems

1. Calculate the cross section for the process

$$\gamma + \text{deuteron} \rightarrow N + P$$

The procedure is the same as that for the photoelectric effect. In the calculation of the matrix element, the final state wave function is again

$$\psi_f(\mathbf{r}) = \frac{1}{\sqrt{V}} \, e^{i\mathbf{p}\cdot\mathbf{r}/\hbar}$$

where $\mathbf{p}$ is the proton momentum. At low energies the wavelength of the radiation is much larger than the "size" of the deuteron, so that $e^{i\mathbf{k}\cdot\mathbf{r}} \approx 1$. To calculate

$$\int d^3r \, e^{-i\mathbf{p}\cdot\mathbf{r}/\hbar} \, \psi_i(\mathbf{r})$$

use

$$\psi_i(\mathbf{r}) = \frac{N}{\sqrt{4\pi}} \, e^{-\alpha(r-r_0)} \qquad r > r_0$$

$$= 0 \qquad r < r_0$$

properly normalized. For what energies would you expect the photon wave-length to be much larger than the range of the potential $r_0 \cong 1.2$ fm?

2. The principle of detailed balance relates the matrix elements for the reactions

$$A + a \rightarrow B + b \qquad \text{I}$$

and

$$B + b \rightarrow A + a \qquad \text{II}$$

Thus

$$\sum |M_\text{I}|^2 = \sum |M_\text{II}|^2$$

where the sum is over both initial and final spin states. Taking into account that in the calculation of a rate or cross section one averages over the initial spin states and sums over the final spin states, show that for the rates

$$\frac{(2J_A + 1)(2J_a + 1)}{p_b^2(dp_b/dE_b)} \frac{dR_\text{I}}{d\Omega_b} = \frac{(2J_B + 1)(2J_b + 1)}{p_a^2(dp_a/dE_a)} \frac{dR_\text{II}}{d\Omega_a}$$

where J_a, J_A, J_b, J_B are the spins of the particles, p_b and p_a are the center-of-mass momenta of particles b and a (I and II must take place at the same total energy), E_b and E_a are the corresponding energies of the particles, and $d\Omega_b$, $d\Omega_a$ are the solid angles in which b and a are observed. Use this result to express the cross section for the radiative capture process

$$N + P \rightarrow D + \gamma$$

in terms of the cross section calculated in Problem 1. Note that the factor $(2J + 1)$ for photons is 2 since there are only two polarization states, and also the spin of the deuteron is 1.

3. The cross section for the reaction

$$\pi^+ + D \rightarrow P + P$$

has been measured for incident π^+ laboratory kinetic energy of 24 MeV, and found to be equal to 3.0×10^{-27} cm^2.

 (a) At what laboratory energy would one be able to carry out a test of detailed balance by measuring the cross section for

$$P + P \rightarrow \pi^+ + D$$

 (The pion mass is $m_\pi c^2 = 140$ MeV; $M_p c^2 = 940$ MeV; $M_D \cong 2M_p$.)

 (b) Given that the spin of the π^+ is 0, what is the predicted cross section for this reaction?

4. What is the radiation length in liquid xenon, for which $Z = 54$, $A = 131$, and $\rho = 3.09$ gm cm^{-3}?

5. Suppose the electron were bound to the nucleus by a square well potential. Calculate the energy dependence of the cross section for the photoelectric effect. Assume that the photon energy is much larger than the binding energy of the electron, and that the potential has a short range.
(*Hint:* See Problem 1.)

References

The mechanisms responsible for the absorption of radiation in matter are discussed in most of the modern physics textbooks e.g., Brehm and Mullin (see References). For a very complete discussion of the experimental techniques used in the measurement of the various effects, see also

E. Segre, *Nuclei and Particles*, W. A. Benjamin, New York, 1964.

THE FOURIER INTEGRAL AND DELTA FUNCTIONS

Consider a function $f(x)$ that is periodic, with period $2L$, so that

$$f(x) = f(x + 2L) \tag{A-1}$$

Such a function can be expanded in a Fourier series in the interval $(-L, L)$, and the series has the form

$$f(x) = \sum_{n=0}^{\infty} A_n \cos \frac{n\pi x}{L} + \sum_{n=1}^{\infty} B_n \sin \frac{n\pi x}{L} \tag{A-2}$$

We can rewrite the series in the form

$$f(x) = \sum_{n=-\infty}^{\infty} a_n e^{in\pi x/L} \tag{A-3}$$

which is certainly possible, since

$$\cos \frac{n\pi x}{L} = \frac{1}{2} (e^{in\pi x/L} + e^{-in\pi x/L})$$

$$\sin \frac{n\pi x}{L} = \frac{1}{2i} (e^{in\pi x/L} - e^{-in\pi x/L})$$

The coefficients can be determined with the help of the orthonormality relation

$$\frac{1}{2L} \int_{-L}^{L} dx \; e^{in\pi x/L} \, e^{-im\pi x/L} = \delta_{mn} = \begin{cases} 1 & m = n \\ 0 & m \neq n \end{cases} \tag{A-4}$$

Thus

$$a_n = \frac{1}{2L} \int_{-L}^{L} dx f(x) \, e^{-in\pi x/L} \tag{A-5}$$

Let us now rewrite (A-3) by introducing Δn, the difference between two successive integers. Since this is unity, we have

$$
\begin{aligned}
f(x) &= \sum_n a_n \, e^{in\pi x/L} \, \Delta n \\
&= \frac{L}{\pi} \sum_n a_n \, e^{in\pi x/L} \, \frac{\pi \, \Delta n}{L}
\end{aligned}
\tag{A-6}
$$

Let us change the notation by writing

$$\frac{\pi n}{L} = k \tag{A-7}$$

and

$$\frac{\pi \, \Delta n}{L} = \Delta k \tag{A-8}$$

We also write

$$\frac{L a_n}{\pi} = \frac{A\,(k)}{\sqrt{2\pi}} \tag{A-9}$$

Hence (A-6) becomes

$$f(x) = \sum \frac{A\,(k)}{\sqrt{2\pi}} \, e^{ikx} \, \Delta k \tag{A-10}$$

If we now let $L \to \infty$, then k approaches a continuous variable, since Δk becomes infinitesimally small. If we recall the Riemann definition of an integral, we see that in the limit (A-10) can be written in the form

$$f(x) = \frac{1}{\sqrt{2\pi}} \int_{-\infty}^{\infty} dk \, A(k) \, e^{ikx} \tag{A-11}$$

The coefficient $A(k)$ is given by

$$
\begin{aligned}
A(k) &= \sqrt{2\pi} \, \frac{L}{\pi} \cdot \frac{1}{2L} \int_{-L}^{L} dx f(x) \, e^{-in\pi x/L} \\
&\to \frac{1}{\sqrt{2\pi}} \int_{-\infty}^{\infty} dx f(x) \, e^{-ikx}
\end{aligned}
\tag{A-12}
$$

Equations A-11 and A-12 define the Fourier integral transformations. If we insert the second equation into the first we get

$$f(x) = \frac{1}{2\pi} \int_{-\infty}^{\infty} dk \, e^{ikx} \int_{-\infty}^{\infty} dy f(y) \, e^{-iky} \tag{A-13}$$

Suppose now that we interchange, without question, the order of integrations. We then get

$$f(x) = \int_{-\infty}^{\infty} dy f(y) \left[\frac{1}{2\pi} \int_{-\infty}^{\infty} dk \, e^{ik(x-y)} \right] \tag{A-14}$$

For this to be true, the quantity $\delta(x - y)$ defined by

$$\delta(x - y) = \frac{1}{2\pi} \int_{-\infty}^{\infty} dk \, e^{ik(x-y)} \tag{A-15}$$

and called the *Dirac Delta function* must be a very peculiar kind of function; it must vanish when $x \neq y$, and it must tend to infinity in an appropriate way when $x - y = 0$, since the range of integration is infinitesimally small. It is therefore not a function in the usual mathematical sense, but it is rather a "generalized function" or a "distribution."[1] It does not have any meaning by itself, but it can be defined provided it always appears in the form

$$\int dx f(x) \, \delta(x - a)$$

with the function $f(x)$ sufficiently smooth in the range of values that the argument of the delta function takes. We will take that for granted and manipulate the delta function by itself, with the understanding that at the end all the relations that we write down only occur under the integral sign.

The following properties of the delta function can be demonstrated:
(i)

$$\delta(ax) = \frac{1}{|a|} \delta(x) \tag{A-16}$$

This can be seen to follow from

$$f(x) = \int dy f(y) \, \delta(x - y) \tag{A-17}$$

If we write $x = a\xi$ and y and $a\eta$, then this reads

$$f(a\xi) = |a| \int d\eta f(a\eta) \, \delta[a(\xi - \eta)]$$

---

[1]The theory of distributions was developed by the mathematician Laurent Schwartz. An introductory treatment may be found in M. J. Lighthill, *Introduction to Fourier Analysis and Generalized Functions,* Cambridge University Press, Cambridge, England, 1958.

On the other hand,

$$f(a\xi) = \int d\eta f(a\eta)\, \delta(\xi - \eta)$$

which implies our result.

(ii) A relation that follows from (A-16) is

$$\delta(x^2 - a^2) = \frac{1}{2|a|}\, [\delta(x - a) + \delta(x + a)] \tag{A-18}$$

This follows from the fact that the argument of the delta function vanishes at $x = a$ and $x = -a$. Thus there are two contributions:

$$\delta(x^2 - a^2) = \delta[(x - a)(x + a)]$$

$$= \frac{1}{|x + a|}\, \delta(x - a) + \frac{1}{|x - a|}\, \delta(x + a)$$

$$= \frac{1}{2|a|}\, [\delta(x - a) + \delta(x + a)]$$

More generally, one can show that

$$\delta[f(x)] = \sum_i \frac{\delta(x - x_i)}{|df/dx|_{x=x_i}} \tag{A-19}$$

where the x_i are the roots of $f(x)$ in the interval of integration.

In addition to the representation (A-15) of the delta function, there are other representations that may prove useful. We discuss several of them.

(a) Consider the form (A-15), which we write in the form

$$\delta(x) = \frac{1}{2\pi} \lim_{L \to \infty} \int_{-L}^{L} dk\, e^{ikx} \tag{A-20}$$

The integral can be done, and we get

$$\delta(x) = \lim_{L \to \infty} \frac{1}{2\pi} \frac{e^{iLx} - e^{-iLx}}{ix}$$

$$= \lim_{L \to \infty} \frac{\sin Lx}{\pi x} \tag{A-21}$$

(b) Consider the function $\Delta(x, a)$ defined by

$$\Delta(x, a) = 0 \qquad\qquad x < -a$$

$$= \frac{1}{2a} \qquad\qquad -a < x < a \tag{A-22}$$

$$= 0 \qquad\qquad a < x$$

Then

$$\delta(x) = \lim_{a \to 0} \Delta(x, a) \tag{A-23}$$

It is clear that an integral of a product of $\Delta(x, a)$ and a function $f(x)$ that is smooth near the origin will pick out the value at the origin

$$\lim_{a \to 0} \int dx f(x) \Delta(x, a) = f(0) \lim_{a \to 0} \int dx \Delta(x, a)$$

$$= f(0)$$

(c) By the same token, any peaked function, normalized to unit area under it, will approach a delta function in the limit that the width of the peak goes to zero. We will leave it to the reader to show that the following are representations of the delta function

$$\delta(x) = \lim_{a \to 0} \frac{1}{\pi} \frac{a}{x^2 + a^2} \tag{A-24}$$

and

$$\delta(x) = \lim_{\alpha \to \infty} \frac{\alpha}{\sqrt{\pi}} e^{-\alpha^2 x^2} \tag{A-25}$$

(d) We will have occasion to deal with *orthonormal polynomials*, which we denote by the general symbol $P_n(x)$. These have the property that

$$\int dx P_m(x) \, P_n(x) \, w(x) = \delta_{mn} \tag{A-26}$$

where $w(x)$ may be unity or some simple function, called the weight function. For functions that may be expanded in a series of these orthogonal polynomials, we can write

$$f(x) = \sum_n a_n P_n(x) \tag{A-27}$$

If we multiply both sides by $w(x)P_m(x)$ and integrate over x, we find that

$$a_m = \int dy w(y) \, f(y) \, P_m(y) \tag{A-28}$$

We can insert this into (A-27) and, prepared to deal with "generalized functions," we freely interchange sum and integral. We get

$$f(x) = \sum_n P_n(x) \int dy w(y) \, f(y) \, P_n(y)$$

$$= \int dy f(y) \left(\sum_n P_n(x) \, w(y) \, P_n(y) \right) \tag{A-29}$$

Thus we get still another representation of the delta function. Examples of the $P_n(x)$ are Legendre polynomials, Hermite polynomials, and Laguerre polynomials, all of which make their appearance in quantum mechanical problems.

Since the delta function always appears multiplied by a smooth function under an integral sign, we can give meaning to its derivatives. For example

$$\int_{-\epsilon}^{\epsilon} dx f(x) \frac{d}{dx} \delta(x) = \int_{-\epsilon}^{\epsilon} dx \frac{d}{dx} [f(x) \delta(x)] - \int_{-\epsilon}^{\epsilon} dx \frac{df(x)}{dx} \delta(x)$$

$$= -\int_{-\epsilon}^{\epsilon} dx \frac{df(x)}{dx} \delta(x) \qquad \text{(A-30)}$$

$$= -\left(\frac{df}{dx}\right)_{x=0}$$

and so on. The delta function is an extremely useful tool, and the student will encounter it in every part of mathematical physics.

The integral of a delta function is

$$\int_{-\infty}^{x} dy\, \delta(y - a) = 0 \qquad x < a$$

$$= 1 \qquad x > a \qquad \text{(A-31)}$$

$$\equiv \theta(x - a)$$

which is the standard notation for this discontinuous function. Conversely, the derivative of the so-called *step function* is the Dirac delta function:

$$\frac{d}{dx} \theta(x - a) = \delta(x - a) \qquad \text{(A-32)}$$

OPERATORS

In this appendix we discuss some topics related to linear operators. The set of admissible wave functions are square integrable functions. Since

$$\psi(x) = \alpha\psi_1(x) + \beta\psi_2(x) \tag{B-1}$$

is square integrable, if $\psi_1(x)$ and $\psi_2(x)$ are square integrable and α, β are arbitrary complex numbers, we say that the ψ's form a *linear space*. An operator A on this space is a mapping:

$$A\psi(x) = \phi(x) \tag{B-2}$$

where $\phi(x)$ is also square integrable. Among all the operators there is a subset called *linear operators*, which have the property that

$$A\alpha\psi(x) = \alpha A\psi(x) \tag{B-3}$$

where α is an arbitrary complex constant, and

$$A[\alpha\psi_1(x) + \beta\psi_2(x)] = \alpha A\psi_1(x) + \beta A\psi_2(x) \tag{B-4}$$

with α, β being complex numbers. A further subset is the *hermitian operators* for which the expectation value for all admissible $\psi(x)$,

$$\langle A \rangle_\psi = \int dx\psi^*(x)\, A\psi(x) \tag{B-5}$$

is real. First we prove that for all admissible ψ_1 and ψ_2

$$\int \psi_2^*(x)\, A\psi_1(x)\, dx = \int [A\psi_2(x)]^*\, \psi_1(x)\, dx \qquad \text{(B-6)}$$

holds.

The reality of $\langle A \rangle$ implies that

$$\int dx \psi^*(x)\, A\psi(x) = \int dx\, [A\psi(x)]^*\, \psi(x) \qquad \text{(B-7)}$$

Now substitute for $\psi(x)$

$$\psi(x) = \psi_1(x) + \lambda \psi_2(x) \qquad \text{(B-8)}$$

This implies that

$$\int dx(\psi_1^* + \lambda^*\psi_2)\, A(\psi_1 + \lambda\psi_2) = \int dx(\psi_1 + \lambda\psi_2)(A\psi_1 + \lambda A\psi_2)^* \qquad \text{(B-9)}$$

Using hermiticity, that is,

$$\int dx \psi_i^* A\psi_i = \int dx \psi_i (A\psi_i)^* \qquad i = 1.2 \qquad \text{(B-10)}$$

we obtain

$$\lambda^* \int \psi_2^* A\psi_1 + \lambda \int \psi_1^* A\psi_2 = \lambda \int \psi_2(A\psi_1)^* + \lambda^* \int \psi_1(A\psi_2)^* \qquad \text{(B-11)}$$

Since λ is an arbitrary complex number, the relations for the coefficient of λ and for the coefficient of λ^* must separately hold. Thus

$$\int dx \psi_2^* A\psi_1 = \int dx (A\psi_2)^* \psi_1 \qquad \text{(B-12)}$$

The next result that we wish to prove is that *eigenfunctions of a hermitian operator corresponding to different eigenvalues are orthogonal.* Consider the two equations

$$A\psi_1(x) = a_1\psi_1(x)$$

and

$$[A\psi_2(x)]^* = a_2\psi_2^*(x) \qquad \text{(B-13)}$$

Note that a_2 is real since the eigenvalues of a hermitian operator are real. Take the scalar product of the first equation with ψ_2^* and the second equation with ψ_1. Thus

$$\int dx \psi_2^* A\psi_1(x) = a_1 \int \psi_2^*(x)\, \psi_1(x)\, dx$$

$$\int dx (A\psi_2)^* \psi_1(x) = a_2 \int \psi_2^*(x)\, \psi_1(x)\, dx \qquad \text{(B-14)}$$

Subtracting, we get

$$(a_1 - a_2) \int \psi_2^*(x)\, \psi_1(x)\, dx = \int dx \psi_2^* A \psi_1 - \int dx (A\psi_2)^* \, \psi_1$$
$$= 0$$

(B-15)

Thus, if $a_1 \neq a_2$, we have

$$\int \psi_2^*(x)\, \psi_1(x)\, dx = 0$$

(B-16)

If we define the hermitian conjugate of the operator A by $A^\dagger$, so that

$$\int dx (A\psi_2)^* \, \psi_1 \equiv \int dx \psi_2^* A^\dagger \psi_1$$

(B-17)

then for a hermitian operator

$$A = A^\dagger$$

(B-18)

We can prove that

$$(AB)^\dagger = B^\dagger A^\dagger$$

(B-19)

To do so, we note that

$$\int \psi_2^* (AB)^\dagger \, \psi_1 = \int (AB\psi_2)^* \psi_1$$
$$= \int (B\psi_2)^* \, (A^\dagger \psi_1)$$
$$= \int \psi_2^* B^\dagger (A^\dagger \psi_1)$$
$$= \int \psi_2^* B^\dagger A^\dagger \psi_1$$

(B-20)

A generalization of this is

$$(ABC \cdots Z)^\dagger = Z^\dagger \cdots C^\dagger B^\dagger A^\dagger$$

(B-21)

Thus, a product of two hermitian operators is only hermitian if the two operators commute:

$$(AB)^\dagger = B^\dagger A^\dagger = BA = AB + [B, A]$$

(B-22)

Another result is that for any operator A, the following

$$A + A^\dagger$$
$$i(A - A^\dagger)$$
$$AA^\dagger$$

(B-23)

will be hermitian.

Next we prove the "uncertainty relations." We define

$$(\Delta A)^2 = \langle A^2 \rangle - \langle A \rangle^2 = \langle (A - \langle A \rangle)^2 \rangle \qquad \text{(B-24)}$$

Let

$$
\begin{aligned}
U &= A - \langle A \rangle \\
V &= B - \langle B \rangle
\end{aligned}
\qquad \text{(B-25)}
$$

and consider

$$\phi = U\psi + i\lambda V\psi \qquad \text{(B-26)}$$

Then

$$I(\lambda) = \int dx \phi^* \phi \geq 0 \qquad \text{(B-27)}$$

With A and B hermitian, so are U and V. We may thus rewrite:

$$
\begin{aligned}
I(\lambda) &= \int dx (U\psi + i\lambda V\psi)^* \, (U\psi + i\lambda V\psi) \\
&= \int dx (U\psi)^* \, (U\psi) + \lambda^2 \int dx (V\psi)^* \, (V\psi) \\
&\quad + i\lambda \int dx [(U\psi)^* \, (V\psi) - (V\psi)^* \, (U\psi)] \\
&= \int dx \psi^* (U^2 + \lambda^2 V^2 + i\lambda[U, V])\psi \\
&= (\Delta A)^2 + \lambda^2 (\Delta B)^2 + i\lambda \int dx \psi^* \, [U, V]\, \psi \geq 0 \\
&= (\Delta A)^2 + \lambda^2 (\Delta B)^2 + i\lambda \, \langle [A, B] \rangle
\end{aligned}
\qquad \text{(B-28)}
$$

The minimum will occur when

$$2\lambda(\Delta B)^2 + i\langle [A, B] \rangle = 0 \qquad \text{(B-29)}$$

Substituting the solution

$$\lambda = -i \, \frac{\langle [A, B] \rangle}{2 \, (\Delta B)^2} \qquad \text{(B-30)}$$

into $I(\lambda)$, we get

$$(\Delta A)^2 - \frac{\langle [A, B] \rangle^2}{4(\Delta B)^2} + \frac{\langle [A, B] \rangle^2}{2(\Delta B)^2} \geq 0$$

that is,

$$(\Delta A^2) \, (\Delta B)^2 \geq \tfrac{1}{4} \, \langle i[A, B] \rangle^2 \qquad \text{(B-31)}$$

Incidentally, the minimum value occurs when ψ is such that $U\psi$ and $V\psi$ are proportional to each other. For the case of the operators x and p, this means that

$$\frac{\hbar}{i} \frac{d\psi(x)}{dx} + i\beta x\psi(x) = 0 \tag{B-32}$$

whose solution is

$$\psi(x) = C\, e^{-\beta(x^2/2\hbar)} \tag{B-33}$$

a ground-state eigenfunction of the harmonic oscillator. It is important to note that the uncertainty relation

$$(\Delta A)^2\, (\Delta B)^2 \geq \tfrac{1}{4}\, (\langle i[A,\, B]\rangle)^2 \tag{B-34}$$

was derived without any use of wave concepts or the reciprocity between a wave form and its Fourier transform. The results depends entirely on the operator properties of the observables A and B.

We conclude the appendix by listing some properties of commutators.

(i)

$$[A,\, B] = -\,[B,\, A] \tag{B-35}$$

(ii)

$$\begin{aligned}
[A,\, B]^\dagger &= (AB)^\dagger - (BA)^\dagger \\
&= B^\dagger A^\dagger - A^\dagger B^\dagger \\
&= [B^\dagger,\, A^\dagger]
\end{aligned} \tag{B-36}$$

(iii) If A and B are hermitian, so is $i[A,\, B]$. This follows directly from the preceding properties.

(iv)

$$\begin{aligned}
[AB,\, C] &= ABC - CAB \\
&= ABC - ACB + ACB - CAB \\
&= A[B,\, C] + [A,\, C]\, B
\end{aligned} \tag{B-37}$$

(v) It may be shown term by term that

$$e^A B e^{-A} = B + [A,\, B] + \frac{1}{2!}\, [A,[A,\, B]] + \frac{1}{3!}\, \big[A,[A,[A,\, B]]\big] + \cdots \tag{B-38}$$

This is known as the Baker-Hausdorff lemma and is of some utility in manipulations of operators.

(vi) It is easily established that

$$[A,[B, C]] + [B,[C, A]] + [C,[A, B]] = 0 \qquad \text{(B-39)}$$

This is called the Jacobi identity.

A more extensive discussion of operators and the linear spaces that they are defined on may be found in J. D. Jackson, *Mathematics for Quantum Mechanics*, W. A. Benjamin, New York, 1962.

RELATIVISTIC KINEMATICS

In this section we summarize some formulas that are useful in simplifying the effects of relativistic transformations from one reference frame to another. A typical application arises in scattering: theory deals with the center-of-mass frame, experiment with the laboratory frame, and the results of the two must be compared. The simplifying technique to be used is based on two results from the theory of special relativity:

(a) The scalar product of two four-vectors, $A_\mu = (A_0, \mathbf{A})$ and $B_\mu = (B_0, \mathbf{B})$, defined by

$$A \cdot B = A_\mu B_\mu \equiv (A_0 B_0 - \mathbf{A} \cdot \mathbf{B}) \qquad \text{(ST 1-1)}$$

is invariant under Lorentz transformations.

(b) The energy and momentum of a particle transform as a four vector

$$p_\mu = \left(\frac{E}{c}, \mathbf{p} \right) \qquad \text{(ST 1-2)}$$

whose square of "length" is given in terms of the rest mass of the particle

$$p^2 = p_\mu p_\mu = \frac{E^2}{c^2} - \mathbf{p}^2 = m^2 c^2 \qquad \text{(ST 1-3)}$$

In general, a collision between two particles, leading to two particles in the final state, for example,

$$A(p_A) + B(p_B) \rightarrow C(p_C) + D(p_D)$$

will be characterized by just two numbers. The reason is that there are $4 \times 4 = 16$ different components of four momenta; these are restricted by four mass conditions (ST 1-3), and four energy and momentum conservation conditions; furthermore, invariance under translation and under rotation implies that six more coordinates, the center of mass momentum, the orientation of the scattering plane in space, and the choice of axes in that plane are irrelevant.

For our two invariants, we take

$$s = (p_A + p_B)^2 = (p_C + p_D)^2 \qquad \text{(ST 1-4)}$$

with the second term following from four-momentum conservation, and

$$t = (p_C - p_A)^2 = (p_D - p_B)^2 \qquad \text{(ST 1-5)}$$

Another possible choice is

$$u = (p_D - p_A)^2 = (p_C - p_B)^2 \qquad \text{(ST 1-6)}$$

These three are not independent, since the reader can easily convince herself that $p_{A\mu} + p_{B\mu} = p_{C\mu} + p_{D\mu}$, the energy-momentum conservation law implies

$$s + t + u = m_A^2 c^2 + m_B^2 c^2 + m_c^2 c^2 + m_D^2 c^2 \qquad \text{(ST 1-7)}$$

The invariants have the following significance.

s: Consider the center-of-mass frame, in which

$$\mathbf{p}_A^* + \mathbf{p}_B^* = 0 \qquad \text{(ST 1-8)}$$

There

$$\begin{aligned} s &= (p_{0A}^* + p_{0B}^*)^2 - (\mathbf{p}_A^* + \mathbf{p}_B^*)^2 \\ &= \left(\frac{E_A^*}{c} + \frac{E_B^*}{c} \right)^2 \\ &= \frac{1}{c^2} (E_A^* + E_B^*)^2 \end{aligned} \qquad \text{(ST 1-9)}$$

that is, it is, within the factor of c^2, the square of the total center-of-mass energy. We follow custom in labeling the center-of-mass coordinates with an asterisk.

t: The significance of t is somewhat clearer in the special (but very common) case that particles A and C, and B and D are the same, as, for example, in the reactions

$$\pi + P \rightarrow \pi + P$$

and

$$\gamma + e \rightarrow \gamma + e$$

In that case, in the center-of-mass frame,

$$\mathbf{p}_B^* = -\mathbf{p}_A^* \qquad \mathbf{p}_D^* = -\mathbf{p}_C^*$$
$$E_A^* + E_B^* = E_C^* + E_D^*$$

(ST 1-10)

and

$$m_A = m_C \qquad m_B = m_D$$

(ST 1-11)

imply that

$$(\mathbf{p}_A^{*2}\, c^2 + m_A^2\, c^4)^{1/2} + (\mathbf{p}_A^{*2}\, c^2 + m_B^2\, c^4)^{1/2} = (\mathbf{p}_C^{*2}\, c^2 + m_A^2\, c^4)^{1/2} + (\mathbf{p}_C^{*2}\, c^2 + m_B^2\, c^4)^{1/2}$$

that is,

$$E_A^* = E_C^* \qquad E_B^* = E_D^*$$

(ST 1-12)

Then

$$t = (p_A - p_C)^2 = \left(\frac{E_A^*}{c} - \frac{E_C^*}{c}\right)^2 - (\mathbf{p}_A^* - \mathbf{p}_C^*)^2$$
$$= -(\mathbf{p}_A^* - \mathbf{p}_C^*)^2$$

(ST 1-13)

that is, it is minus the square of the momentum transfer in the center-of-mass frame. Note that t is related to the center-of-mass scattering angle. The preceding yields

$$t = -\mathbf{p}_A^{*2} - \mathbf{p}_C^{*2} + 2\mathbf{p}_A^* \cdot \mathbf{p}_C^*$$
$$= -\mathbf{p}_A^{*2} - \mathbf{p}_C^{*2} + 2|\mathbf{p}_A^*|\,|\mathbf{p}_C^*|\cos\theta^*$$

(ST 1-14)

The laboratory frame is characterized by $\mathbf{p}_B^L = 0$

$$p_{B\mu} = (m_B c, \mathbf{0})$$

(ST 1-15)

Thus

$$s = (p_A + p_B)^2 = p_A^2 + p_B^2 + 2p_A \cdot p_B$$
$$= m_A^2 c^2 + m_B^2 c^2 + 2m_B E_A^L$$

(ST 1-16)

and

$$t = (p_D - p_B)^2$$
$$= m_D^2 c^2 + m_B^2 c^2 - 2m_B E_D^L$$
$$= (p_A - p_C)^2$$
$$= m_A^2 c^2 + m_C^2 c^2 - 2E_A^L E_C^L/c^2 + 2\mathbf{p}_A^L \cdot \mathbf{p}_c^L$$
$$= m_A^2 c^2 + m_C^2 c^2 - 2E_A^L E_C^L/c^2 + 2|\mathbf{p}_A^L|\,|\mathbf{p}_C^L|\cos\theta^L$$

(ST 1-17)

This, with the help of

$$E_A^L + m_B c^2 = E_C^L + E_D^L \tag{ST 1-18}$$

and *the invariance of s and t*, that is, the fact that s and t have the same values in the center-of-mass and the laboratory frames (or any other frames) allows us to compute the relation between center-of-mass scattering angle and laboratory scattering angle, and between the energies in the two frames.

The transformation properties of differential cross sections, $d\sigma/d (\cos \theta)$ are obtained from the statement, which can be established when one formulates scattering theory relativistically, that $d\sigma$ *is an invariant.* Hence

$$\frac{d\sigma}{dt} \tag{ST 1-19}$$

is invariant and, to exhibit the cross section in a form in which the transformations from one frame to another are most easily done, it is best to write it as a function s and t. We will not do this here. As a final peripheral comment we note that the expression for the

$$\int \frac{d^3\mathbf{p}}{(2\pi\hbar)^3}$$

is not relativistically invariant. However, the manifest invariance of

$$\int \cdots \int d^4p \; \delta(p^2 - m^2c^2)$$
$$= \int d^3\mathbf{p} \int_{p_0>0} dp_0 \delta(p_0^2 - \mathbf{p}^2 - m^2c^2) \tag{ST 1-20}$$
$$= \int d^3\mathbf{p} \frac{1}{2\sqrt{\mathbf{p}^2 + m^2c^2}} = \frac{c}{2} \int \frac{d^3\mathbf{p}}{(\mathbf{p}^2c^2 + m^2c^4)^{1/2}}$$

shows that

$$\int \frac{d^3\mathbf{p}}{E} \frac{1}{(2\pi\hbar)^3} \tag{ST 1-21}$$

is invariant. Matrix elements in relativistic theories always have the particles normalized not according to

$$\frac{1}{\sqrt{V}} e^{i\mathbf{p}\cdot\mathbf{r}/\hbar}$$

but according to

$$\frac{1}{\sqrt{V}} \frac{1}{\sqrt{E}} e^{i\mathbf{p}\cdot\mathbf{r}/\hbar}$$

so that the necessary factors emerge from the square of the matrix element.

THE DENSITY
OPERATOR

In all of our discussions we have dealt with the time development of physical systems, whose initial states were of the form

$$|\psi\rangle = \sum_n C_n |u_n\rangle \qquad \text{(ST 2-1)}$$

Often such initial states are not the ones that are provided by the method of preparing the states. It may be that instead of a single ensemble consisting of identical states $|\psi\rangle$ we may be presented with a number of different ensembles on which measurements are to be performed. We may have a set of ensembles of the form

$$|\psi^{(i)}\rangle = \sum_n C_n^{(i)} |u_n\rangle \qquad \text{(ST 2-3)}$$

and all we know is that the probability of finding an ensemble characterized by (i) is p_i, with

$$\sum_i p_i = 1 \qquad \text{(ST 2-4)}$$

For example, we may have a beam of hydrogen atoms in an excited state, with fixed energy and orbital angular momentum l, but completely unpolarized, so that all m-values $-l \leq m \leq l$ are equally probable. In that case $p_m = 1/(2l + 1)$,

444

independent of m. It is *not correct* to say that the beam is described by the wave function

$$|\psi\rangle = \sum_m C_m |Y_{lm}\rangle$$

with $|C_m|^2 = 1/(2m + 1)$, since the physical situation represents $2m + 1$ independent beams, so that there is no phase relationship between different m-values.

The *density operator* formalism allows us to deal with both of these cases.

PURE STATE

Consider a pure state first. Define the density operator ρ by

$$\rho = |\psi\rangle\langle\psi| \qquad \text{(ST 2-5)}$$

We can write this in the form

$$\rho = \sum_{m,n} C_n C_m^* |u_n\rangle\langle u_m| \qquad \text{(ST 2-6)}$$

The matrix elements of ρ in the u_n basis are

$$\rho_{kl} = \langle u_k| \rho |u_l\rangle = \langle u_k| \sum_{m,n} C_n C_m^* |u_n\rangle\langle u_m|u_l\rangle$$
$$= C_k C_l^* \qquad \text{(ST 2-7)}$$

We observe that

(a)

$$\rho^2 = |\psi\rangle\langle\psi|\psi\rangle\langle\psi| = |\psi\rangle\langle\psi| = \rho \qquad \text{(ST 2-8)}$$

(b)

$$\text{Tr } \rho = \sum_k \rho_{kk} = \sum_k |C_k|^2 = 1 \qquad \text{(ST 2-9)}$$

(c) We can also write the expectation value of some observable as

$$\langle A\rangle = \langle\psi|A|\psi\rangle = \sum_{m,n} C_m^* \langle u_m|A|u_n\rangle C_n$$
$$= \sum_{m,n} C_m^* C_n A_{mn} = \sum_{m,n} A_{mn}\rho_{nm} \qquad \text{(ST 2-10)}$$
$$= \text{Tr } (A\rho)$$

The results of equations (ST 2-8)–(St 2-10) are independent of the choice of the complete set of basis vectors $|u_n\rangle$. To see this, consider the set $|v_n\rangle$. By the general expansion theorem, we can write

$$|v_n\rangle = \sum_m T_m^{(n)} |u_m\rangle$$

with

$$T_m^{(n)} = \langle u_m | v_n \rangle \equiv T_{mn}$$

Note that

$$\sum_n T_{mn}(T^+)_{nk} = \sum_n T_{mn}T_{kn}^* = \sum_n \langle u_m | v_n \rangle \langle u_k | v_n \rangle^*$$

$$= \sum_n \langle u_m | v_n \rangle \langle v_n | u_k \rangle = \delta_{mk}$$

so that the matrix T is unitary. Then

$$|\psi\rangle = \sum_k D_k |v_k\rangle$$

$$= \sum_k D_k T_{kl} |u_l\rangle$$

so that

$$C_l = D_k T_{kl} = (T^{\text{tr}})_{lk} D_k \equiv U_{lk}D_k$$

Since T is unitary, so is $U \equiv T^{\text{tr}}$, the transpose of the matrix T. Thus

$$\rho_{kl} = C_k C_l^* = (U)_{km} D_m (U)_{ln}^* D_n^*$$

or

$$= U \rho_D (U)^+$$

where ρ_D is the density operator in the v-basis. Thus

$$\rho_D = U^+ \rho U$$

It follows from the unitarity of U that the properties of ρ also apply to ρ_D.

Since $\rho = \rho^+$ it follows that ρ may be diagonalized by a unitary transformation. This means that it is possible to choose a basis $|v_n\rangle$ such that ρ is diagonal. Since $\rho^2 = \rho$, this means that the eigenvalues can only be 1 and 0, and since $\text{Tr}\rho = 1$, only one eigenvalue can be 1, and all the others must be zero. Thus only one of the D_k can be nonvanishing. This means that in a suitably chosen basis, *a pure state is a state that is an eigenstate of a maximally commuting set of observables, whose eigenfunctions are the set* $|v_n\rangle$.

MIXED STATE

For a mixed state we define the density operator by

$$\rho = \sum_i |\psi^{(i)}\rangle p_i \langle \psi^{(i)}| \qquad \text{(ST 2-11)}$$

In the $|u_n\rangle$ basis, this takes the form

$$\rho = \sum_{im,n} C_n^{(i)} C_m^{(i)*} p_i |u_n\rangle\langle u_m|$$

so that

$$\rho_{kl} = \langle u_k|\rho|ul\rangle = \sum_i p_i\, C_k^{(i)} C_l^{(i)*} \qquad\qquad \text{(ST 2-12)}$$

Note that $\rho_{kl} = \rho_{lk}^*$ so that ρ is hermitian. Since

$$\sum_n |C_n^{(i)}|^2 = 1$$

it follows that

$$\text{Tr } \rho = \sum_k \rho_{kk} = \sum_i p_i = 1 \qquad\qquad \text{(ST 2-13)}$$

as before. Also

$$
\begin{aligned}
\langle A \rangle &= \sum_i p_i \, \langle \psi^{(i)}|A|\psi^{(i)}\rangle \\
&= \sum_i \sum_{mn} p_i \, \langle \psi^{(i)}|u_n\rangle\langle u_n|A|u_m\rangle\langle u_m|\psi^{(i)}\rangle \\
&= \sum_i \sum_{mn} p_i \, C_m^{(i)} C_n^{(i)*}\, A_{nm} \\
&= \sum_{mn} \rho_{mn} A_{nm} = \text{Tr } (\rho A)
\end{aligned}
\qquad\qquad \text{(ST 2-14)}
$$

as for pure state. On the other hand, it is no longer true that $\rho^2 = \rho$. In fact

$$\rho^2 = \sum_j \sum_i |\psi^{(i)}\rangle p_i \langle \psi^{(i)}|\psi^{(j)}\rangle p_j \langle \psi^{(j)}| = \sum_i |\psi^{(i)}\rangle p_i^2 \langle \psi^{(i)}|$$

and, since $\sum_i p_i = 1$,

$$\text{Tr } \rho^2 = \sum_i p_i^2 < 1 \qquad\qquad \text{(ST 2-15)}$$

for a mixed state.

It follows from the Schrödinger equation

$$\frac{d}{dt}\,|\psi^{(i)}\rangle = -\frac{i}{\hbar}\, H|\psi^{(i)}\rangle$$

and (since $H = H^+$)

$$\frac{d}{dt}\,\langle \psi^{(i)}| = \frac{i}{\hbar}\,\langle \psi^{(i)}|H$$

that

$$\frac{d}{dt}\rho = -\frac{i}{\hbar}H\rho + \frac{i}{\hbar}\rho H = -\frac{i}{\hbar}[H, \rho] \qquad \text{(ST 2-16)}$$

Note that the sign is opposite to the expression for the time rate of change of a general operator, which reads

$$\frac{d}{dt}A = \frac{i}{\hbar}[H, A]$$

The simplest application of the formalism is in the description of a beam of electrons or any other particles of spin 1/2. Here ρ is a 2 × 2 Hermitian matrix. The most general form of such a matrix is

$$\rho = \frac{1}{2}(a\mathbf{1} + \mathbf{b} \cdot \boldsymbol{\sigma}) \qquad \text{(ST 2-17)}$$

with a and $\mathbf{b}$ real. The condition tr$\rho = 1$ implies that $a = 1$. We can calculate ρ^2 as follows:

$$\rho^2 = \frac{1}{4}(1 + \mathbf{b} \cdot \boldsymbol{\sigma})(1 + \mathbf{b} \cdot \boldsymbol{\sigma}) = \frac{1}{4}(1 + \mathbf{b}^2 + 2\mathbf{b} \cdot \boldsymbol{\sigma})$$
$$= \frac{1}{2}\left(\frac{1 + \mathbf{b}^2}{2} + \mathbf{b} \cdot \boldsymbol{\sigma}\right) \qquad \text{(ST 2-18)}$$

The density matrix ρ will describe a pure state only if $\rho^2 = \rho$, that is, if $\mathbf{b}^2 = 1$. For a mixed state, it follows from (ST 2-15) that $\mathbf{b}^2 < 1$.

Let us now obtain a physical interpretation for $\mathbf{b}$. Consider a mixture of spin 1/2 beams. Each of the beams will have electrons aligned along either the z- or x- or y-axis. The fraction of particles that are in an eigenstate of σ_z with eigenvalue $+1$ will be denoted by $f_3^{(+)}$. The fractions that are in an eigenstate of σ_z with eigenvalue -1 will be denoted by $f_3^{(-)}$ and so on, so that

$$f_3^{(+)} + f_3^{(-)} + f_1^{(+)} + f_1^{(-)} + f_2^{(+)} + f_2^{(-)} = 1 \qquad \text{(ST 2-19)}$$

The eigenstates, with eigenvalues ± 1 of σ_z, σ_x, and σ_y are

$$\begin{pmatrix}1\\0\end{pmatrix}\begin{pmatrix}0\\1\end{pmatrix} \quad \begin{pmatrix}1/\sqrt{2}\\1/\sqrt{2}\end{pmatrix}\begin{pmatrix}1/\sqrt{2}\\-1/\sqrt{2}\end{pmatrix} \quad \begin{pmatrix}1/\sqrt{2}\\i/\sqrt{2}\end{pmatrix}\begin{pmatrix}1/\sqrt{2}\\-i/\sqrt{2}\end{pmatrix}$$

Thus the density matrix has the form

$$\rho = f_3^{(+)}\begin{pmatrix}1\\0\end{pmatrix}(1 \;\; 0) + f_3^{(-)}\begin{pmatrix}0\\1\end{pmatrix}(0 \;\; 1) + f_1^{(+)}\begin{pmatrix}1/\sqrt{2}\\1/\sqrt{2}\end{pmatrix}(1/\sqrt{2} \;\; 1/\sqrt{2})$$
$$+ f_1^{(-)}\begin{pmatrix}1/\sqrt{2}\\-1/\sqrt{2}\end{pmatrix}(1/\sqrt{2} \;\; -1/\sqrt{2}) + f_2^{(+)}\begin{pmatrix}1/\sqrt{2}\\i/\sqrt{2}\end{pmatrix}(1/\sqrt{2} \;\; -i/\sqrt{2})$$
$$+ f_2^{(-)}\begin{pmatrix}1/\sqrt{2}\\-i/\sqrt{2}\end{pmatrix}(1/\sqrt{2} \;\; i/\sqrt{2})$$

A little algebra, with the use of (ST 2-19) shows that this can be written in the form

$$\rho = \frac{1}{2} + \frac{1}{2}\boldsymbol{\sigma} \cdot \mathbf{P} \qquad\qquad \text{(ST 2-20)}$$

where $P_i = f_i^{(+)} - f_i^{(-)}$. The fraction of particles in a mixture that is aligned in the $+z$ direction minus the fraction that is aligned in the $-z$ direction is called the *polarization* in the z direction, and we denote it by P_3. Similarly for the other directions. Thus by comparing (ST 2-20) with (ST 2-18) we can interpret $\mathbf{b}$ as the net polarization vector $\mathbf{P}$ of the beam. In the case of beams of atoms of angular momentum l, the most general ρ is a $(2l + 1) \times (2l + 1)$ Hermitian matrix, and the interpretation of the elements is more complicated. Further discussion of the density matrix is beyond the scope of this book.

THE WENTZEL–KRAMERS–BRILLOUIN APPROXIMATION

This approximation method is particularly useful when one is dealing with slowly varying potentials. Exactly what this means will become clear later. One wants to solve the equation

$$\frac{d^2\psi(x)}{dx^2} + \frac{2m}{\hbar^2}\,[E - V(x)]\,\psi(x) = 0 \tag{ST 3-1}$$

and to do so, it is useful to write

$$\psi(x) = R(x)\,e^{iS(x)/\hbar} \tag{ST 3-2}$$

Then

$$\frac{d^2\psi}{dx^2} = \left[\frac{d^2R}{dx^2} + \frac{2i}{\hbar}\frac{dR}{dx}\frac{dS}{dx} + \frac{i}{\hbar}R\frac{d^2S}{dx^2} - \frac{1}{\hbar^2}R\left(\frac{dS}{dx}\right)^2\right]e^{iS(x)/\hbar} \tag{ST 3-3}$$

so that the differential equation splits into two, by taking the real and imaginary part of (ST 3-1) after (ST 3-3) has been substituted. The imaginary part gives

$$R\frac{d^2S}{dx^2} + 2\frac{dR}{dx}\frac{dS}{dx} = 0 \tag{ST 3-4}$$

that is,

$$\frac{d}{dx}\left(\log \frac{dS}{dx} + 2 \log R\right) = 0$$

whose solution is

$$\frac{dS}{dx} = \frac{C}{R^2} \tag{ST 3-5}$$

The real part reads

$$\frac{d^2R}{dx^2} - \frac{1}{\hbar^2} R \left(\frac{dS}{dx}\right)^2 + \frac{2m[E - V(x)]}{\hbar^2} R = 0$$

which, when (ST 3-5) is substituted, becomes

$$\frac{d^2R}{dx^2} - \frac{C^2}{\hbar^2} \frac{1}{R^3} + \frac{2m[E - V(x)]}{\hbar^2} R = 0 \tag{ST 3-6}$$

At this point we make the approximation that

$$\frac{1}{R} \frac{d^2R}{dx^2} \ll \frac{C^2}{\hbar^2} \frac{1}{R^4} = \frac{1}{\hbar^2} \left(\frac{dS}{dx}\right)^2 \tag{ST 3-7}$$

so that the equation becomes

$$\frac{C^2}{R^4} = 2m[E - V(x)] \tag{ST 3-8}$$

Thus

$$\frac{C}{R^2} = \frac{dS}{dx} = \sqrt{2m[E - V(x)]} \tag{ST 3-9}$$

and hence

$$S(x) = \int_{x_1}^{x} dy \sqrt{2m[E - V(x)]} \tag{ST 3-10}$$

The condition for the validity can be translated into a statement about the variation of $V(x)$. It will be satisfied if $V(x)$ varies slowly in a wavelength, which varies from point to point, but which for slowly varying $V(x)$ is defined by

$$\lambda(x) = \frac{\hbar}{p(x)} = \frac{\hbar}{\{2m[E - V(x)]\}^{1/2}} \tag{ST 3-11}$$

At the points where

$$E - V(x) = 0 \tag{ST 3-12}$$

special treatment is required, because in the approximate equation (ST 3-8) $R(x)$ appears to be singular. This cannot be, and this means that the approximation (ST 3-7) must be poor there. The special points are called *turning points* because it is there that a classical particle would turn around: It can only move where $E - V(x) \geq 0$. The way of handling solutions near turning points is a little too technical to be presented here. The basic idea is that we have a solution to the left of the turning point [where $E > V(x)$, say], of the form

$$\psi(x) = R \, e^{i \int_{x_1}^{x} dy \sqrt{(2m/\hbar^2)[E-V(y)]}} \tag{ST 3-13}$$

and a solution to the right of the turning point [where $E < V(x)$], and what we need is a formula that interpolates between them. In the vicinity of the turning point one can approximate $\sqrt{(2m/\hbar^2)[E - V(x)]}$ by a straight line over a small interval, and solve the Schrödinger equation exactly. Since it is a second-order equation, there are two adjustable constants, one of which is fixed by fitting the solution to (ST 3-13) and the other by fitting it to

$$\psi(x) = R \, e^{- \int_{x_1}^{x} dy \sqrt{(2m/\hbar^2)[V(y) - E]}} \tag{ST 3-14}$$

the solution to the right of the turning point.[1] The preceding solution decreases in amplitude as x increases. The total attenuation at the next turning point, when $E \geq V(x)$ again, is

$$\frac{\psi(x_{II})}{\psi(x_I)} \simeq e^{- \int_{x_1}^{x_{II}} dy \sqrt{(2m/\hbar^2)[V(y) - E]}} \tag{ST 3-15}$$

which is just the square root of the transmission probability that we found in Chapter 5.

[1]For more details, see almost any of the more advanced books on quantum mechanics, for example, J. L. Powell and B. Crasemann, *Quantum Mechanics*, Addison-Wesley, Reading, Mass., 1961; L. I. Schiff, *Quantum Mechanics*, McGraw-Hill, New York, 1968.

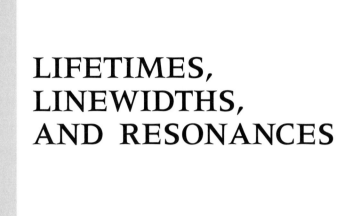

LIFETIMES, LINEWIDTHS, AND RESONANCES

In this section we do three things:

(a) We discuss a somewhat improved treatment of transition rates that shows how the exponential decay behavior comes about, following the general approach of V. Weisskopf and E. P. Wigner.

(b) We show how the Lorentzian shape for the linewidth comes about.

(c) We show that the scattering amplitude for a photon by the atom in its ground state peaks strongly when the energy of the incident photon is equal to the (shifted) energy of the excited state.

To simplify the problem as much as possible, we consider an atom with just two levels, the ground state, with energy 0, and a single excited state, with energy E. The two states are coupled to the electromagnetic field, which we will take to be scalar, so that no polarization vectors appear. We will only consider the subset of eigenstates of H_0 consisting of the excited state ϕ_1, for which

$$H_0\phi_1 = E\phi_1 \tag{ST 4-1}$$

and of the ground state + one photon, $\phi(\mathbf{k})$, for which

$$H_0\phi(\mathbf{k}) = \epsilon(\mathbf{k})\ \phi(\mathbf{k}) \tag{ST 4-2}$$

and limit ourselves to these in an expansion of an arbitrary function. This is certainly justified when the coupling between the two states, ϕ_1 and $\phi(\mathbf{k})$ through the potential V, is small, as in electromagnetic coupling, since then the influence of two-, three-, . . . , photon states is negligible. Note that

$$\langle \phi_1 | \phi(\mathbf{k}) \rangle = 0 \qquad \text{(ST 4-3)}$$

even when the $\mathbf{k}$ is such that the energies $\epsilon(\mathbf{k})$ and E are the same. The states are orthogonal because one has a photon in it and the other does not, and because for one of them the atom is in an excited state, and for the other it is not.

The solution of the equation

$$i\hbar \frac{d\psi(t)}{dt} = (H_0 + V) \, \psi(t) \qquad \text{(ST 4-4)}$$

may be written in terms of the complete set

$$\psi(t) = a(t) \, \phi_1 \, e^{-iEt/\hbar} + \int d^3k b(\mathbf{k}, t) \, \phi(\mathbf{k}) \, e^{-i\epsilon(\mathbf{k})t/\hbar} \qquad \text{(ST 4-5)}$$

When this is inserted into (ST 4-4),

$$i\hbar \frac{da}{dt} e^{-iEt/\hbar} \, \phi_1 + Ea \, e^{-iEt/\hbar} \, \phi_1$$

$$+ i\hbar \int d^3k \frac{db(\mathbf{k}, t)}{dt} e^{-i\epsilon(\mathbf{k})t/\hbar} \, \phi(\mathbf{k}) + \int d^3k\epsilon(\mathbf{k}) \, b(\mathbf{k}, t) \, e^{-i\epsilon(\mathbf{k})t/\hbar} \, \phi(\mathbf{k})$$

$$= Ea(t) \, e^{-iEt/\hbar} \, \phi_1 + \int d^3k\epsilon(\mathbf{k}) \, b(\mathbf{k}, t) \, e^{-i\epsilon(\mathbf{k})t/\hbar} \, \phi(\mathbf{k})$$

$$+ a(t) \, e^{-iEt/\hbar} \, V\phi_1 + \int d^3k b(\mathbf{k}, t) \, e^{-i\epsilon(\mathbf{k})t/\hbar} \, V\phi(\mathbf{k})$$

results. If we take the scalar product with ϕ_1, we get

$$i\hbar \frac{da}{dt} = a(t)\langle \phi_1 | V | \phi_1 \rangle + \int d^3k b(\mathbf{k}, t) \, e^{-i[\epsilon(\mathbf{k}) - E]t/\hbar} \, \langle \phi_1 | V | \phi(\mathbf{k}) \rangle$$

Since V, acting on a state, is supposed to change the photon number by one, $\langle \phi_1 | V | \phi_1 \rangle = 0$. With the notation

$$\epsilon(\mathbf{k}) - E = \hbar\omega(\mathbf{k})$$
$$\langle \phi_1 | V | \phi(\mathbf{k}) \rangle = M(\mathbf{k}) \qquad \text{(ST 4-6)}$$

the equation becomes

$$i\hbar \frac{da(t)}{dt} = \int d^3k b(\mathbf{k}, t) \, e^{-i\omega(\mathbf{k})t} \, M(\mathbf{k}) \qquad \text{(ST 4-7)}$$

If we take the scalar product with $\phi(\mathbf{q})$, and again use photon counting to set $\langle \phi(\mathbf{q}) | V | \phi(\mathbf{k}) \rangle = 0$, we get, after a little manipulation, using a normalization that

$$\langle \phi(\mathbf{q}) | \phi(\mathbf{k}) \rangle = \delta(\mathbf{k} - \mathbf{q}) \qquad \text{(ST 4-8)}$$

the equation

$$i\hbar \frac{db(\mathbf{q}, t)}{dt} = a(t) \, e^{i\omega(\mathbf{q})t} \, M^*(\mathbf{q}) \tag{ST 4-9}$$

Since $b(\mathbf{k}, 0) = 0$ if the excited state is occupied at $t = 0$, a solution of this equation is

$$b(\mathbf{k}, t) = \frac{1}{i\hbar} M^*(\mathbf{k}) \int_0^t dt' \, e^{i\omega(\mathbf{k})t'} \, a(t') \tag{ST 4-10}$$

We now insert this into (ST 4-7) to get

$$\frac{da(t)}{dt} = -\frac{1}{\hbar^2} \int d^3k |M(\mathbf{k})|^2 \, e^{-i\omega(\mathbf{k})t} \int_0^t dt' \, a(t') \, e^{i\omega(\mathbf{k})t'} \tag{ST 4-11}$$

Let us make an ansatz for the solution of this equation that holds for large t. We propose that

$$a(t) = a_0 \, e^{-zt} \tag{ST 4-12}$$

When this is substituted in (ST 4-11) we get

$$\begin{aligned} -ze^{-zt} &= -\frac{1}{\hbar^2} \int d^3k \, |M(\mathbf{k})|^2 \, \frac{e^{-i\omega(\mathbf{k})t}}{i\omega(\mathbf{k}) - z} \, (e^{i\omega(\mathbf{k})t} \, e^{-zt} - 1) \\ &= -\frac{1}{\hbar^2} \int d^3k \, |M(\mathbf{k})|^2 \, \frac{1}{i\omega(\mathbf{k}) - z} \, (e^{-zt} - e^{-i\omega(\mathbf{k})t}) \end{aligned} \tag{ST 4-13}$$

The second term vanishes for large t. The reason is that the integrand consists of a product of a smoothly varying function of $\mathbf{k}$ multiplied by a bounded, rapidly varying one. The integral can be shown to vanish faster than any power of $1/t$ according to the Riemann-Lesbegue lemma. Accepting this result, we are left with

$$z = \frac{1}{\hbar^2} \int d^3k \, |M(\mathbf{k})|^2 \, \frac{1}{i\omega(\mathbf{k}) - z} \tag{ST 4-14}$$

Let us write

$$z = \frac{\gamma}{2} + \frac{i\Delta}{\hbar} \tag{ST 4-15}$$

and since $|M(\mathbf{k})|^2$ is small, so are both γ and $\Delta/\hbar$. The equation then yields, dropping the smaller product of $\Delta/\hbar$ and $|M(\mathbf{k})|^2$,

$$\frac{\gamma}{2} + \frac{i\Delta}{\hbar} = \frac{1}{\hbar^2} \int d^3k \, |M(\mathbf{k})|^2 \, \frac{-i\omega(\mathbf{k}) + \gamma/2}{(\gamma/2)^2 + (\omega(\mathbf{k}) - \Delta/\hbar)^2} \tag{ST 4-16}$$

If we now make use of the relation

$$\lim_{\lambda = 0^+} \frac{\lambda}{\sigma^2 + \lambda^2} = \pi\delta(\sigma) \tag{ST 4-17}$$

we get

$$
\begin{aligned}
\gamma &= \frac{2\pi}{\hbar} \int d^3\mathbf{k}\ |M(\mathbf{k})|^2\ \delta(\hbar\omega(\mathbf{k}) - \Delta) \\
&= \frac{2\pi}{\hbar} \int d^3\mathbf{k}\ |\ M(\mathbf{k})|^2\ \delta(\epsilon(\mathbf{k}) - (E + \Delta))
\end{aligned}
\tag{ST 4-18}
$$

and

$$
\begin{aligned}
\Delta &= -\int d^3\mathbf{k}\ |M(\mathbf{k})|^2\ \frac{1}{\hbar\omega(\mathbf{k}) - \Delta} \\
&= \int d^3\mathbf{k}\ |M(\mathbf{k})|^2\ \frac{1}{E + \Delta - \epsilon(\mathbf{k})}
\end{aligned}
\tag{ST 4-19}
$$

In both of these integrals, the Δ in the delta function and the energy denominator can be left out to the order of accuracy of the calculation. We leave it to draw attention to the fact that it is $E + \Delta$ that appears everywhere, that is, the energy of the excited state is shifted by Δ. Thus the coefficient of ϕ_1 in $\psi(t)$, with the initial condition

$$
a(0) = a_0 = 1
\tag{ST 4-20}
$$

is

$$
a(t) = e^{-\gamma t/2}\ e^{-i(E+\Delta)t/\hbar}
\tag{ST 4-21}
$$

and the probability of finding $\psi(t)$ in the initial state ϕ_1 after a (long) time t is

$$
|a(t)|^2 = e^{-\gamma t}
\tag{ST 4-22}
$$

We note from (ST 4-18) that γ is the decay rate calculated in perturbation theory. Also, the phase factor involves the energy *shifted by the second-order perturbation energy shift*, as comparison with (16-16) shows. The only difference is that the intermediate states summed over here form a continuum. Since the energy denominator in (ST 4-19) can vanish, the prescription

$$
i\Delta/\hbar = \underset{\lambda \to 0}{\mathrm{Lim}}\ \frac{1}{\hbar^2} \int d^3\mathbf{k}\ |M(\mathbf{k})|^2\ \frac{-i\omega(\mathbf{k})}{\lambda^2 + (\omega(\mathbf{k}) - \Delta/\hbar)^2}
\tag{ST 4-23}
$$

must be used.

Another quantity of interest is the probability that the state $\psi(t)$ ends up in the state $\phi(\mathbf{k})$ at $t = \infty$. This is given by $|b(\mathbf{k}, \infty)|^2$, where, according to (ST 4-10) and (ST 4-12), we have

$$
\begin{aligned}
b(\mathbf{k}, \infty) &= \frac{1}{i\hbar} M^*(\mathbf{k}) \int_0^\infty dt'\ e^{-[z - i\omega(\mathbf{k})]t'} \\
&= \frac{M^*(\mathbf{k})}{i\hbar}\ \frac{1}{z - \omega(\mathbf{k})}
\end{aligned}
$$

Thus

$$b(\mathbf{k}, \infty) = \frac{M^*(\mathbf{k})}{\epsilon(\mathbf{k}) - E - \int d^3k' \dfrac{|M(\mathbf{k}')|^2}{E - \epsilon(\mathbf{k}')} + i\hbar\gamma/2} \qquad \text{(ST 4-24)}$$

and the absolute square of this,

$$|b(\mathbf{k}, \infty)|^2 = \frac{|M(\mathbf{k})|^2}{[\epsilon(\mathbf{k}) - E - \Delta]^2 + (\hbar\gamma/2)^2} \qquad \text{(ST 4-25)}$$

yields the Lorentzian shape for the linewidth, that is, the photon energy is centered about the (shifted) energy of the excited level, with the width described by $\hbar\gamma/2$. The energy shift is small, and usually ignored.

The same form appears in the scattering problem. Consider the scattering of a "photon" of momentum $\mathbf{k}_i$ by the atom in the ground state. The state of the system is again described by (ST 4-1), (ST 4-2), (ST 4-7), and (ST 4-9), except that initially, which here means at $t = -\infty$, the state is specifically given as $\phi(\mathbf{k}_i)$, so that

$$b(\mathbf{q}, t) = \delta(\mathbf{q} - \mathbf{k}_i) \qquad \text{at } t = -\infty \qquad \text{(ST 4-26)}$$

Hence the integration of (ST 4-9) gives

$$b(\mathbf{q}, t) = \delta(\mathbf{q} - \mathbf{k}_i) + \frac{1}{i\hbar} M^*(\mathbf{q}) \int_{-\infty}^{t} dt' a(t') e^{i\omega(\mathbf{q})t'} \qquad \text{(ST 4-27)}$$

The quantity of interest is the amplitude for a transition into a final state in which the photon has momentum $\mathbf{k}_f$ at $t = +\infty$, that is, it is

$$\langle \phi(\mathbf{k}_f) | \psi(+\infty) \rangle = b(\mathbf{k}_f, +\infty)$$

$$= \delta(\mathbf{k}_f - \mathbf{k}_i) - \frac{i}{\hbar} M^*(\mathbf{k}_f) \int_{-\infty}^{\infty} dt' a(t') e^{i\omega_f t'} \qquad \text{(ST 4-28)}$$

$$(\omega_f \equiv \omega(\mathbf{k}_f))$$

using the previous equation.

Substituting (ST 4-27) into (ST 4-7) yields the equation

$$\frac{da(t)}{dt} = \frac{1}{i\hbar} e^{-i\omega_i t} M(\mathbf{k}_i) - \frac{1}{\hbar^2} \int d^3k |M(\mathbf{k})|^2 e^{-i\omega(\mathbf{k})t} \int_{-\infty}^{t} dt' a(t') e^{i\omega(\mathbf{k})t'} \qquad \text{(ST 4-29)}$$

This may be integrated, taking into account that $a(-\infty) = 0$, to give

$$a(t) = \frac{M(\mathbf{k}_i)}{i\hbar} \int_{-\infty}^{t} dt' e^{-i\omega_i t'}$$

$$- \frac{1}{\hbar^2} \int d^3k |M(\mathbf{k})|^2 \int_{-\infty}^{t} dt' e^{-i\omega(\mathbf{k})t'} \int_{-\infty}^{t'} dt'' a(t'') e^{i\omega(\mathbf{k})t''} \qquad \text{(ST 4-30)}$$

Now the integral $\int_{-\infty}^{t} dt' \, e^{-i\omega_i t'}$ is not well defined. The standard procedure is to write it in the form

$$\underset{\epsilon \to 0}{\text{Lim}} \int_{-\infty}^{t} dt' \, e^{-i(\omega_i + i\epsilon)t'} = \underset{\epsilon \to 0}{\text{Lim}} \, i \, \frac{e^{-i(\omega_i + i\epsilon)t}}{\omega_i + i\epsilon} \qquad \text{(ST 4-31)}$$

The use of a convergence factor, which is then allowed to vanish in a well-defined way, is somewhat similar to the treatment of the Coulomb potential as the limiting case of a screened Coulomb potential, which was addressed in Chapter 23. Next, as can be seen from Figure ST 4-2,

$$\int_{-\infty}^{t} dt' \, e^{-i\omega(\mathbf{k})t'} \int_{-\infty}^{t'} dt'' a(t'') \, e^{i\omega(\mathbf{k})t''} = \int_{-\infty}^{t} dt'' a(t'') \, e^{i\omega(\mathbf{k})t''} \int_{t''}^{t} dt' \, e^{-i\omega(\mathbf{k})t'}$$

$$= \frac{i}{\omega(\mathbf{k})} \int_{-\infty}^{t} dt'' a(t'')[e^{-i\omega(\mathbf{k})(t-t'')} - 1]$$

so that

$$a(t) = \frac{M(\mathbf{k}_i) \, e^{-i\omega_i t}}{h(\omega_i + i\epsilon)} - \frac{i}{\hbar^2} \int d^3k \, \frac{|M(\mathbf{k})|^2}{\omega(\mathbf{k})} \int_{-\infty}^{t} dt'' a(t'')[e^{-i\omega(\mathbf{k})(t-t'')} - 1] \qquad \text{(ST 4-32)}$$

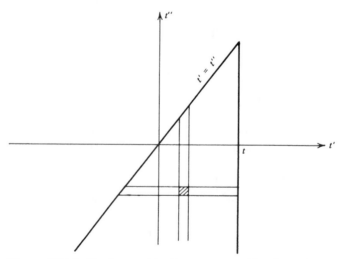

Figure ST4-1. The integral in (ST4-30) can either be written as a "sum" of the vertical strips, as in the equation, or, a sum of the horizontal strips, as in (ST4-32). The same interchange of orders is used in (ST4-33) except that the vertical line at t'' is shifted to $+\infty$.

According to (ST 4-28) the quantity of interest for nonforward scattering (so that the first term can be ignored) is

$$
\int_{-\infty}^{\infty} dt\, a(t)\, e^{i\omega_f t} = \frac{M(\mathbf{k}_i)}{\hbar(\omega_i + i\epsilon)} \int_{-\infty}^{\infty} dt\, e^{i(\omega_f - \omega_i)t}
$$

$$
- \frac{i}{\hbar^2} \int d^3 k\, \frac{|M(\mathbf{k})|^2}{\omega(\mathbf{k})} \int_{-\infty}^{\infty} dt\, e^{i\omega_f t} \int_{-\infty}^{t} dt''\, a(t'') [e^{-i\omega(\mathbf{k})(t-t'')} - 1]
$$

$$
= \frac{2\pi M(\mathbf{k}_i)}{\hbar(\omega_i + i\epsilon)}\, \delta(\omega_f - \omega_i)
$$

$$
- \frac{i}{\hbar^2} \int d^3 k\, \frac{|M(\mathbf{k})|^2}{\omega(\mathbf{k})} \int_{-\infty}^{\infty} dt''\, a(t'') \int_{t''}^{\infty} dt\, \{ e^{i\omega(\mathbf{k})t''} e^{i[\omega_f - \omega(\mathbf{k})]t} - e^{i\omega_f t} \}
$$

(ST 4-33)

where in the last term we again rewrote the integral with the help of Figure ST 4-1. The t-integration can again be done with the help of the convergence factor trick, so that (ST 4-33) now reads

$$
\int_{-\infty}^{\infty} dt\, a(t)\, e^{i\omega_f t} = \frac{2\pi M(\mathbf{k}_i)\, \delta(\omega_f - \omega_i)}{\hbar(\omega_i + i\epsilon)} + \frac{1}{\hbar^2} \int d^3 k\, \frac{|M(\mathbf{k})|^2}{\omega(\mathbf{k})} \int_{-\infty}^{\infty} dt''\, a(t'')\, e^{i\omega_f t''}
$$

$$
\times \left[\frac{1}{\omega_f - \omega(\mathbf{k}) + i\epsilon} - \frac{1}{\omega_f + i\epsilon} \right]
$$

which is in the form of an equation for the unknown. This may be solved to give

$$
\int_{-\infty}^{\infty} dt\, a(t)\, e^{i\omega_f t} = \frac{2\pi M(\mathbf{k}_i)\, \delta(\omega_f - \omega_i)}{\hbar(\omega_i + i\epsilon)}
$$

$$
\times \frac{1}{1 - \dfrac{1}{\hbar^2 \omega_f} \displaystyle\int d^3 k\, \dfrac{|M(\mathbf{k})|^2}{\omega_f - \omega(\mathbf{k}) + i\epsilon}}
$$

(ST 4-34)

Hence, in the nonforward direction

$$
b(\mathbf{k}_f, \infty) = -\frac{i}{\hbar} M^*(\mathbf{k}_f) \cdot 2\pi M(\mathbf{k}_i)\, \delta(\omega_f - \omega_i)
$$

$$
\times \frac{1}{\hbar\omega_i + i\epsilon - \displaystyle\int d^3 k\, \dfrac{|M(\mathbf{k})|^2}{\hbar\omega_f - \hbar\omega(\mathbf{k}) + i\epsilon}}
$$

$$
= \frac{-2\pi i\, \delta(\hbar\omega_f - \hbar\omega_i)\, M(\mathbf{k}_i)\, M^*(\mathbf{k}_f)}{\epsilon(\mathbf{k}_i) - E - \displaystyle\int d^3 k\, \dfrac{|M(\mathbf{k})|^2}{\epsilon(\mathbf{k}_i) - \epsilon(\mathbf{k})} + i\pi \displaystyle\int d^3 k\, |M(\mathbf{k})|^2\, \delta[\epsilon(\mathbf{k}_i) - \epsilon(\mathbf{k})]}
$$

(ST 4-35)

The amplitude peaks strongly when the incident (and final) energy $\epsilon_i\ (= \epsilon_f)$ is near the energy of the excited state of the atom, shifted to $E + \Delta E$, as in (ST 4-25). This justifies the comments made at the end of Chapter 18.

PHYSICAL CONSTANTS[1]

N_0 (Avogardro's number)	$6.0221367(36) \times 10^{23}$ mol^{-1}
c (speed of light—definition)	$2.99792458 \times 10^{10}$ cms^{-1}
e (electron charge)	$1.60217733(49) \times 10^{-19}$ C
	$4.80653199(15) \times 10^{-10}$ e.s.u.
1 MeV	$1.60217733(49) \times 10^{-6}$ erg
$\hbar$ (Planck constant/2π)	$1.05457266(63) \times 10^{-27}$ erg-s
	$6.5821220(20) \times 10^{-22}$ MeV-s
α (fine structure constant $e^2/\hbar c$)	$1/137.0359895(61)$
k (Boltzmann constant)	$1.380658(12) \times 10^{-16}$ erg-K^{-1}
m_e (electron mass)	$9.1093897(54) \times 10^{-27}$ g
	$0.51099906(15)$ MeV/c^2
m_p (proton mass)	$1.6726231(10) \times 10^{-24}$ g
	$938.27231(28)$ MeV/c^2
m_n (neutron mass)	$1.6749286(1) \times 10^{-24}$ g
	$939.56563(28)$ MeV/c^2
1 a.m.u. ($m(C^{12})/12$)	$1.6605402(10) \times 10^{-24}$ g
	$931.49432(28)$ MeV/c^2
a_0 ($\hbar/m_e c\alpha$)	$0.529177249(24) \times 10^{-8}$ cm
R_∞ ($= m_e c^2 \alpha^2/2$)	$13.605698(40)$ eV
G (gravitational constant)	$6.67259(85) \times 10^{-8}$ cm^{3}g^{-1}s^{-2}
μ_{Bohr} (Bohr magneton $e\hbar/2m_e c$)	$0.578838263(52) \times 10^{-14}$ MeV G^{-1}

[1]The numbers are taken from a very interesting article by E. R. Cohen and B. N. Taylor, *Phys. Today*, BG 9–14 (August 1994).

REFERENCES[1]

G. Baym, *Lectures on Quantum Mechanics*, W. A. Benjamin, New York, 1969.

This is a very appealing book, with just the right mixture of formalism, intuitive arguments, and applications. It should be considered an advanced book, accessible to the student who has covered the material in this book.

H. A. Bethe and R. W. Jackiw, Intermediate Quantum Mechanics (2nd edition), W. A. Benjamin, New York, 1968.

This book contains detailed discussions of calculational methods applicable to the theory of atomic structure, multiplet splittings, the photoelectric effect, and atomic collisions. Much of the material is not to be found in any other textbook. The book is thus an advanced text, as well as an exhaustive reference book.

H. A. Bethe and E. E. Salpeter, *Quantum Mechanics of One- and Two-Electron Atoms*, Springer-Verlag, Berlin/New York, 1957.

This reprint of the authors' article in the *Handbuch der Physik* is an elaborate, detailed, definitive treatment of the problem at hand. It is a book about atoms and not about quantum mechanics, and the level is high. It is an excellent reference book.

[1]Many books have been written about quantum mechanics. I have studied from some of them, read others, glanced at a few, and probably missed some others. The list given here is not meant to be exhaustive, and no book is criticized by its omission. In particular, no books on quantum chemistry are listed.

D. Bohm, *Quantum Theory*, Dover Publ., New York, 1989.

The book is discursively written, on a level comparable to the present book. The author pays much attention to the principles of quantum theory and gives an excellent discussion of the quantum theory of the measurement process. There are few applications and not many problems.

S. Borowitz, *Fundamentals of Quantum Mechanics*, W. A. Benjamin, New York, 1967.

This is a well-written book, about half of which is devoted to the theory of waves and to classical mechanics. The level is comparable to that of the present book.

J. J. Brehm and W. J. Mullin, *Introduction to the Structure of Matter*, John Wiley & Sons, New York, 1989.

This is a very thorough, advanced book that covers much of modern physics in a very readable style. An excellent reference book for anybody wishing to get more qualitative detail for any of the topics that are mentioned cursorily in *Quantum Physics*.

E. U. Condon and G. H. Shortley, *The Theory of Atomic Spectra*, Cambridge University Press, Cambridge, England, 1959.

This is a very detailed reference book on all aspects of atomic spectra, although it does not make use of the recent techniques that depend on group theory. The book is very advanced, and thus its shortcomings in the technical developments are not important for anybody but the specialist. It is very useful for the student.

S. Brandt and H. D. Dahmen, *Quantum Mechanics on the Personal Computer* (3rd edition), Springer-Verlag, New York, 1994.

This is the only book that I have seen in which personal computers are a central tool in the teaching of quantum mechanics. To the extent that PCs become more and more available, and as students become more and more adept at their use, inevitably the exploration of solutions of the Schrödinger equation will be much aided by this tool.

C. Cohen-Tannoudji, B. Diu, and F. Laloe, *Quantum Mechanics*, John Wiley & Sons, New York, 1977.

This is an encyclopedic book of more than a thousand pages. The student will find detailed coverage of many aspects of atomic physics. The mathematical level is quite a bit above that of *Quantum Physics*.

R. H. Dicke and J. P. Wittke, *Introduction to Quantum Mechanics*, Addison-Wesley, Reading, Mass., 1960.

I enjoyed this book very much. It is on a level comparable to the present book, and discusses a few topics, notably quantum statistics, that are not treated here. The problems are excellent.

P. A. M. Dirac, *The Principles of Quantum Mechanics* (4th edition), Oxford University Press (Clarendon), Oxford, 1958.

This is a superb book by one of the major creators of quantum mechanics. The student who has studied the material in this book will have no trouble with Dirac; if he is at all serious about mastering quantum mechanics, he should sooner or later go through Dirac's book.

R. P. Feynman and A. R. Hibbs, *Quantum Mechanics and Path Integrals*, McGraw-Hill, New York, 1965.

In 1948, R. P. Feynman proposed a different formulation of quantum mechanics. In this book the equivalence of this formulation to the standard theory is demonstrated, and the "path integral" expression for the general amplitude is exploited in a number of calculations. The selection of material is very interesting, and the point of view is different from the one developed by the author. Thus this somewhat more advanced book presents an excellent complement to this book.

R. P. Feynman, R. B. Leighton, and M. Sands, *The Feynman Lectures on Physics*, Vol. 3, *Quantum Mechanics*, Addison-Wesley, Reading, Mass., 1965.

In this introduction to quantum mechanics, Feynman abandons the path integral and approaches the subject from the point of view of state vectors. A large number of fascinating examples are discussed with the minimum of formal apparatus. A superb complementary book, whose only shortcoming is the absence of problems.

K. Gottfried, *Quantum Mechanics*, Vol. 1, *Fundamentals*, W. A. Benjamin, New York, 1966.

This is a very advanced book, distinguished by the care with which the various topics are discussed. The treatment of the measurement process and of invariance principles is excellent. The student who has mastered the material in this book should be able to read Gottfried's book, provided he or she has acquired the necessary mathematical equipment.

D. Griffiths *Introduction to Quantum Mechanics*, Prentice-Hall, Englewood Cliffs, N.J. 1995.

This well-written, attractive book is roughly on the level of Dicke and Wittke or Saxon. It contains a nice selection of topics, including a discussion of the geometric phase.

W. Heisenberg, *The Physical Principles of the Quantum Theory*, Dover, New York, 1930.

This reprint of some 1930 lectures given by Heisenberg on the physical significance of the quantum theory still makes good reading. The discussion of the uncertainty relations is particularly useful.

H. A. Kramers, *Quantum Mechanics*, Interscience, New York, 1957.

This book by one of the founders of the subject is at its best in the discussion of spin and the introduction to relativistic quantum theory, both rather advanced subjects. The student who is comfortable with quantum mechanics will find browsing through this book enjoyable and rewarding.

L. D. Landau and E. M. Lifshitz, *Quantum Mechanics (Nonrelativistic Theory)* (2nd edition), Addison-Wesley, Reading, Mass., 1965.

The book by Landau and Lifshitz is one of a series of superb books covering all of theoretical physics. It is hard to think of this as a textbook for any but the most sophisticated students. Any student, however, once he reaches the advanced level, will find much that is useful in this book. There is an assumed mathematical facility on the part of the student.

Richard L. Liboff, *Introductory Quantum Mechanics*, Holden-Day, San Francisco, 1980.

This is a very attractive book, which is written on roughly the same mathematical level as this book. It covers in more detail, and somewhat differently, much of the material in the first two-thirds of *Quantum Physics*, and is an excellent additional reference.

Harry J. Lipkin, *Quantum Mechanics—New Approaches to Selected Topics*, North-Holland, Amsterdam, 1973.

Lipkin's book deals with a number of advanced topics in the application of quantum mechanics, in a simple way. The physics is always at the forefront of the discussion, and a student who has a good command of *Quantum Physics* will get much benefit and pleasure out of this book.

A. Messiah, *Quantum Mechanics* (in 2 volumes), John Wiley & Sons, New York, 1968.

This book gives a complete coverage of quantum theory from the treatment of one-dimensional potentials through the quantization of the electromagnetic field and the relativistic wave equation of Dirac. It is an advanced book, and it assumes a mathematical sophistication that few first year graduate students possess. It is an extremely worthwhile book.

E. Merzbacher, *Quantum Mechanics* (2nd edition) John Wiley & Sons, New York, 1970.

Together with the book by Schiff, this is the standard first year graduate textbook, and deservedly so. The complete range of concepts and phenomena is treated with economy and taste. The book should be available to the student who has gone through the material is this book.

R. Omnes *The Interpretation of Quantum Mechanics*, Princeton University Press, Princeton, N.J., 1994.

An important book dealing with recent work on the extension of the standard interpretation of quantum mechanics.

D. Park, *Introduction to the Quantum Theory*, (3rd edition) McGraw-Hill, New York, 1984.

This attractive book is written on the same level as the present book. Among the topics discussed by Park, and absent in this volume, is the subject of quantum statistics, which is treated with clarity.

W. Pauli, *Die Allgemeinen Prinzipien der Wellenmechanik, Handbuch der Physik*, Vol. 5/1, Springer-Verlag, Berlin/New York, 1958.

The advanced student who reads German will find in this reprint of a 1930 article by Pauli a concise definitive discussion of quantum mechanics. There are no applications, but all of the important matters are there.

P. J. E. Peebles *Quantum Mechanics*, Princeton University Press, Princeton, N.J. 1992.

This is a nicely written textbook at the undergraduate level. It differs from other books at this level (except that by Bohm) by a detailed discussion of what is really meant by the measurement process in quantum mechanics.

A. B. Pippard, *The Physics of Vibration*, Cambridge University Press, Cambridge, England, 1978.

This book covers all kinds of oscillators, classically and quantum mechanically. It is not a textbook in the usual sense. It is a delight to read.

J. L. Powell and B. Crasemann, *Quantum Mechanics*, Addison-Wesley, Reading Mass., 1961.

The strength of this book is in the painstaking working out of all of the mathematical details of wave mechanics and matrix mechanics. Probably all of the mathematical aspects of these subjects that have been bypassed in this book can be found here. There is a good discussion of the WKB approximation and of the general properties of second-order differential equations. There are relatively few applications, and there are more exercises than problems.

M. E. Rose, *Elementary Theory of Angular Momentum*, John Wiley & Sons, New York, 1937.

An advanced treatment of angular momentum and the many applications to atomic and nuclear physics.

J. J. Sakurai, *Modern Quantum Mechanics* (S. F. Tuan, Editor), Addison-Wesley, Reading, Mass., 1994.

This excellent book by the late J. J. Sakurai is written at a somewhat more advanced level than *Quantum Physics*, like the books by Merzbacher and Schiff. This first year graduate textbook really does have a modern flavor with a selection of topics that leads quite naturally into the areas of advanced quantum mechanics of interest to particle physicists. The book has an excellent collection of problems.

D. S. Saxon, *Elementary Quantum Mechanics,* Holden-Day, San Francisco, 1968.

This book is on the same level as the present one, and it is a useful reference, since the selection of topics is just a little different, as is the emphasis and the choice of applications. The book contains an excellent set of problems.

L. I. Schiff, *Quantum Mechanics* (3rd edition), McGraw-Hill, New York, 1968.

This is one of the standard first year graduate textbooks. It is perhaps a little too compact, and thus most suitable for the well-prepared student. The level of mathematical sophistication assumed is above that of the reader of the present book.

F. Schwabl, *Quantum Mechanics,* Springer-Verlag, New York, 1992.

This is a more advanced book with an interesting selection of topics.

R. Shankar, *Principles of Quantum Mechanics,* Plenum Press, New York, 1980.

This is a more sophisticated and mathematically advanced book than the present book. It contains alternative treatment of some of the topics discussed in this book, and is a good reference.

M. P. Silverman, *And Yet it Moves: Strange Systems and Subtle Questions in Physics,* Cambridge University Press, New York, 1993.

This book deals in a qualitative way with a number of topics that involve the basic principles of quantum physics. It is fun to read and quite accessible to the student who has covered a good part of *Quantum Physics.*

INDEX